SPRINGER

LAB MANUAL

Springer
Berlin
Heidelberg
New York
Barcelona
Budapest
Hong Kong
London
Milan
Paris
Santa Clara
Singapore
Tokyo

Angel Cid-Arregui Alejandro García-Carrancá (Eds.)

Microinjection and Transgenesis

Strategies and Protocols

With 90 Figures, Including 10 Color Plates

Springer

Dr. Angel Cid-Arregui

Neurobiologie
Universität Heidelberg
Im Neuenheimer Feld 364
D-69120 Heidelberg
Germany

Dr. Alejandro García-Carrancá

Instituto de Investigaciones Biomédicas
Universidad Nacional Autónoma de México
04510 Mexico City
Mexico

ISBN 3-540-61895-3 Springer-Verlag Berlin Heidelberg New York

Library of Congress Cataloging-in-Publication Data

Microinjection and transgenesis : strategies and protocols / Angel Cid-Arregui, Alejandro García-Carrancá (eds.). p. cm. -- (Springer lab manual) Includes bibliographical references and index. ISBN 3-540-61895-3 (wire-o-binding alk. paper) 1. Transgenic animals--Laboratory manuals. 2. Genetic engineering--Laboratory manuals. 3. Microinjections--Laboratory manuals. I. Cid-Arregui, Angel, 1959- . II. García-Carrancá, Alejandro, 1955- . III. Series.
QH442.6.M53 1997
572.8'1'0724--dc21

Production: PRO EDIT GmbH, D-69126 Heidelberg
Cover Design: Design & Production, D-69121 Heidelberg
Typesetting: Mitterweger Werksatz GmbH, D-68723 Plankstadt

SPIN 10530162 31/3137 5 4 3 2 1 0 – Printed on acid free paper –

Foreword

The establishment of microinjection protocols about 20 years ago for cultured cells and shortly thereafter for the generation of transgenic mice by microinjection of DNA into fertilized mouse eggs greatly influenced many fields of biology. Not only have the data generated using these approaches contributed to a large extent to our present understanding of gene regulation and cellular function of higher eukaryotic cells, but current knowledge and future developments in this area will certainly have a great impact on basic and applied research for many years to come.

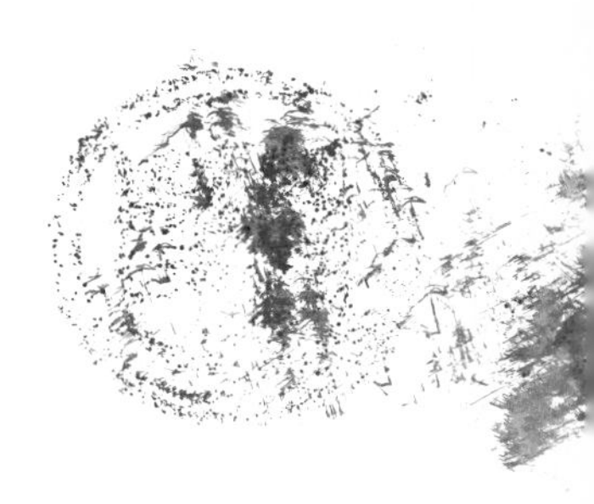

This laboratory manual describes the current state of the art in this research area and focuses primarily on both the experimental strategies with an extensive bibliography and the detailed procedures. A large number of studies are presently being performed and a great variety of new experimental designs are rapidly being developed. The book contains protocols on injection of somatic cells as well as on injection of embryos, the use of similar equipment being a common feature. In the articles dedicated to somatic cells, full descriptions of the manual and automatic injection systems are given as well as the methods for the analysis of injected cells by video-microscopy, electron microscopy or in situ hybridizations. In addition, comprehensive protocols are given for injection experiments with very different purposes, such as to study signal transduction or microtubule dynamics. A large section deals with the principles, the new strategies and various applications of transgenic technology primarily employing mice, but also using rabbits and *Drosophila*.

Following an introduction to mouse husbandry with a description of the requirements for transgenic mouse studies, we learn about the culture of male germ cells and the attempts to transfer sperm-mediated genes. The different strategies for generating germline chimaeras, as well as the production of YAC transgenic mice are outlined, and the numerous possible applications for using the mouse embryo as a „test tube" to identify gene activities are described. Finally, the various ways to apply histological markers, such as *lacz*, for multiple purposes are presented and an overview is given on the application of transgenic mouse studies to analyze the immune system and on virus research. No book on transgenic approaches can now be published without reviewing the possibilities and strategies for conditionally altering gene activities using inducible site-specific recombinases. On the other hand, the cloning of new genes from transgenic insertion sites still appears to provide a valuable tool for mouse geneticists for the study of developmental control genes.

The book contains a large amount of useful information. The advantages of microinjection protocols as well as their limitations are extensively discussed and I am convinced that the reader of this book will find useful suggestions, novel ideas as well as surprising facts on a relatively new but fast-growing research area. The numerous novel approaches in transgenic research with the constantly improving techniques, such as the transfer of nuclei from differentiated cells into enucleated sheep oocytes, together with the advancement of better cryopreservation techniques for mouse sperm and fertilized eggs, will indeed revolutionize modern biology. Some of the excitement of those working in the field is described here and is passed on to the reader of the book.

Vienna, March 1997 Erwin F. Wagner

Preface

"... Even granting that the genius subjected to the test of critical inspection emerges free from all error, we should consider that everything he has discovered in a given domain is almost nothing in comparison with what is left to be discovered"

Santiago Ramón y Cajal (1825–1934)
Precepts and Counsels on Scientific Investigation: Stimulants of the spirit, 1951

The interest of man in manipulating and modifying everything he discovers must obey an ancestral instinct which is indispensable for evolution. Once the adequate tools were available such interest focused on the functional unit of living organisms, the cell.

Despite the enormous complexity of their molecular organization, it is possible to microinject eukaryotic cells so that they recover perfectly from the perturbations caused by the manipulation. Indeed, careful intranuclear and/or intracytoplasmic microinjections neither alter significantly the proliferation capacity of somatic cells in culture, nor interfere with the development of fertilized eggs. This fact, together with the advantages of the direct manipulation of cells, have made microinjection the technique of choice for single cell experiments in vitro and for producing genetically modified animals.

This manual is intended to provide the reader with the techniques and methods needed to perform experiments using microinjection in studies on single cells or on transgenic embryos and animals, in particular mice, but also rabbits and flies. We are aware of the fact that the success of such experiments relies strongly on their feasibility, the proper design and correct interpretation of results. Therefore, special emphasis has been placed on providing a comprehensive introduction to each strategy described, besides the necessary working protocols. In addition, extensive references provide greater detail or alternative methods. In order to accomplish these goals within the limited space of a manual, repetition of earlier books has been carefully avoided wherever possible without affecting the comprehensiveness of the protocols.

Basically the same equipment is required for microinjection of somatic cells in culture and of oocytes and fertilized eggs. For this reason, both methodologies are described in this book. Complete microinjection setups for manual/automatic and computer assisted microinjection are described in Part I. The first 10 chapters include not only some complementary techniques that must often be combined with microinjection for single cell experiments, but also a compilation of representative applications of these techniques to cell biology, gene expression and signal transduction studies. Part II, consisting of the remaining 20 chapters, deals with a great variety of aspects ranging from mouse husbandry to the production of transgenic animals and the targeted inactivation of genes using the emerging techniques of site-specific recombination. These chapters have been arranged according to subject following ontogenic criteria: germ cells, oocytes, pre-implantation embryos, post-implantation embryos, targeting inactivated mutants and transgenics. All chapters deal with techniques using mice, except for Chapters 29 and 30 which describe transgenic rabbits and the microinjection of *Drosophila* embryos, respectively.

We would like to thank all the authors whose efforts and collaboration made this book possible. We also thank our colleagues for many helpful suggestions. In addition, we are grateful to Dieter Czeschlik and Jutta Lindenborn at Springer-Verlag, and Constanze Sonntag at PRO EDIT GmbH for their suggestions and help. Finally, the support of the Internationales Büro of the German BMBF and the Mexican CONACYT is gratefully acknowledged.

September 1997

Angel Cid-Arregui
Alejandro García-Carrancá

Contents

Part I

Strategies and Protocols for Microinjection Experiments Using Somatic Cells in Culture

Chapter 1

Protocols for the Manual and Automatic Microinjection of Somatic Cells in Culture and for the Analysis of Microinjected Cells

ANGEL CID-ARREGUI[1*], CARLA SANTANA[2], MIRIAM GUIDO[2], NESTOR MORALES-PEZA[2], MARIA VICTORIA JUAREZ[3], ROGER WEPF[3], AND ALEJANDRO GARCIA-CARRANCA[2]

Introduction

Micromanipulation of living cells was first used in electrophysiology studies during the first half of this century. Reliable techniques for intracellular recording were developed in the late 1940s. Less than one decade later, similar techniques were applied to perform direct transfer of biological material into cells (see Chambers and Chambers 1961). Initially, microinjection was designed for transplantation of mammalian nuclei (Graessmann 1970) and chromosomes (Diacumakos 1973) into recipient cells in culture. Further refinement of materials and instruments, however, permitted intracellular transfer of solutions containing macromolecules of various origins, such as viral RNA (Graessmann and Graessmann 1976; Stacey et al. 1977), purified proteins (Mabuchi and Okuno 1977; Tjian et al. 1978), and viral DNA (Anderson et al. 1980; Capecchi 1980). Likewise, microinjection was used for DNA transfer into fertilized eggs and a cloned gene was soon reported to be expressed in mouse somatic tissues (Gordon et al. 1980). Later, recombinant genes were stably introduced into the mouse germ line (Brinster et al. 1981; Wagner et al. 1981). Since then, microinjection has been used routinely to generate transgenic animals, and applied to a wide variety of studies using living cells, such as gene expression, signal transduction, or cytoskeleton studies.

* corresponding author: phone: +49-6221-548 321, fax: +49-6221-548 301, e-mail: angel.cid-arregui@urz.uni-heidelberg.de

[1] Dept. of Neurobiology, University of Heidelberg, Im Neuenheimer Feld 364, D-69120-Heidelberg, Germany

[2] Department of Molecular Biology, Instituto de Investigaciones Biomédicas, UNAM, 04510, Mexico City, Mexico

[3] European Molecular Biology Laboratory, Meyerhofstrasse 1, 69012 Heidelberg, Germany

Microinjection allows direct access to the two main intracellular compartments, the nucleus and the cytoplasm. Thus, this technique makes possible single cell experiments in which one or both intracellular milieu can be precisely modified. Moreover, the injected cells can be treated conveniently with appropriate changes in the composition of the culture medium. Table 1.1 summarizes all variables of a microinjection experiment with some illustrative examples. As indicated, a great variety of strategies can be devised by unique combinations of the different variables injection product, target compartment (cytoplasmic, nuclear or both), single or multiple injections, specific cell treatments (before or after microinjection), and the different analytical techniques.

Proper design of microinjection experiments requires careful consideration of the advantages and limitations of this technique. The following are some major advantages of microinjection:

- It can be applied to a wide variety of cell types from either established cell lines or primary cultures. Studies using the latter may benefit especially from microinjection because such cells are usually difficult to grow and very sensitive to changes in the culture medium. They survive in culture for a few days or weeks (e.g., neurons) or a few passages (e.g., fibroblasts, keratinocytes). Therefore, microinjection may be a unique method for DNA transfer experiments with these cells.
- A small amount of injection sample is required, eliminating the need for large-scale preparation and purification of injection products, such as DNA, antibodies, or recombinant proteins, which are often expensive and time-consuming to obtain.
- A great variety of synthetic and biological products can be injected into cells, from small oligonucleotides to yeast artificial chromosomes (YACs), and from peptides to purified organelles.
- The volume of sample injected into each cell, and hence the number of molecules of the injection product, can be estimated.
- Two or more different products can be injected, simultaneously or consecutively, into the same cell. In fact, injected cells can be retrieved and injected a second time with the same or a different sample.
- Microinjection can be combined with previous, simultaneous, and/or subsequent treatments of cells to help provide optimal conditions for the experiment.

- In combination with sensitive analytical techniques, such as immunofluorescence, in situ hybridization, or PCR, the effect of the injected product can be evaluated at the single-cell level.
- Speed of results. For most purposes, injected cells can be analyzed shortly after injection.

Microinjection has also some limitations:

- It requires special equipment and some expertise.
- It demands manipulation of cells one by one. Therefore, only a limited number of cells (up to a few hundred) can be studied at a time.
- Microinjection causes some degree of damage to the cell when the tip of the microneedle penetrates through the plasma membrane and the nuclear envelope. This is due to the mechanical force it exerts, but also to the fluid released from its opening, which increases the intracellular pressure and, depending on the nature of the sample being injected, may cause focal changes in ionic strength, osmolarity and pH. These effects may be more obvious on the cytoskeleton but can also reach some intracellular compartments, thus complicating the interpretation of the result.

All these aspects have to be considered carefully when planning microinjection experiments to ensure that the appropriate strategy is chosen. Thus, for instance, microinjection is not normally used for DNA transfer because it is laborious and time-consuming, as compared to the calcium precipitate, electroporation or lipofection methods. However, for some experiments (e.g., using cells from primary explants, or YAC vectors) the advantages of microinjection make it the method of choice for DNA transfer.

This chapter describes the equipment and methods to perform manual and automatic microinjection into cultured cells. In addition, working protocols for the analysis of cells microinjected with common reporter plasmids are given.

Materials

Microinjection equipment

The simplest microinjection setup consists of an inverted microscope, a micromanipulator, and a puller with which to prepare suitable micropipettes (see Fig. 1.1). For best results, the use of an accurate pressuring

Table 1.1. Parts of a microinjection experiment with examples of different experimental approaches

Intranuclear/ Intracytoplasmic Microinjection	Intracellular processing	Effects of the injected compound(s)	Analysis of the microinjected cells
• DNA: cellular, viral, plasmids, YACs • RNA: cellular, viral *in vitro* transcribed • Antisense: RNA, oligonucleotides • Ribozyme: synthetic, *in vitro* transcribed • Antibodies: unlabeled, conjugated • Proteins: cellular, recombinant • Purified organelles: Nuclei, vesicles • Other natural or synthetic products • Mixtures of compounds	• Transcription • Translation • Annealing • Binding • Modification (Phosphorilation, etc.) • Transport • Degradation, etc.	• Interference with cellular functions: Cell cycle arrest Block of intracellular signaling pathways Block of intracellular membrane traffic etc. • Involvement in cellular functions: Regulation of gene expression Signal transduction Cytoskeleton assembly etc.	• Observation of phenotype: changes in cell morphology, growth arrest, apoptosis, multinucleation, cell fusion, etc. • Biochemical analyses: CAT assay, spectrophotometric measurements of enzyme activities • Immunofluorescence, immunohistochemistry • *In situ* hybridization • Immunoelectron microscopy • Molecular biology analyses: Southern and Northern blots, RNase protection, PCR etc.
• Additional cell treatment: ^{35}S-Methionine (for metabolic labeling), BrdU (labeling of newly synthesized DNA), FITC-transferrin (marker for receptor-mediated endocytosis), horse radish peroxidase (marker for fluid phase endocytosis), actinomicyn D (blocks RNA synthesis), cycloheximide (blocks protein synthesis), brefeldin A (blocks ER to Golgi vesicle traffic), nocodazole (depolymerizes microtubules), G-418 (for selection of neomycin resistence), etc.			

device (see below) is highly recommended. In addition to this, the automated and the automated computer-assisted systems developed by Eppendorf and Zeiss require some additional sophisticated devices (see below; see also Ansorge and Saffrich, Chap. 2, this Vol.). It is advisable to install the microinjection setup in a room used exclusively for this purpose and located in the vicinity of the cell culture facilities. In addition, microscope and micromanipulator should be on a metal platform, which is is placed onto an antivibration table.

- Leitz heavy base, or similar, where the microscope and micromanipulators are mounted
- Inverted microscope: e.g., Nikon TMS and Diaphot 300/2; Zeiss Axiovert 100 or 135, or Axioskop FS (Fig. 1.1-A)
- Micromanipulators (Fig. 1.1A, B): Leitz (manual system, should be fixed to the platform) or Eppendorf 5171 (automatic system for axial microinjection, it is fixed to the microscope stage)
- Microinjector (Fig. 1.1B): device which provides the pressure required to release sample solution from the micropipette for injection. Eppendorf 5242. Alternatively, a syringe (ideally of glass, with a screw-plunger, available from Leitz) fixed to the platform
- Glass micropipettes: there are two types of microinjecting pipettes that can be used:

  Ready-to-use prepulled micropipettes from Eppendorf. They are mounted on a plastic screw which fits into the Leitz micropipette holder

  Micropipettes produced in the laboratory using a puller device (see below and Fig. 1.1C). It is recommended to use capillaries made of borosilicate glass (available from Clark Electromedical Instruments, Pangbourne, Berks, UK, cat. no. GC120TF-10). These capillaries have an outer diameter of 1.2 mm and a wall 0.13 mm thick. Spanning the entire length of the capillary there is a filament of 0.1 mm∅ adhering to its inner wall, which supports displacement of the sample solution to the tip of the micropipette (see Fig. 1.2). See Schnorf et al. (1994) for a detailed characterization of micropipettes pulled from different types of capillaries and a method to determine the diameter of the tip opening by measurement of the threshold bubble pressure.

  Microinjection pipettes can be treated with silane to avoid interaction of sample and/or culture medium components with the hydrophilic glass surface. The most commonly used silanization protocols involve a chemical reaction between hydrophilic groups at the glass surface and an organic silane that contains a hydrophilic glass-reactive moiety

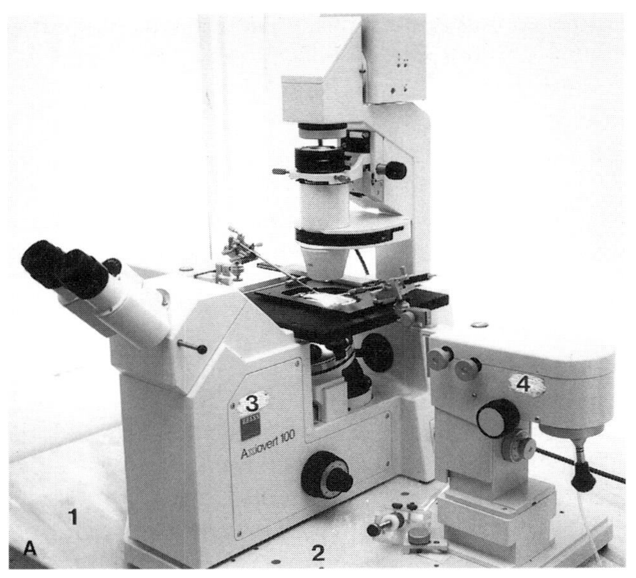
3
Axiovert 100
4
1
2
A

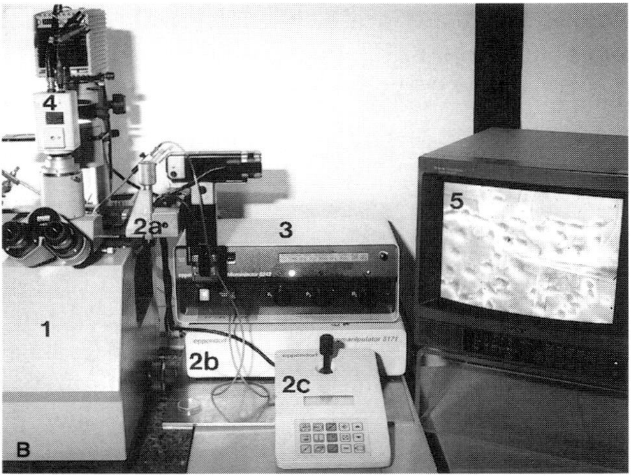
4
2a
3
5
1
2b
2c
B

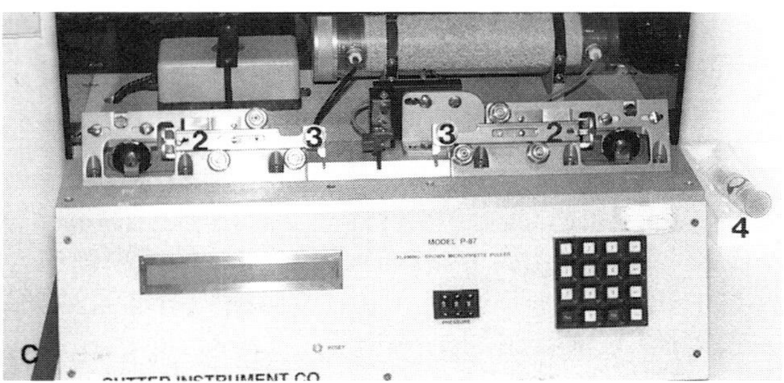
2
3
3
2
4
MODEL P-87
C

◀ **Fig. 1.1A–C.** Assembled instruments for manual and automatic microinjections (see Materials). **A** Manual microinjection setup: *1* antivibration table; *2* Leitz heavy base; *3* Zeiss Axiovert inverted microscope; *4* right-side Leitz micromanipulator. **B** Automatic microinjection setup: *1* Zeiss inverted microscope; *2* Eppendorf automatic micromanipulator (Model 5171): *a* micromanipulating unit with three motorized subunits assembled to provide movement in the x, y, and z axis; *b* power unit; *c* small computer unit with keyboard and joystick; *3* Eppendorf microinjector (Model 5242); *4* TV camera; *5* TV monitor showing on the screen cells on a glass coverslip being microinjected by the micropipette which is on the right side pointing to the center of the screen. **C** Horizontal puller from Sutter Instrument (Model P-87): *1* filament holder; *2* right and left slides; *3* clamps; *4* glass capillaries

and a hydrocarbon chain which confers hydrophobicity (see Leyden 1985; Proctor 1992 and refs. therein). Both the inside and outside surfaces of the micropipette can be treated. However, in general, there is no need to treat the inner surface of the micropipette, because binding of nucleic acids and proteins to untreated glass does not occur at the physiological ionic strength of commonly used injection solutions (Thomas et al. 1979). Furthemore, single-stranded DNA and proteins bind to silanized glass even more strongly than to untreated glass (see Proctor 1992). Silanization of the outside surface of the micropipette may be more advisable, because it prevents adhesion of cellular hydrophilic material to the glass which often blocks the micropipette opening and impairs ejection. However, silane treatment is not easy to perform, and, in general, adhesiveness does not impair injection of a reasonable number of cells with the same micropipette before requiring its change. Therefore, we do not recommend routine silanization of the micropipettes.

– Pipette puller
  Horizontal puller: Sutter Instrument Co. (Novato, USA), Model P-87 (Fig. 1.1C). Load a capillary, fix both sides by tightening the clamps on the right and left slides, so that it is centered with respect to the heating filament. Set current, force and time: e.g., Heat = 640, Pull = 150, Vel = 100, Time = 135. Press "start" button and wait until the capillary has been pulled into two micropipettes suitable for microinjection (see Fig. 1.2)
  Vertical puller: David Kopf Instruments (Tujunga, California, USA), Model 720. Load a glass capillary in the upper clamp, center it with respect to the filament and tighten clamp. Rise lower slide and tighten clamp. Adjust heat and solenoid range (e.g., heat = 11 and solenoid = 3.6). Press "start" button

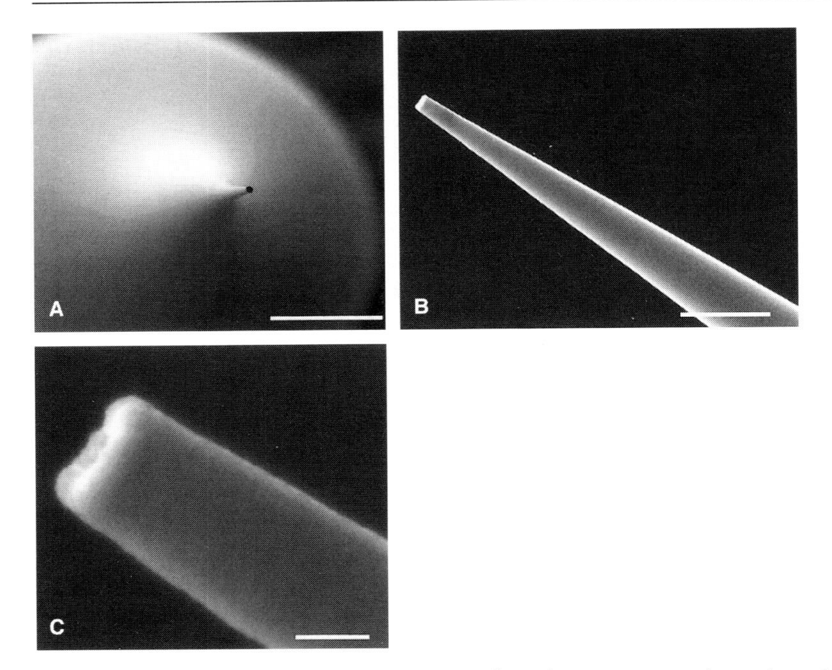

Fig. 1.2A–C. Scanning electron micrographs of a representative micropipette tip. The micropipette was produced by pulling a borosilicate glass capillary (see Materials) in a horizontal puller device like that shown in panel **C** of Fig. 1.1. The inner diameter of the tip is about 150 nm and the outer roughly 350 nm. Sharpness and shape of the tip edges may vary slightly from one micropipette to another. Micropipettes were sputter-coated with 10 nm Au/Pd, which may enlarge the overall appearance and smoothen the tip edges. The micropipettes were imaged in a FESEM (Philips XL-30FEG) at 30 keV. *Bar* **A** 200 μm, **B** 2 μm, **C** 200 nm

Cell culture equipment

- Cell culture incubator and laminar flow hood
- Plastic dishes and glass coverslips. Plate the cells to be microinjected on 60-mm∅ tissue culture dishes, each containing several glass coverslips (up to six, 10–15 mm∅). When the cells are ready for injection (for most purposes at 20–40 % confluency), transfer one coverslip to a new 60 mm∅ dish containing 4–5 ml of buffered medium

Two days before plating cells for microinjection, treat an appropriate number of glass coverslips as follows:

1. Place coverslips, separated and in vertical array, on ceramic or metal racks.

2. Rinse briefly by immersion of racks into a glass jar filled with tap water.
3. Drain racks on paper towels and transfer them to another jar containing 0.1 N HCl. Incubate overnight.
4. Rinse coverslips in tap water 5×10 min each.
5. Immerse racks in 70 % alcohol and incubate for 60 min.
6. Rinse briefly in deionized water.
7. Drain on paper towels.
8. Let dry for 30 min at room temperature.
9. Bake at 220–250 °C for 6 h.

- Diamond-point pencil. It is used to mark small circles or crosses on the coverslips. Cells inside or around this mark will be injected. This facilitates retrieval of injected cells. Alternatively, coverslips with an etched grid are commercially available (CellLocate coverslips, Eppendorf).

Buffered culture medium

In addition to the medium required for culturing the cell line or primary culture under study, which is maintained at physiological pH values with the aid of the 5 % CO_2 atmosphere provided by the incubator, it is necessary to use a buffered medium to keep the cells during microinjection. The most convenient medium should be found attending the characteristics of the cells and the nature of the experiment. Suitable buffered media can be prepared as follows:

- Culture medium with 25 mM HEPES pH 7.2 (stock 1 M, sterile filtered and kept at room temperature) substituted for an equivalent amount of the bicarbonate to maintain the pH in the physiological range
- Culture medium without serum and buffered with 25 mM HEPES, pH 7.2
- Hank's balanced saline solution (HBSS), buffered with 25 mM HEPES pH 7.2

These media should contain penicillin and streptomycin. Antifungal antibiotics, however, are not recommended as they may damage the cells and are not necessary if microinjection is performed in a clean room used only for this purpose.

Injection buffer

There are many possible buffer solutions that can be used to prepare samples for injection. The following are the buffers in most widely used:

1. Tris-EDTA buffer: 10 mM Tris-HCl pH 7.4, 0.25 mM EDTA. This is the commonest buffer used in nucleic acid solutions to be injected into the nucleus of cultured cells and eggs (see Wassarman and DePanphilis, 1993; Hogan et al. 1994).
2. Phosphate buffers. Phosphate is the main intracellular anion and this makes logical its use in injection buffers. However, when used at physiological concentrations (e.g., 48 mM K_2HPO_4, 14 mM NaH_2PO_4, 4.5 mM KH_2PO_4, pH 7.2, Graessmann et al. 1980), it may precipitate as calcium phosphate at the opening of the microneedle, where the injection buffer mixes with the culture medium, thus blocking ejection of the sample through the tip. Therefore, some authors prefer to use buffers with a lower concentration of phosphate compensated with a concentration of chloride above physiological levels, which allows maintaining Na^+ and K^+ concentrations close to the intracellular values (e.g., 10 mM each $H_2PO_4^-$ and HPO_4^{2-}, pH 7.2, 84 mM Cl^-, 100 mM K^+, 17 mM Na^+, 1 mM EDTA, see Proctor 1992).

Fluorescent markers

- FITC-Dextran (Sigma FD-70S or FD-150S)
- Rhodamine-dextran (70S, Sigma R-9379)
- Rhodamine-dextran, 40,000 MW, lysine fixable (Molecular Probes D-1843)

Prepare 10 % (w/v) stock solutions of these markers in injection buffer, store at −20 °C protected from the light. Working solutions are prepared by diluting stocks in injection buffer to a final concentration of 0.25–1 % (see Sect. 1.2). These solutions should be filtered through 0.22 μm filters and stored at 4 °C protected from the light.

Enzyme markers

- Luciferase (Sigma L-5256)
- HRP (Sigma P-8250). Peroxidase substrate TMB and peroxidase solution (Pierce 1854050 and 1854060, respectively)

Immunofluorescence and histochemistry reagents

- Fixatives. In general, injected cells can be readily fixed with one of the fixatives described below. However, for each particular antibody the most appropriate fixative should be found
  Methanol. For fixation, immerse the coverslips in 100 % methanol at −20 °C for 2–5 min

Buffered 4 % paraformaldehyde/0.2 % glutaraldehyde (PFA/GA). Prepare as follows:

1. Dissolve 0.5 g of paraformaldehyde in 25 ml of PBS by heating to 60 °C while adding 10 M NaOH (1 small drop per min) until PFA is dissolved and the solution clears (it takes 5–10 min).
2. Let cool to room temperature.
3. Add 20 μl of 25 % glutaraldehyde.
4. If necessary, adjust pH to 7–7.2 with 1 M NaOH.

Fix the cells as follows:

1. Incubate the coverslips in PFA/GA for 10–15 min.
2. Rinse briefly in PBS.
3. Quench autofluorescence by incubating in 100 mM NH_4Cl in PBS 2×10 min each.
4. For immunofluorescence, permeabilize the cells by incubating in 0.2 % Triton X-100 in PBS at room temperature for 5 min. Alternatively, permeabilize by passing the coverslips through an ethanol series: 25, 50, 70, 90, 100, 90, 70, 50, and 25 % ethanol, 1 min each. Finally transfer to PBS and proceed for immunofluorescence.

- Blocking solution: 2 % fetal calf serum, 2 % BSA (Sigma A-6793) and 0.2 % fish gelatin (Sigma G-7765) in PBS. For blocking use undiluted. For diluting first and secondary antibodies use 1:10 blocking solution in PBS
- Mounting medium. We use routinely Mowiol mounting medium prepared as follows:

1. Add 10 g of Mowiol 4–88 (Calbiochem no. 475904) to 40 ml PBS.
2. Stir for 2 h.
3. Add 20 ml glycerin and stir for an additional 10 min.
4. Centrifuge at 3000 rpm for 15 min.
5. Make aliquots of 1 ml and store at −20 °C. This solution is very dense and should not be vortexed but mixed gently by inversion after thawing to avoid bubble formation. Do not re-freeze aliquots, keep them at 4 °C until new use.

- N-acetyl cysteine (NAC, Sigma A-8199). Prepare a stock solution 500 μM in MEM, neutralize as needed with NaOH, store at −20 °C. For use: dilute 1:5000 in the culture medium

1.1 Microinjection

Procedures

This section describes the preparation of cells, samples and microneedles for injection, and the procedures for the manual and automatic micromanipulation of cultured cells.

Cells Virtually, any cell in culture can be microinjected. Cell lines and primary cultures from most tissues attach firmly to tissue culture dishes and also to glass coverslips. Moreover, most cell types are large and robust enough to allow easy and efficient microinjection. Some cells, however, are very small or attach weakly to the glass which makes microinjection very difficult and time-consuming. In addition, there are cells that grow in suspension and need to be held using a holding pipette placed opposite the injecting needle. The protocols given here refer to the first type of cells, as it is the most common, and would need to be adapted to the peculiarities of the other two types.

Prepare cells for microinjection as follows:

1. One day before injection, plate the cells on glass coverslips (10–15 mm∅) to a density of 250–1000 cells per coverslip, depending on the experiment.
2. Just before starting injection, transfer a coverslip with the cells to be injected to a new tissue culture dish (35 mm∅), quickly mark a cross (or one or more circles) on the coverslip using a diamond-point pencil washed with 70 % ethanol and flamed briefly under the hood, and add 3 ml of buffered culture medium (see Materials, above).
3. Inject cells for 30 min. In general, injection sessions should last no longer than 30 min to minimize cell damage due to deleterious effects of the microscope light and the culture conditions during injection.
4. After injection, remove the buffered medium, wash gently with PBS, transfer coverslip to a dish containing preincubated (30 min) fresh culture medium, and return it to the incubator.

Injection samples

DNA. For most purposes, circular (intact) plasmids are chosen for microinjection because they provide good levels of expression. However, for some experiments, like those involving homologous recombination, DNA fragments devoid of vector sequences will need to be microinjected.

Oligonucleotides. For microinjection, oligonucleotides should be dissolved in water or in the same buffer as used for DNA (see Materials, above). It is advisable to use HPLC-purified oligonucleotides. It may be also very convenient to use phosphorothioate derivatives, which are more resistant to degradation and hence have a more intense and prolonged effect on the cells (see Güimil García and Eritja, Chap. 5, this Vol.).

RNA. RNA for microinjection is commonly obtained by in vitro transcription of plasmids designed to express the gene of interest. The enzyme used in this step should be inactivated and, preferably, removed (e.g., by purifying the RNA using glass milk) because it might damage the cells. For some experiments, total or poly A(+) RNA from cells or tissues has to be microinjected into host cells. In these cases, we recommend purifying the RNA through CsCl gradients.

Antibodies and purified proteins. Dialyze protein solutions to be injected against microinjection buffer. In general, use solutions whose protein concentration is in the range of 1 mg/ml.

Note: Consult Sambrook et al. (1989), Proctor (1992), Wassarman and DePanphilis (1993), and Hogan et al. (1994) for methods on the preparation of DNA, RNA, oligonucleotide, and protein solutions for microinjection.

Micropipettes

Micropipettes are prepared from glass capillaries as described in Materials, above. Once pulled, they should be handled carefully to avoid breakage. Contact with the part of the capillary close to the tip should be avoided to prevent further contamination of the culture medium, where it will be immersed during injection. For better reproducibility of experiments it is preferable, whenever possible, to use the same micropipette in the entire injection session.

1. Load the micropipette with 0.5–1 μl of injection sample by pipetting it from the rear using a sterile microloader (commercially available from Eppendorf or self-made from a normal glass capillary flamed and pulled).

2. The inner filament in the micropipettes pulled from capillaries GC120TF-10, which runs along the capillary wall, supports displacement of sample solution from the rear to the tip of the micropipette. Therefore, a brief immersion of just 1 mm of the back side of the capillar, which must not have been in contact with the fingers nor anything else, into the sample solution is enough to ensure that a small volume of sample reaches the tip.

Centering and positioning micropipettes

1. Attach a loaded micropipette to the micromanipulator by introducing its free side into the holder adapter, which is then screwed in hardly to fix the pipette. The adapter with the pipette is then fixed to the holder tube attached to the micromanipulator.
2. Transfer a coverslip containing the cells to be injected to a new dish containing buffered medium (see Materials, above) so that it is centered in the dish.
3. Place the dish on the microscope stage. Try to center roughly one of the circles on the coverslip (see Materials, above) in the center of the illuminated field, in front of the lens, preferably where the light is more intense.
4. Focus on the cells using a low magnification objective (e.g., 5 or 10× for 50 or 100× magnification, respectively).
5. Center roughly the tip of the micropipette, which appears as a shadow, in the optic field.
6. Lower the pipette slowly until it enters the medium, then stop.
7. Looking through the microscope, move the pipette, adjusting the micromanipulator as necessary until you see its shadow over the field of view. Then try to distinguish where the tip is and center it roughly.
8. Lower the pipette slowly until you see it more sharply, but still not in focus.
9. Change to the working magnification (e.g., 320x). Focus the cells and look for the tip of the pipette. If you do not find it, return to the lower magnification and try to center it. Repeat this step.
10. Lower the tip carefully until it is in focus.

Pressure Systems. Setting Injection Pressures

Stand by and ejection pressures applied to the injecting pipette may be provided by any of the following systems:

Pressure systems

- The simplest device consists of a 50-ml glass syringe connected to the micropipette through the back opening of the holder by plastic tubing. The entire system is filled with air. In some cases this syringe is fixed to a metal support to which its plunger is screwed so that it can be moved up and down. Turning right, the plunger goes down and raises the pressure applied to the pipette, while turning left, the plunger goes up and decreases pressure. Care must be taken, however, not to exert negative pressure because this may cause cultured medium to be sucked back into the pipette. When using a normal syringe, only when pressure is exerted on it is the injecting sample forced out of the pipette, while the flow of sample ceases when pressure is released.
- More sophisticated automatic microinjectors are available from Eppendorf. This device permits accurate control of the pressure applied to the pipette throughout the experiment. It provides a constant positive pressure that generates a continuous flow of injecting sample, thereby resulting in less clogging of the pipette and eliminating backflow problems. The device is connected to a bottle of compressed gas which provides the input pressure. In addition, it is connected to the pipette holder by plastic tubing. The output pressure can be modulated at three different levels by regulating the corresponding valves:

- **P3** is the lowest level of pressure. It is applied continuously to the pipette (holding pressure) to avoid aspiration of culture medium. Its value can be adjusted ranging from 0 to 700 hPa, depending on the size of the opening of the micropipette tip. We advise setting P3 at $\leq$100 hPa for injection pressures (P2, see below) $\leq$300 hPa, and at 100–200 for higher P2 values. If manual microinjection is performed, P3 may be adjusted so that it can be used as injection pressure.
- **P2** is the intermediate level of pressure, which provides the force needed for microinjection (injection pressure). It can be set between 45 and 2000 hPa. Common values required for injection using GC120TF-10 capillaries with a tip opening of about 150 nm (Fig. 1.2) are between 200 and 400 hPa. The time of injection during which this pressure will be applied is selected on the control panel ranging from 0.1 to 9.9 s. If manual microinjection is performed, P2 is applied by the

operator pressing the footpad briefly. In this case, care should be taken that the switch from P3 to P2 occurs when the tip of the pipette is inside the cell. When using the automated microinjection device, it will recall automatically the P2 pressure during each injection. Each change from P3 to P2 is recorded on the control panel, which shows the total number of cells injected.

- **P1** is the highest output range (3000–10,000 hPa). The commonest setting is 5000–6000 hPa. P1 is applied manually and its use is restricted to cleaning the capillary when it is obstructed by impurities that may be present in the injection solution itself or inside the capillary.

Setting optimal injection pressures

Once the tip of the pipette is positioned and focused in the plane of the cells, and just before starting a round of injections, it should be verified that the tip of the pipette is open. Then, the next step is to adjust the optimal injection pressure, i.e., that providing ejection of a reasonable volume of sample in a short time (0.5–1 s), which ensures minimal damage to the cells.

1. Confirm that the tip of the microneedle is open by applying full pressure for 1 or 2 s. If it is open, the flow of injection solution will be seen clearly coming out of the pipette as a brighter front penetrating the surrounding medium. If ejection is not observed when applying full pressure, but the pipette slides on the coverslip instead, the tip is closed. Try to open it by applying full pressure for 3–5 s several times and, if necessary, beating gently with a finger on the micromanipulator at the same time, so that the tip breaks against the glass surface. If this does not suffice to open the tip, the micropipette should be replaced by a new one. If it has the same problem, revise the settings of the puller and, if necessary, change the conditions by increasing the current in steps of 0.01 A until open needles are obtained.
2. Select a cell in the center of the optical field and place the tip of the micropipette just above the nucleus.
3. Set microinjection time to 0.5–1 s in the front panel of the microinjector.
4. Lower the micropipette until it is inside the nucleus and apply the P2 pressure by pressing briefly on the footpad. Take the tip back to the initial position. In this step can be assessed if the injection pressure is either:
   a) appropriate, if a small volume of sample was left inside the nucleus which did not cause apparent enlargement

b) high, if the nucleus was slightly enlarged
c) excessive, if the nucleus was clearly enlarged
d) too low, if no ejection was observed. Note that compression on the cell surface by the micropipette causes local changes in optical refringency which may give the impression that the cell was injected while it was not. Only experience helps distinguish real from apparent injection.

5. Modify the P2 value according to the result of the previous step and test it in a new cell. Repeat until the appropriate value is found.
6. The pressure test described above should be performed in the perinuclear cytoplasm if intracytoplasmic injection is intended. The same considerations as in step **4** to set the appropriate injection pressure are valid here. Although the cytoplasm admits injection of higher volumes of sample, yet these should never be so high as to enlarge the cell.
7. If a syringe is used to provide pressure, appropriate injection pressures are achieved by pressing with more or less intensity on the pestle of the syringe according to the changes in appearance of the cells as described above. Obviously, such device will never permit an accurate control of the injection pressure as the microinjector does, but in experienced hands it provides reliable injections.

Microinjection Performance

According to the micromanipulation technique used, three types of microinjection can be distinguished: manual, in which micropipette movements are operated manually; automatic, and automatic computer-assisted microinjection. The last two use a device to hold the micropipette that performs automatic movements in the x, y, and z axis, and which is operated from a control board or a computer, respectively.

Microinjection experiments can be performed by any of these methods with no significant difference in the results (see Table 1.2 for a comparison). Indeed, a manual system may well fulfill the needs of virtually every experiment. However, automatic micromanipulation can make injection easier and faster, and result in less damage to the cells. For experiments requiring injection of many cells (i.e., several hundreds), the automatic systems may be ideal. Nevertheless, certain types of cells are easier to inject with a manual device because it allows the finest micromanipulation.

Manual system. In order to perform successful and reliable injections within a reasonable time by manual micromanipulation, it is essential to use high-quality equipment, like that described under Materials, above. The Leitz micromanipulator is the best manual device available (Fig. 1.1A) as it is simple to handle and highly precise. The microneedle, which is fixed to the holder of the micromanipulator, can be moved in the x, y, and z axis by operating three knobs and, for the finest movements, a joystick (Fig. 1.1A). Once the tip of the needle is centered with respect to the cellular compartment to be injected, it is moved downwards in the y axis until it enters the cell, at which moment injection starts (Fig 1.3). This is in contrast with the automatic system, where the needle moves axially (i.e. along its own axis) towards the cell surface during injection (see below). In experienced hands, manual performance is as reliable and fast as automated microinjection.

Table 1.2 Comparison of the different microinjection systems

Microinjection system	Optimal rate of micro-injection[a]	Reproducibility[b,f]	Double injection of the same cell[c,f]	Applicability to biochemical studies[d,f]	Applicability to singe cell studies[e,f]
Manual	100–200	++	+	++	+++
Automated	200–300	++	++	++	+++
Automated-computerized	500–700	++	+++	+++	+

[a] These values correspond to numbers of cells microinjected in 30 min (an interval of time considered reasonably safe, in terms of cell viability, for cells being microinjected at room temperature in an appropriate medium buffered with HEPES). The optimal number of cells microinjected per time unit depends largely on the quality of the equipment and the experience and ability of the experimentator. It depends also on the cell type (e.g., smaller cells are more difficult to inject), the characteristics of the injection solution (some products favor stickiness at the needle tip), and the culture conditions. For further information on the efficiency of the three systems see Ansorge 1982; Pepperkok et al. 1988).

[b] Reproducibility of microinjection experiments is taken here as a function of the viability of the injected cells and the accuracy of injection. Technically, these two parameters are specified by the damage caused by injection and the volume of sample injected, which are in turn determined by the following factors: (1) diameter of the opening of capillary tip, (2) flow of injection solution, and (3) time of microinjection. The use of a precise puller to produce needles of constant tip diameter and a microinjector to provide accurate pressure and injection time values will help maintain those factors in the optimal range, which varies from cell type to cell type. The system of microinjection will also influence reproducibility. In general, automatic systems cause less damage to the cells than the manual method because the needle is pushed by the micromanipulator in a plane which forms a 45° angle with the plane of the dish surface (axial microinjection), whereas in the manual system the needle moves only in the vertical plane and, therefore the mechanical pressure it needs to cross the cell surface is higher (see Fig. 2); in addition, the possibility that cellular components attach to the capillary tip on its way back out of the cell is also higher by this method due to its vertical displacement. Finally, the time that the needle tip remains into the cell can be precisely controlled only when using the automatic systems.

[c] Double injections of cells are easier using the automated-computerized system because the machine keeps in the memory the positions of previously injected cells. By the other two methods the experimentator seeks himself the cells injected in the first round of microinjection. However, viability of double injected cells is considered lower by the manual method due to the reasons stated in [b].

[d] For optimal results these studies require the injection of hundreds of cells, what is easier and faster to perform by the fully automatic method.

[e] For studies which require injection of a reduced number of cells the three systems are suitable, although the manual and semiautomatic methods are advisable because of their lower cost.

[f] Classification: (+++) very good, (++) good, (+) not recommended.

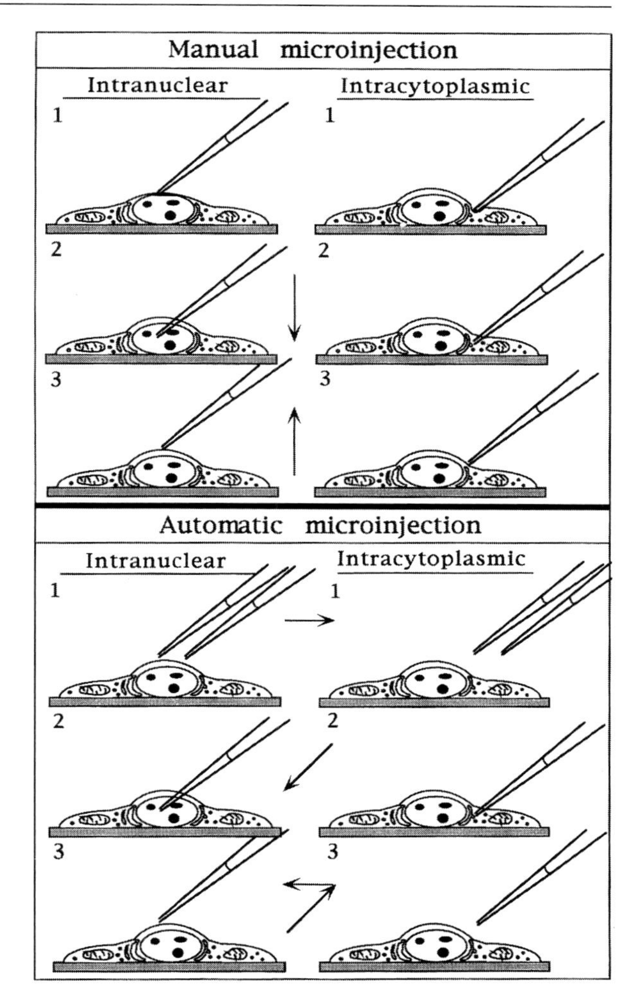

Fig. 1.3. Schematic representation of cell injections using manual or automatic micromanipulation. Note the different approach of the micropipette tip to the cell in each system: vertical in the manual, axial in the automatic

Manual intranuclear injection

1. Using the lowest magnification (50×), focus on the plane of the cells situated in the area of the coverslip labeled for injection. Center the micropipette by eye in the illuminated area and lower it until it enters the culture medium. Center the tip of the micropipette in the optical field and lower it until it is in focus. Then, change to the working magnification (320×) and focus on the cell surface. Center and lower the tip until it is in focus. Move the microscope stage until the tip points to the nucleus of a cell.

2. Lower the tip carefully until it enters the nucleus (see Fig. 1.3, upper left panel).

3. Apply injection pressure: press on the footpath switch of the microinjector to apply P2 for the selected time (see previous section). If a syringe is being used, ensure that enough pressure is applied at this moment.
4. Move the micropipette upwards carefully until it leaves the cell.
5. Move the microscope stage to find a new cell, center its nucleus with respect to the tip and repeat steps **2** to **4.** If Cellocate coverslips are used (see Materials, above), injected cells can be recorded in the book provided by the manufacturer.

Manual intracytoplasmic injection

1. Place the tip above the perinuclear cytoplasm. This is the most convenient part for injection because in it the cell is thicker and there is room for the segment of the tip that stays inside the cell during injection.
2. Lower the tip until it touches the plasma membrane and enters the cell (see Fig. 1.3, upper right panel).
3. Apply pressure P2 for 1–2 s.
4. Move the micropipette upwards until the tip is outside the cell and several μm above its surface.
5. Move the microscope stage, find a new cell, and proceed as above.

Automatic System. This type of microinjection requires an automated micromanipulator like the 5171 of Eppendorf (see Materials, above and Fig. 1.1B), which has three components (see Eppendorf 5171 Operating Manual):

- Micromanipulating unit, with three modules for the x, y, and z coordinates, each equipped with a drive unit and a stepper motor
- Control unit. It consists of a small computer with a keyboard and a joystick
- Power unit

Automatic intranuclear and intracytoplasmic injections

The Eppendorf 5171 micromanipulator performs automatically all micropipette movements required for axial injection (see Fig. 1.3 lower panel). In a first step, the coordinates of the micropipette positioned inside the nucleus or the cytoplasm of a test cell are registered and stored in the control unit computer. Then, injection proceeds by centering each new cell with respect to the micropipette tip and recalling the coordinates using the joystick.

1. Center the micropipette tip in the optic field as described above for the manual system, but operating the micromanipulating unit automatically from the control board.
2. At the working magnification (320×), lower the tip to touch a point of the cell surface region above the nucleus or above the perinuclear space, until a slight depression is seen.
3. Fix position parameters for intranuclear or intracytoplasmic injection in the control unit.
4. Lift the tip a few μm above the cell surface. It should be slightly out of focus but still visible. Select a new cell and microinject it by pressing on the button placed above the joystick. The micropipette describes first a retrograde trajectory on the horizontal plane (toward the micromanipulator unit) followed by a rapid axial movement toward the cell, which ends exactly when the tip reaches the coordinates fixed in the previous step (see Fig. 1.3). At this moment, the control unit of the micromanipulator, which is connected to the microinjector, activates the injection pressure (P2) for the preselected time. Immediately after this, the micropipette returns to the initial position.
5. Correct the parameters in the control board as necessary. Because the coverslip and dish surfaces are not perfectly flat, it is necessary to adjust the intranuclear or intracytoplasmic coordinates when injection progresses from one region to the other of the coverslip.

1.2 Protocols for the Analysis of Microinjected Cells

Estimation of Ejection Flow and Average Injection Volume

The ejection flow (EF, volume of sample solution released from the micropipette per time unit) can be easily estimated using a radioactive marker. The micropipette is loaded with a stock solution containing a radioactive isotope (^{14}C or ^{35}S) and ejected at selected conditions of pressure and time into a drop of water. The sample is measured in a scintillation counter. Comparison to a standard curve obtained with a series of dilutions of the stock solution allows calculate the volume of sample ejected, which divided by the time of ejection gives the EF. The EF is determined by the same factors listed below for the injection volume. Typical ejection flow values fall in the range of 0.5–1 μl/s (Ansorge and Pepperkok 1988).

The volume of sample that is injected into each cell (AIV, average injection volume) can be estimated using reporter enzymes like luciferase or horseradish peroxidase (HRP).

Procedure

AIV estimation using luciferase

1. Inject a precise number of cells (50–100) with a 2 mg/ml stock solution of luciferase in PBS or injection buffer.
2. Rinse the coverslip gently in PBS and lyse the cells with 50 µl of 2 % NP-40/0.2 % SDS in PBS. Use the entire cell extract to measure luciferase activity in a luminometer. Make a standard curve using 50-µl samples from a series of dilutions of the luciferase stock solution.
3. Calculate the amount of luciferase in the cell lysate and estimate the volume of solution injected. To calculate the AIV, divide this value by the number of cells injected.

AIV estimation using HRP

1. Inject a precise number of cells (50–100) with a 20 mg/ml stock solution of HRP in PBS or injection buffer.
2. Rinse the coverslip gently in PBS and lyse the cells with 50 µl of 2 % NP-40/0.2 % SDS in PBS.
3. Pipette lysate into a spectrophotometer disposable cuvette containing 50 µl PBS.
4. Prepare a series of dilutions of the HRP standard solution and pipette 50 µl of each dilution into cuvettes containing 50 µl of 2 % NP-40/0.2 % SDS in PBS.
5. Add 200 µl of a mixture (1:1, v/v) of peroxidase substrate TMB and peroxidase solution to the cuvettes of steps **3** and **4.** Incubate at room temperature for 5–30 min (depending on the intensity of the blue color developed by the samples during the reaction).
6. Stop reaction by adding 700 µl of 1M PO_4H_3 (the samples become yellow) and read absorbance at 450 nm. Estimate the total volume of HRP injected and from this the AIV.

Analysis of Cells Injected with Fluorescent Markers

Fluorescent markers, like FITC-dextran, are frequently used as coinjection markers to visualize injected cells. At concentrations between 0.25

and 1 % they exhibit very low toxicity and remain visible for 2–4 days in proliferating cells, or longer in quiescent and postmitotic cells. Living cells microinjected with these markers can also be visualized; however, exposure to UV light for just a few seconds may cause irreversible damage to the cells. The same applies to injection products labeled with a fluorochrome, like, for instance, antisense oligonucleotides coupled to FITC or rhodamine (see Fig. 1.4A,B).

1. Rinse injected cells gently twice in PBS.
2. Fix briefly with methanol (2 min) or PFA/GA (5 min) (see Materials, above).
3. Rinse the coverslip briefly in deionized water.
4. Mount on a slide with a drop of Mowiol.

If immunofluoresce or any other treatment of the injected cells is to be performed, it is advisable to use a fixable dextran (see Materials, above) to avoid losses of marker by diffusion during the washing steps.

Cells Injected with Reporter Plasmids

X-Gal staining of cells expressing the *E. coli* β-galactosidase

1. Inject a convenient number of cells with a LacZ expressing plasmid. Allow cells express the plasmid for a convenient time.
2. Wash coverslip briefly and gently twice in PBS pH 7.3/2 mM $MgCl_2$.
3. Fix in PFA/GA/2 mM $MgCl_2$ at room temperature for 10 min.
4. Wash twice with PBS pH 7.3 / 2 mM $MgCl_2$.
5. Transfer coverslip to one of a 12-multiwell dish and add 0.5 ml of staining solution containing: 1 mg/ml X-Gal (Biomol, Hamburg, cat. no. 02249, stock 40 mg/ml in dimethyl-formamide), 0.02 % NP-40, 2 mM $MgCl_2$, 5 mM $K_3Fe(CN)_6$ (stock 50 mM in PBS pH 7.3), and 5 mM $K_4Fe(CN)_6$ (stock 50 mM in PBS pH 7.3).
6. Incubate at 37 °C. The staining is usually visible after 1–2 h of incubation, depending on the promoter driving β-Gal expression. Nevertheless, intensity continues to increase over 12–24 h.
7. Rinse twice in PBS. Cover with PBS for examination or, preferably, rinse in deionized water and mount with Mowiol on a glass slide (Fig. 1.4C).

Note: For details of X-Gal staining and quantitation of galactosidase activity see Cid et al. (1993), see also Cid-Arregui et al., Chapter 26, this volume.

Visualization of cells injected with plasmids expressing the green fluorescent protein (GFP)

The GFP (238 amino acids) was isolated from the bioluminiscent jellyfish *Aequorea victoria* (Prasher et al. 1992) as a monomer which emits green fluorescent light upon excitation with ultraviolet or blue light. The GFP retains its fluorescence capacity when expressed recombinantly in eukaryotic cells (Chalfie et al. 1994). Its visualization, however, is limited by the low amount of light that it emits (less than 1 photon per molecule), as compared with most fluorescent compounds (several thousands of photons per molecule). Nevertheless, GFP is a very useful tool that can be utilized as a reporter in gene expression studies, for the subcellular localization of fusion proteins, and in organelle traffic studies.

1. Inject cells with a GFP-expressing plasmid (100 μg/ml in injection buffer). The incubation time for the cells permitting an appropriate level of GFP expression, depends largely on the transcriptional activity of the promoter from which it is expressed. Using the cytomegalovirus (CMV) promoter we have observed visible amounts of GFP as early as 6 h after injection (see Fig. 1.4D). Prior to fixation, some authors recommend incubating the cells at 30 °C for 2 h to enhance the fluorescence capability of the protein.
2. Fix the cells as described in Materials, above. GFP fluorescence persists after methanol and PFA/GA fixation.
3. For some experiments it may be interesting to visualize the GFP in living cells. In this case, it is convenient to supplement the culture medium with N-acetyl cysteine (NAC) to protect cells from damage caused by UV light (Mayer and Noble 1994). Prepare a stock of 500 mM NAC (Sigma A-8199) in MEM, neutralize as needed with NaOH. For use: dilute 1:5000 in culture medium. After having photographed or recorded with a videocamera the expressing cells for a few seconds, rinse the coverslip briefly in PBS, transfer into a dish containing appropriate culture medium (without NAC), return it to the incubator and incubate as required for the experiment.

Cells injected with plasmids expressing chloramphenicol acetyl transferase (CAT)

The CAT assay is very useful for monitoring transcriptional activity of eukaryotic promoters and enhancers (see Gorman et al. 1982). A major advantage of the method is that the activity values obtained are usually proportional to the activity of the promoter. In addition, no endogenous CAT activity is detected in mammalian cells (see Fig 1.5).

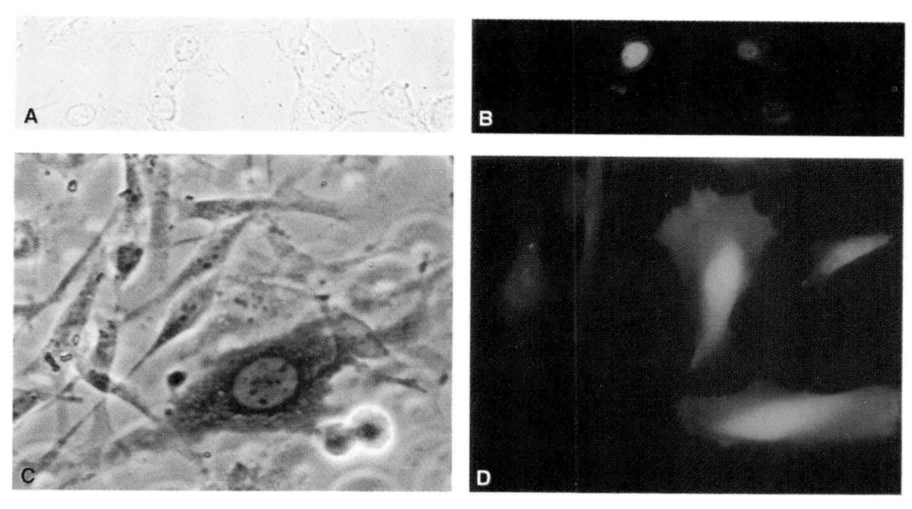

Fig. 1.4A–D. Visualization of fluorescent samples and reporter gene products in injected cells. **A** and **B** Phase contrast (**A**) and fluorescence microscopy (**B**) of FITC-conjugated antisense oligonucleotides in living cells photographed immediately after intranuclear injection. **C** X-gal staining of a cell microinjected with the bacterial LacZ gene expressed from the HCMV promoter. **D** fluorescence microscopy of cells expressing GFP. After injection, the cells were incubated at 37 °C/5 % CO_2 for 6 h, then fixed in methanol for 5 min at −20 °C, air-dried and mounted with Mowiol

CAT Assay

1. Inject at least 100 cells (ideally 300–500) with a solution of 100 mg/ml of a plasmid expressing the CAT gene. Incubate the injected cells for the appropriate time.
2. Rinse in PBS, drain.
3. Scrape the cells with the tip of a pipette loaded with 100 µl of TGD buffer (250 mM Tris pH 8, 15 % glycerol, 5 mM dithiotreitol) while pipetting up and down. Transfer cell suspension to a 1.5-ml microfuge tube.
4. Prepare crude extracts either by sonicating in a water bath or by performing five to six freeze-thaw cycles in a dry-ice/ethanol bath and a 37 °C water bath.
5. Centrifuge at 4 °C for 10 min to eliminate debris.
6. Use the entire cellular extract to determine CAT activity after completing volume to 300 µl with 250 mM Tris pH 8, 15 % glycerol.

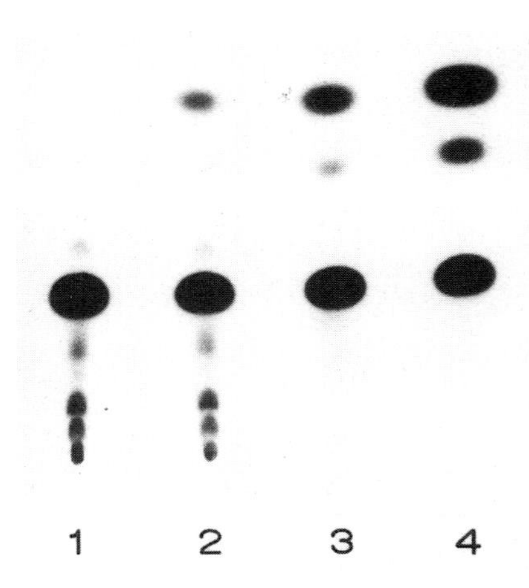

Fig. 1.5. Biochemical analysis (CAT assay) of injected cells. About 100 cells were injected with a reporter plasmid carrying the CAT gene expressed from RSV promoter (Cid et al. 1993). After injection, cells were incubated overnight to allow reporter gene expression, lysed, and used for CAT assay as described in Section 1.2. *1* Control non-injected cells; *2* cells microinjected with 100 µg/µl of RSV-CAT plasmid, the whole extract was used for the assay; *3*, *4* control assay performed with 20 and 100 µg of protein extract, respectively, from cells stably transfected with the same RSV-CAT plasmid used for microinjection

7. Add 8 µl of acetyl CoA (20 mM), 2.5 µl of ^{14}C-chloramphenicol (25 µCi/ml; 0.0625 µCi). Vortex vigorously. Incubate at 37 °C for 45 min.
8. Stop reaction by adding 250 µl of ethylacetate (99.5 %).
9. Centrifuge 10 min at 4 °C in a microfuge.
10. Take upper organic phase containing labeled chloramphenicol and its acetylated forms.
11. Dry in Speed-Vac for 1 h or let stand overnight in a fume hood.
12. Resuspend in 10 µl of ethylacetate.
13. Spot samples in small drops on a silica plate. Samples should be separated at least 1.5 cm from each other, and 2 cm from the edge of the plate.
14. Chromatography. Place the plate vertically into a glass recipient containing an adequate volume of chloroform/methanol (19:1, v/v). Let chromatography run for 45–60 min.
15. Let the plate stand at room temperature for a few min and expose on an X-ray film.

Analysis of Cells Injected with Antibodies

1. After microinjection, incubate cells for at least 30 min.
2. Rinse coverslip gently in PBS.
3. Fix in cold methanol or PFA/GA as described in Materials, above. The optimal conditions for each antibody should be tested previously (see Harlow et al. 1988; Hockfield et al. 1993).
4. Block for 30 min in blocking solution (see Materials, above).
5. Add a FITC- or rhodamine-conjugated secondary antibody diluted in blocking solution 1:10 in PBS. Incubate at room temperature for 1 h, or at 4 °C overnight, in a humid chamber.
6. Wash 5× with PBS, rinse in deionized water and mount with Mowiol.

References

Anderson WF, Killos L, Sanders-Haigh L, Kretchmer PJ, Diakumakos EG (1980) Replication and expression of thymidine kinase and human globin genes microinjected into mouse fibroblasts. Proc Natl Acad Sci 77:5399–5403

Ansorge W, Pepperkok R (1988) Performance of an automated system for capillary microinjection into living cells. J Biochem Biophys Meth 16:283–292

Brinster RL, Chen HY, Trumbauer M, Senear AW, Warren R, Palmiter RD (1981) Somatic expression of herpes thymidine kinase in mice following injection of a fusion gene into eggs. Cell 27:223–231

Capecchi M (1980) High efficiency transformation by direct microinjection of DNA into cultured mammalian cells. Cell 22:479–488

Chalfie M, Tu Y, Euskirchen G, Ward WW, Prasher DC (1994) Green fluorescent protein as a marker for gene expression. Science 263:802–805

Chambers R and Chambers E L (1961) Exploration into the nature of living cell. Harvard Univ. Press, Cambridge, Massachusetts

Cid A, Auewarakul P, García-Carrancá A, Ovseiovich R, Gaissert H, Gissmann L (1993) Cell-type-specific activity of the human papillomavirus type 18 upstream regulatory region in transgenic mice and its modulation by tetradecanoyl phorbol acetate and glucocorticoids. J Virol 67: 6742–6752

Diacumakos E G (1973) Methods for micromanipulation of human somatic cells in culture. Methods Cell Biol 7:287–311

Gordon JW, Scangos GA, Plotkin DJ, Barbosa JA, Ruddle FH (1980) Genetic transformation of mouse embryos by microinjection of purified DNA. Proc Natl Acad Sci 77:7380–7384

Gorman CM, Moffat LF, Howard BH (1982) Recombinant genomes which express chloramphenicol acethyltransferase in mammalian cells. Mol Cell Biol **2:** 1044–1051

Graessmann A (1970) Mikrochirurgische Zellkerntransplantation bei Säugetierzellen. Exp Cell Res 60:373–382

Graessmann A, Graessmann M (1976) 'Early' simian-virus-40-specific RNA contains information for tumor antigen formation and chromatin replication. Proc Natl Acad Sci 73:366–370

Graessmann A, Graessmann M, Müller C (1980) Microinjection of early SV40 DNA fragments and T antigen. Meth Enzymol 65:816–825

Harlow E, Lane D (1988) Antibodies: A laboratory manual. Cold Spring Harbor Laboratory Press. Cold Spring Harbor New York

Hogan B, Beddington R, Costantini F, Lacy E (1994) Manipulating the mouse embryo, a laboratory manual, 2nd edition. Cold Spring Harbor Laboratory Press, Cold Spring Harbor, New York

Hockfield S, Carlson S, Evans C, Levitt P, Pintar J, Silberstein L (1993) Selected methods for antibody and nucleic acid probes. Cold Spring Harbor Laboratory Press. Cold Spring Harbor, New York

Leyden DE (ed) (1985) Silanes, surfaces and interfaces. Gordon and Breach Science Publishers, New York

Mabuchi I, Okuno M (1977) The effect of myosin antibody on the division of starfish blastomeres. J Cell Biol 74:251–263

Mayer M, Noble M (1994) N-acetyl-cystein is a pluripotent protector against cell dead and enhancer of trophic factor-mediated cell survival in vitro. Proc Natl Acad Sci 91:7496–7500

Prasher DC, Eckenrode VK, Ward WW, Prendgast FG, Cormier MJ (1992) Primary structure of the *Aequorea victoria* green-fluorescent protein. Gene 111:229–233

Proctor GN (1992) Microinjection of DNA into mammalian cells in culture: theory and practice. Methods Mol Cell Biol 3:209–231

Sambrook J, Fritsch EF, Maniatis T (1989) Molecular cloning: a laboratory manual, 2nd edition. Cold Spring Harbor Laboratory Press, Cold Spring-Harbor, New York

Schnorf M, Potrycus I, Neuhaus G (1994) Microinjection technique: routine system for characterization of microcapillaries by bubble preasure measurements. Exp Cell Res 210:260–267

Stacey, DW Allfrey VG, Hanafusa H (1977) Microinjection analysis of envelope-glycoprotein messenger activities of avian leucosis viral RNAs. Proc Natl Acad Sci 74:1614–1618

Thomas CA, Saigo K, Mcleod E, Ito J (1979) Then separation of DNA segments attached to proteins. Anal Biochem 93:158–166

Tjian R, Fey G, Graessmann A (1978) Biological activity of purified Simian virus 40 T antigen proteins. Proc Natl Acad Sci 75:1279–1283

Wagner EF, Stewart TA, Mintz B (1981) The human beta-globin and a functional thymidine kinase gene in developing mice. Proc Natl Acad Sci 78:5016–5020

Wassarman, PM, DePamphilis ML (1993) Guide to techniques in mouse development. Methods in Enzimology, vol. 225. Academic Press, New York

Chapter 2

Automated Computer-Assisted Microinjection into cultured somatic cells

Wilhelm Ansorge[1] and Rainer Saffrich[1*]

Introduction

Capillary microinjection into cultured somatic cells growing on a solid support developed rapidly after the introduction of the technique by Graessmann (for an overview see Celis 1986), and is now established as one of the most versatile methods of introducing substances into such living cells. Microinjection made it possible to use single cells as objects to study complex cellular processes, structure and function in vivo. A large variety of molecules like dyes, proteins, and nucleic acids can be injected and their activity studied. Applications for microinjection experiments are found in cytology, cell biology, physiology, molecular biology, developmental biology, and pharmacology for functional analysis of enzymes and enzyme systems, cellular structural proteins, hormonal control mechanisms, intra- and extracellular control mechanisms, signal transduction and cell cycle studies, transport processes, protein biosynthesis, translation and transcription of genetic information, etc.

The most significant disadvantage of the technique is the limitation in the number of cells that can be injected within a given time. This can be a problem if the time for injection is limited, which depends on the experiment; for example, when a large number of cells has to be injected for a biochemical assay or when microinjection is used to produce stable transfected cell lines.

An important improvement in this respect was the introduction of automation in the micromanipulation and microinjection procedures as well as the control and standardization of experimental conditions, like

[1] European Molecular Biology Laboratory, Meyerhofstrasse 1, D-69117 Heidelberg, Germany
* corresponding author: phone: +49-6221-387 535, fax: +49-6221-387 306, e-mail: saffrich@embl-heidelberg.de

preparation of cells, or the production of reproducible injection capillaries (Ansorge 1982; Ansorge and Pepperkok 1988).

The development of a computer-assisted and microprocessor-controlled injection system provides high injection rates with optimum reproducibility of injections and makes quantitative microinjection possible (Pepperkok et al. 1988). The AIS (*a*utomated *i*njection *s*ystem, Zeiss, Germany) permits reliable and simple microinjection into living cells by a system configuration of various hardware and software components operating together to control the injection procedure. The software reliably controls the precise injection process and allows fast and precise axial injections at a rate of up to 1500 cells per hour. Axial injections ensure the highest viability of the injected cells. Retrieving injected cells for analysis, and multiple injections into the same cells are possible with the system. The user friendly and easy-to-operate software program, in addition, shortens the training period for the experimenter.

With the AIS-time consuming test series and experiments, as well as biochemical assays on injected cells, can now be carried out rapidly with outmost precision and reproducibility at an exceptionally high injection rate with less damage to the cells. New biochemical approaches may be possible with this system.

The system has been extended in our lab for fast automated analysis and quantitation of gene expression by computerized microimaging of injected cells (Pepperkok 1993).

The degree of automation described here is still incomplete. Automation efforts have been concentrated on where it is really useful in terms of balancing effort and results. Probably complete automation will be impossible to achieve, because of the complexity and variety of the problems involved.

Outline

Description of the System and the Components

The basic setup of the automated microinjection system is built around an inverted microscope (Zeiss Axiovert) equipped with long-distance phase-contrast objectives for observation (5×-32×). The micromanipulator consists of a motorized stage for accurate positioning of the cells and motorized z-axis drive with the microcapillary holder moving the injection capillary. Injection pressure is supplied from an Eppendorf 5242 pneumatic microinjector. A video system with a CCD camera and a monitor for observation of cells together with the microprocessor control unit and the required software for controlling the microinjection procedure via graphics overlay on the video system, complete the setup. As additional component, epifluorescence excitation equipment should be added to the microscope. A heating stage and a CO_2 incubation chamber are useful and may be necessary for special applications. The working system in our laboratory is shown in Fig. 2.1.

2.1 Operation Principles and Modes of Operation

A menu guides the user through the operation of the AIS system. The basic routines are:

- MARK/INJECT for marking and injection of cells
- FIND CELLS retrieval of injected cells for further injections or analysis of the cells
- SYSTEM SETUP for setting of parameters and calibration of the system

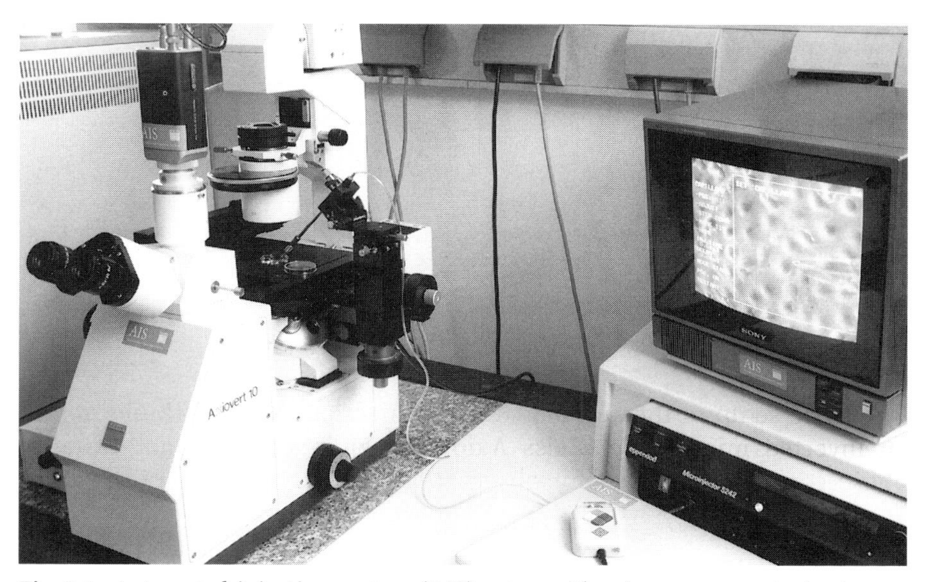

Fig. 2.1. Automated injection system (AIS) setup with microscope, motorized stage and z-micromanipulator, pressure microinjector, and video camera system

The injection parameters such as injection pressure, injection time, injection angle and depth, injection speed, and observation magnification can be selected as required. The reproducible precision of the injections is mainly limited by the stepping motors, it is 0.25 μm with the AIS setup.

The initial system setup and calibrations (camera alignment, stage/video calibration, stage limits) will not be described here. For reference see the AIS operating instructions manual (Zeiss 1989).

First, the principles of the automated injection are described. The injection procedure is performed and controlled via a graphics overlay on the image of the cells acquired with the video system from the microscope and displayed on the monitor. The software is menu-driven and assists the user in marking the cells to be injected with a cursor. The overlay system on the video monitor is important for simultaneous observation of the cells and operation and control of the motorized stage, the micromanipulator, and the injections. No tedious looking through the eyepiece of the microscope to perform injections, but relaxed working conditions, makes the microinjections more successful.

Let us assume that we want to inject the cell in a plane z_c at position x_1/y_1 and the capillary tip is at position x_2/y_2 in a plane z_w above the cell.

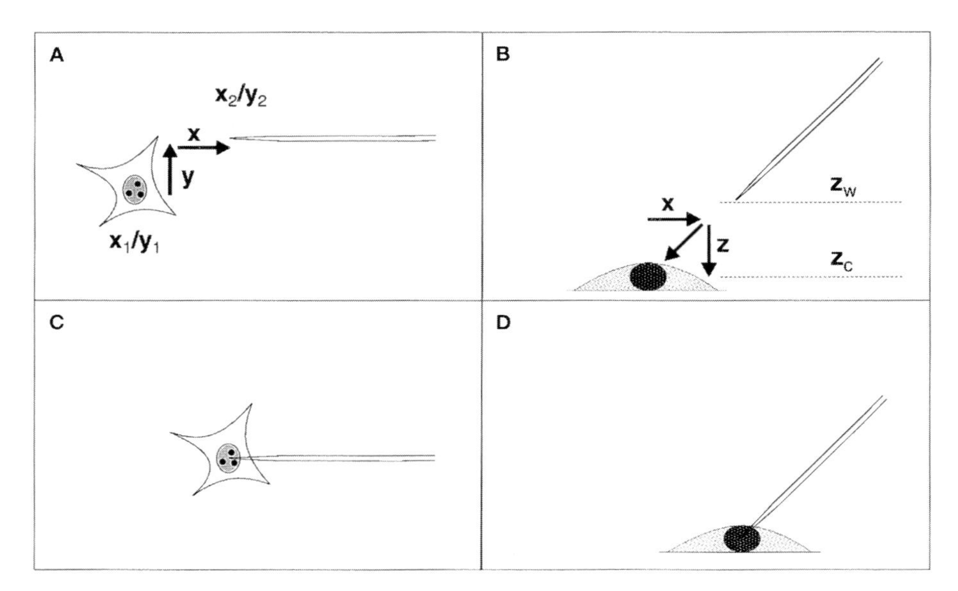

Fig. 2.2 A–D Schematic drawing of the automated injection process. **A** and **C** Top view of capillary and cell before injection and when injecting. **B** and **D** Side view of capillary and cell before injection and when injecting. Details are described in the text

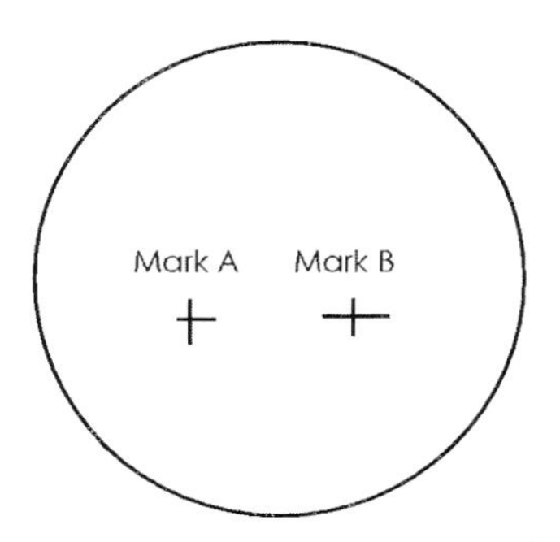

Fig. 2.3. Glass coverslip with *two crosses* as reference marks for storing and retrieving the coordinates of injection positions

This is shown schematically in Fig. 2.2A–D, where the left panel shows a view from top, the right panel a view from the side. The injection point into the cell at position x_1/y_1 is marked via the graphics overlay on the monitor. Now the system moves the cell in x–y direction to the x_2/y_2 position of the injection capillary. The initial movement is in y-direction from y_1 to y_2. Then the capillary is lowered in z-direction from z_w to z_c, together with the stage movement in x-direction from x_1 to x_2 to inject the cell. When the cell is moved with the motorized scanning stage to the location of the capillary tip, the system performs an axial injection according to the defined angle (45° in Fig. 2.2).

Depending on the chosen mode of operation, the injection data are stored in the computer and this allows the continuation of interrupted injection experiments. Intermittent observation of injected cells is also possible by storing their position coordinates on the coverslip in the computer.

Procedure

Operations Before and at the Beginning of the Program Run

Marking of glass coverslips

Each coverslip with cells to be injected has to carry two marks defining a fixed cell coordinate system which is stored on the computer hard disk in a file and allows later retrieval of the injected cells. The marks may be two crosses (see Fig. 2.3) scratched on the glass coverslips using a diamond-tip pen followed by sterilization in an autoclave oven. Orient the coverslip in the petri dish on the microscope stage in such a way that the connecting line between the two marks is parallel to the x-axis of the microscope stage. These marks are entered into the system in an initial dialog menu where the user has to find the two marks and enter their exact position with the cursor. The first mark defines the origin of the coordinate system for the cell coordinates and the second mark the direction of the x-axis.

Filling micropipettes with injection sample

1. To avoid micropipette clogging during microinjection, the sample should always be centrifuged for at least 10 min at 10,000 *g* at 4 °C.
2. Deliver the sample through the rear open end of the micropipette down near to the tip using a sterile microloader from Eppendorf. Typically loaded volumes are 0.2–1 μl.
3. Insert the filled micropipette quickly into the capillary holder of the microinjector and lower it rapidly until the micropipette tip reaches the medium of the culture dish containing cells to be injected as described below.

The speed of step (**3**) is important, because the sample in the micropipette tip may easily dry if it remains too long in the air, and micropipette clogging occurs.

Finding the capillary tip

The following procedure helps to prevent breakage of the micropipette tip during its centering in the microscope field. All steps should be performed in the capillary position CELL chosen in the AIS program menu.

1. Switch to a low objective magnification, i.e., 5×.
2. Focus on the cells.
3. Fix the capillary holder on the z-micromanipulator and position the tip of the capillary approximately straight above the objective.
4. Choose ADJUST on the graphics overlay menu and lower the capillary until it just enters the culture medium in the petri dish.
5. Move the capillary in the x- and y-directions with the screws for manual movement and fine adjustment on the z-micromanipulator until you see the tip centered on the video screen (in the case of problems try to focus on the capillary tip with the objective).
6. Lower capillary and carefully, but not all the way, to its focus to avoid breakage of the micropipette.
7. Change to the next highest magnification.

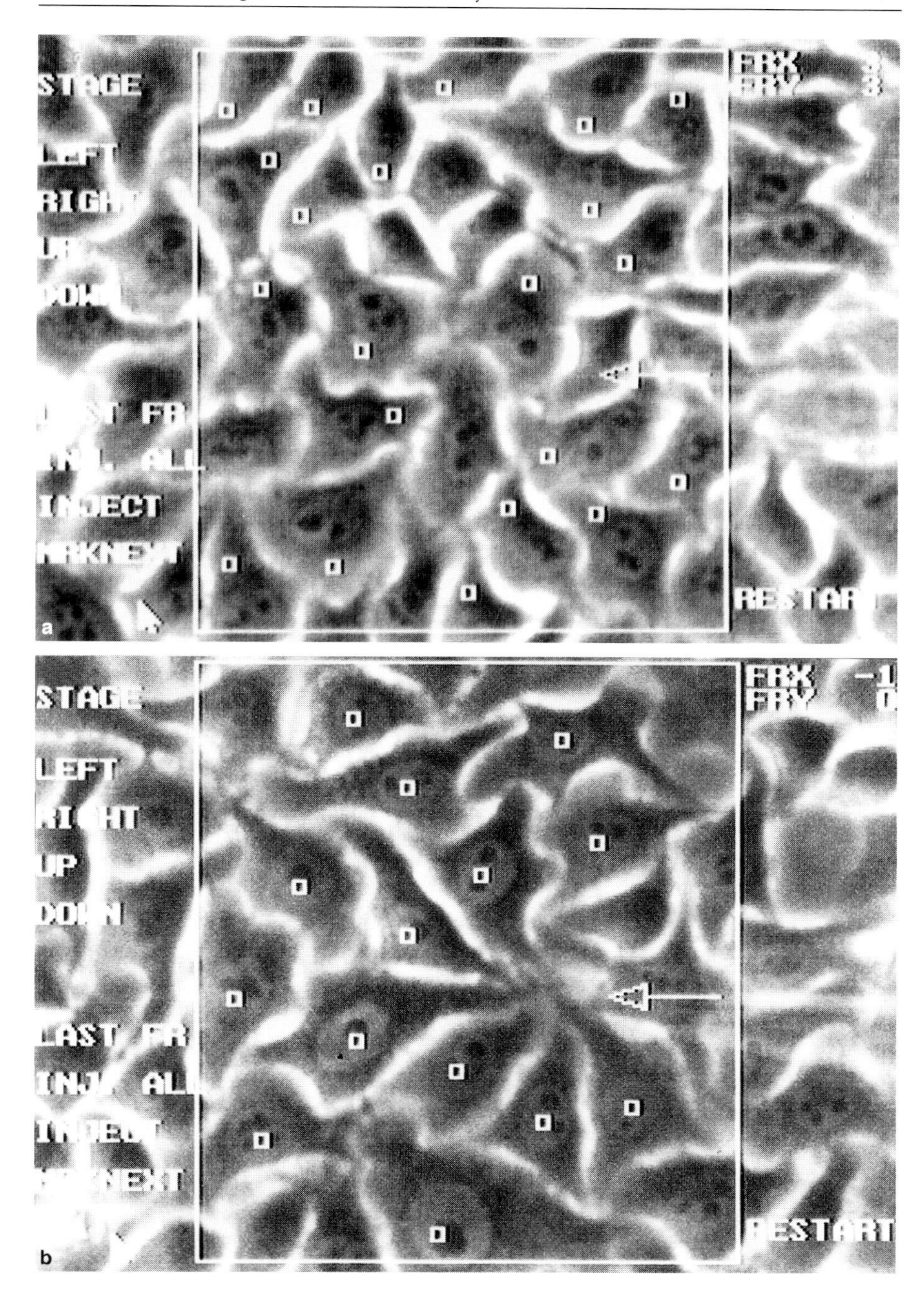

Fig. 2.4a, b. Display of cells and the graphics overlay on the video

8. Repeat steps **5–7** until you reach the working magnification (32×). Now the cells should be in focus and the micropipette should be centered, but not in focus.
9. Lower capillary carefully further down until cells and tip are both in focus. This is the case when the tip touches the cell. Alternatively, lower it until the tip touches the glass surface of the coverslip in the petri dish. Both cells and tip are now in the same focal plane.
10. Mark tip position with the MARK TIP menu item.
11. Change to WAIT position and the capillary is lifted 30 μm (default value) above the plane of the cells.
12. Perform some test injections to adjust the proper injection pressure and injection time by clicking with the cursor on the cells at the desired injection position (cytoplasm or nucleus).
13. Some slight readjustment of the capillary z-position may be necessary and is done with STEP DOWN or ADJUST.
14. Microinjection of cells can now be continued in this way.

The pressure applied to the micropipette remains constant during injection. It is adjusted at the beginning of the session by the microinjector according to the cell type and injection sample used.

Automated Microinjection Procedure

After the initial steps **1–13** described in Section 2.3.1 have been carried out, the system is ready for the next level of automated microinjection, which provides further automation and storage of injection positions for later retrieval of injected cell.

1. Switch the capillary position to WAIT.
2. Choose the program menu POS. OK. This will change to a new graphics overlay menu.
3. Select a frame with cells to inject using the menu items NEXT FRAME or STAGE, then choose MARK NEXT.
4. Now mark all the cells to be injected by pointing with the cursor to the exact position of injection on the cells displayed on the video monitor. Nuclear injections should be performed in the center of the cell

nucleus and cytoplasmic injections are performed just next to the nucleus (see Fig. 2.4a and b).

5. Choose the program menu INJECT. All the marked cells in the field are injected automatically under computer control and their stage coordinates are stored on computer disk.
6. Switch to a neighboring field containing noninjected cells with NEXT FRAME or STAGE and repeat steps **4** to **5.**

Retrieving Microinjected Cells

Every injected cell can be retrieved by its stage coordinates stored on the computer disk during microinjection. For this purpose, the glass coverslip on which the cells are growing must contain two distinguishable orientation marks as explained above (see also Fig. 2.3). In the FIND CELLS program menu, the position of these orientation marks must be entered again into the computer with the cursor as described above to define the cell coordinate system a second time. The program now recalculates the injection positions from the relative orientation of this coordinate system and the first one from the injection session stored on the computer hard disk.

Microinjecting the Same Cells Twice

The procedure to retrieve injected cells is also used to microinject the same cells a second time with, for example, another sample. Therefore the microinjected cells are retrieved and microinjected as described in the above procedures.

2.2 Experiments for Learning and Training Microinjection

To learn microinjection technique with the automatic system, brief experiments, which are useful to get used to the features of the automated microinjection system, and also to control the progress in the technical skills of the experimenter, are shown. The suggested experiments are, in principle, not restricted to automatic microinjection but can also be done with manual and semiautomatic systems; but they are easier to perform with the AIS. First steps in microinjection should be as easy as possible to avoid frustration to the user.

The experiments are simple and do not need special preparation. To start with, choose cells growing flat and which are well spread like HeLa cells, NIH 3T3 cells, or CHO cells. They are most easily injected and are frequently used in injection experiments. As injection material, fluorescent dyes like FITC dextrans are suitable and also a plasmid DNA which can be used as a reporter. In our teaching courses, pSV1 and a GFP vector (Chalfie et al. 1994) are used for this purpose. For a very detailed review on DNA microinjection see Proctor (1992).

Procedure

First Training Steps in Microinjection

The microinjection system has to be equipped for epifluorescence excitation.

Training microinjection

1. Grow cells on coverslips having two marks (crosses) as described above or on coverslips with an etched grid (CELLocate, Eppendorf).
2. Pull capillary and fill 1–2 μl FITC-dextran solution (1 % in PBS, sterile filtered) into the capillary.
3. Place petri dish with cells on coverslip on the microscope stage.
4. Start AIS, define file for storing positions of injected cells, and define the two reference marks or two positions on the gridded coverslips for later retrieval of the cells.
5. Fix capillary in the holder and mount it on the micromanipulator.
6. Focus on cells and bring the capillary into the field of observation as described above.
7. Do some test injections to select proper injection pressure and injection time.
8. Switch to automatic mode.
9. Inject 100 cells into the cytoplasm with the FITC solution.

10. Then, restart and remove petri dish, put back into incubator, incubate for 1 h. Well-injected cells will survive, cells damaged too much by the microinjection will die and usually detach from the coverslip

11. After 1-h incubation, bring the cells back to the AIS , reload file of inject positions, and retrieve the cells with the FIND CELLS menu. Observe cells with fluorescence excitation and count the positive cells in fluorescence and check in phase contrast if they are alive and look healthy. Determine the microinjection success rate (this equals the number of positive cells found again at the retrieved position). The success rate should be >90 % after a few training sessions with the automated system. The same procedure should be carried out for nuclear injection.

The success of injections is determined not only by survival rate of the injected cells but depends also on the type of experiment to be performed. Survival rates can be high, if in most injected cells only a small amount is injected, but the result of the experiment may not be satisfactory. The same may happen if the amount of sample injected is quite high but just not high enough to kill the cells. In this case, the time the cells need to recover from the injection may be too long in that particular experiment or may interfere with the effect to be investigated. Therefore one has also to gain experience for the right amount of sample injected in a certain experiment. This has also to be trained and can only be learned by experimental experience. The next protocol gives some suggestions as to how to gain this experience. This special experiment is most useful to inject DNA to transfect cells or express proteins.

Improving Microinjection Skills

The microinjection system has to be equipped for epifluorescence excitation. Use a reporter plasmid of your choice or pSV1 or a GFP vector (the later has the advantage that the staining steps needed for the other reporter vectors are not necessary). Coinjection of a fluorescent dye should be carried out to control also the survival rate of the cells. The coinjection dye has to be of a color different from the color of the staining used to observe the transfected and expressing cells. In addition, the coinjection dye has to be fixable, otherwise it will be washed out during the later staining procedure (not necessary with GFP, where living cells can be observed).

Improving microinjection

1. Grow cells on coverslips having two marks (crosses) as described above or on coverslips with an etched grid (CELLocate, Eppendorf).
2. Mix plasmid solution with dye solution in a proper ratio to get the appropriate final concentrations of the DNA (i.e..100 ng/μl for pSV1 with a SV40 promotor) and the dye (0.5–1 % dye concentration).
3. Pull capillary and fill 1–2 μl DNA/coinjection marker solution into the capillary.
4. Place petri dish with cells on microscope stage.
5. Start AIS and define the two marks or two positions on the gridded coverslips for later retrieval of the cells.
6. Fix capillary in the holder and mount it on the micromanipulator.
7. Focus on cells and bring the capillary into the field of observation as described above.
8. Perform some test injections to select proper injection pressure and injection time.
9. Switch to automatic mode.
10. Inject 100 cells.
11. Then restart and remove petri dish, put back into incubator, incubate for at least 6 h. Well-injected cells will survive, cells damaged too much by the microinjection will die and usually detach from the coverslip.
12. After incubation, fix and stain cells as necessary, and bring the cells back to the AIS and retrieve the cells with the FIND CELLS mode. Observe cells with fluorescence excitation, count cells that have survived using the fluorescence of the coinjection marker, and check in phase contrast if they are alive and look healthy. Look then for positively transfected cells. Determine microinjection survival rate (this equals the number of coinjection marker positive cells found) and transfection efficiency (the rate of positively stained cells to number of cells that have survived multiplied by 100).

The success rate should be >90 % after a few training sessions with the microinjection system. The survival rate may probably be sometimes higher than 100 % taking into account that injected cells will divide during the incubation time after injection.This is already a positive indication that the cells were not too much disturbed in their cell cycle by the injections with the capillary.

Comments

Several other factors and conditions are of importance for successful and reproducible microinjection like the preparation of the cells or the reproducible production of injection capillaries. Their control and standardization can improve the automated microinjection approach.

- **Production and Handling of Capillaries**
  Good reproducibility of the volumes injected into living cells is achieved with the automatic microinjection system. It is based on the availability of reproducibly prepared micropipettes in shape and tip size. With computer controlled pullers like the Sutter Instruments P87 or commercially available injection capillaries this became possible. For a certain diameter of the capillary opening at the tip, the ejection rate from the capillary depends linearly on the pressure applied and decreases linearly with increasing viscosity of the sample. Capillaries are pulled immediately prior to injection. The tips are sterile directly after pulling due to the high temperature needed to melt the glass and pull the capillaries. No further treatment of the capillaries is carried out. In our experience, this may only introduce additionally problems for the microinjection. Problems arising during microinjection can in most cases be attributed to the sample solution directly and/or mutual interference between the sample and the medium in which the cells are injected. Easy filling of the sample near the tip from the rear open end of the capillary is achieved with microloaders.

- **Characterization of Tip Diameter by Bubble Pressure Measurement**
  The protocol described below allows characterization of the tip diameter by bubble pressure measurement (Schnorf et al. 1994). It is a nondestructive method for determining the tip size and the micropipettes can still be used for microinjection experiments without any restrictions. This is not the case when micropipettes are characterized by electron microscopy.

1. Insert the micropipette into the capillary holder connected to the pressure microinjector (Eppendorf 5242) as for microinjection.
2. Dip the tip into a beaker filled with 100 % ethanol (should be sterile filtered through a 0.22 μm filter).

3. Increase the pressure of the microinjector until air bubbles from the micropipette tip appear, indicating that the tip is open. Typical pressure values for good micropipettes used in our laboratory for nuclear and cytoplasmic injections are between 2000 and 4000 hPa, indicating a tip size of 0.3–0.7 μm. Lower values indicate a tip size above 1 μm. At higher pressure values the micropipette tip will be very small and the risk of clogging during microinjection is rather high.

- **Coinjection Marker Solutions**
  Stock solutions of fluorescent labeled dextrans as coinjection markers for microinjection are prepared as 2 % solutions in either pure H_2O or PBS depending on the application. PBS solutions are used for direct injection or for mixing with the sample solution to inject when used as coinjection marker; H_2O solutions are used for mixing with DNA solutions in coinjection experiments. Final concentration of coinjection marker dextran solutions for injection should be 0.5–1 % for easy detection.

Solutions should always be centrifuged just prior to injection for at least 10 min at 10,000–15,000 U/min to minimize clogging of injection capillaries by small particles and/or aggregates in the sample. Always pipette the solution to fill into the injection capillary from the surface of your sample.

- **Cell preparation**
  Cells are growing on small glass coverslips (∅ 11–15 mm or 10 × 10 mm square) to maintain better planarity of the solid support compared to plastic petri dishes. In addition, breakage of the capillary tip when touching the glass is reduced as compared to plastic material. Place the coverslips with cells into the center of the petri dish. Coverslips resting at the edge of the petri dish tend tostay less plain, and readjustment of the capillary height has to be made more often when moving to a different field on the coverslip. To attain a flat position of the coverslip, press the coverslip slightly down to the bottom with a forceps. Best results with microinjection are obtained with cells which are well spread and do not vary much in their flatness.

References

Ansorge W (1982) Improved system for capillary microinjection into living cells. Exp Cell Res 140:31–37

Ansorge W, Pepperkok R (1988) Performance of an automated system for capillary microinjection into living cells. J Biochem Biophys Methods 16:283–292

Celis E, Graessmann A, Loyter A (1986) Microinjection and organelle transplantation techniques. Academic Press, San Diego

Chalfie M, Tu Y, Euskirchen G, Ward WW, Prasher DC (1994) Green fluorescent protein as a marker for gene expression. Nature 263:802–805

Pepperkok R, Schneider C, Philipson L, Ansorge W (1988) Single cell assay with an automated capillary microinjection system. Exp Cell Res 178:369–376

Pepperkok R, Herr S, Lorenz P, Pyerin W, Ansorge W (1993) System for quantitation of gene expression in single cells by computerized microimaging: application to c-fos expression after micronjection of anti-casein kinase II antibody. Exp Cell Res 204:278–285

Proctor GN (1992) Microinjection of DNA into mammalian cells in culture: theory and practice. Methods Mol Cell Biol 3:209–231

Schnorf M, Potrykus I, Neuhaus G (1994) Microinjection technique: routine system for characterization of microcapillaries by bubble pressure measurement. Exp Cell Res 210:260–267

Zeiss (1989) AIS operating Instructions, Jena

Suppliers

List of Equipment, Materials, and Suppliers

- FITC-Dextran, 70,000 Mw, cat. no. FD-70S, SIGMA (St. Louis, MO, USA)
- FITC-Dextran, 150,000 Mw, cat. no. FD-150, SIGMA (St. Louis, MO, USA)
- TRITC-Dextran, 70,000 Mw, cat. no. R9379, SIGMA (St. Louis, MO, USA)
- FITC-Dextran, fixable, 70,000 Mw, cat. no. D-1822, Molecular Probes (Eugene, OR, USA)
- Petri dishes, 35 mm∅, gamma irradiated, cat .no. 153066, NUNC (Roskilde, DK)
- Petri dishes, 35 mm∅, gamma irradiated, with glass coverslip insert, cat .no. P35G-0-14-C, MatTek (Ashland, MA, USA)
- Glass coverslips, round, 11 mm∅, Menzel Gläser (Braunschweig, Germany).
- Gridded glass coverslips CELLocate, round, 12 mm∅, cat. no. 5245 951.002, Eppendorf (Hamburg, Germany).

- Microloader, cat. no. 5242 956.003, Eppendorf (Hamburg, Germany)
- Prepulled and sterile ready to use micropipettes (Femtotips) cat. no. 5242 952.008, Eppendorf (Hamburg, Germany)
- Raw glass capillaries, cat. no. GC120TF-10, Clark Electromedical Instruments (Reading, UK)
- Pipette Puller model P-87, Sutter Instrument (Novato, CA, USA),
- AIS system, ZEISS (Jena, Germany)

Chapter 3

Immunofluorescence and Immunoelectron Microscopy of Microinjected and Transfected Cultured Cells

GUSTAVO EGEA[1, 2*], TERESA BABIA[1], ROSER PAGAN[2], ROSER BUSCA[2], INMACULADA AYALA[1], FERRAN VALDERRAMA[1], MANUEL REINA[2], AND SENEN VILARO[2]

Introduction

Immunofluorescence microscopy is the most common method to analyze expression and localization of a given protein in cells that have been microinjected or transfected previously with the appropriate DNA constructs. It offers the advantage that it is quick, easy to perform, and allows examination of a large number of cells within a short time. However, illumination with UV light is often damaging for the cells, and the fluorescence tends to bleach as a result of the excitation of the fluorochrome by the UV light. Nowadays, these disadvantages have been overcome by sophisticated systems such as video intensification cameras and confocal microscopy. In addition, the information that can be obtained by immunofluorescence microscopy is also restricted by the limited resolution of the optical lens. Moreover, some fixation conditions can induce artifactual changes in the intracellular localization of a significant number of molecules (Melan and Sluder 1992; Griffiths et al. 1993). Therefore, if the precise localization of a protein is being studied, it will always be necessary to look at the fine structure of the cell, using immunoelectron microscopy techniques. Nevertheless, whenever possible, the utilization of immunofluorescence in combination with immunoelectron microscopy is preferable. In this chapter, we describe in detail protocols for immunofluorescence and immunoelectron microscopy analyses which are particularly suited for cells in culture.

* corresponding author: phone: +34-3-4021909; fax: +34-3-4021907;
e-mail: egea@medicina.ub.es
[1] Dept. Biologia Cel. i Anatomia Patològica, Facultat de Medicina, Fundació Clínic-Universitat de Barcelona, Barcelona, Spain
[2] Dept. Biologia Cel. Animal i Vegetal. Facultat de Biologia. Universitat de Barcelona, Barcelona, Spain

3.1 Immunofluorescence Protocols

We describe here protocols to perform indirect immunofluorescence microscopy. This methodology can be applied to microinjection experiments in which antibodies and intact or epitope-tagged proteins or their respective cDNAs inserted into appropriate shuttle vectors are being microinjected or transfected.

Materials

Reagents

- Ammonium chloride (NH_4Cl), bovine serum albumin (BSA), hydrochloric acid (HCl), glycine, methanol, paraformaldehyde, Tris, Triton X-100, sodium azide (NaN_3), sodium chloride (NaCl), saponin, immunofluorescence mounting medium.

Solutions

- PBS (150 mM NaCl, 0.1 % NaN_3, 0.01 M phosphate buffer, pH 7.2)
- TBS (150 mM NaCl, 0.1 % NaN_3, 50 mM Tris-HCl, pH 7.3)
- Fixative: 4 % paraformaldehyde (PFA) in PBS. To prepare 200 ml of fixative, 8 g of PFA are disolved in 100 ml distilled water and heated to 60–65 °C under constant stirring. The resultant solution has a cloudy appearance. Add 6 µl of HCl 10 N and the fixative solution becomes clear (store this solution at −20 °C in aliquots of 5 ml). Subsequently, the fixative solution is filtered and cooled down at room temperature. Mix the fixative solution with 100 ml of PBS (2×) and adjust the pH to 7.4. This fixative can be mantained at 4 °C for several days, although the pH should be checked before use.
- 0.1 % saponin/1 % BSA in PBS (can be stored as frozen aliquots)
- Methanol cooled at −20 °C
- 50 mM NH_4Cl in PBS
- 100 mM glycine in PBS
- 0.2–0.5 % Triton X-100 in PBS

Procedure

Localization of Microinjected Proteins or Proteins Expressed from Microinjected cDNAs

Localization protocol

1. Rinse coverslips (∅10–12 mm) in PBS.
2. Fix cells in 4 % paraformaldehyde in PBS for 10–15 min at room temperature.
3. Rinse in PBS, 2×5 min.
4. Treat coverslips with 50 mM NH_4Cl or 100 mM glycine in PBS for 30 min to block free aldehyde groups (some investigators miss out this step, but we consider it important to prevent virtual background associated to unspecific binding of antibodies to negatively charged groups).
5. Rinse in PBS, 2×5 min.
6. Transfer coverslips (with cells face down) onto a 10–20-μl drop of 0.1 % saponin/1 % BSA in PBS or TBS for 10 min in a Parafilm sheet in a moist chamber for 15 min. Saponin is used to permeabilize the fixed cells. A 10-min treatment with Triton X-100 in PBS is often used to permeabilize the membranes. Although structure integrity is less preserved, this treatment offers the advantage of having a permanent effect and, thus, does not require continued treatment as occurs with saponin.
7. Drain the excess of permeabilization solution onto paper wipes and transfer each coverslip directly onto a 6–10-μl drop of primary antibody diluted in 0.1 % saponin/1 % BSA/PBS or TBS placed in a moist chamber. Incubate for 60 min at room temperature or overnight at 4 °C.
8. Rinse in PBS or TBS, 2×5–10 min under gentle stirring.
9. Incubate with appropriate fluorescein or rhodamine-labeled secondary antibody also diluted in 0.1 % saponin/1 % BSA/PBS or TBS for 30–45 min at room temperature. We suggest performing this step in the dark by covering the moist chamber with aluminium foil.
10. Rinse in PBS or TBS, 2×5–10 min.
11. Drain the excess of PBS or TBS and mount the coverslip with a small drop of ImmunoFluore mounting medium (ICN) or similar reagent (Mowiol, Gelvatol, etc) onto a microscope slide.

12. Preparations can be visualized immediately and can be stored at 4 °C in the dark for several weeks.

Fixation with methanol

Most samples to be visualized by immunofluorescence microscopy can also be fixed by using methanol. The protocol is then more simple, and is performed as follows:

1. Rinse the coverslips with PBS.
2. Immerse the coverslips in methanol (previously cooled at −20 °C) for 2–3 min only. Methanol induces both fixation and permeabilization of the cells. Thus, the addition of a permeabilization reagent (saponin or Triton X-100) is not necessary. This fixative has the disadvantage that solubilization of membrane components is higher than with aldehyde fixatives.
3. Follow the above protocol from step 5, but using 1 % BSA in PBS or TBS instead of saponin.

Detection of Microinjected Antibodies or Fluorescent-Labeled Molecules

When a polyclonal or monoclonal antibody or a fluorescent-conjugated molecule is microinjected, the experimental procedure for its detection is shorter. Basically, the above protocol can be applied with the exception that incubation with a primary antibody is not necessary. If fluorescein-labeled molecules are microinjected, only the fixation and mounting steps are performed. However, depending on the experimental conditions, it may be appropriate to further incubate with a primary antibody although it had already been microinjected previously (e.g., see Mangeat and Burridge 1985).

3.2 Immunoelectron Microscopy Methodology

The localization of extracellular and cell surface antigens by electron microscopy using antibodies and electro-dense markers has proved to be a powerful tool over the past twenty years. However, localization of intracellular components has been more difficult, since it is essential to provide access for antibodies and markers without significantly altering the fine structure of the cell. There are two major strategies to overcome this problem:

- Permeabilize fixed cells with detergents or polar solvents. Consequently, cell membranes are partially (or completely) destroyed and the entry of antibodies and markers is allowed. Subsequently, the cells are postfixed and processed for common epoxy resin. This technical procedure is known as pre-embedding immunocytochemistry.
- Embed fixed cells, section them, and subsequently incubate with antibodies and markers. This strategy is known as post-embedding immunocytochemistry. Embedding could be performed at low temperatures in a partially water-soluble resin or in a frozen solution containing cryoprotectants.

The most widely used markers for immunoelectron microscopy are:

- horseradish peroxidase (HRP) and
- colloidal gold particles

HRP is practically restricted to pre-embedding procedures and needs diaminobenzidine (DAB) as a substrate to give rise to an electron-dense material which precipitates. The product of this reaction cannot be quantitated and can diffuse to other subcellular compartments. Consequently, the interpretation of the results may be controversial. On the other hand, HRP can provide useful qualitative information in certain studies. However, in microinjection and transfection experiments, colloidal gold has become a very powerful marker, and it is used in studies focused on intracellular trafficking and the fine localization of macromolecules. Particles of colloidal gold exhibit high contrast when they are visualized at the electron microscope. Furthermore, they can be prepared with different diameters (from 1 to 40 nm) and be easily conjugated to numerous macromolecules (i.e., nucleic acids, antibodies, lectins, enzymes, lipids). It is beyond the scope of this chapter to describe the preparation of colloidal gold particles of different size and their conjugation to macro-molecules (for this see Handley 1989; Egea 1993; Griffiths 1993). Moreover, there are numerous commercial available gold-labeled reagents (BioCell, Janssen, Nanoprobes, etc). Thus, only step-by-step immunocytochemical protocols that use colloidal gold particles will be described in detail here.

Finally, cells to be microinjected and studied at EM level are usually seeded onto small sheets of Thermanox plastic (Nunc) that have been previously sterilized by immersion in methanol for 10 min. Thereafter, they are placed at the bottom of petri dishes containing culture medium. Subsequently, cells are microinjected and subjected to the appropriate experimental conditions, then fixed and processed according to the pre-embedding or post-embedding immunolabeling protocols described below.

3.2.1 Pre-embedding procedures using colloidal gold probes

Labeling of cell surface components for endocytosis studies

Labeling of external exposed cell surface components of cultured cells or isolated organelles is probably one of the simplest methods in immunoelectron microscopy. The surface of the cell/organelle is usually accessible to antibodies or molecular probes. In addition, after immunolabeling cell surface components with gold particles, cells can be maintained under proper conditions, so that it is possible to follow the intracellular fate of endocytosed components. Thus, this technique is useful not only for the study of cell surface components, but also for endocytosis studies in cultured cells (see Fernàndez-Borja et al. 1996). The method used in such studies combines immunocytochemistry in live or lightly fixed cells and conventional electron microscopy techniques. As a result, it is possible to observe the subcellular localization of antigens preserving at the same time the fine structure of the cell (see Fig. 3.1). After immunolabeling, samples are fixed and embedded in resin for ultramicrotomy. The embedding procedures vary with respect to chemical compositions of resins, protocols for dehydration, procedures for infiltration with resins, as well as properties of the embedded tissue. Epoxy resins are the most commonly used embedding media in biological electron microscopy. They combine good ultrastructural preservation of the sample, ease of sectioning, reproducibility, and relative ease of handling. Here, we shall describe the method for cell surface immunolabeling as well the embedding procedure that we use for electron microscopy analyses.

Fig. 3.1 A–G. Immunoelectron microscopy of cell surface components and its application to endocytosis studies. Cells expressing the human placental alcaline phosphatase (PLAP, a GPI-linked protein) were processed as indicated in Section 3.2.1 using polyclonal antibodies to PLAP and protein A-gold (15-nm particles) as a marker. The figure shows a gallery of representative intracellular endocytic structures (*arrows*) into which colloidal gold particles can be seen after different times of incubation at 37 °C. Note that PLAP is internalized by nonchlatrin-coated vesicles (**A–C**). Subsequently, gold particles are accumulated in early endosomes (**D–E**) and in tubular-like structures (**E–F**). In addition, gold particles are also located within multivesicular bodies (**G**). *mp* Plasma membrane. *Bar* **A**, **B** 0.3 μm, **C**, **D** 0.15 μm, **E**, **F**, **G** 0.2 μm. ▶

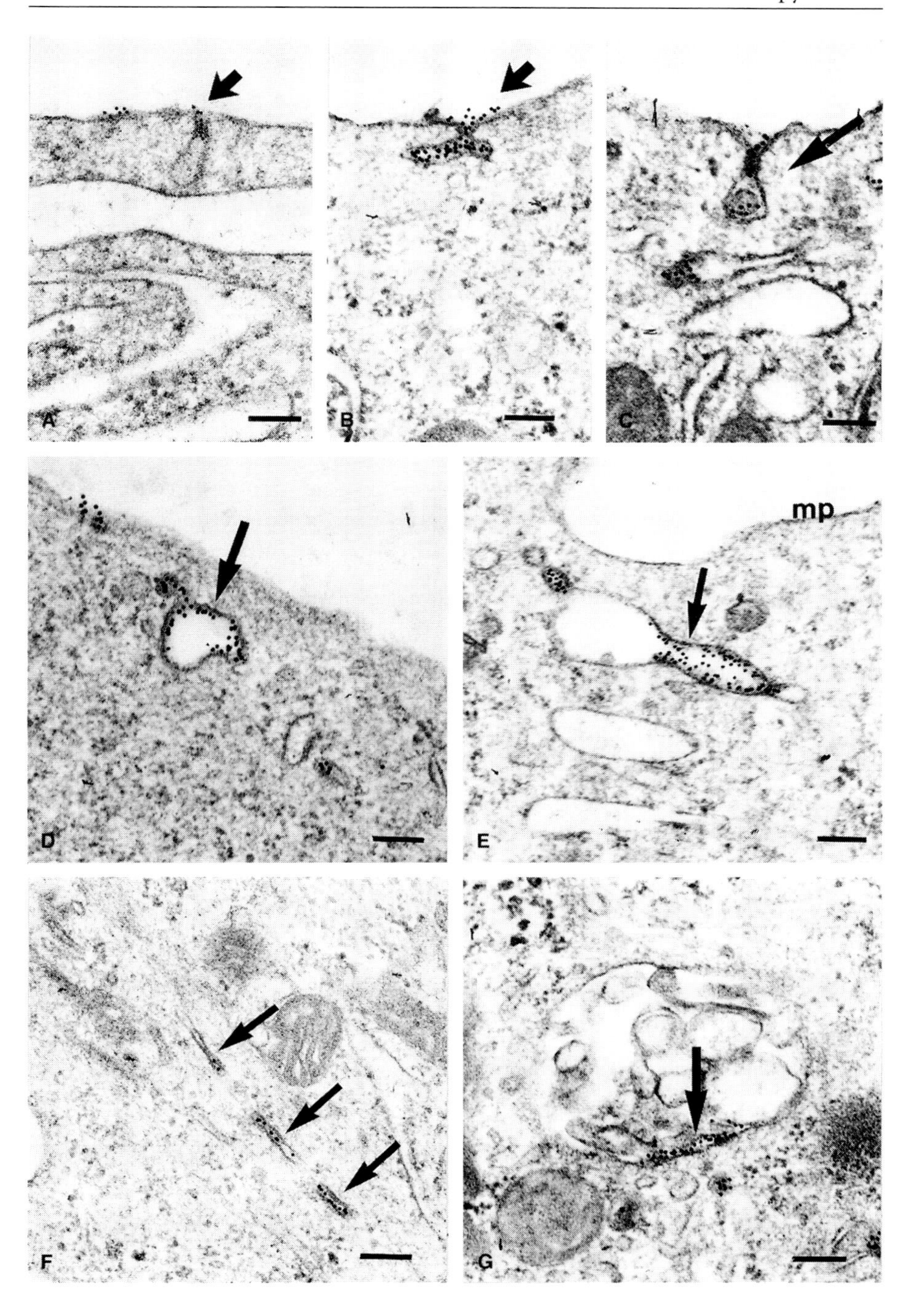
A
B
C
D
E
mp
F
G

Materials

Reagents

- Bovine serum albumin (BSA), cell culture medium, hydrochloric acid, dodecenyl succinic anhydride (DDSA), epoxy resin kit (TO24, TAAB Laboratories Equipment Ltd) containing epoxy resin (TAAB 812), ethanol, glutaraldehyde, gold-conjugated probes (protein A, gold-conjugated secondary antibodies); HEPES, maleic acid ($C_4H_4O_4$), methyl nadic anhydride (MNA), paraformaldehyde, propylene oxide (C_3H_6O), sodium hydroxide, uranyl acetate dihydrate [$(CH_3COO)_2UO$-$2H_2O$], and 2,4,6-tri(dimethylaminomethyl) phenol (DMP-30).

Solutions

- Incubation medium. Prepare 50–100 ml of the same culture medium that is used to culture the cells to be labeled (e.g., DMEM) containing 20 mM HEPES, pH 7.4, and 1 % BSA (incubation medium). Do not use fetal calf serum. Dilute antibodies and gold probes in incubation medium. Antibody should be used at a dilution between 2 and 20 µg/ml. If a large amount of primary antibody is available, it is advisable to prepare Fab fragments conjugated to gold particles to avoid cell surface antigen dimerization or oligomerization (see Harlow and Lane 1988 for general protocols).
- PBS
- 0.1 and 0.2 M cacodylate buffer, pH 7.2
- Fixative: 2 % paraformaldehyde, 2.5 % glutaraldehyde in 0.1 M sodium cacodylate buffer. To make 250 ml fixative solution, mix 25 ml of 25 % aqueous glutaraldehyde stock solution and 25 ml of 20 % paraformaldehyde stock solution (prepared as indicated previously; adjust the pH to 7.2 with NaOH (0.1 N) or HCl (0.1 N). Complete volume to 250 ml with distilled water. The fixative can be maintained at 4 °C for several days. Check pH before use
- Ethanol dilutions. Prepare 70, 90, and 95 % ethanol in water
- Maleate buffer (0.05 M). To make 100 ml, dissolve 0.58 g maleic acid in 80 ml water and adjust pH to 5.2 with NaOH (1 N). Fill up to 100 ml with distilled water
- Uranyl acetate (0.5 %) in 0.05 M sodium maleate buffer. To make 100 ml, dissolve 0.58 g maleic acid in 80 ml water and adjust pH to 6.0 with NaOH (1 N). Dissolve 0.5 g uranyl acetate dihydrate in this solution and adjust pH (if necessary) to 5.2 with NaOH. Fill up to 100 ml with water
- Epoxy resin. To make 100 g resin mix, 48 g TAAB 812, 10 g DDSA, and 33 g MNA. Stir for 5 min. Add 2 g DMP-30 and stir for another 5 min. The complete epoxy mixture should be used for initial infiltration

within the next few hours as it will slowly start to polymerize also at room temperature. The freshly mixed complete resin can be stored at (−20 °C) for months in closed vials. The vials must attain room temperature before being opened and used for embedding

Procedure

Labeling of cell surface proteins

An example where this protocol was used is shown in Figure 3.1.

1. Establish cells cultures using 35-mm diameter dishes.
2. Precool dishes, medium, and antibodies at 4 °C.
3. Wash dishes with incubation medium.
4. Incubate cells for 15 min to 1 h at 4 °C with the appropriate dilution of the primary antibody.
5. Rinse twice in incubation medium at 4 °C.

When primary antibodies or probes conjugated to gold particles are not available, repeat steps **4** and **5**, incubating the dishes at 4 °C with the apropiate dilution of gold markers.

6. Add prewarm incubation medium to the dishes and tranfer them to a 37 °C cell culture incubator and take one dish at each different incubation time (e.g., 0, 15, 30, and 60 min).
7. After each time period, fix cells with a mixture of 2 % paraformadehyde and 2.5 % glutaraldehyde in phosphate or cacodylate buffer for 1 h at room temperature.
8. Rinse dishes with PBS.
9. Scrape cells in a small volume of PBS (i.e, 200 μl) and collect them by centrifugation.
10. Cell pellets are embedded in 10 % gelatin and postfixed overnight in 2 % paraformaldehyde at 4 °C. Fixed gelatin with embedded cells can be easily cut into blocks; this facilitates further handling of the sample.
11. Postfix in 1 % OSO_4 for 1 h at room temperature, rinse, and proceed as indicated below for plastic embedding.

Embedding

The following procedure results in embedded blocks with good cutting properties from a variety of cell types. It includes in-bloc staining, which improves contrast in the subsequent section staining. The same procedure can be used for embedding tissues, cells, or organelle samples.

12. Rinse fixed cells for 2×30 min in buffer (e.g., same buffer as used for fixation). Perform subsequent steps at 4 °C until step **17.** If in-bloc staining is not desired, proceed directly to step **16.**

13. For in bloc staining rinse for 2×15 min with sodium maleate buffer. As in all following steps up to **21**, remove fluid with a Pasteur pipette. The sample must never be allowed to dry.

14. Stain for 60 min in uranyl acetate in maleate buffer.

15. Rinse for 2×15 min in sodium maleate buffer.

16. Dehydrate for 2×15 min in 70 % ethanol.

17. Dehydrate for 2×15 min in 90 % ethanol. This and following steps of dehydration and infiltration are carried out at room temperature.

18. Dehydrate for 2×15 min in 95 % ethanol.

19. Dehydrate for 2×15 min in absolute ethanol.

20. Place sample for 2×15 min in propylene oxide. Take particular care that the sample does not dry, as propylene oxide evaporates very rapidly. Because propylene oxide is toxic and very volatile, this and all subsequent steps should be carried out in a well-ventilated hood with gloves.

21. Infiltrate the specimens for 60 min in a mixture of 50 % propylene oxide and 50 % completely mixed epoxy resin.

22. Transfer the specimens to the surface of the epoxy resin in a clean vial containing 100 % resin. Use fine forceps or a wooden stick. Leave the specimens in the epoxy resin overnight at room temperature.

23. Fill the flat embedding mold with epoxy resin.

24. Polymerize the specimens at 60 °C for 2 days.

25. Obtain ultrathin sections acording to conventional procedures.

Labeling of intracellular components

As indicated previously, one of the problems to overcome in pre-embedding procedures is the different barriers which impair diffusion of immunoreagents. However, for microinjected or transfected cells in culture, this does not represent an important problem as it occurs with tissue sections, because extracellular matrix components are practically absent. Furthermore, in some experimental approaches it is possible and even convenient to microinject gold-conjugated products (i.e., proteins or other macromolecules). In these cases, no further immunocytochemical reactions are necessary, and the common fixation and embedding protocols can be followed (Dworetzky 1991).

Materials

Reagents

Ammonium chloride (NH_4Cl), bovine serum albumin (BSA), di-sodium phosphate anhydrous (Na_2HPO_4), ethanol, epon, gelatin, glutaraldehyde (25 or 50 % in water), gold-labeled reagents (protein A, streptavidin, anti-biotin, secondary antibodies), methanol, osmium tetroxide (OsO_4), paraformaldehyde, potassium ferrocyanide ($K_3Fe[CN]_6$), saponin, skimmed milk, sodium cacodylate, sodium di-hydrogen-phosphate monohydrate ($NaH_2PO4\text{-}H_2O$) Triton X-100, uranyl acetate.

Solutions

- PBS
- 0.1 and 0.2 M phosphate buffer, pH 7.4
- 0.1 and 0.2 M cacodylate buffer, pH 7.4
- Fixatives: 4 or 8 % paraformaldehyde in 0.1 M phosphate or cacodylate buffer (prepare as described previously); 1 % glutaraldehyde in 0.1M phosphate or cacodylate buffer (add 1 ml of glutaraldehyde (–25 % in water) to 24 ml of buffer); 1 % uranyl acetate in PBS or distilled water; 1–2 % OsO_4 in PBS or distilled water; 1 % OsO_4/0.8 % $K_3Fe[CN]_6$ in PBS or distilled water
- 50 mM NH_4Cl in PBS
- 0.2–0.5 % Triton X-100 in PBS
- 5 % skimmed milk
- 0.1 % gelatin in PBS
- 1 % BSA in PBS
- 0.1 % saponin/1 % BSA in PBS

Procedure

Common protocol for localization of intracellular molecules

The fixation and extraction steps of this protocol are very similar to those described for immunofluorescence.

1. Fix cells or plastic sheets with paraformaldehyde (4–8 % in PBS or 0.1 M phosphate or cacodylate buffer).
2. Rinse in PBS.
3. Quench free aldehyde groups with 50 mM NH_4Cl in PBS.
4. Extract the cells with 0.2–0.5 % Triton X-100 in PBS for 10 min. This procedure is very appropriate when cytoskeletal proteins are being studied (Kouklis et al. 1991; Shelden and Wadsworth 1992; Takeda et al. 1994, Tomkiel et al. 1994), but when a fine structure should be preserved, it is advisable to follow the saponin method as described above. Note that saponin should be present in all subsequent incubation steps.
5. Rinse in PBS.
6. Treat the cells with 5 % skimmed milk solution or 0.1 % gelatin or 1 % BSA to block nonspecific reaction with antibodies.
7. Incubate with the primary antibody properly diluted in the blocking solution for 3 h at room temperature.
8. Rinse in PBS.
9. Incubate with protein A-gold or gold-labeled secondary antibody diluted in PBS for 1 h at room temperature or overnight at 4 °C.

Note: To facilitate a compromise between a good preservation of the fine cell structure and a good epitope accesibility, reagents complexed to ultrasmall gold particles (1.4 to 3 nm) should be used. In this case, it is necessary to perform a silver amplification of these gold particles. It is known that gold particles can nucleate silver deposition from an appropriate silver solution. This silver-coated gold particle catalyzes more silver deposition and the silver grains grow in size. Nowadays, there are commercial silver amplification kits (Amersham intenSE, Janssen, Nanoprobes HQ Silver, BioCell) and the amplification reaction can be used either just after immunocytochemical reaction with the gold-labeled reagent preceding Epon embedding (Kouklis et al.1991; Campbell and Gorbsky 1995), or directly onto Epon or Lowicryl ultrathin sections (Paterson et al. 1995).

10. Rinse thoroughly in PBS.

11. Fix in 1 % glutaraldehyde in 0.1 M phosphate/cacodylate buffer (sometimes 0.1 % tannic acid is added to improve membrane contrasting) for 15 min at room temperature.

12. Rinse 3×5 min, in 0.1 M phosphate/cacodylate buffer.

13. Postfix with 1 % osmium tetroxide (OsO_4) in PBS for 1–2 h at 4 °C in the dark.

14. Rinse in distilled water and stain with 1 % uranyl acetate, dehydrate in increased sequential ethanol grade baths, and embed in Epon (as described previously).

Alternative protocol for pre-embedding

A succesful simple pre-embedding protocol for immunogold electron microscopy that combines dry methanol fixation with anti-biotin antibodies has been described by Miller et al. (1993). These authors microinjected a biotinylated protein and proceeded as follows:

1. Fix the injected cells with methanol (−20 °C) for 3 min.
2. Rinse in PBS for 1 min.
3. Incubate with goat anti-biotin and normal donkey serum at 1:20 dilution in 0.1 M cacodylate buffer for 30 min at 37 °C.
4. Rinse in cacodylate buffer 3×9 min.
5. Incubate with rabbit anti-goat secondary antibody conjugated with 5 nm gold particles at 1:5 dilution in cacodylate buffer for 30 min at 37 °C.
6. Rinse in cacodylate buffer.
7. Fix again with 1 % glutaraldehyde in cacodylate buffer for 1 h.
8. Rinse three times, for 20 min each, in cacodylate buffer.
9. Postfix with 1 % OsO_4, 0.8 % potassium ferrocyanide ($K_3Fe[CN]_6$) in cacodylate buffer for 2–6 h at 4 °C in the dark.
10. Rinse in distilled water and counterstain with 1 % uranyl acetate, dehydrate with graded ethanols and embed in Epon as above.

Labeling of Intracellular Components. Whole-mount Immuno Cytochemistry

Within the cell or the tissues, the filamentous structures of cytoskeleton are distributed in three dimensions. However, the spatial distribution of the cytoskeleton could not be viewed by the conventional methods of electron microscopy on ultrathin sections. For this, electron microscopy in whole-mount preparation is the best choice to study the whole cytoskeleton distribution of cultured cells (Pagan et al. 1996). This technique takes advantage of the detergent/solvent insolubility of the cytoskeleton components as well as their associated proteins. The same method can also be used for detection of cytoskeleton-associated molecules. In this method, cells are plated onto electron microscope grids (200–300 mesh) as for coverslips, and the density chosen at the subconfluent state to give one or two cells per grid square after attachment and spreading. In general, cells spread more slowly on plastic films than on glass. To improve spreading, prior to plating, the filmed grids are incubated overnight with a drop of serum.

Materials

Solutions

- Cytoskeleton buffer (CB): 10 mM PIPES, 100 mM NaCl, 300 mM sucrose, 3 mM $MgCl_2$ $6H_2O$, 1 mM EGTA, 1.7 mM phenylmethylsulfonyl fluoride (PMSF), pH 6.8. To make 1 l, weigh 3.46 g PIPES, 5.8 g NaCl, 102.6 g sucrose, 0.608 g $MgCl_26H_2O$, 0.076 g EGTA. Dissolve in 900 ml H_2O, adjust pH to 6.8 with NaOH (1 N), and fill up to 1 l. Store at 4 °C. Add PMSF (1.7 nM) before use
- 0.5 % Triton X-100 in CB. Add 500 μl of Triton X-100 in 100 ml of CB. Prepare before use.
- Paraformaldehyde solution; 2 % solution of paraformaldehyde in 0.5 % Triton X-100 in CB, pH 6.8. Prepare just before use
- Glutaraldehyde solution; 2.5 % solution of glutaraldehyde made up in 10 mM phosphate buffer (PB). Prepare just before use.

Procedure

Protocol for whole-mount immunocyto-chemistry

The application of this protocol is illustrated in Fig. 3.2.
Cells should be grown on sterilized gold grids covered with Formvar and coated with carbon. To prepare Formvar-coated grids, the reader is addressed to more specialized EM books (e.g., Hayat 1989).

1. Rinse grids gently in PBS or CB at room temperature (5 s).
2. Extract the cells with 0.5 % Triton X-100 in CB for 3 min at 4 °C.
3. Fix cells in 2 % paraformaldehyde solution for 30 min at 4 °C.
4. Rinse 3×5 min in 20 mM glycine-10 mM PB, at room temperature.
5. Block free aldehyde groups in 1 % BSA in 20 mM glycine-10 mM PB, 20 min.
6. Incubate cells with 10–12 µl of primary antibody or primary antibody mixture (if double labeling is desired) diluted in 20 mM glycine-10 mM PB for 30 min.
7. Rinse 3×5 min in 20 mM glycine-10 mM PB.
8. Incubate the cells with gold-labeled secondary antibody mixture also diluted in 20 mM glycine-10 mM PB for 20 min.
9. Rrinse 3×5 min in 20 mM glycine-10 mM PB.
10. Rinse 3×5 min in 10 mM PB.
11. Rinse 3×5 min in 10 mM PB.
12. Postfix cells in glutaraldehyde solution for 1 h.
13. Rinse cells 3×5 min in 10 mM PB.
14. Postfix cells with 1 % OsO4 in 10 mM PB 1 h.
15. Rinse 3×5 min in PBS.
16. Rinse 3×5 min in distilled water.
17. Dehydrate grids through a series of increasing ethanol concentrations:
    a) 50 % ethanol (2×5 min)
    b) 70 % ethanol (2×5 min)
    c) 90 % ethanol(2×5 min)
    d) absolute ethanol (2×5 min).

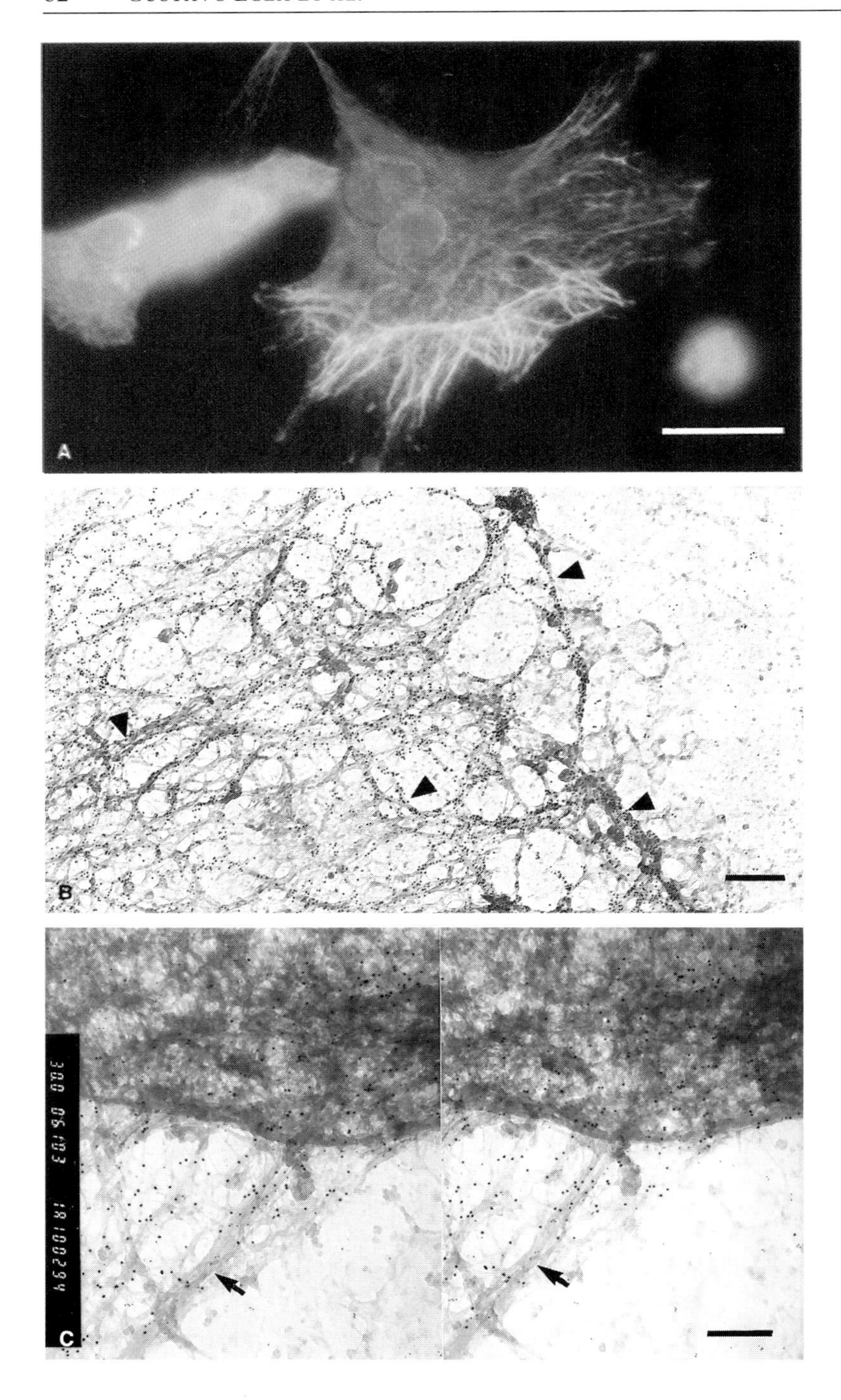
A
B
C

◀ **Fig. 3.2A–C.** Labeling of intracellular components using the whole-mount procedures. The figure shows cells coexpressing two intermediate filaments (vimentin and cytokeratin) and their distribution. **A** Hepatocytes were fixed in methanol (−20 °C) for 2 min, rinsed in PBS, and processed for indirect double immunofluorescence, as indicated in Section 3.2. Primary antibodies were a mouse anti-vimentin, which was detected by a FITC-conjugated goat anti-mouse (green), and a rabbit anti-cytokeratin 18 which was detected by a TRITC-conjugated goat anti-rabbit (red). Hoechst 33342 was used for nuclear staining (blue). The picture was obtained by triple exposure. **B** and **C** Samples were obtained following the whole-mount and embedded-free procedures described in the text. Cells were grown on gold grids covered with Formvar and coated with carbon. Following extraction in cytoskeleton buffer, cells were fixed and stained with the polyclonal antibody against cytokeratin 18 (**B**) and with the same mixture of primary antibodies used in **A** (**C**). Gold-conjugated anti-rabbit (15 nm) and anti-mouse (10 nm) immunoglobulins were used as secondary antibodies to detect cytokeratin and vimentin, respectively. Electron and stereo-electron micrographs were obtained in a Hitachi electron microscope. Stereoscopic images were made at 10° total tilt angle. Note in **B** the complex network of cytokeratin filaments (*arrowheads*). **C** shows the simultaneous localization of vimentin (10 nm) and cytokeratin (15 nm) in the same bundle (*arrow*). *Bar* **A** 20 μm, **B** 0,2 μm, **C** 0,1 μm

18. Before EM observation cells must be dried through the CO_2 critical point. It is necessary to place the grids in a transition fluid, amyl acetate. To this end, perform the following steps:
a) place the grids in a mixture of ethanol-amyl acetate 3:1 (2×5 min)
b) ethanol-amyl acetate 2:2 (2×5 min)
c) ethanol-amyl acetate 1:3 (2×5 min)
d) pure amyl acetate.

19. Dry cells through the CO_2 critical point, coat lightly with carbon, and examine in a transmission electron microscope.

Intracellular Labeling. DNP Immunogold Method

This is a relatively novel method (Pathak and Anderson 1989, 1991; Rothemberg et al. 1992) for localizing sparse antigens in thin sections. It combines the sensitivity of the horseradish peroxidase with the quantitative capability of the gold particles and consists of a pre-embedding and a post-embedding procedure. The pre-embedding method uses low concentration of permeabilization detergents and a secondary antibody that has multiple covalently attached dinitrophenol (DNP; for conjugation of DNP molecule see Little and Eisen 1967 and Pathak and Anderson 1989). After plastic embedding, the DNP residues are readily detected with a tertiary anti-DNP antibody followed by protein A-gold labeling or gold-conjugated antibodies following the common methods of post-embedding in plastic sections (see Sect. 3.2). The method offers good sensitivity in addition to very good morphology. However, this method requires time and good experience in EM methods.

Materials

Reagents

Albumin bovine (BSA Sigma Faction V), ammonium chloride (NH_4Cl), DNP-goat anti-mouse or DNP-goat anti-rabbit according to primary antibody used, dodecenyl succinic anhydride (DDSA), ethanol absolute, eponate, Formvar-carbon coated grids, glutaraldehyde (50 %), goat anti-rabbit IgG-10 nm gold conjugated, lead citrate, magnesium chloride (Mg Cl_2), monoclonal anti-DNP (Oxford Biomedical Research), nickel grids, osmium tetroxide (OsO_4), paraformaldehyde (PFA), potassium chloride (KCl), propylene oxide (PO), rabbit anti-mouse IgG, saponin, sodium azide (NaN_3), sodium chloride, sodium hydroxide (NaOH), sodium metaperiodate ($NaIO_3$), sodium phosphate (dibasic) (Na_2HPO_4-$7H_2O$), sodium, phosphate (monobasic) (NaH_2PO_4-H_2O), tannic acid, Tris, uranyl acetate, 2,4,6-tri (dimethyl-aminomethly)phenol (DNP-30).

Stock solutions

- Sodium phosphate buffer (sPB) 0.2 M, pH 7.4. Solution A: 0.2 M sodium phosphate (monobasic). Solution B: 0.2 M sodium phosphate (dibasic). Add sol A to sol B (until pH=7.4, or to the desired value)
- Paraformaldehyde 16 %
- Potassium chloride (KCl), 1 M
- Magnessium chloride ($MgCl_2 6H_2O$), 1 M
- Ammonium chloride NH_4 Cl, 2 M
- 2,4,6-trinitrophenol (picric acid) 50 nM
- Uranyl acetate 4 %. Keep refrigerated, protected from light and filtrate before use
- Sodium metaperiodate (saturated solution). Dissolve to 1 g/ml in distilled water
- Tannic acid 1 %. Keep refrigerated, protected from light and filtrate before use
- Lead citrate. In a 50 ml tube: to 1.33 g lead nitrate add 1.76 g sodium citrate. Add 30 ml boiled, cooled, filtered distilled H_2O. Shake for 1 min. Let stand 30 min with intermittent shaking. Add 8 ml 1 N NaOH

(made fresh). Make volume to 50 ml with boiled, cooled, filtered H_2O. Invert to mix gently. Store at 4 °C, tightly covered. Filter before use.

Working solutions

- Buffer A. 0.1 M sPB, 3 mM$MgCl_2$, 3 mM KCl, pH 7.4. To prepare 1 l, add 500 ml sPB (0.2 M), 3 ml $MgCl_2$ (1 M), 3 ml KCl (1M), and fill up with distilled water
- Buffer B. Buffer A with 0.2 % BSA and 0.01 % saponin. Dissolve 0.2 g of BSA (faction V) and 10 mg saponin in 100 ml of buffer A
- Buffer C. 20 mM Tris, 3 mM NaN_3, 225 mM NaCl, pH 8
- Fixative I. 3 % PFA, 3 mM TNP, 3 mM $MgCl_2$, 3 mM KCl, 0.1 M sodium phosphate buffer, pH 7.4. For 30 ml (enough for ten dishes of 60 mm^2) add 15 ml sPB 0.2 M, 5.6 ml PFA 16 %, 90 µl $MgCl_2$ 1 M, 90 µl KCl 1 M, 1.8 ml TNP 50 mM, and fill up to 30 ml with distilled water
- Fixative II. 1.33 % gluraraldehyde in buffer A. To prepare 30 ml add 15 ml sPB 0.2 M, 0.798 ml gluraraldehyde 50 %, 90 µl $MgCl_2$ 1 M, 90 µl KCl 1 M, and made up to 30 ml water
- Fixative III. 1 % OsO_4, 1.5 % potassium ferrocyanide in buffer A. To prepare 20 ml add 10 ml sPB 0.2 M, 60 µl $MgCl_2$ 1 M, 60 µl KCl 1 M, 300 mg potasium ferrocyanide, 5 ml OsO_4 4 %, and make up to 20 ml with distilled water
- Blocking solution. 0.1 M NH_4Cl in Buffer A. Prepare 60 ml for two blocking steps as follows: 30 ml sPB 0.2 M, 3 ml NH_4Cl 2 M, 180 µl $MgCl_2$ 1 M, 180 µl KCl 1 M and fill up with distilled water
- Tannic acid 0.02 % in buffer A. To prepare 30 ml add 15 ml PB 0.2 M, 600 µl tannic acid 1 %, 90 µl $MgCl_2$, 90 µl KCl 1 M, and fill up with distilled water.
- Epon mixture. Epon 12 (15.3 g), DDSA (8.67 g), NMA (7.31 g), DMP-30 (435 µl). Prepare as indicated previously.

Procedure

Intracellular Labeling Using DNP Immunogold (see Fig. 3.3)

Pre-embedding for labeling with DNP immunogold

1. Fix cells, grown in culture dishes, with fixative I; 3 ml per dish, 1 h, at room temperature.
2. Rinse in buffer A, 3×5 min.
3. Block free aldehyde groups with 0.1 M NH_4Cl in buffer A; 3 ml per dish, 30 min at 4 °C.

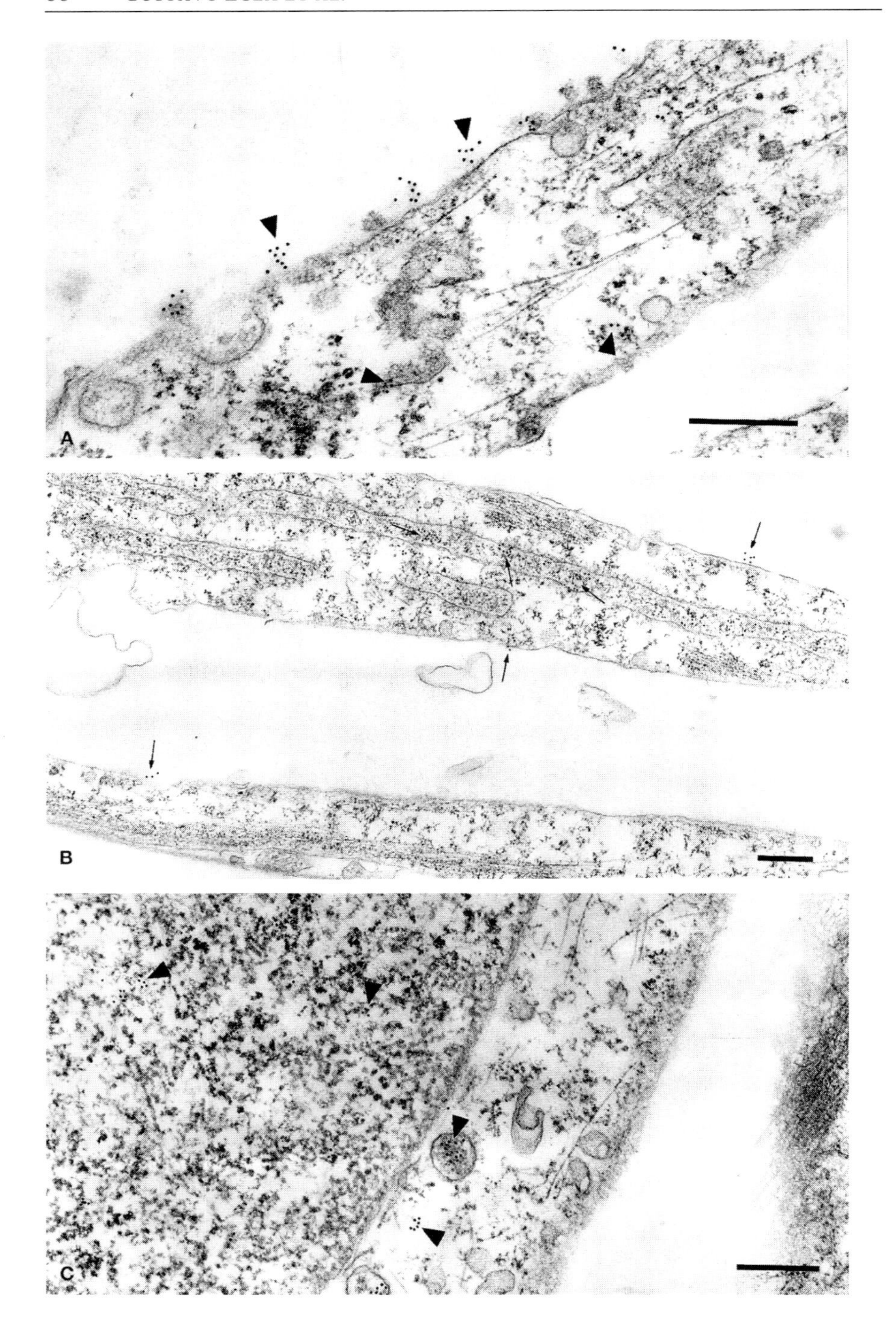
A
B
C

◀ **Fig. 3.3A–C.** Intracellular labeling using the DNP immunogold method. Three representative examples of immunolabeling obtained with fibroblast cells by using the dinitrophenol-immunogold method. **A** A cell surface-bound protein (lipoprotein lipase) was detected using monoclonal antibodies against this protein (*arrowheads*). **B** A rough endoplasmic reticulum resident protein (the receptor-associated protein, also known as RAP) was detected using polyclonal antibodies (*arrows*). **C** A recycling cell surface protein (the low-density lipoprotein receptor related protein/a2M receptor or LRP) was detected using polyclonal antibodies (*arrowheads*). Gold particles used were of 10 nm. *Bar* 0.4 mm

4. Rinse in buffer A, 3×5 min.
5. Incubate primary antibody diluted in buffer B (if you use crystalline BSA use 0.15 %) for 1–2 h at 4 °C. 1 ml of antibody solution is enough for ∅60-mm dishes. Dishes should be kept on agitation throughout the incubation time, otherwise some areas of them might dry.
6. Rinse in buffer B, 4×5 min.
7. Incubate cells with DNP-conjugated secondary antibodies. Diluted in buffer B at 15–20 µg/ml. Use 2 ml/dish, 1–2 h at 4 °C.
8. Rinse in buffer B 4×5 min.
9. Rinse in buffer A 3×5 min.
10. Fix with fixative II; 3 ml/dish for 1 h at room temperature.
11. Rinse in buffer A 3×5 min. First overnight stop.
12. Block free aldehyde groups with 0.1 M NH_4Cl in buffer A for 30 min.
13. Rinse in buffer A, 3×5 min.
14. Fix with fixative III, 2 ml/dish, 1 h at room temperature.
15. Rinse in buffer A, 3×5 min.
16. Contrast with 0.02 % tannic acid in buffer A for 30 min. Protect from light by covering dishes with aluminum foil.
17. Rinse in buffer A, 3×5 min.
18. Rinse in distilled water, 3×5 min.
19. Ethanol 30 %, 2×7.5 min.
20. Ethanol 50 %, 2×7.5 min.

21. Ethanol 70 %, 2×7.5 min. Second overnight step at 4 °C.
22. Stain with 0.25 %, uranyl acetate in 70 % ethanol for 1 h. Protect from the light.
23. Ethanol 90 %, 2×25 min.
24. Ethanol 100 %, 2×30 min.
25. Detach cells from the dishes with propylene oxide (PO) at 4 °C, and transfer the cell suspension to Eppendorf tubes. **Note:** Be careful in this step because long-term treatments with propylene oxide can dissolve the plastic of the culture dish. Pellet cells by centrifugation (1 min at 12,000 rpm). Take off the supernatant, add new PO, and resuspend the cells. Centrifugate and repeat the step once more.
26. Tranfer pellets to a mixture of 1 vol Epon:2 vol PO for 1 h. Resuspend the pellets and centrifuge in a Beckman microfuge (do this in all subsequent steps).
27. Add a mixture of 1 vol Epon:1 vol PO, for 1 h.
28. Add a mixture of 2 vol Epon:1 vol PO for 1 h.
29. Epon overnight at room temperature.
30. Epon, 2 h at room temperature.
31. Epon , 12–24 h at 60 °C.
32. Perform ultrathin sections according to conventional procedures.

Post-embedding

33. Float gold or silver-colored sections on Formvar-carbon-coated nickel grids in satured $NaIO_3$ for 40 min.
34. Rinse sections by jet wash (2 ml/s) with distilled water, 1 min/grid (see Fig. 3.4c).
35. Air-dry.
36. Float on a drop of 1 % BSA (or 1 % ovalbumin) in buffer C for 60 min. (Fig. 3.4a,b).
37. Incubate sections with 6–10 ml monoclonal anti-DNP (1–3 μg/ml) in buffer C, containing 0.1 % BSA. Overnight at 4 °C or 2 h at room temperature.
38. Rinse in buffer C/ 0.1 % BSA, 4×5 min.

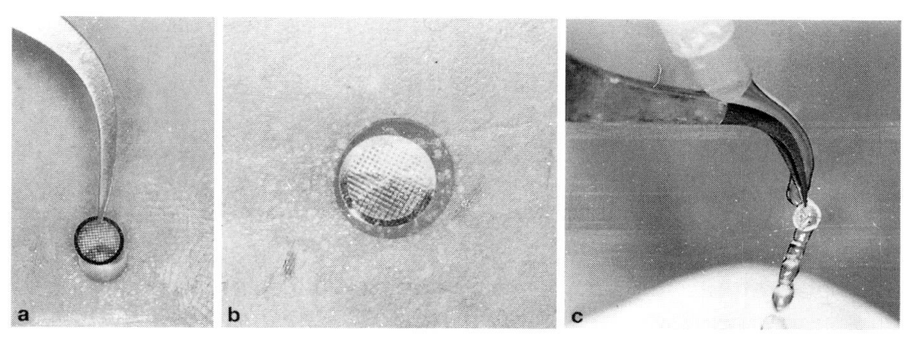

Fig. 3.4a–c. Handling of electron microscope grids for post-embedding immunolabeling. Sequences for inmunogold labeling of Lowicryl or Epon ultrathin sections. See text for details. (Egea 1993)

39. Incubate sections with rabbit anti-mouse (5 μg/ml) in buffer C/0.1 % BSA, 2 h at room temperature.

40. Rinse in buffer C/ 0.1 % BSA, 4×5 min.

41. Incubate sections with goat anti-rabbit IgG 10 nm gold at 1/40 dilution in buffer C/0.1 % BSA for 1 h at room temperature.

42. Jet wash with distilled water, 1 min/grid.

43. Air-dry.

44. Stain with 4 % uranyl acetate for 15 min. Protect from light.

45. Rinse in water and let sections air-dry.

46. Stain with 2 % lead citrate for 10 min under nitrogen atmosphere.

47. Rinse in distilled water and let air-dry.

48. EM observation.

3.2.2 Post-embedding Procedures Using Colloidal Gold Probes

The post-embedding technique is the most widely used immunocytochemical methodology for electron microscopy. Basically, cells are directly frozen (after cryoprotection) or embedded in a resin (mainly Epon, Araldite, LR White, Lowicryls, or Unicryl) and immunocytochemical reactions are performed on ultrathin cryosections or on ultrathin sections, respectively. On the other hand, most protocols followed for immunofluorescence are extended to the immunoelectron microscope. In general, this is the case, but it should be pointed out that higher concentrations of antibody are needed for EM. This is the way to obtain a better signal-to-noise ratio. Optimal antibody concentrations are usually about ten fold higher than those used for immunofluorescence on permeabilized culture cells (Griffiths 1993).

At present, the most popular resin-embedding methods use the low-temperature embedding in low cross-linked hydrophylic resins like Lowicryl (Villiger 1991) and the epoxy resin Epon. Thin sections prepared from resin-embedded tissues or cells have been succesfully used for post-embedding immunolabeling (Roth 1989). Here, we described a complete and detailed protocol of how to handle ultrathin sections of resin-embedded microinjected (or transfected) cells to localize macromolecules at the ultrastructural level. However, detailed description of the different protocols for embedding with low-temperature resins and to obtain ultrathin sections are beyond the scope of this chapter. The reader is advised to consult more specialized EM books (Hayat 1989; Griffiths 1993) or an experienced electron microscopist or EM service. Whichever embedding medium is used, ultrathin sections should be collected on Formvar/Parlodion-coated nickel/gold grids (no copper grids!).

Materials

Reagents

Ammonium chloride (NH_4Cl), BSA, glutaraldehyde (25 or 50% in water), disodium hydrogen-phosphate anhydrous (Na_2HPO_4), fetal calf serum (FCS), gelatin, gold-labeled reagents (i.e, protein A, streptavidin, antibiotin, secondary antibodies), lead acetate, osmium tetroxide, PFA, skimmed milk, sodium cacodylate, sodium di-hydrogen-phosphat monohydrate (NaH_2PO_4-H_2O), sodium metaperiodate, Triton X-100, Tween-20, uranyl acetate.

Solutions

- PBS
- 0.1 and 0.2 M phosphate buffer pH 7.4
- 0.1 and 0.2 M cacodylate buffer pH 7.4
- 0.2–1% BSA in PBS
- 0.1% gelatin in PBS
- 2–5% skimmed milk

- 5 % FCS
- 1 % BSA, 0.1 % Triton X-100, 0.1 % Tween-20 in PBS (PBS+++).
- 2 % uranyl acetate in double distilled water (filter before use)
- 1 % lead acetate (filter the whole solution and carefully add parafilm sheet onto the surface)
- 50 mM NH4Cl in PBS
- Fixatives. The fixatives described below are those used routinely because they provide a good compromise between morphological preservation and antigen recognition capability of the antibody. These fixatives are the following:
  4–8 % paraformaldehyde in 0.1 M phosphate or cacodylate buffer.
  4–8 % paraformaldehyde with 0.01–0.5 % glutaraldehyde in 0.1 M phosphate or cacodylate buffer. **Note:** Remember to quench the free-aldehyde groups with ammonium chloride (see previous sections).
- Saturated aqueous solution of sodium metaperiodate (only for Epon ultrathin sections).

Procedure

Immunogold Labeling on Ultrathin Sections of Cells Embedded in Low-Temperature Resins (Lowicryl)

1. Float the grids, sections down, for 5–10 min on a large drop of PBS alone, or containing either 0.5–1 % BSA, or 0.1 % gelatin, or 2–5 % skimmed milk, or 5 % fetal calf serum (FCS) as blocking agents.

2. Transfer grids (Fig. 3.4a) onto a droplet (5–10 μl) of the specific antibodies (properly diluted in one of the above blocking solutions) for 60 min at room temperature or overnight at 4 °C in a moist chamber (Fig. 3.4b). This step is not necessary if cells were microinjected with antibodies. In this case, go directly to step **4.**

3. Jet wash the grids by a mild spray of PBS (Fig. 3.4c), and then immerse them in PBS for 2 min. This process should be repeated once. Alternatively, grids can be floated onto the surface of a large volume of PBS in an ELISA microwell plate and maintained under constant stirring with a magnetic stirring device for 3×30 s. Finally, the side of the grid without sections is blotted dry on a filter paper, but never let the sections dry!

4. Place the grids onto a droplet (10 μl) of protein A-gold complex (pAg) for 45 min at room temperature. The appropriate dilution of pAg is according the size of the gold particles. We use the following dilutions in PBS containing 1 % BSA, 0.1 % Triton X-100, and 0.1 % Tween-20

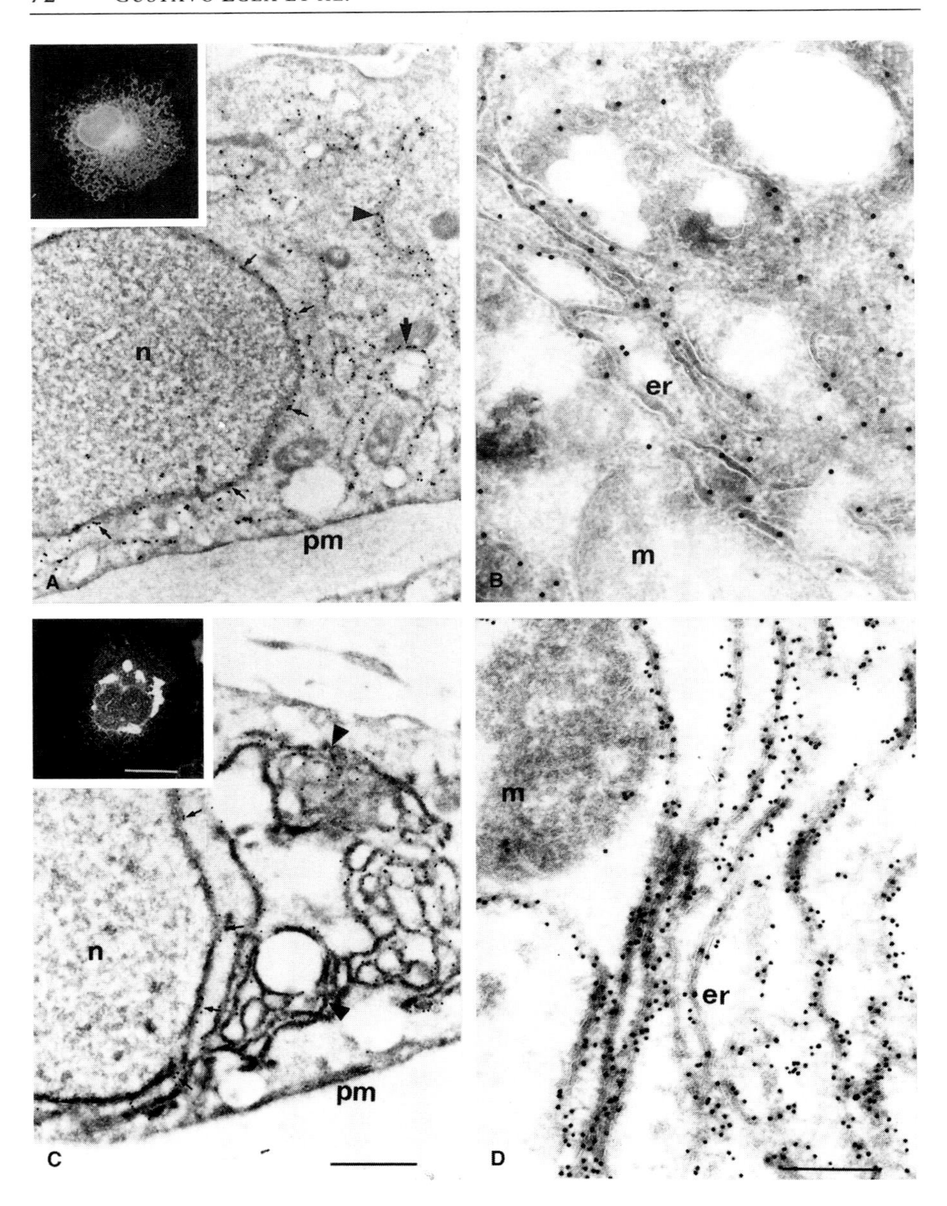
n
pm
A
er
m
B
m
n
pm
C
er
D

◀ **Fig. 3.5A–D.** Postembedding in cryoultrathin sections. Intracellular distribution of wild-type human lipoprotein lipase (hLPL) and mutant N43A hLPL protein in transfected COS1 cells. The mutant consists of the substitution of Asn43 for a Ala to study the role of glycosylation in the intracellular processing of LPL protein. COS1 cells were transfected with either wild-type (**A, B**) or mutant (**C, D**) cDNAs. They were processed for immunoelectron microscopy (48 h posttransfection) following the procedure for immunolabeling of cryoultrathin sections as indicated in the text. Immunolabeling was performed using polyclonal affinity-purified chicken anti-LPL, rabbit anti-chicken immunoglobulins, and protein A-gold (15 nm). *Inserts* show the immunofluorescent microscopy of wild-type (**A**) and mutant LPL (**C**) transfected cells. Note that mutant N43A hLPL accumulates in the lumen of the endoplasmic reticulum. *er* Endoplasmic reticulum; *n* nucleus; *m* mitochondria; *pm* plasma membrane. *Bar* **A, C** 1 mm, insert 20 mm, **B, D** 0.3 mm (Buscà et al. 1995)

(abbreviated PBS+++) measured to 525 nm spec trophotometric optical density (OD_{525}):

a) OD=0.40 for pAg 15 nm
b) OD=0.08 for pAg10 nm
c) OD=0.06 for pAg5 to 8 nm

5. Rinse grids twice in PBS as described in step **3** and finally also briefly jet wash in distilled water and allow to air dry (30 min) before counterstaining.

6. Contrasting of sections is done with 2 % uranyl acetate for 6 min, rinsed with distilled water and then with 1 % lead acetate for 45 s under a nitrogen atmosphere. Finally, grids are rinsed in distilled water and allowed to dry before being examined by electron microscopy.

Note: This is the protocol followed when using a polyclonal antibody. However, if an antigen is being localized with monoclonal antibodies, the protein A reacts weakly. In this case use:

a) an intermediate step between the primary antibody (step **2**) and the protein A-gold step (step **4**) with an affinity-purified rabbit anti-mouse IgG and/or IgM at a final concentration of 25 μg/ml in PBS containing 0.1 % Tween -20 for 45 min at room temperature, or
b) commercial gold-labeled secondary IgG or IgM antibodies diluted in PBS +++ at an OD_{525} of 0.13 for 45 min. See also protocol followed by Meyer et al. (1992).

In addition, the use of biotinylated monoclonal antibodies is quite common. In this case, the biotin molecules are detected with an anti-biotin gold labeled secondary antibody (at an OD_{525} of 0.15 in PBS+++, for 45 min), or with an anti-biotin secondary antibody (at 20 μg/ml in PBS, for 45–60 min) followed by pAg (at the dilutions mentioned above), or with streptavidin-gold (at OD_{525} 0.2 in 0.2 % BSA in PBS for 60–90 min).

Immunogold Labeling of Ultrathin Sections of Epon-Embedded Cells

The above protocols apply when ultrathin sections of Lowicryl-embedded cells are used. The same methods can be used if thin sections are obtained from Epon-embedded material, but an etching step is necessary before the cytochemical reactions are performed. Etching consists of the chemical extraction of resin to make antigens accessible to the immunocytochemical reactions. Diverse procedures can be used before step **1** but the most effective is a saturated aqueous solution of sodium metaperiodate for 10–60 min (see Sect. 3.5). This method is particularly useful if epoxy-embedded cells and tissues were postfixed previously with osmium tetraoxide (Bendayan and Zollinger 1983).

After the etching step, grids are extensively rinsed in distilled water and the immunocytochemical reactions are performed as described above. Sections are contrasted with 2 % uranyl acetate for 10 min and 1 % lead acetate for 2 min under nitrogen atmosphere.

Post-embedding of Cryoultrathin Sections

The cryosectioning technique (see Fig. 3.5), also called the Tokuyasu technique, offers a number of advantages over most other methods for high resolution immunolabeling (for more details see Griffiths 1993). This method has the following advantages:

- It is the most sensitive technique for immunolabeling because the antigen can be affected only by the initial aldehyde fixation. Freezing and thawing of the specimen do not seem to affect its antigenicity.
- As the sections are not embedded in resin, they offer higher access of the antibody to the antigen as compared with other techniques.
- The possibilities for enhancing the contrast of membrane structures are greater than by any other method.
- The entire procedure can be performed in one working day.

However, the critical and more difficult step of this method is obtaining cryoultrathin sections. Description of the technical requirements for cryoultrathin sectioning is beyond the scope of this chapter.

In this method, it is necessary to fix the cells or tissues (Buscà et al. 1995). Usually, 4–8 % PFA in phosphate, Hepes, or Pipes buffer is used. To improve preservation of the fine structure, the fixative is often supplemented with 0.1–0.5 % glutaraldehyde or 0.5–1 % acrolein, both strong cross-linkers. Cryoprotected, vitrified biological specimens can easily be sectioned at low temperatures. The most widely used cryoprotectant is sucrose, used at concentrations between 2 and 2.3 M. High concentrations of sucrose give softer tissue blocks which have to be sectioned at lower temperatures. For difficult specimens, an alternative cryoprotectant mixture developed by Tokuyasu consisting of 1.8 M sucrose and 20 % (w/v) polyvinylpyrolidone (PVP, MW 10,000) is recommended.

Materials

Solutions

- Aqueous uranyl acetate (UA), 3 g UA in 100 ml ddH2O. Filter before use (is a saturated solution). Store at 4 °C
- Uranyl acetate in maleate buffer, (can use 1 % to 3 %). 1 g (or 3 g) UA in 100 ml maleate buffer (50 mM, pH 7.4). Filter and store at 4 °C
- Neutral uranyl acetate. Mix equal parts 0.3 M oxalic acid with 4 % aqueous UA. Adjust the pH to 7.5. Use pH-indicator strips, since heavy metals will block the pH-meter electrode
- Lead citrate. In a 50-ml plastic tube (1.33 g lead nitrate plus 1.76 g sodium citrate). Add 30 ml boiled, cooled, filtered double distilled (ddH_2O). Shake for 1 min. Let stand for 30 min with intermittent shaking. Add 8 ml 1 N NaOH (made fresh). Raise volume to 50 ml with boiled, cooled, filtered ddH_20. Invert to mix gently. Store at 4 °C, tightly closed
- Methyl cellulose. 2 g methyl cellulose (25 centipoises) layered on top of 100 ml ddH_20. Leave at 4 °C for 3 days (MC dissolves better at 4 °C). Centrifuge using Ti70 rotor for 30 min or more at 60,000 rpm, Store at 4 °C
- 10 % FCS/PBS. 10 ml of FCS brought to 100 ml with PBS. Aliquot in 5-ml bottles and store at −20 °C
- 1 % Fish skin gelatin (FSG)/PBS. 1 ml FSG in 100 ml with. Aliquot in 5-ml bottles and store at −20 °C.

Procedure

Immunolabeling of cryoultrathin sections from microinjected cells (see Fig. 3.5)

1. Collect grids by floating them on a solution to block nonspecific binding of antibodies. 1–5 % fetal calf serum (FCS)/PB, 1 % FSG and 2 % gelatin can be used. Use 5 % gelatin if you wish to keep the grids for several days before processing.
2. Rinse in PBS once.
3. If the cells have been fixed with glutaraldehyde, free aldehyde groups should be quenched in 0.02 M glycine in 5–10 % FCS/PBS for 10 min. Rinse in PBS for a total of 5 min.
4. Centrifuge antibody solution (1 min at 13,000 *g*) diluted in 1–5 % FCS/PBS.
5. Rinse in PBS 6×2 min each.
6. Incubate grids in protein-A gold for 20–30 min. Dilute protein-A gold in 1–5 % FCS/PBS. The concentration is critical. Too-high concentrations give nonspecific binding.
7. Rinse in PBS 6×4 min each.
8. Rinse in distilled water, 4×1 min each.
9. Incubate in 2 % methyl cellulose solution (25 cps) containing 0.1–0.4 % uranyl acetate, 3×6 min each.
10. Pick the grid up with a 3-mm loop. Remove the excess fluid with filter paper. **Note:** The thickness of the methyl cellulose film determines the contrast and the extent of drying artifacts.
11. Air-dry the grid suspended in the loop.
12. Examine the grids at the electron microscope.

For double labeling, after step **7**, the grids are floated on 1 % glutaraldehyde in PBS for 5 min, followed by many rinses in PBS (Slot et al. 1991). Steps **3** to **7** are then repeated using as secondary antibody labeled gold beads of different size, and then followed by steps **8** to **12**.

Acknowledgments. The skilled technical assistance of Ana Orozco, and Susana Castel (Serveis Cientifico-Tècnics, Universitat Barcelona) is acknowledged. Work done in the authors' laboratory is supported by grants to G.E. (FIS 94/0468 SAF 97-0016, and TeleMarató TV3 de Catalunya) and to S.V. (ClCYT PB94-1548 and TeleMarató TV3 de Catalunya). T.B. is a recipient of a ClRIT postdoctoral fellowship; I.A. is a predoctoral fellow from the Spanish Ministry of Education and Science; F.V. is a predoctoral fellow from Universitat de Barcelona.

References

Bar-Sagi D (1991) Phospholipase A2: microinjection and cell localization techniques. Methods Enzymol. 197:269-279

Bendayan M, Zollinger M (1983) Ultrastructural localization of antigenic sites on osmium-fixed tissues applying the protein A-gold technique. J Histochem Cytochem 31:101-109

Buscà R, Pujana MA, Pognonec P, Auwerx J, Deeb SS, Reina M, Vilaró S (1995) Absence of N-glycosylation at asparagine 43 in human lipoprotein lipase induces its accumulation in the rough endoplasmic reticulum and alters this cellular compartment. J Lipid Res 36:939-951

Campbell MS, Gorbsky GJ (1995) Microinjection of mitotic cells with 3F3/2 anti-phosphoepitope antibody delays the onset of anaphase. J Cell Biol 129:1195-1204

Dworetzky SI (1991) Microinjection of colloidal gold. In: Hayat MA (ed) Colloidal gold. Principles, methods and applications, vol 3. Academic Press, New York, pp 265-279

Egea G (1993) Lectin cytochemistry using colloidal gold methodology. In: Gabius H-J, Gabius S: Lectins and glycobiology. Springer, Berlin Heidelberg New York, pp 217-233

Fernández-Borja M, Bellido D, Vilella E, Olivecrona G, Vilaró S (1996) Lipoprotein-lipase mediated uptake of lipoproteins in human fibroblasts. Evidence for a receptor-independent internalization pathway. J Lipid Res 37:464-481

Griffiths G (1993) Fine structure immunocytochemistry. Springer, Berlin Heidelberg New York

Griffiths G, Parton RG, Lucocq C, van Deurs B, Craun D, Slot JW, Geuze HJ (1993) The immunofluorescent era of membrane traffic. Trends Cell Biol 3:214-219

Handley DA (1989) The development and application of colloidal gold as a microscope probe. In: Hayat MA (ed) Colloidal gold. Principles, methods and applications, vol 1. Academic Press, San Diego, pp 1-12

Harlow E, Lane D (1988) Antibodies. A laboratory manual. Cold Spring Harbor Laboratory, Cold Spring Harbor

Hayat MA (1989) Principles and techniques of electron microscopy. Biological applications. McMillan Press, London

Kouklis PD, Papamarcaki T, Merdes A, Georgatos S (1991) A potential role for the COOH-terminal domain in the lateral packing of type III intermediate filaments. J Cell Biol 114:773-786

Little JR, Eisen HN (1967) Preparation of immunogenic 2,4-dinitrophenyl and 2,4,6-trinitrophenyl proteins. Methods Immunol Immunochem 1:128-135

Mangeat PH, Burridge K (1985) Microinjection and immunofluorescence microscopy as tools to study cytoskeletal organization in tissue culture cells. In: Bullock GR, Petrusz P (eds) Techniques in immunocytochemistry, vol 3. Academic Press, New York, pp 79–97

Melan MA, Sluder G (1992) Redistribution and differential extraction of soluble proteins in permeabilized cultured cells. Implications for immunofluorescence microscopy. J Cell Sci 101:731–743

Meyer RA, Laird DW, Revel J-P, Johnson RG. (1992) Inhibition of gap junction and adherens junction assembly by connexin and A-CAM antibodies. J Cell Biol 119:179–189

Miller RK, Khuon S, Goldman RD (1993) Dynamics of keratin assembly: exogenous type I keratin rapidly associates with type II keratin in vivo. J Cell Biol 122:123–135

Pagan R, Martin I, Alonso A Llobera M, Vilaró S (1996) Vimentin filaments follow pre-existing cytokeratin network during epithelial-mesenchymal transition of neonatal hepatocytes. Exp Cell Res 222: 333–344

Paterson H, Adamson P, Robertson D (1995) Microinjection of epitope-tagged Rho family cDNAs and analysis by immunoblotting. Methods Enzymol 256:162–173

Pathak RK, Anderson RGW (1989) Use of dinitrophenol-IgG conjugates to detect sparse antigens by immunogold labeling. J Histochem Cytochem 37:69–74

Pathak RK, Anderson, RGW (1991) Use of dinitrophenol IgG conjugates: immunogold labeling of cellular antigens on thin section of osmificated and Epon-embedded speciments. In: Hayat MA (ed) Colloidal gold. Principles, methods and applications, vol 3. Academic Press, New York, pp 223–241

Roth J (1989) Postembedding labeling on Lowicryl K4M tissue sections: detection and modification of cellular components. In: Tartakof AM (ed) Methods in cell biology, vol 31. Academic Press, London, pp 514–553

Rothemberg KG, Heuser JE, Donzell WC, Ying Y-S, Glenney JR, Anderson RGW (1992) Caveolin, a protein component of caveolae membrane coats. Cell 68:673–682

Shelden E, Wadsworth P (1992) Microinjection of biotin-tubulin into anaphase cells induces transient elongation of kinetochore microtubules and reversal of chromosome-to-pole motion. J Cell Biol 116:1409–1420

Slot JW, Geuze HJ, Gigengack S, Lienhard GE, James DE (1991) Inmunolocalization of the insulin-regulatable glucose transporter in brown adipose tissue of the rat. J Cell Biol 113:123–135

Takeda S, Okabe S, Funakoshi T, Hirokawa N (1994) Differential dynamics of neurofilament-H protein and neurofilament protein in neurons. J Cell Biol 127:173–185

Tomkiel J, Cooke CA, Saitoh H, Bernat RL, Earnshaw WC (1994) CENP-S is required for maintaining proper kinetochore size and for a timely transition to anaphase. J Cell Biol 125:531–545

Villiger W (1991) Lowicryl resins. In: Hayat MA (ed) Colloidal gold. Principles, methods and applications, vol 3. Academic Press, New York, pp 59–73

Chapter 4

Videomicroscopy of Living Microinjected Hippocampal Neurons in Culture

Frank Bradke[1] and Carlos G. Dotti[1]*

Introduction

Videomicroscopy is the appropriate method to study many morphological and intracellular changes of individual cells. In this chapter the two principal types of videomicroscopy will be described: videomicroscopy with normal phase contrast, and video enhanced contrast differential interference contrast (VECDIC) microscopy. The first is especially suited to study morphological changes of the cell such as cell movement, retraction, outgrowth, filopodia dynamics (Nobes and Hall, 1995), and mitochondria movement. VECDIC, in contrast, is based on Normarski microscopy. With this method it is possible to track moving vesicles in neurites (Allen et al. 1982; see also Fig. 4.1) and lamellipodia (Fisher et al. 1988), microtubule polymerization (Cassimeris et al. 1988), and actin dynamics in growth cones (Forscher and Smith 1988). Beside these methods, organelle transport can be also specifically visualized and documented by fluorescent microscopy using specific fluorescent probes (Lee and Chen 1988).

The chapter is divided into four parts: the first deals with the practical problem of how to keep cells alive and in good condition for long observation periods. The workstation we use is described in the second part. The third part deals with a brief theoretical background on VECDIC. Finally, we describe how to set the microscope for VECDIC and how to obtain and process images.

* corresponding author: phone: +49-6221-387 322; fax: +49-6221-387 306;
e-mail: dotti@embl-heidelberg.de

[1] European Molecular Biology Laboratory, Meyerhofstrasse 1, D-69117-Heidelberg, Germany

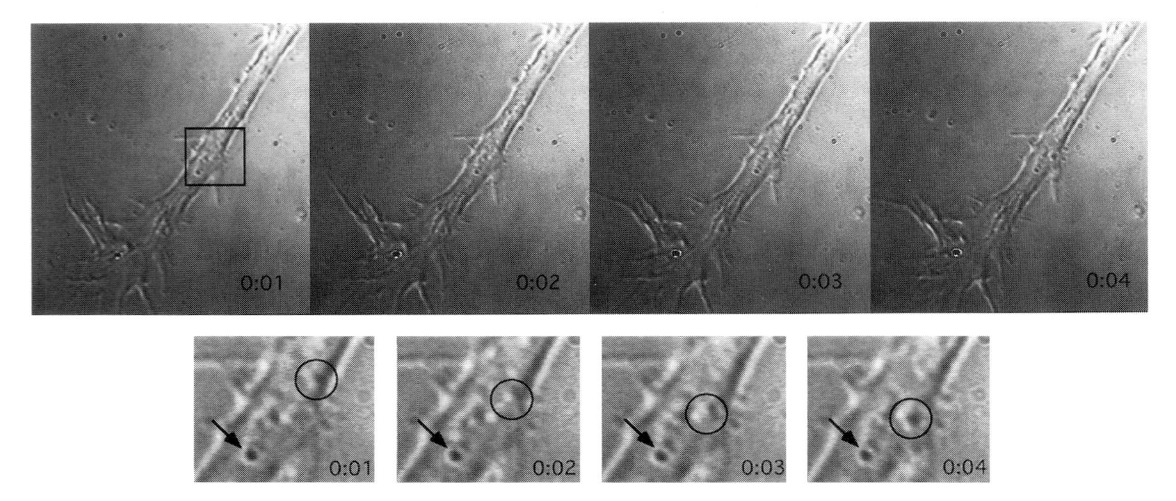

Fig. 4.1. Organelle and growth cone dynamics in a neurite of a hippocampal neuron visualized by VECDIC. Filopodia movement is visible in real time (see long *left* filopodia in the upper panel) as well as organelle movement (see magnification of the squared part in the lower panel, time 1–4 s): a nonmoving object (*arrow*) and a moving organelle (*circle*) are shown

Materials

General Considerations for Handling Cells

Different cell types respond very differently to stress situations such as microinjection. Swiss 3T3 cells are an example of a robust cell line. They tolerate microinjection and can be used for videomicroscopy directly afterwards (Nobes and Hall 1995). Primary culture cells are, in general, more sensitive. Our treatment of rat embryo hippocampal neurons after microinjection and during videotaping will be presented as an extreme example which is probably exaggerated for most other cell types. We stress technical aspects which could be eliminated for more robust cells.

Microinjection is a stress for cells. It is therefore important to limit changes in their environment, such as temperature and pH. However, even in carefully planned experiments, changes in medium, small temperature shifts, and moving the cell culture dishes are unavoidable. We keep these changes to a minimum and ensure that the cells have time to adapt to their new environment as well as to recover in their original medium in the incubator.

For microinjection, hippocampal neurons are transferred from their original neuronal medium to a buffer solution, i.e., Hank's basic salt solution (HBSS). We microinject cells using the Zeiss AIS automatic microinjector. After microinjection, the neurons are returned to the original culture dish containing the growth medium and left in the incubator for 2 h. For videomicroscopy we use sealed chambers and medium which buffers physiological pH at atmospheric concentration of gases for a short time. Thus, it is possible to assemble and seal the chamber without pH change. The possibilities here are multiple. As chamber medium we use growth medium / air medium (1/1) prewarmed to 37 °C (see recipes of the media at the end of this paragraph). These media are specifically used for rat hippocampal neurons (for details of this culture, see Goslin and Banker 1991). Air medium is the same as growth medium, also in terms of osmolarity, except that it is buffered with Hepes instead of $NaHCO_3$. To modify media for other cell types, one possibility is to add a final concentration of Hepes, which doubles the molarity of $NaHCO_3$. Example: bring Earle's salt-based MEM (EMEM) to 50 mM Hepes (EMEM contains 26 mM $NaHCO_3$). However, hippocampal neurons show arrested growth under these conditions. For long-term recordings we also add 0.1–1 mM N-acetyl-L-cysteine, a light antioxidant which has been reported to inhibit apoptosis and keep the cells healthier under stress conditions (Mayer and Noble 1994). The observation video chamber is prepared and warmed in the incubator for 10 min. The microscope stage can be heated with an air fan controlled by a thermometer. We use an objective heater ring which warms the objective and also warms the specimen through the thermal conductance of the oil on the lens. Usually, the stage is heated 1 to 2 h after the cells have been located at the microscope. Heating the stage is necessary to keep the cells in growth conditions. Special care has to be taken with the temperature. Neurons are very sensitive to temperatures slightly above 37 °C but are not harmed by temperatures around 25–30 °C.

Media

- Growth medium (N_2)
  10 ml 10× N_2 supplement
  90 ml N-MEM
  100 mg egg albumin

Store at 4 °C no longer than 1 week

- Air medium:
  80 ml MEM (8 ml 10× MEM, bring to 80 ml with water)
  10 ml 10× N_2
  3.2 ml 20 % glucose
  1 ml glutamine (from Gibco cat.no.: 25030–024)
  1.4 ml 5.5 % $NaHCO_3$
  1.5 ml hepes
  1 ml 1.1 % pyruvic acid
  100 mg egg albumin
  Bring to 100 ml with deionized H_2O, aliquot in 50-ml tubes, store at 4 °C
- N-MEM:
  50 ml 10× MEM (GIBCO Cat.No.21430–012)
  5 ml 1.1 % pyruvic acid (SIGMA Cat.No.P-2256)
  5 ml 200 mM glutamine (GIBCO Cat.No.25030–024)
  15 ml 20 % glucose
  20 ml 5.5 % $NaHCO_3$
  Bring to 500 ml with sterile water
- 10× N_2 supplement:
  1 ml 5 mg/ml insulin (Sigma I-5500)
  1 ml 20 μM progesterone (Sigma PO-1301)
  1 ml 100 mM putrescine (Sigma P-7505)
  1 ml selenium dioxide (Sigma S-9379)
  100 mg transferrin (Sigma T-2252)
  Bring to 100 ml with N-MEM. Aliquote and store at −20 °C

The Video Chambers

To observe living cells under the microscope, a chamber is necessary which gives the cells a constant environment in terms of hygroscopy, pH, and temperature. We describe three different chambers which fit different tasks: the simple, the flexible, and the long-term chamber. All of the chambers are temperature-controlled by an objective heating ring from Bioptechs, Inc., Butler, PA, USA. The different chambers, as well as the objective heating ring, are shown in Fig. 4.2.

The Simple Chamber, the Petri Dish. The simplest solution to observe cells under the microscope is a special petri dish with a hole in the bottom with a glass coverslip attached (Glass Bottom Microwells plastek cultureware from the MatTek corporation, Ashland, MA; Part No. P35G-1.5–14-C., uncoated No. 1.5.) The cells are grown in the same dish in which the experiment will be performed. The glass surface can be treated like normal glass coverslips. To coat the surface of the glass, we use the method described in Goslin and Banker (1991). In case the glass has to be acid-treated before coating, the acid must not come into contact with the plastic edge. Otherwise, the dishes become leaky. To prevent air exchange during manipulation, the lid can be sealed with parafilm. The dishes are well suited for phase-contrast and Normaski microscopy using an inverted microscope and a long-distance condensor. However, this chamber cannot be used for high-resolution Normaski because the oil condensor must come very close to the cells to illuminate the specimen properly. For recording times of less than 1 h, addition of Hepes to the medium is not required. Sealing the petri dish is enough.

The Flexible Chamber, the Metal Slide. This chamber has the shape and size of a microscopy glass slide but is made out of aluminum (see Fig. 4.3). A hole of 8 mm is drilled in the middle. The edges around the hole are milled from both sides. The milling diameter depends on the coverslip used (milling diameter = coverslip diameter + 1 mm). The hole can also be adjusted to different sizes but there should be enough milled area to seal the coverslip to the slide properly. The metal slides can also have a small microaqueduct leading from the outside to the interior of the chamber to treat cells with drugs and change media while observing. This chamber can be used for both phase contrast as well as VECDIC microscopy because it permits optimal optical working distance. This is the only one of the three chambers which can be used with a non-inverted microscope. Moreover, this chamber can also be used with different coverslips, e.g. gridded coverslips (CELLocate coverslips, No. 5245 963.000–00, Eppendorf GmbH, Hamburg). This is not possible for the other two chambers.

The major drawback of the flexible chamber is that the small volume of medium in the chamber limits observation time.

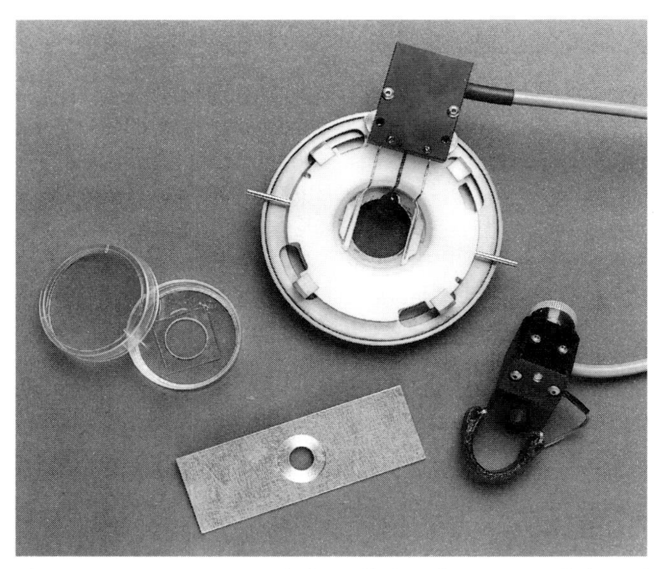

Fig. 4.2. Heating ring and the cell chambers: petri dish with a coverslip bottom, the metal slide, and the FCS 2 chamber

Procedure

How to Use the Chamber

1. Put a thin film of Vaseline on one side of the milled edge using a syringe.
2. Place a clean coverslip onto the hole and press it gently to allow the Vaseline to form a film between the glass and the metal.
3. To avoid movement of the coverslip as well as air exchange of the medium a layer of warmed, fluid VALAP (Vaseline/lanolin/paraffin: 1/1/1) can be used to seal the border coverslip-aluminum.
4. Put a thin film of Vaseline on the milled/side edge of the opposite of the metal slide. Mark this side. (This is the side where the coverslip containing cells is placed.)
5. Leave the metal chamber in the incubator to warm.
6. Mount the coverslip containing cells:

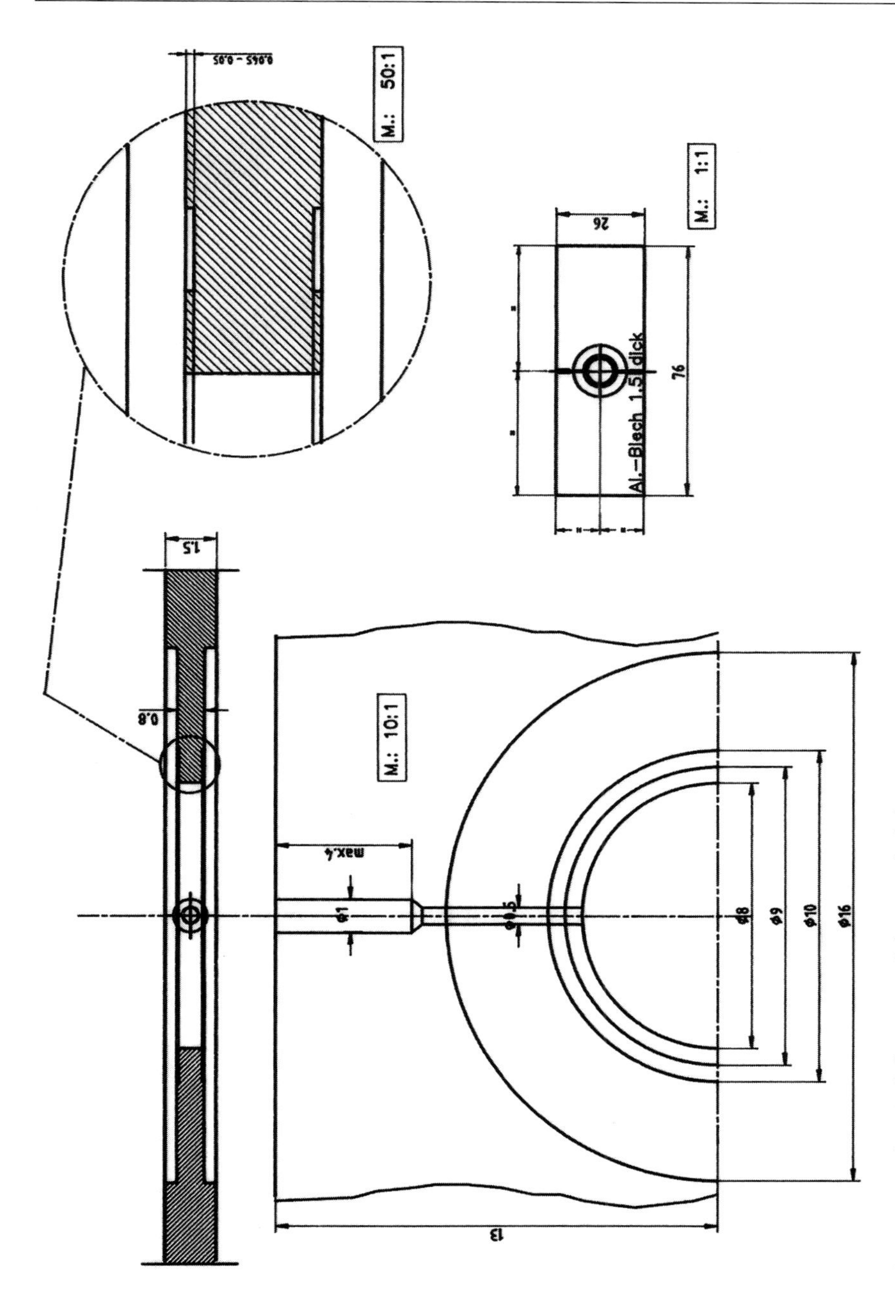

Fig. 4.3. Mechanical drawing of the metal slide

a) Take your dish containing the coverslip with the microinjected cells as well as the prewarmed metal chamber and place them in a laminar flow hood. A tissue placed beneath the chamber will stop heat exchange.
b) Pipette 2 μl of 50 mM Hepes into the chamber, take 100 μl of medium from the petri dish and add it into the chamber. Alternatively, pipette 50 μl of Air-MEM and add another 50 μl of medium derived from the petri dish. The well of the chamber is now completely covered with medium.
c) Take the coverslip with the microinjected cells from the dish using a sterile forceps, flip it so that the cells face the medium, and place it into the free milled side.
d) Press the coverslip gently to the milled edge and remove excess medium with a tissue. If necessary, seal with VALAP.
e) Cover the metal chamber in another dry tissue to reduce temperature shifts and then go to the microscope. Put a drop of warm oil on the coverslip and place the chamber on the microscope stage with this side facing the lens.

The Long-Term Observation Chamber. For long-term observations (i.e., 24–48 h) a more sophisticated system is necessary. We use the FCS-2-chamber from Bioptechs, Inc., Butler, PA, USA, which allows temperature control and perfusion of medium. It consists of a metal ring which contains the heat regulatory element. Coverslips of 40 mm diameter fit into this chamber. Due to its special perfusion system, the chamber allows exchange of medium in a closed system with a minimum of current. Medium can be slowly exchanged by a peristaltic pump. We use this chamber with an objective heater ring also from Bioptechs. To assemble the chamber follow the manufacturer's instructions.

Major drawback is that the cells have to be grown on 40-mm coverslips.

Use of the FCS 2 System

1. Sterilize the plastic spacer and the microaqueduct in 70 % ethanol for 15 min.
2. Let them dry under sterile conditions.
3. Rinse in sterile water and, finally, in Air-MEM.

4. Fit the upper gasket on the plastic ring.
5. Put the wet microaqueduct on top.
6. Put the plastic spacer (0.1, 0.25, or 1 mm) on top of the microaqueduct.
7. Add 0.4 ml Air-MEM into the chamber.
8. Take the dish with the microinjected cells.
9. Remove 1.2 ml of the culture medium and add it into the chamber.
10. Take the 40 mm coverslip out of the petri dish and put it gently on top of the spacer with the cells facing the medium. Try to avoid air bubbles.
11. Close the chamber as described in the instruction manual provided by the manufacturer.
12. Cover the chamber with some tissues and bring it to the microscope.

We have kept neurons alive for more than 4 days under these conditions, choosing a 1-mm spacer and a temperature of 36 °C for the objective heating.

Description of the Videomicroscopy Workstation

Microinjection and videorecording are performed at different microscopes. This is convenient, since the microinjection workstation is too frequently used to allow long observation times. Here we describe our videorecording setup (see also Fig. 4.4).

The Zeiss Axiovert 135 inverted microscope (Fig. 4.4A) is equipped with 40× (NA 1.0), 63× (NA 1.4), and 100× (NA 1.4) Plan Apochromat objectives. The microscope has a 100 W, 12 V light bulb. A green filter is installed in the light path to reduce photo damage to the cells. To tape phase-contrast and low resolution Normaski images, we record using a long-distance condensor and reduced light but increase the gain with the CCD camera control. For VECDIC we work with the maximum of light but with the green filter. For VECDIC some additional microscope equipment is necessary: Polarizer, analyzer, 1.6× optovar, appropriate condensors and Wollaston prisms specific to the numeric apertures of the objectives.

For recording we use a Sony Hyper HAD camera (Fig. 4.4B), in connection with a Hammatsu CCD camera C 2400 control panel (Fig. 4.4C).

Fig. 4.4. Workstation: inverted microscope (*A*), video camera (*B*), camera control panel (*C*), black and white monitors (*D*, *G*), Macintosh (*E*), video recorder (*F*) and optical drive (*H*)

For alternate recording of phase, Normaski, and fluorescence, we use a Cohu camera (Fig. 4.4B) from the 4910 series (4913–5100/000) with the same CCD camera control. VECDIC images made with this camera are of good quality. The Sony camera produces images with slightly higher contrast. However, the advantage of the Cohu camera is that fluorescent signals can also be captured because it can integrate the low fluorescent signals directly on a camera chip.

From the control panel, the live image is projected into a black and white video monitor (Fig. 4.4D). From the monitor the signal is grabbed into a Power Macintosh 8500 equipped with an image grabber (LG3 image grabber, Scion Co.; Fig. 4.4E) and to a Panasonic time lapse video recorder AG6720 (Fig. 4.4F) in parallel. Another monitor (Fig. 4.4G) is connected to the output of the video recorder to monitor the recordings. Time lapse video recording is essential to record slowly occurring events, e.g., movement of organelles. Images are collected at a rate slower than video rate and played back at video rate (34 frames per s). Normally, VHS tapes are used, but for some critical parts of the experiment we change to higher-quality tapes, i.e., SVHS tapes. This improves the image only if the video recorder is SVHS-equipped. Through the grabber, single pictures as well as series can also be processed on the computer using NIH image software. Pictures can be stored on an optical drive (Fig. 4.4H). We use a 230 MB storage drive (Fujitsu M2512 A).

Theoretical Background to Normaski

Differential interference contrast (DIC) light microscopy is especially well suited to observe objects which have distinct borders. Objects can be resolved which are normally not visible by transmission light microscopy, e.g., microtubule polymerization and depolymerization can be followed (Cassimeris et al. 1988) as well as organelle movement along neurites (Allen et al. 1982; see also Fig.4.1).

DIC microscopy works by splitting a polarized light beam into two orthogonally polarized light beams using a Wollaston prism located in the condensor. The split beams shear from each other, but become two parallel light trains converted by the condensor lens. The two light trains derived from one light beam are in the nanometer range away from each other, i.e., below the resolution of the light microscope's optics, and have a well-defined phase difference, i.e., are coherent. While going through the specimen, one light train is, in general, exposed to optically denser material than the other, causing the light train which passes through the optically denser zone to decrease its wavelength in comparison with the other train. Thus the phase relationship of the two coherent light trains has changed after passing through the specimen. They are then recombined by a second Wollaston prism located at the base of the objective, where they will interfere with each other. Destructive and constructive interferences are visible in the microscope as the dark and bright edges of the specimen. The resulting contrast can be adapted to the specimen by moving the second Wollaston prism. The shadowing effect is maximal when the edge of the object is parallel to the orientation of the Wollaston. No shadowing can be observed when the edge of a specimen is orthogonal in the orientation of the Wollaston prism.

All the information of the specimen extracted by the light beams is now in the image. However, the signal-to-noise ratio is very low because all the light beams which passed the specimen unchanged are also present in the picture. (Just try to remove the analyzer: the image is very bright, but the specimen can hardly be observed.) A second polarizer orthogonally oriented to the first polarizer is installed in the light path to reduce these unchanged light beams close to extinction. With the second polarizer in place, the image observed through the occular comprises only light beams which have changed their polarization when passing through the specimen and the Wollaston prism.

For more detailed information about the theoretical background of differential interference contrast light microscopy, we recommend the original article of Allen et al. (1969). The review by Salmon et al. (1989)

on VECDIC provides a good introduction to the possibilities and technical background of videorecording in the Normaski microscopy.

How to Adjust the Normaski

(For both metal chamber and FCS2 chamber on an inverted microscope)

1. Put oil on the side of the coverslip where the cells are.
2. Fix specimen on the stage, the cells facing the objective. Approach the objective to the specimen.
3. Put a drop of oil on the top coverslip and approach the condensor.
4. Focus.
5. Close the aperture of the bulb as well as of the condensor.
6. Koehler the condensor aperture. In the ideal case the aperture of the light source should be in focus now as well. Often, they are slightly mismatched, which decreases the light yield.
7. Switch to TV observation.
8. Turn the gain to a middle setting which allows a reasonable picture. Later, the gain may be changed according to the tasks: higher contrast requires maximum gain, which will also increase the noise.
9. Move the Wollaston until the objects show a minimum of shadow. In the ideal case you should see a small shadowing from both sides (see Fig. 4.5). Point objects should appear as airy disks. This state is called extinction.
10. Now move the Wollaston slightly. The point object will not show shadows from both sides, but has only one shadowy side (see Fig. 4.5). The more the prism is tilted, the stronger the shadowing will be. The increasing contrast reduces the resolution. Again, the right compromise between contrast and resolution depends on what is to be observed.
11. The image shows different light intensities caused by the tilting of the prism. The shadowing function of the camera control helps to compensate for this and gives a well illuminated image.

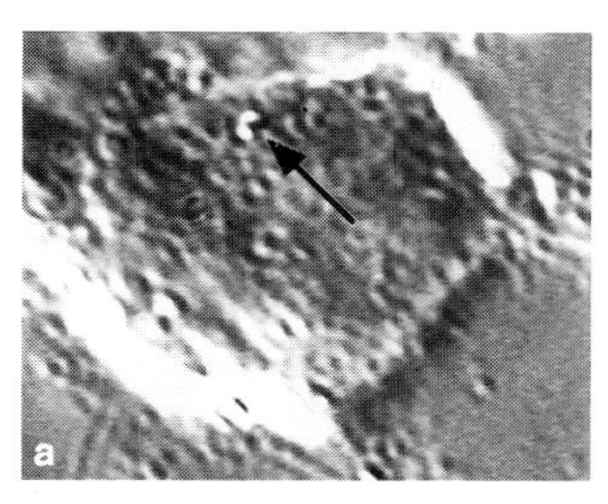

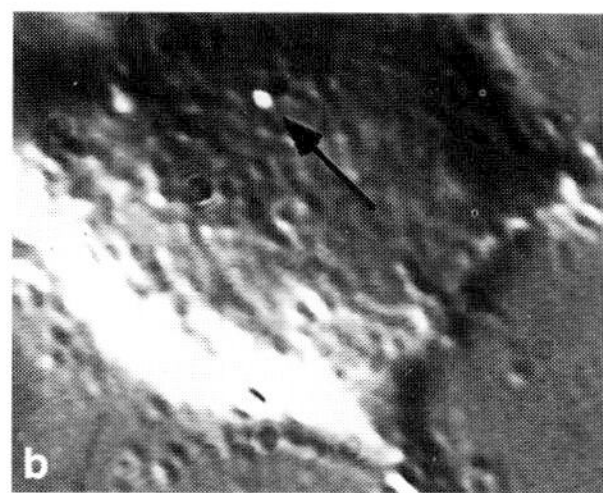

Fig. 4.5a, b. Moving the Wollaston prism: position of extinction (**a**): illumination from both sides leading at point objects to the appearance of airy disks (**a**, *arrow*). Slight movement out of this position (**b**) causes shadowing of one side. The point objects become shaded from one side(**b**, *arrow*)

Troubleshooting

Chambers

- The medium in the chamber turns basic, the cells look unhealthy

- There is too much gas exchange between the medium and the outside of the chamber. There could be two problems. Either the chamber is not sealed well enough. In this case, use parafilm for sealing the simple chamber or seal the aluminum chamber with warm VALAP. It is possible to clog the metal tubes of the FCS2 chamber with hot wax (this can be removed easily with water pressure.). On the other hand, the medium could be exposed to normal air too long before being sealed in the chamber. This will be no problem after practicing the assembly procedure for a while. To keep the pH constant add more AIR-medium.

- The cells look unhealthy and die

- Check the temperature. Did you calibrate your heating system? Reduce the temperature. Cells are often sensitive to small increases in temperature but quite robust at low temperature.
- Check the pH. If the color of the medium changed to basic, proceed as described before.
- Do you illuminate too much? Reduce the light. Compensate with increasing the gain of the camera control which will unfortunately cause a noisier image. Alternatively, reduce your observation time.
- Do you use a green filter? Add a heat filter into the light pass. Do you use N-acetyl-L-cysteine?

VECDIC

- The image is bright and the specimen is hardly observable
- Reduce the light. Make sure by phase contrast light microscopy that you are near the focal plane and that a specimen is spotted. Change again to DIC microscopy: change the condensor, insert the polarizer and analyzer and the Wollaston prism in the light path. Koehler again.
- The image is too dark
- Increase the light and/ or the gain of the camera control
- Move the Wollaston prism
- Oil condensor: is enough oil between coverslip and condensor?

Acknowledgments. We are very grateful to Sigrid Reinsch and Ernst Stelzer for helpful discussions and comments on the manuscript.

References

Allen RD, David GB, Normaski G (1969) The Zeiss-Normaski differential interference equipment for transmitted-light microscopy. Z Wiss Mikrosk Mikrosk Tech 69:193–221
Allen RD, Metuzals J, Tasaki I, Brady TS, Gilbert SO (1982) Fast axonal transport in the squid giant axon. Science 218:1127–1129
Cassimeris L, Pryer NK, Salmon ED (1988) Real-time observations of microtubule dynamic instability in living cells. J Cell Biol 107:2223–2231
Fisher GW, Conrad PA, DeBiasio RL, Taylor DL (1988) Centripetal transport of cytoplasm, actin, and the cell surface in lamellipodia of fibroblasts. Cell Motil Cytoskel 11:235–247
Forscher P, Smith SJ (1988) Actions of cytochalasins on the organization of actin filaments and microtubules in a neuronal growth cone. J Cell Biol 107:1505–1516
Goslin K, Banker G. (1991) Rat hippocampal neurons in low-density culture. In: Banker G, Goslin K (1991) Culturing nerve cells. MIT Press, Cambridge, pp 251–281
Lee C, Chen LB (1988) Dynamic behaviour of endoplasmatic reticulum in living cells. Cell 54:37–46
Mayer M, Noble M (1994) N-actyl-L-cysteine is a pluripotent protector against cell death and enhancer of trophic factor-mediated cell survival in vitro. PNAS 91:7496–7500
Nobes CD, Hall A (1995) Rho, Rac and Cdc42 GTPases regulate the assembly of multimolecular focal complexes associated with actin stress fibers, lammelipodia, and filopodia. Cell 81:53–62
Salmon T, Walker RA, Pryer NK (1989) Video-enhanced differential interference contrast light microscopy. BioTechniques 7:624–33

Preparation, Purification, and Analysis of Synthetic Oligonucleotides Suitable for Microinjection Experiments

RAMON GÜIMIL GARCIA[1] AND RAMON ERITJA[1*]

Introduction

Synthetic oligonucleotides are being used in a wide variety of applications. Recently, interest has grown to utilize oligonucleotides and their analogs as antisense inhibitors of gene expression (Crooke and Lebleu 1993). Although antisense oligonucleotides can be administrated directly to the culture media, microinjection is also used. Microinjection of oligonucleotides into cells permits the choice of the injection site and the delivery of the desired amount of oligonucleotides inside the cells. Oligonucleotides are injected into cells to inhibit gene expression (Kleuss et al. 1994; Kola and Sumarsono 1995; Moulds et al. 1995) and to analyze the fate of antisense oligonucleotides in cells (Chin et al. 1990; Dagle et al. 1991; Leonetti et al. 1991).

Oligonucleotide synthesis is performed usually on solid supports. Recent advances in chemical methods have greatly reduced the amount of time and effort to produce oligonucleotides and their analogs. Automation of the synthetic process has been possible and oligonucleotide synthesizers have been available since 1980. The purpose of this chapter is to provide experimental protocols for the preparation of oligonucleotides, it is divided into two parts. In the first part the assembly of the monomers using the syringe method is described (Tanaka and Letsinger 1982). Different protocols for the purification of synthetic oligonucleotides are provided in the second part.

* corresponding author: phone: +49-6221-387210; fax: +49-6221-387306; e-mail: eritja@embl-heidelberg.de

[1] European Molecular Biology Laboratory, Meyerhofstrasse 1, D-69117-Heidelberg, Germany

5.1
Synthesis of Oligonucleotides Using the Syringe Method

The procedure that follows is designed to prepare a small oligonucleotide (up to 30 bases) in a scale of 1–2 μmol with minimal equipment using the phosphoramidite approach (Caruthers et al. 1987). The synthetic process will be performed inside a syringe equipped with a filter. The different chemicals are aspirated to the syringe and expelling the solutions. Every addition of a monomer consists in four reactions (Fig. 5.1).

1. Detritylation: elimination of the protective group of the 5'-OH. Perform with a 2–3 % solution of dichloroacetic or trichloroacetic acid in dichloromethane.
2. Coupling: the 5'-OH of the nucleoside attached to the solid support is condensed with the DMT deoxyribo 3'-nucleoside phosphoramidite activated with tetrazol.
3. Capping: unreacted 5'-OH are converted to acetate esters by reaction with a solution of acetic anhydride and N-methylimidazole.
4. Oxidation: phosphite-triester is converted to phosphate-triester with a iodine solution. The oxidation step could be also used to prepared phosphate modified oligonucleotides such as nucleoside phosphorothioates.

After the assembly of the sequence, cleavage of the linkage between the oligonucleotide and the support and elimination of the protective groups is performed with concentrated ammonia.

Materials

Reagents and Equipment

- 2 ml polypropylene/ polyethylene syringe with polypropylene plunger equipped with a porous polypropylene frit (B. Braun Diessel Biotech, Melsungen, Germany, cat. no. 920 152 / 1 or 920 252 / 8, ABIMED, Langefeld, Germany, cat no. 80270)
- Hamilton Gastight syringe (5 ml) with a Luer lock needle
- Screwcap vial (5 ml)
- TLC chamber
- TLC plates (Silica gel, 60 F254, 5×10 cm, Merck, Darmstadt, Germany, cat. no. 1.16834)
- Standard DNA synthesis reagents (see suppliers at the end of the chapter)

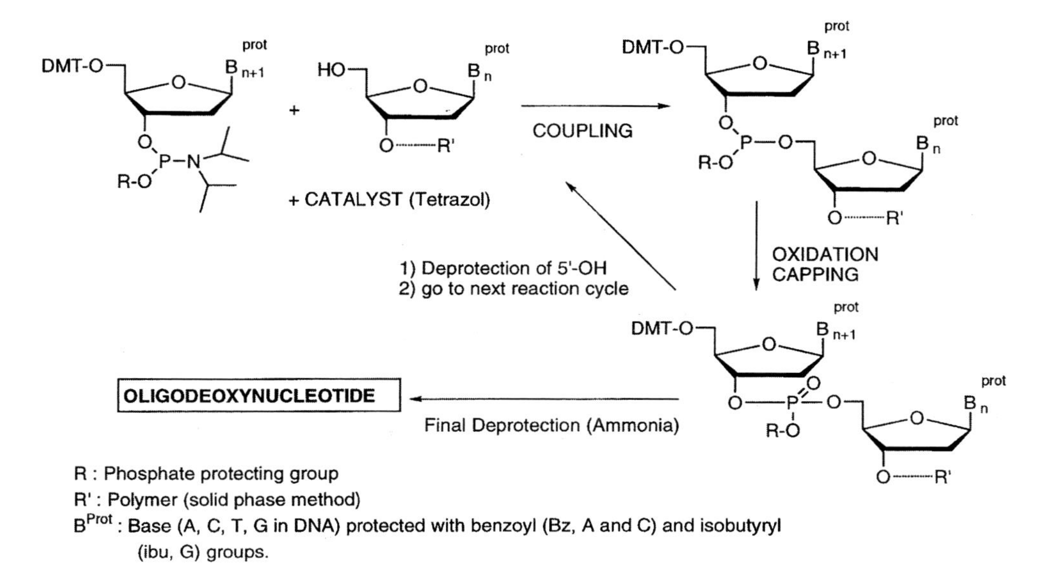

Fig. 5.1. Outline of the reactions involved on one addition of a nucleoside unit using phosphoramidite derivatives

- Solid support loaded with the first nucleoside
- DMT nucleoside phosphoramidites
- Concentrated ammonia (25–32 %)
- UV spectrophotometer
- Oven or heating plate at 50–60 °C
- UV lamp (254 nm)
- Speed-vac or Rotovap

Solutions for the synthesis

- Dichloromethane
- 2–3 % dichloroacetic or trichloroacetic in dichloromethane (deblocking solution)
- Acetonitrile HPLC grade
- Anhydrous acetonitrile packed in a bottle with septum
- 0.4 M tetrazol solution in acetonitrile
- Acetic anhydride in tetrahydrofuran, lutidine (Cap A)
- N-methylimidazole in tetrahydrofuran (Cap B)
- Iodine solution (0.1–0.02 M) in pyridine, tetrahydrofuran, water

Other solutions

- 0.1 M 4-toluensulfonic acid in acetonitrile
- 1-propanol / concentrated ammonia / water (55:35:10)

  Note: Use gloves, lab coat, and protective glasses. Work under a ventilated fume-hood is advisable.

Procedure

Assembly of the Sequence

The oligonucleotide sequence is prepared from the 3'-end by addition of the appropriate phosphoramidite derivative following the 3' to 5' direction.

1. Fill a series of small bottles with one of the solvents or reagents required in the synthesis. Dissolve the DMT nucleoside phosphoramidites with the appropriate amount of dry acetonitrile to make a 0.1 M solution (use the Gastight syringe to transfer the acetonitrile).
2. Set up a syringe with the appropriate frit. Weigh the appropriate amount (30–60 mg for 1–2 μmol) of solid support loaded with the first nucleoside (the nucleoside present at the 3'-end) in the syringe. Set a needle on the syringe.
3. Draw 1 ml of acetonitrile into the syringe, mix the CPG with the solution by gentle agitation for 15 s and expel the solvent. Repeat the operation a total of three times.
4. Wash the support three times with 1 ml of dichloromethane as indicated in step **3.**
5. Draw 1 ml of deblocking solution and mix the CPG with the solution by gentle agitation for 1 min. A strong orange color will develop. Collect the orange solution in a separate culture tube.
6. Wash the support twice with dichloromethane (1 ml), collect the washings together with the previous orange solution.
7. Repeat step **5** and observe if more color is produced. If the solution is orange, mix the solution and the CPG for 1 min and collect the orange solution together with the previous washings.
8. Wash with dichloromethane (1 ml) twice.
9. Wash with acetonitrile HPLC grade (1 ml) twice.

10. Wash with anhydrous acetonitrile (1 ml) twice.

11. First draw 0.5 ml of 0.4 M tetrazol solution followed by 0.5 ml of the appropriate phosphoramidite solution (0.1 M). Add a plastic cap on the tip of the needle to preserve the solution from ambient humidity. Mix for 2 min with gentle agitation. Then discard the solution.

12. First aspirate 0.5 ml of Cap A (acetic anhydride solution) and then 0.5 ml of Cap B (N-methylimidazole). Mix for 30 s with gentle agitation and discard the solution.

13. Draw the iodine solution, mix for 30 s and discard the solution. The oxidation step could be used to introduce modified phosphate bonds (Crooke and Lebleu 1993). For example, replacing iodine solution with solutions with sulfurizing agents (see below), it is possible to introduce phosphorothioate linkages.

14. Wash with acetonitrile until the complete removal of the brown color of the iodine. Usually it needs four washings of 1 ml each.

15. Repeat the synthetic cycle (steps **4** to **14**) until completion of the sequence.

16. After the addition of the last base, and depending of the method used for the purification (see below), you can either remove the last DMT group to have a 5'-OH free or leave the DMT group at the 5'-end. If the fully deprotected oligonucleotide is desired, you should follow steps **5** to **9** after the addition of the last phosphoramidite. To obtain the 5' protected DMT-oligonucleotide, you should stop the synthetic cycle in step **14** of the addition of the last phosphoramidite.

Ammonia Deprotection

1. Remove the plunger for the syringe and allow to dry the solid support. Remove the solid support from the syringe and place it in a screwcap vial. Add 1–1.5 ml of concentrated ammonia and close the vial tightly. Place the vial at 50–60 °C for a minimum of 6 h (is usually left overnight).

2. Remove the vial from the oven and allow the vial to cool at room temperature or in the refrigerator. Remove the solid support by centrifugation or by filtration using an HPLC filter or a glass pipette with a cotton plug. The resulting ammonia solution could be used directly if the oligonucleotide will be purified by reversed-phase cartridge (the oligonucleotide should have the DMT group on). Otherwise, ammonia should be eliminated. If the number of oligonucleotides is small, the best way is to transfer the ammonia solution to a round-bottomed flask and concentrate the solution with a Rotovap. Alternatively a Speed-vac could be used, but it is important to make sure that the vacuum is not very high at the beginning to avoid sample losses when ammonia is applied to the vacuum.

3. The residue is dissolved in water and the absorption at 260 nm of the solution is measured. Usually a 1/100 dilution is needed. For an oligonucleotide of 15–20 bases, usually 50–80 OD units at 260 nm of crude are obtained in a 1 μmol synthesis.

Analysis of the DMT Cation Absorbance

Coupling efficiency could be measured by the absorbance of the orange solutions collected in test tubes during the synthesis of the oligonucleotide.

1. For a standard 1 μmol synthesis, transfer the effluent from each detritylation to a 10-ml volumetric flask. Rinse the test tube with 0.1 M 4-toluensulfonic acid in acetonitrile and transfer the contents to the volumetric flask. Dilute to the mark, stopper the flask, and mix gently.
2. Measure the absorbance of the DMT cation in a UV-visible spectrophotometer at 500 nm. Usually the absorbance of the solution is higher than 2 OD units and a 1:10 or 1:100 dilution should be made. Alternatively, the solutions could be measured at 520–530 nm where the absorbance of the undiluted solutions is close to 1 OD unit.
3. The overall yield of the synthesis is obtained by dividing the absorbance of the solution obtained in the last detritylation (last tube) by the absorbance from the first coupling reaction (second tube). The yield of each particular coupling could be obtained by dividing the absorbance of the DMT cation solution before and after the coupling.

Analysis by TLC

The quality of the newly synthesized oligonucleotides can be controlled by gel electrophoresis (described on the purification part) and by thin layer chromatography (if the oligonucleotide is shorter than 30 bases).

1. Set up a TLC chamber with the running solution 1-propanol-concentrated ammonia-water (55:35:10). The solution should be around 1 cm deep.
2. Draw a line with pencil at 2 cm from the bottom of the plate.
3. Apply 1 μl of the ammonia solution onto the TLC plate at the line drawn before. Dry the sample with the help of a hairdryer or let it dry.
4. Put the plate in a glass chamber until the liquid front reaches most of the TLC plate.
5. Dry the plate either with the hairdryer or at room temperature.
6. Analyze the TLC under UV-light (254 nm). Typically a 20-mer have a retention factor (Rf) of 0.4.

Results and Comments

The protocol described above allows the preparation of approx. 3 mg of crude oligonucleotide (20 bases long, 1 μmol scale). DNA synthesizers can perform similar synthesis cycles automatically and they are common instruments in most molecular and cellular biology laboratories. The standard scales that an automatic synthesizer can produce are : 0.04, 0.2, 1, and 10 μmol.

A similar cycle is used for the preparation of oligonucleotides having phosphorothioate linkages. Specifically, the change of the iodine solution for a 0.5 M TETD solution in acetonitrile or a 50 mM 3H-1,2-benzodithiol-3-one 1,1-dioxide (Beaucage reagent) solution in acetonitrile will transform the phosphite-triesters to phosphorothioate. This reaction could be performed at any position of the oligonucleotide. The reaction time for sulfurization using TETD is 15 min and 30 s if Beaucage reagent is used. The phosphoramidite method allows the introduction of a large variety of different modifications that prevents exonuclease degradation. For further information see the following reviews: Crooke and Lebleu (1993); Beaucage and Iyer (1993).

The progress of the synthesis is controlled by the color of the DMT cation released during the acid treatment. The colored solutions are collected during the synthesis and they can be measured at 500 nm. A good synthesis should have a similar intensity of color throughout the synthesis.

The homogeneity of the product obtained could be checked after ammonia deprotection using TLC, or ion-exchange HPLC or electrophoresis. A major product (>80 % for a 20-mer) is observed together with several small products (less than 1 %) corresponding to the truncated sequences.

5.2 Purification and Analysis of Synthetic Oligonucleotides

There are different methods to purify synthetic oligonucleotides. Depending on the available instrumentation, it is possible to use either gel electrophoresis or HPLC techniques (Fig. 5.2). Ion-exchange HPLC provides a good resolution of the full-length product from failure sequences. HPLC on reversed-phase columns is also used if the DMT group of the last addition is left on the oligonucleotide during the ammonia treatment (DMT on synthesis) because the full-length oligonucleotide containing the DMT group is easily separated from truncated sequences. A similar result could be obtained using reversed-phase cartridges. The final product is less pure than the product obtained by HPLC or electrophoresis, but a good recovery is obtained in less time and no special equipment is needed for cartridge purification. Also protocols to desalt and obtain the sodium salts are described.

Materials

For gel electrophoresis

- Electrophoresis power supply (0–3000 Volts)
- Electrophoresis glass plates (20×20), Teflon spacers (0.4–1.5 mm), and comb
- UV lamp (254 nm)
- Urea
- 10× TBE: Tris base 105 g; boric acid 55.6 g and EDTA disodium salt. 2 H_2O. Dissolve in water and adjust to 1 l of solution (pH should be 7.8–8)

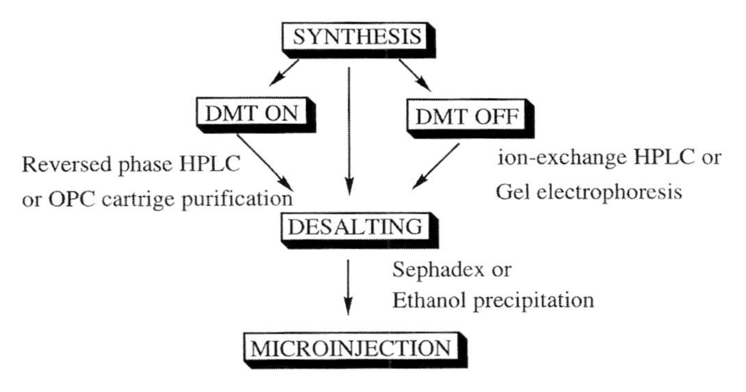

Fig. 5.2. Schematic flow sheet showing the different steps for oligonucleotide purification

- Acrylamide/ bisacrylamide stock solution: 38 g of acrylamide and 2 g of bisacrylamide dissolve in water and adjust to 100 ml water.
  Warning: The unpolymerized form of acrylamide and bisacrylamide are highly toxic. Wear gloves when preparing the solution.
- N,N,N',N'-tetrametylethylenediamine (TEMED)
- 20 % ammonium persulfate (100 mg) in water (0.4 ml).
- TLC plate (silica gel 60F254, 20×20 cm, Merck, Darmstadt, Germany, cat. no. 1.16484)
- Saran-Wrap

For HPLC

- HPLC or FPLC apparatus
- Plastic syringes (5 ml)
- HPLC filters (Nylon Acrodisc, 0.2 μm)
- Reversed-phase HPLC column (PRP-1, Hamilton or Nucleosil 10C18 or similar)
- Ion-exchange HPLC column (Mono Q, Pharmacia)
- HPLC solutions (reversed-phase) : A: 5 % acetonitrile in 0.1 M triethylammonium acetate
- pH 6.5; B: 70 % acetonitrile in 0.1 M triethylammonium acetate pH 6.5
- HPLC solutions (ion exchange): A: 10 mM NaOH, 0.3 M NaCl pH 12.5; B: 10 mM
- NaOH, 0.9 M NaCl pH 12.5
- Detritylation solution: 80 % acetic acid in water

For reversed-phase cartridge purification

- Reversed-phase cartridges (COP, Cruachem Ltd, Glasgow, Scotland or OPC, Applied Biosystems-Perkin Elmer, Foster City, California, USA or Poly-Pak, Glen Research, Sterling, Virginia, USA).
- Syringe (5 ml)
- UV spectrophotometer
- Acetonitrile
- 2 M triethylammonium acetate pH 7.5
- Diluted ammonia (1/10)
- Water
- 2 % trifluoroacetic acid in water
- 20 % acetonitrile in water

For desalting and obtaining of sodium salt

- Sephadex G10 or G-25
- Dowex 50 W×4–200 (100–200 mesh)
- UV spectrophotometer
- NaOH 1 M
- Water

Procedure

Polyacrylamide Gel Electrophoresis (PAGE)

Polyacrylamide gel electrophoresis may be used for analysis and purification of oligonucleotides. The thinness of the gel may range from 0.2–0.4 mm (for analytical purposes) to 1–2 mm (for preparative purposes); 15–20 % acrylamide and denaturing conditions (7 M urea) are optimal.

1. Prepare the appropriate glass plates, spacers and comb. (Sambrook et al. 1989).
2. Prepare the solution of urea-polyacrylamide. Fifty ml are needed for analytical purposes and 150 ml for preparative purposes. For analytical gels: weight 20.9 g of urea and add 5 ml of 10× TBE and 20 ml of the 40 % acrylamide/ bisacrylamide stock solution. Add water to a final volume of 50 ml and stir to dissolve the solids with gentle warming.
3. Add 12.5 µl of TEMED and 125 µl of the ammonium persulfate. Mix thoroughly and pour the solution immediately into the gel cast and add the comb. After polymerization of the gel (5–15 min), install the

glass plates in the electrophoresis apparatus and run the gel at 750 V for 2 h.

4. After the run, the gel is carefully transferred to a TLC plate wrapped with Saran wrap.
5. The oligonucleotides are visualized with UV light (254 nm).
6. If the gel is preparative, excise the desired gel band with a razor blade and put the gel fragments with 2–3 ml of water overnight in the refrigerator. Then filter, wash with water, and discard the gel fragments.
7. Desalt the oligonucleotide solution using Sephadex G-10 (see below) or HPLC or by ethanol precipitation.

Ion-Exchange HPLC

Rapid analysis and /or purification may be accomplished using anion-exchange HPLC. Silica gel or polymeric columns carrying chemically bonded quaternary ammonium groups are used. Oligonucleotides are separated because negatively charged phosphate groups on the DNA interact with the tetraalkylammonium cations contained on the column. A gradient of increased ionic strength is used to elute oligonucleotides in order of increasing chain length (Fig. 5.3). The protocol described uses Pharmacia Mono Q column. The size of the column is a function of the amount of product to inject. Analytical columns (HR 5/5) give good results for 0.1–5 OD (40-nmols scale). For 1 µmol scale synthesis, a larger column is needed (HR 10 /10).

1. Equilibrate the column with solution A.
2. Inject the sample and run a 60-min linear gradient from 0 % B to 70 % B.
3. Collect the last eluting peak and quantify product from the UV absorption of the solution.
4. Neutralize the solution and desalt.

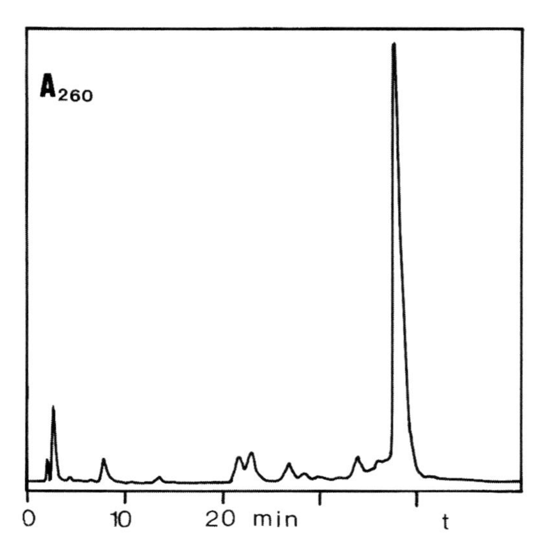

Fig. 5.3. Analytical ion-exchange HPLC chromatogram of a crude dodecamer (5'-CACCGACGGCGCC 3') prepared on 10-μmol scale on an Applied Biosystems 394 DNA synthesizer (DMT off). Column Mono Q HR 5 / 5. Flow rate 0.5 ml/ min. Solvent A: 10 mM NaOH, 0.3 M NaCl pH 12.5; Solvent B: 10 mM NaOH, 0.9 M NaCl pH 12.5. A 50-min gradient from 10 to 60 % B. Absorbance at 260 nm

Reversed-Phase HPLC

The separation of the desired sequence from truncated sequences is obtained due to the hydrophobic character of the DMT group (Fig. 5.4). The oligonucleotide should be obtained from a synthesis in which the DMT of the last phosphoramidite has not been removed (DMT on). After separation of the DMT oligonucleotide, the DMT group is removed with acetic acid and usually a second purification is performed. A 200×4 mm column is used for analytical purposes (0.1–5 OD). Repeated injections or a larger column are needed for 1 μmol scale.

1. Equilibrate the column with 10 % B.
2. Inject the sample and run a 30-min linear gradient form 10 % B to 70 % B.
3. Collect the peak eluting near 50 % B.
4. Repeat the injections with the remainder of the crude mixture and collect the appropriate peak. Quantify the product from the UV absorption of the solution and evaporate the combined fractions.
5. Treat the residue with 1 ml of 80 % acetic acid at room temperature for 30 min.

6. Add 1 ml of water and extract the acetic acid and the dimethoxytrityl alcohol with ethyl ether (3×6 ml).

7. Add one drop of ammonia to the aqueous solution and concentrate the solution.

8. Dissolve the residue in 0. 5 ml of water or buffer A, filter the sample and repurify the sample by HPLC. A 30-min gradient from 0 % B to 50 % B is used.

9. Collect the main peak and quantify the product from the UV absorption.

Reversed-Phase Cartridge Purification

Similarly to the reversed-phase HPLC, the purification is based on the presence of the DMT group at the 5' end. Cartridge purification is relatively inexpensive because no special instrumentation is needed. It is important to process oligonucleotides with G at the 5' end as soon as possible because they have the tendency to lose the DMT group. The following protocol is given for COP cartridges.

1. Wash the reversed-phase cartridge with 2 ml of acetonitrile.

2. Equilibrate the cartridge with 2 ml of 2 M triethylammonium acetate.

3. Dilute the ammonia-deprotection solution containing the DMT oligonucleotide with 1–2 ml of water. Pass the solution through the cartridge collecting the eluate. Pass the solution a second time. Keep the eluate.

4. Wash the cartridge with 3 ml of 1.5 M aqueous solution (water / concentrated ammonia 9:1).

5. Wash the cartridge with 2 ml of water.

6. Fill the syringe barrel with 3–4 ml of 2 % TFA in water and pass half of the solution. Let stand for 2–3 min and flush the remainder to waste.

7. Wash with 2 ml of water

8. Elute the desired, fully deprotected, oligonucleotide from the cartridge with 1 ml of 20 % acetonitrile and collect the sample in a 1.5–2-ml Eppendorf tube. In the case of phosphorothioate DNA analogs,

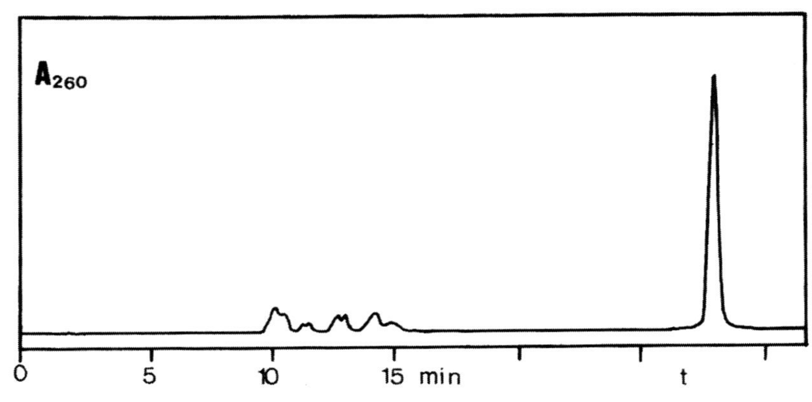

Fig. 5.4. Analytical reversed-phase HPLC chromatogram of the same crude dodecamer prepared using the DMT on protocol. Column Radial-Pak (Bondapak C-18). Flow rate 2 ml / min. Solvent A: 5 % acetonitrile in 0.1 M triethylammonium acetate pH 6.5. Solvent B: 70 % acetonitrile in 0.1 M triethylammonium acetate pH 6.5. A 30 min gradient from 0 to 50 % B. Absorbance at 260 nm

fluorescein-labeled oligonucleotides, or 2'-O-methyl RNA analogs, a 35 % acetonitrile/water solution is recommended due to higher hydrophobicity.

9. For 1 μmol synthesis repeat the process two to three times. The same cartridge could be used for the same oligonucleotide, giving up to 20 OD units per cycle.

Desalting

Oligonucleotides coming from HPLC (especially ion exchange) and electrophoresis should be desalted before use. Molecular sieve columns are useful for these purposes, but ethanol precipitation can also be used.

1. Equilibrate a Sephadex G-10 or G-25 column with water. A prepacked column can be used (NAP-10, Pharmacia).
2. Dissolve the sample with 1 ml of water.
3. Load the sample to the column and elute with water.
4. Collect 1 ml samples and localize the first UV-absorbing material, usually the first 2–4 ml. Combine the fractions and concentrate to dryness.

Preparation of Oligonucleotides in the Sodium Salt Form

Oligonucleotides coming from reversed-phase purification contain triethylammonium salts and counterions of the phosphates. A large amount of these counterions can be toxic to the cells. Repeated liophylization removes a large part of these volatile salts but it is convenient also to exchange the counterion using either ion-exchange or ethanol precipitation.

1. Weigh 1–2 g of Dowex 50×4 per μmol of oligonucleotide.
2. Soak the ion-exchange resin with 1 M NaOH solution and poured on a column. Wash the column with 20–50 ml of 1M NaOH solution and afterwards with 200 ml of water. Check the pH of the eluate to be neutral.
3. Dissolve the sample with 1 ml of water and load the sample to the column. Elute with water.
4. Collect 1-ml samples and localize the first UV-absorbing material, usually the first 2–4 ml. Combine the fractions and concentrate to dryness.

Ethanol Precipitation

1. Dissolve the sample with 0.3 ml of water (0.02–0.03 ml per OD unit of oligonucleotide) and add 0.05 ml of 3 M sodium acetate.
2. Add 1 ml of ethanol and mix.
3. Store in the freezer for at least 30 min and centrifuge at high speed for 5 min. For oligonucleotides of less than 15 bases, isopropanol may be substituted for ethanol to ensure complete precipitation.
4. Remove the supernatant with a pipette and dry the pellet with a Speed-vac.

Results and Comments

Short oligonucleotides can be used for microinjection experiments directly after synthesis and ethanol precipitation (Kleuss et al. 1994). In this case, small amounts of truncated sequences will be also present on the experiment. In order to remove the truncated sequences, PAGE, reversed-phase, or ion-exchange HPLC techniques can be used. Recovery

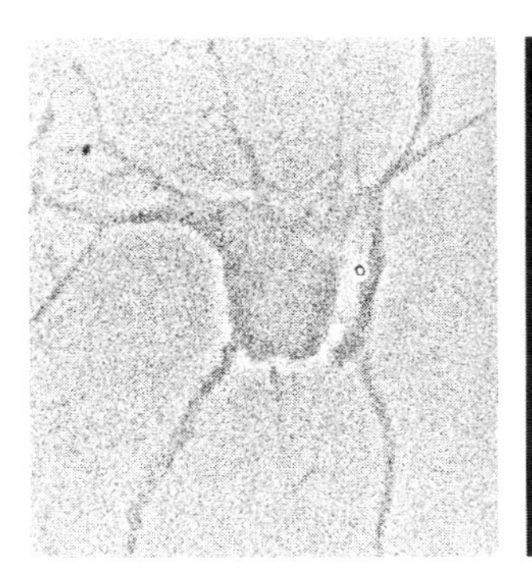
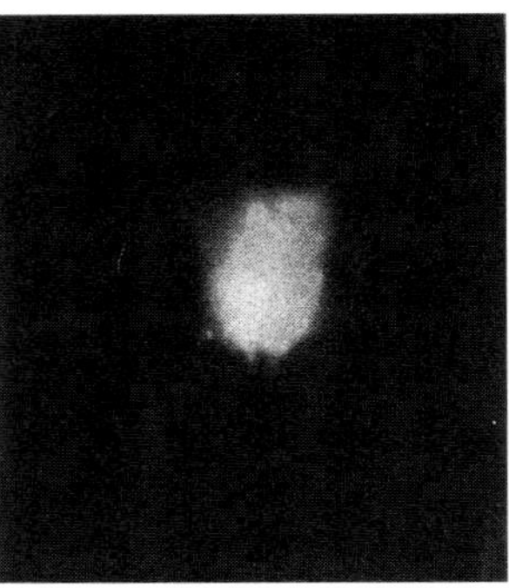

Fig. 5.5. A fully polarized hippocampal neuron in culture photographed immediately after intranuclear microinjection of fluorescein-labeled oligodeoxynucleotides designed for antisense inhibition of a heat shock protein

of the full-length product varies from 20 to 50 %, depending on the length and the method used. Also, reversed-phase cartridges can be useful for rapid removal of the truncated sequences.

Natural phosphodiester oligonucleotides can be used for microinjection experiments, but they are degraded inside the cells by nucleases (Dagle et al. 1991). To avoid nuclease degradation, oligonucleotide-containing phosphorothioate bonds are frequently used (Chin et al. 1990; Leonetti et al. 1991; Beaucage and Iyer 1993; Crooke and Lebleu 1993). Other modifications used in microinjection experiments are 5-propynyl pyrimidine oligonucleotides (Moulds et al. 1995), 2'-O-allyl oligoribonucleotides (Moulds et al. 1995), and phosphoramidates (Dagle et al. 1991). In order to analyze the intracellular distribution of microinjected oligonucleotides, fluorescent compounds such as fluoresceine or rhodamine can be incorporated into the oligonucleotides (Chin et al. 1990, Leonetti et al. 1991). Figure 5.5 shows a neuronal cell that has been microinjected with a phosphorothioate oligonucleotide labeled with fluoresceine.

For microinjection experiments, oligonucleotides are dissolved in water or phosphate-buffered saline solutions at a concentration ranging from 1 to 100 μM. Before microinjection, the oligonucleotide solution is centrifuged for 10 min at 13,000 g to remove materials in suspension. An injection volume of about 20 femtoliters of a 5-μM solution is used for microinjection into nucleus (Kleuss 1994); 50–500 femtoliters of a 0.1 mM oligonucleotide solution is used for microinjection into cytoplasm (Chin et al. 1990). Oocytes are injected with 10–50 nanoliters at a 0.1-M concentration (Dagle et al. 1991).

Acknowledgments. We thank Marten Wiersma for technical assistance. Fig. 5.5 was taken by Dr. A. Cid-Arregui.

References

Beaucage SL, Iyer RP (1993) The synthesis of modified oligonucleotides by the phosphoramidite approach and their applications. Tetrahedron 49:6123–6194

Caruthers MH, Barone AD, Beaucage, SL, Dodds, DR, Fischer, EF, McBride, LJ, Matteucci, M, Stabinsky, Z, Tang, J-Y (1987) Chemical synthesis of deoxyoligonucleotides by the phosphoramidite method. Methods Enzymol 154:287–313

Chin DJ, Green GA, Zon G, Szoka FC, Straubinger RM (1990) Rapid nuclear accumulation of injected oligodeoxyribonucleotides. New Biol 2:1091–1100

Crooke ST, Lebleu B (eds) (1993) In: Antisense research and applications. CRC Press, Boca Raton

Dagle, JM, Weeks, DL, Walder, JA (1991) Pathways of degradation and mechanism of action of antisense oligonucleotides in *Xenopus laevis* embryos. Antisense Res Dev 1:11–20

Kleuss C, Schultz G, Witting B (1994) Microinjection of antisense oligonucleotides to asses G-protein subunit function. Methods Enzymol 237:345–353

Kola I, Sumarsono SH (1995) Microinjection of in vitro transcribed RNA and antisense oligonucleotides in mouse oocytes and in early embryos to study the gain- and loss-of-function of genes. Methods Mol Biol 37:135–149

Leonetti JP, Mechti N, Degols G, Gagnor C, Lebleu B (1991) Intracellular distribution of microinjected antisense oligonucleotides. Proc Natl Acad Sci USA 88:2702–2706

Moulds, C, Lewis, JG, Froehler, BC, Grant, D, Huang, T, Milligan, JF, Matteucci, MD, Wagner, RW (1995) Site and mechanism of antisense inhibition by C5-propyne oligonucleotides. Biochemistry 34:5044–5053

Sambrook, J, Frisch, EF, Maniatis, T (1989) Molecular cloning: a laboratory manual, 2nd edn, vols 1, 2, 3. Cold Spring Harbor Laboratory, Cold Spring Harbor

Tanaka, T, Letsinger, RL (1982) Syringe method for stepwise chemical synthesis of oligonucleotides. Nucleic Acids Res 10:3249–3260

Suppliers

Suggested Suppliers for Oligonucleotide Synthesis Products
Applied Biosystems-Perkin Elmer
850 Lincoln Centre Drive
Foster City, CA 94404; USA
Phone (415) 570–6667; (800) 345–5224; Fax (415) 572–2743

PerSeptive Biosystems, Inc.
500 Old Connecticut Path
Framingham, MA 01701; USA
Phone (508) 383–7700; (800) 899–5858

Bio Genex
4600 Norris Canyon Rd.
San Ramon, CA 94583; USA
Phone (510) 275-0550; (800) DNA-PURE; Fax (510) 866-2594

Glen Research
44901 Falcon Place
Sterling, Virgina 20166; USA
Phone (703) 437-6191; (800) 327-GLEN; Fax (703) 435-9774

Eurogentec
Parc scientifique du Sart Tilman
Rue Bois Saint Jean 14
4102 Seraing; Belgium
Phone 32-41.66.01.50; Fax 32-41.65.51.03.

Pharmacia Biotech Norden
Djupdalsvägen 20-22; Box 776;
191 27 Sollentuna; Sweden
Phone 46-(0)8-623-8500; Fax 46-(0)8-6230069.

Cruachem Ltd; Tood Campus
West of Scotland Science Park
Acre Road
Glasgow G20 0UA; Scotland
Phone 41-945-0055; Fax 41-946-6173

Clontech Laboratories, Inc.
1020 East Meadow Circle
Palo Alto, CA 94303-4230; USA
Phone (415) 424-8222; (800) 662-CLON; Fax (415) 424-1064 and (800) 424-1350

Peninsula Laboratories, Inc.
611 Taylor Way
Belmont, CA 94002-4041M; USA
Phone (415) 592-5392; (800) 922-1516; Fax (415) 595-4071

Sigma Chemical Co
P.O. Box 14508
St Louis, Missouri 63178; USA
Phone (314) 771-5750; (800) 521-8956; Fax (314) 771-5757 and (800) 325-5052

Chapter 6

Colorimetric in Situ Hybridization

Jaime Berumen[1]

Introduction

In situ hybridization, the demonstration of genetic information within a morphological context, is a powerful technique which is being used with increasing confidence and frequency both in experimental and diagnostic pathology. The development of colorimetric methods amenable to routinely processed formalin-fixed tissues has the potential to further increase its use.

Specific DNA or RNA is visualized by this combination of the molecular biology technique of nucleic acid hybridization and the cytochemical technique of immuno- or affinity detection. An appropriately labeled DNA or RNA probe is allowed to bind to its DNA or RNA target sequence. For the microscopic detection of the bound probe, enzyme-labeled antibodies or affinity molecules such as avidin are used. The topological positions of the genes and their expression, at the RNA level, are thus maintained.

Nonradioactive methods of hybridization and detection are essential for clinical adoption of this technology. Nonradioactive probes are stable reagents and result in reduced environmental hazards. Brigati and coworkers first demonstrated the feasibility of using biotin-labeled nucleotides with an avidin-enzyme detection system for the demonstration of viral genetic information in formalin-fixed paraffin-embedded tissue sections in 1983 (Brigati et al. 1983). Since that time, the technique has been refined and shortened, and when fully optimized, shown to have sensitivity comparable to ^{35}S-autoradiography (Unger et al. 1991). Other affinity labels such as digoxigenin and sulfone are available and direct labeling of probes with enzymes or fluorescent tags has been developed.

[1] Laboratorio Multidisciplinario de Investigation, Escuela Militar de Graduados de Sanidad-Escuela Medico Militar, Universidas del Ejercito y Fuerza Aerea. Apdo. Postal 35–556, 11649 Mexico City, Mexico; phone: 5–25–2020303; fax: 5–25–2020303; e-mail: escuelam@solar.sar.net

The procedure contains aspects related to both the maintenance of the morphology of the material under investigation and the achievement of specificity with the hybridization reaction. In addition, the efficiency needed for both the hybridization and the detection steps impose separate requirements to the entire procedure.

Probe Preparation and Labeling

The probe is the key reagent that determines the ultimate specificity and, in part, the sensitivity of the hybridization assay. The size and complexity are the most important features of probes to achieve a high level of sensitivity and specificity. Probes with higher genetic complexity provide increased sensitivity because a significant portion of the target is represented in the probe and available for hybridization. Probe size should also be such as to favor the kinetics of the hybridization reaction and allow for maximal tissue penetration, thus allowing the time of hybridization to be reduced, and reducing the nonspecific background signal. Optimal probe fragment size ranges from 50 to 200 nucleotides.

Probes used for in situ hybridization are generally prepared using standard molecular biological techniques. The most widely used are double-stranded DNA (dsDNA) probes consisting of the DNA of interest cloned into a plasmid DNA vector. Also used are single-stranded DNA and RNA probes isolated using, for example, the M13 phage system or asymmetric PCR and transcribed from fragments cloned into a plasmid containing the bacteriophage SP6 or T7 promotor, respectively. Synthetic oligonucleotides are also being used; however, the amount of genetic information represented in these short sequences limits the sensitivity when compared to the genetically complex recombinant probes.

Probes can be labeled either enzymatically or chemically. The enzymatic methods incorporate appropriately modified nucleotides, such as the biotin-11-dCTP, using the techniques generally adopted for radioactive labeling in molecular biology. Nick translation and random primer extension are the enzymatic methods most commonly used to label DNA probes (Sambrook et al. 1989). Random primer extension gives a higher specificity labeling (the number of incorporated labeled nucleotides per μg of DNA probe) than nick translation procedure; even the template DNA is not labeled and at the end of the reaction only half of the DNA probe is labeled. Besides, with the random priming labeling, considerable net DNA synthesis occurs, resulting in a 10–40-fold amplification of the probe.

The chemical methods utilize a variety of techniques such as photobiotinylation, sulfonation, mercuration, and N-acetoxy-2-aacetylaminofluorene labeling. Before, during, or after labeling the probe, size should be controlled such that the probe penetrates easily into the fixed tissues and cells.

Prehybridization Treatment of Cell and Tissue Slides

The morphology of the tissue under investigation will be determined by the fixation, pretreatment, and hybridization conditions. Tissue fixation and pretreatment are important to maintain the topological position of the target DNA and RNA and for penetration of probe and detection reagents. Penetration of probe and detection reagent can be improved by pepsin or proteinase K pretreatment. For most tissues, pepsin-HCl treatment is most satisfactory, but for cell smears, proteinase K gives better results.

Optimal digestion conditions are essential for success of in situ hybridization. The adequacy of tissue preparation for probe penetration and hybridization is determined by the results of hybridization with a positive control, like the human placental DNA probe. If the cells are well preserved and adequately prepared for hybridization, the human genomic material within each nucleus should generate a signal with this probe. Thus, human placental DNA serves as a simple but effective endogenous positive control probe.

The degree of sensitivity is achieved by careful optimization of the digestion conditions without loss of fine nuclear morphology. In fact, the use of this positive control was essential for developing colorimetric technology with a sensitivity comparable to that of autoradiographic methods (Unger et al. 1991). Decreased digestion (less time, acid, pepsin or proteinase K) gives better morphology, but less sensitivity.

Optimal digestion may not always be attainable in tissue sections because of extreme variations in fixation and processing. Without recourse to an endogenous positive control as a guide towards maximizing the signal, digestion conditions would be difficult to estimate. Use of this control permits poorly preserved tissues to be eliminated from further consideration, and reduces false-negative hybridizations due to inadequate protease digestion. Concentration of pepsin/HCl and proteinase K should be optimized for different tissue and cellular types, and the positive control probe should be included in every set of experiments along with the tested probes (Unger et al. 1991).

Hybridization and Posthybridization Reaction Conditions

Molecular hybridization is a reaction in which a single-stranded target sequence in solution, on filters, or within tissues, and a complementary molecular probe anneal to form double-stranded hybrid molecules. In the case of DNA hybridization, both target sequences and DNA probe are originally double-stranded, and before hybridization they must be denatured.

The two complementary DNA strands can be separated by agents that break the hydrogen bonds, such as heat, extremes of acid or alkaline pH, and hydrogen bond-disrupting chaotropic agents like formamide. The energy needed to break the bonds between the A-T pairs (with two hydrogen bonds) is lower than for G-C pairs (with three hydrogen bonds). The backbone phosphate groups are negatively charged, and this electrotastic repulsion assists in the separation of the two strands once the hydrogen bonds are broken. For RNA hybridization, denaturation of target sequences is not theoretically necessary, but is advantageous because RNA molecules assume a secondary structure. The melting temperature (T_m) of a duplex DNA molecule is defined as the temperature at which 50 % of the duplex molecules are denaturated, and it is determined by the proportion of guanidine and cytidine nucleotides (% G+C), the length of the molecules in base pairs (L), the molar concentration of monovalent cations (M), and the amount of formamide used in the reaction mixture. These variables are linked by an equation derived by experiments of DNA association in solution which can be used to calculate the melting temperature in degrees Celsius.

$$T_m = 81.5 + 16.6\ (\log M) + 0.41\ (\%\ G{+}C) - 0.72\ (\%\ \text{formamide}) - 600/L$$

The T_m of double-stranded molecules is also influenced by the presence of mismatches of bases in the complementary strands due to small sequence differences and decreased 1 °C per 1 % of sequence mismatches in the duplex. The melting temperature of dsRNA is higher than that of RNA-DNA hybrids, which in turn is higher than that of dsDNA.

Melting temperature properties govern the reaction conditions for optimal hybridization. Single-stranded DNA denatured by heat will reform double-stranded hybrids when the temperature is lowered again. Probe molecules binding to target sequences that do not have a full sequence homology will form hybrids having a high percentage of base with mismatches (% mismatch). The temperature of renaturation or hybridization depends on the perfection or stringency required for hybrid formation. The temperature of hybridization for perfect base pairing between the probe and target DNA is close to the T_m of the hybrid molecule. Thus, the stringency of hybridization is a function of the difference between the T_m and the hybridization temperature, and determines the degree to which mismatched hybrids are permitted to form.

Since the rate of the reannealing is maximum at 25 °C below the T_m (T_m-25) and falls as the temperature of hybridization approaches the T_m, hybridization conditions between T_m-15 to T_m-25 are recommended to allow rapid reannealing. Higher stringency is then achieved by post-hybridization washes to values closer to the T_m, according to the equation above.

The reaction rate is a function of the concentrations of probe and target sequences. The effective concentration of probe DNA in the hybridization solution can be increased by the addition of water-capturing polymers like dextran sulfate. More detailed information about these hybridization conditions can be found in Hames and Higgins (1985).

Specificity of Signals

Sensitive detection of small target sites is best achieved when high signals are obtained in the absence of any background reaction. Background reaction can be derived from nonspecific probe binding and nonspecific detector binding. Probe binding specificity is determined by the stringency of the hybridization and washing conditions. Specificity of detections is determined by the type of detector molecules and incubation conditions used. Background reduction can be achieved by staining under highsalt conditions (such as with 4× SSC), which prevents ionic binding of detector molecules. This is most effective for highly charged molecules such as avidin conjugates. The plasmid pBR322 is commonly used as the negative control probe, in the same conditions as the tested probe, to monitorize the unspecific signal.

Only when the negative control is absolutely negative can a delicate colorimetric signal be interpreted with confidence. This confidence in visual interpretation also increase with experience. In some tissue sections these requirements for optimal sensitivity can be hard to achieve. Background can be an especially difficult problem, with liver and kidney exhibiting quite high levels of endogenous avidin binding activity. In such situations, alternative labeling and detection system can be applied.

Outline

The entire in situ hybridization procedure consist of the following steps:

- Preparation of the labeled probe
- Preparation of the slides, fixation, and pretreatment of the material on the slide
- Denaturation of target DNA (not for mRNA)
- The hybridization reaction and the removal of nonbound probe
- The detection steps of the bound probe.

Each of the separate steps is discussed in detail below and full protocols for tissue sections and cultured cells are given in the subsequent sections.

Materials

Reagents

- Nick translation kit (Boehringer 976776)
- BioPrime DNA labeling system (BRL 18094–011)
- Bio-11-dUTP 0.4 mM (BRL 9507SA)
- Mini-spin columns Sephadex G-50 (Boehringer Mannheim 100406)
- DNase I (BRL 8047SA)
- BSA (BRL 5561UA)
- Avidin-alkaline phosphatase conjugate (Dako D–365)
- T4 DNA polymerase 1 U/μl (BRL 8005SA)
- CrystalMount (Biomeda)
- Permount (Fisher Scientific)
- Rubber cement (Oncor).
- Brij-35 sol. 30 % w/v (Sigma 430AG-6)
- NBT (Sigma N-68786)
- BCIP (Sigma B-6149)
- 3-aminopropyltriethoxysilane (Sigma A-3648)
- Nuclear fast red (Sigma N-8002)

Solutions

DNase activation buffer: 10 mM TrisCl pH 7.6, 5 mM $MgCl_2$, 1 mg/ml nuclease free BSA. Store at –20° C

10× nick translation buffer: 500 mM TrisCl pH 7.8, 50 mM $MgCl_2$, 100 mM 2 Mercaptoethanol, 100 μg/ml nuclease free BSA. Store at –20° C

10× T4 DNA polymerase buffer: 0.33 M Tris-acetate pH 8.0, 0.66 M potassium acetate, 0.1 M magnesium acetate, 5 mM DTT, 1 mg/ml BSA

10× gel buffer: 0.3 M NaCl, 30 mM EDTA

10× running buffer: 0.3 M NaOH, 30 mM EDTA

1× neutralizing solution: 3 M NaCl, 0.5M Tris-Cl pH 7.5

10× dye solution: 50 mM Tris-Cl pH 7.5, 20 % Ficoll, 0.9 % xylene cyanol FF, 0.9 % bromophenol blue

Proteinase K solution: 100 mM Tris-HCl pH 7.6, 50 mM EDTA, 50 μg/ml proteinase K

100× Denhardth's solution: 2 % Ficoll, 2 % Polyvinyl pyrrolidone, 2 % BSA

Hybridization cocktail: 45 % formamide (BRL), 5× SSC, 1× Denhardt's solution, 25 mM $NaPO_4$ pH 6.5, 10 % dextran sulfate (Sigma), 500 μg/ml Salmon sperm DNA (500 pb), 3 μg/ml of biotinylated probe

Tris-saline pH 7.5 Brij: 0.1 M Tris-Cl pH 7.5, 0.1 M NaCl, 5 mM $MgCl_2$, 0.25 % Brij-35–30 % solution

Tris-saline pH 9.5 Brij: 0.1 M Tris-Cl pH 9.5, 0.1 M NaCl, 50 mM $MgCl_2$, 0.25 %Brij-35–30 % solution

McGadey reagent: 67 μl NBT 50 mg/ml 50 % DMF, 33 μl BCIP 50 mg/ml DMF

10 ml Tris-saline pH 9.5 Brij

Nuclear fast red solution: nuclear fast red 0.1 %, aluminum sulfate 5 %. Heat the solution until the reagents dissolve, filter, and add crystal of thymol.

6.1 Preparation of Biotinylated DNA Probes

Sequential Nick Translation and Random Primer Labeling

In this sequential labeling protocol the DNA may be conveniently labeled with biotin first by nick translation and then by random primer extension reactions in the presence of biotin-11-dCTP and the other three unlabeled deoxynucleotides (dATP, dTTP, dGTP). This sequential protocol improves the specificity labeling compared to the random priming procedure, because the template DNA for random priming is already biotinylated by nick translation before the new extended DNA strands are labeled. At the end of this sequential protocol most DNA strands are biotinylated.

Procedure

1. To a sterile microcentrifuge tube on ice add:
   μl DNA to be labeled (volume to equal 1 μg DNA)
   2.0 μl 10× nick translation buffer
   8.0 μl nucleotide buffer solution (200 μM each of dATP, dGTP, dCTP and Bio-11-dUTP)
   2 μl enzyme solution (5 U DNA polymerase, 100 pg DNase I)
   μl sterile distilled water
   To bring final volume to 20.0 μl
2. Mix gently, pop-spin to concentrate reagents in bottom of tube and incubate at 15 °C in ice-water bath for 2 h.
3. Heat in boiling water for 10 min.
4. Take 5 μl of the nick translation mixture (250 ng of DNA) and put in a new sterile microcentrifuge tube, then add 19 μl of sterile water and 20 μl of 2.5× random primers solution. Mix gently.
5. Denature DNA by heating for 7 min in a boiling water bath; then immediately cool on ice and incubate for 5 min. Centrifuge for 15 s.
6. Add 5 μl 10× dNTPs mix (2 mM each of dATP, dTTP and dGTP, 1 mM of biotin-14-dCTP and dCTP) and 1 μl of Klenow fragment DNA polymerase (40 U/μl). Mix gently but thoroughly. Centrifuge for 15 s.
7. Incubate at 37 °C for 2 h, then stop the reaction by heating for 10 min in a boiling water bath and add 50 μl of 1× SSC-0.1 % SDS to make a final volume of 100 μl. Avoid stopping the reaction with EDTA because this impairs the use of DNase I in case the probe needs to be shortened.
8. Remove nonincorporated deoxyribonucleotides triphosphate by chromatography on a 1-ml mini-spin column (Sephadex G-50).
9. Collect eluate and spot 1 to 2 μl onto nitrocellulose paper, dry in oven at 80 °C for 30 min and to detect the biotinylated probe proceed from step **4** to **10** of blot and detection procedure (see below).
10. The final concentration of the biotin-labeled DNA probe is approximately 3 μg in 100 μl. It may be stored in solution at −20 °C for at least 1 year.

Re-Nick Protocol

Before use, the probe size should be determined by alkaline agarose gel electrophoresis. For in-situ-hybridization probe size should average 200 bases. If probes are not below 500 bases, more DNase needs to be added to the nick translation/random primer mixture. The amount will have to be titrated for each lot of enzyme, 1–2 μl of 10^{-6} dilution is good starting point. Dilutions should be prepared immediately prior to use in DNase activation buffer. DNase does not survive freeze-thawing too well, so aliquoting is needed.

1. To sterile tube on ice add:
   μl biotinylated DNA (to equal 1 μg)
   5 μl 10× nick translation buffer
   1–2 μl diluted DNase I (dilute to 10^{-6} immediately prior to use in DNase activation Buffer)
   μl sterile distilled water,
   to bring final volume to 50 μl.
2. Mix gently, pop-spin and incubate in 15 °C ice-water bath for 2 h.
3. Stop reaction and separate product on spin column as described in the labeling procedure.

Procedure for Sizing Biotinylated Labeling Probes

End-labeled DNA markers

1. Perform the digestion of 5 μg of pBR322 with HinfI in a microcentrifuge tube:
   33 μl DD H_2O
   5 μl 10× T_4 DNA polymerase reaction buffer
   10 μl pBR322 (0.5 μg/μl)
   2 μl HinfI (10 U/μl)
   50 μl total volume
2. Mix, pop-spin, incubate in 37 °C water bath, 2 h.
3. Add to total DNA digested above:
   5 μl 2 mM dNTPs mix (dCTP, dATP, dGTP, and Bio-11-dUTP)
   10 μl 10× T_4 DNA polymerase reaction buffer
   12.5 μl T_4 DNA polymerase (12.5 U)
   22.5 μl DD H_2O
   100 μl total volume

4. Mix and incubate for 15 min in a 12 °C ice-water bath.
5. Add 10 µl 500 mM EDTA. Mix and heat in 70 °C water bath for 5 min. The final concentration of marker DNA is 50 ng/µl

2 % agarose alkaline denaturing gel

1. Prepare 2 % agarose mini-gel in 1× gel buffer. Heat to dissolve agarose, cool to about 70 °C prior to pouring gel.
2. Mix 1 µl of 10× running buffer, 1 µl of dye solution, 2 µl labeled DNA (100 ng) and 6 µl of DD H_2O.
3. Cover the gel with 1× running buffer and leave the gel to equilibrate for 15 min.
4. Load the samples into gel through the running buffer.
5. Electrophorese at 75 mA until bromphenol blue dye front is 3/4 down the gel (about 2.5 h).

Blot and detection

1. Neutralize gel in 3 M NaCl, 0.5 M Tris-Cl pH 7.5 until pH is less than 9. This requires about four changes of 30 ml buffer, 10 min/change. Rinse gel in H_2O.
2. Set up Southern blot with nitrocellulose filter using 10× SSC. Blot at room temperature for 2 h (overnight will work if more convenient).
3. Wash filter in 2× SSC 5 min, air dry and bake in 80 °C oven between filter paper for 1 h to overnight (may be stored at this point in refrigerator).
4. Block in 3 % BSA/Tris-saline pH 7.5 for 30 min in 37 °C water bath.
5. Bake in 80 °C oven between filter paper 30 min. (May be stored at this point in refrigerator).
6. Block in fresh 3 % BSA/Tris-saline pH 7.5 solution at 37 °C for 30 min.
7. Incubate with avidin-alkaline phosphatase conjugate (1:500 in 1 % BSA/Tris-saline pH 7.5) at room temperature for 10 min.
8. Wash nitrocellulose 4× 3 min/wash in Tris-saline pH 7.5 at room temperature. (Washes should occur in vessel which has not been exposed to conjugate).
9. Wash in Tris-saline 9.5, 2× 3 min at room temperature.
10. Develop in McGadey reagent, 1 h at room temperature, or until signal reaches desired intensity. Reaction is faster at 37 °C.

6.2 In Situ Hybridization for Target DNA in Tissues and Cultured Cells

Procedure

Silanization of Slides

Freshly prepare 2 % solution of 3-aminopropyltriethoxysilane in acetone. Dip good quality clean glass slides in solution for 2 min. Rinse twice in distilled water and air dry. Slides may be stored in dust-free box at room temperature until use.

Prehybridization Treatment of Tissue Slides

1. Pick up 3 μm sections of formalin-fixed paraffin-embedded tissue block.
2. Incubate the slides in the oven at 70 °C for 15 min to overnight in a metallic tray to firmly adhere tissues to the slides.
3. Dewax the tissue sections incubating the slides in Xylol (Coplin Jar) for 2× 5 min at 37 °C.
4. Incubate in absolute ethanol for 2×5 min (Coplin Jar) and air-dry the slides.
5. Digest the tissue section with pepsin 0.125 mg/ml in 0.025 N HCl. Apply 100 μl of the solution, cover with plastic coverslip and incubate at 37 °C for 20 min in a humidified chamber.
6. Wash the slides 3× 5 min in TS-Brij pH 7.5.
7. Dehydrate the tissue sections incubating for 2 min sequentially in 70 %, 95 %, and absolute ethanol.

Prehybridization Treatment of Cultured Cells

1. Collect cells from culture flask in 2 ml of fresh PBS. Pellet the cells at room temperature, 5 min at 1000 *g* in the microcentrifuge.
2. Resuspend and wash the cells gently but thoroughly in 2 ml of fresh PBS by using a 1-ml Eppendorf pipette. Pellet the cells as above. Repeat two more times.

3. Resuspend the cells in 200 μl of fresh PBS and spread 10 μl in a glass sibanized slide.
4. Incubate the slides in the oven at 70 °C for 30 min.
5. Fix in 95 % ethanol for 15 min.
6. Digest the cells with 50 μg/ml proteinase K solution in the same conditions as above. Then follow steps **6** and **7** of the tissue pretreatment procedure. Concentration of pepsin/HCl and proteinase K should be optimized for different tissue and cellular types. For both tissue and cell experiments, the human placental biotinylated DNA is used as the endogenous positive control probe to control for optimization of digestion conditions.

Hybridization (Metallic tray)

1. Apply 50 μl of the hybridization cocktail to each slide over tissue (avoid forming bubbles), then cover with a 22×40 coverslip. When all slides are finished, seal with rubber cement.
2. Place the slides on a metallic tray preheated at 100 °C, and probe and target DNA are simultaneously denatured at 100–105 °C for 15 min in a convection oven.
3. Perform the hybridization by incubating the slides at 37 °C in a moist chamber for 2 h.

Posthybridization Washes (Coplin Jar)

1. Remove glue from slides carefully and soak slides in 2× SSC-0.25 % Brij. Coverslip should fall off, if not, slide coverslip off carefully.
2. The nonspecifically bound probe is removed with a series of graded salt washes: 2×3 min each in 2× SSC-0.25 % Brij and 0.2× SSC-0.25 % Brij at room temperature; in 0.1× SSC-0.25 % Brij at 50 °C and in 2× SSC-0.25 % Brij at room temperature.

The temperature and salt concentration of stringent washes depend on the T_m of the hybrid probe/target DNA and the stringency required. It is necessary to consider that the T_m of biotinylated probes is 5 °C below the T_m of unlabeled DNA.

The plasmid pBR322 is used as negative control probe, under the same conditions as the tested probe, to monitor the unspecific signal.

Colorimetric Detection (Humidified Chamber/Coplin Jar)

1. Apply 200 μl 3 % BSA solution on the tissue slide, cover with a plastic coverslip and incubate for 5 min at 37 °C in a moist chamber.
2. Mix 4 μl avidin-alkaline phosphatase conjugate in 1 ml of 1 % BSA solution (final dilution 1:200). Apply 20 μl of the conjugate, cover with a plastic coverslip, and incubate for 20 min at 37 °C in the moist chamber.
3. Wash the slides 3×3 min each in Tris-saline pH 7.5 Brij and 2× in Tris-saline pH 9.5 Brij.
4. To develop the signal, apply 30 μl McGadey reagent and a plastic coverslip, incubate for 1 h at 37 °C in the moist chamber.
5. Wash 3×1 min each in Tris-saline pH 7.5.
6. Counterstain the tissue, applying 20 μl of the nuclear fast red solution and incubate for 20 s at room temperature. Wash in Tris-saline pH 7.5, cover with CrystalMount, air-dry, and mount with Permount.

Results

The sensitivity of this colorimetric in situ hybridization protocol with biotinylated probes is comparable to radioactive methods, with the colorimetric method having the advantages of speed, probe stability, and improved signal localization (Unger et al. 1991). Target DNA detection with colorimetric in situ hybridization resulted in a very clear purplish-blue nuclear signal contrasted with a soft pink of the counter-stain and negligible background. The lack of background permits analysis of the signals even at 1000× magnification (Fig. 6.1). This lack of background signal is not achieved with autoradiographic methods, and background grain counts must be subtracted in cases where the signal is low.

The in situ hybridization assay is considered satisfactory when the tissue slide gives a uniform, even nuclear reaction with the endogenous positive control probe (human placental DNA) and give no signal with the negative control probe (pBR322 DNA).

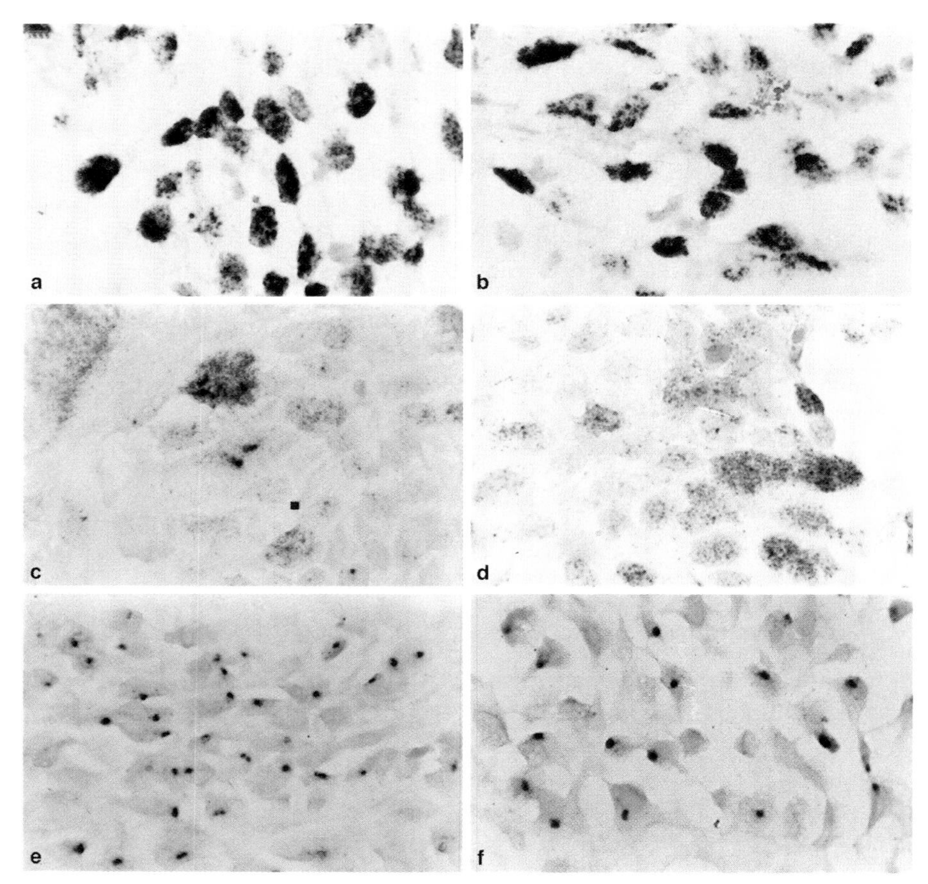

Fig. 6.1a–f. Signal patterns of HPV16 DNA in cervical cancers detected by colorimetric in situ hybridization with biotinylated probes. (BCIP/NBT signal development with nuclear fast red counterstain; original magnification 1000×). **a** and **b** Diffuse pattern. **c** and **d** Mixed pattern. **e** and **f** Dot pattern. Signal is dark and confined to the nucleus, which is pink

Hybridization of tissues and culture cells with biotinylated probes may produce a variety of nuclear signals, ranging from a discrete dot to a diffuse signal covering the whole nucleus, depending on the amount, status, and kind of target and probe DNA. Single genes or integrated viral DNAs give a dot signal whose intensity and size depend on the number of copies present in tandem in the cellular genome. On the other hand, multiple repeated DNA sequences give a diffuse pattern, as is the case of the HPG probe hybridized over normal nucleus or viral probes in the case of tissues containing a high copy number of episomal viral forms, such as the human papilloma virus (HPV) type 11 in condyloma acuminata.

One of the first in situ hybridization assays that may be incorporated into the pathology research laboratories is that for the detection and typing of human papillomaviruses (HPV). Many studies have linked some HPV to cervical carcinomas and their precursor lesions. In situ studies have been useful in depicting the virus within neoplastic cells. In addition to providing accurate detection and typing of the virus, in situ assays can give information about the distribution, amount, and integration state of viral genomes within individual tumor cells. HPV exists as an episome in most condylomas and cervical dysplasias, while in most carcinomas, HPV is integrated onto a host chromosome. Thus, integration has been postulated to be an important step in oncogenesis.

The hybridization of cervical cancers tissues with the HPV16 produce a variety of nuclear signals (see Fig. 6.1) and represent a very nice example of the powerful information that may be uncovered by the colorimetric in situ hybridization technique. Tumors containing high copy number of episomal forms show a diffuse pattern in the cellular nucleus (Fig. 6.1a,b), whereas those with viral DNA integrated into the cellular genome show one dot in the nucleus (Fig. 6.1e,f). In some tumors, the entire spectrum of signals is observed, at times consisting of different patterns in separate areas of the tumor (not shown) and at other times within a single focus with adjacent nuclei showing diffuse and single dot pattern (Fig. 6.1c,d), suggesting that a combination of viral episomal amplification and integration may be occurring during tumor progression (Berumen et al. 1995).

This colorimetric in situ hybridization protocol may be useful to monitorize the incorporation of transfected vectors into the genomic DNA of somatic cells transfected permanently in vitro or derived from transgenic mice. The hybridization of these cells with the biotinylated transfected vector probes produced a fine dot signal in the nucleus, indicating that the transfected genes are integrated into the cellular genome (data not shown).

References

Berumen J, Unger ER, Casas L, Figueroa P (1995) Amplification of human papillomavirus types 16 and 18 in invasive cervical cancer. Hum Pathol 26:676–681

Brigati DJ, Myerson D, Learry JJ, Spalholz B, Travis SZ, Fong CKY, Hsiung GD, Ward DC (1983) Detection of viral genomes in cultured cells and paraffin embedded tissue sections using biotin-labeled hybridization probes. Virology 126:32–50

Hames BD, Higgins SJ (1985) Nucleic acid hybridization, a practical approach. IRL Press, Oxford

Sambrook J, Fritsch EF, Maniatis T (1989) Molecular cloning, a laboratory manual. Cold Spring Harbor Laboratory, Cold Spring Harbor

Unger ER, Hammer ML, Chenggis ML (1991) Comparison of ^{35}S and biotin labels for in-situ hybridization. Use of an HPV model system. J Histochem Cytochem 39:145–150

Suppliers

Boehringer Mannheim Corporation
Biochemical Products
9115 Hague Road
PO Box 50414
Indianapolis, IN 46250-0414
USA

DAKO Corporation
22, North Milpas Street
Santa Barbara California 93103
USA

Fisher Scientific
Communications Dept. 710
711 Forbes Avenue
Pittsburgh PA 15219
USA

Life Technologies, Inc.
Producer of GIBCO BRL Products
Latin America Office
8451 Helgerman Court
Gaithersburg, MD 20877
USA

Oncor, Inc.
209 Perry Parkway
Gaithersburg, MD 20877
USA

SIGMA Chemical Company
PO Box 14508
St. Louis, MO, U.S.A. 63178-9916
Biomeda
Foster City, CA
USA

Manual Microinjection: Measuring the Inhibitory Effect of HIV-1-Directed Hammerhead Ribozymes in Tissue Culture Cells

ROBERT HORMES[1] AND GEORG SCZAKIEL[1*]

Introduction

The acquired immunodeficiency syndrome (AIDS; Barre-Sinoussi et al. 1983; Popovic et al. 1984) is caused by infection with the human immunodeficiency virus type 1 (HIV-1). To investigate the potential of in vivo active inhibitors of HIV-1, we established a transient test system in tissue culture cells. Test substances were comicroinjected together with infectious proviral HIV-1 DNA into human colon-carcinoma cells (SW 480). The suppression of HIV-1 replication was quantified by measurement of the release of infectious HIV-1 particles after amplification of HIV-1 produced in initially microinjected SW480 cells by cocultivated $CD4^+$ MT-4 cells (Sczakiel et al. 1990).

The adherently growing SW480 cells were spread out on glass slides. When grown to semiconfluence, cells were microinjected using glass capillaries which allow the transfer of water soluble material, including the biomacromolecules such as proteins and nucleic acids and even virus particles, into the nuclei or the cytoplasm of culture cells (Graessmann and Graessmann 1986). Here we describe a protocol for investigating the inhibitory effectiveness of synthetic hammerhead ribozymes directed against HIV-1 in living cells. The ribozymes were generated by in vitro transcription with T7- or T3-RNA polymerase from linearized plasmid DNA templates. The plasmids contain the transcribed ribozyme-coding sequence between a T7- or T3-promoter and a restriction site (Homann et al. 1993). Length requirements of the antisense flanks as well as the subcellular delivery of the ribozymes were analyzed. The flowsheet shown in Fig. 7.1 schematically summarizes the steps of the microinjec-

* corresponding author: phone: +49-6221-424939; fax: +49-6221-424932; e-mail: sczakiel@dkfz-heidelberg.de

[1] Forschungsschwerpunkt Angewandte Tumorvirologie, Deutsches Krebsforschungszentrum, Im Neuenheimer Feld 242, D-69120 Heidelberg, Germany

tion assay which was used to determine ribozyme-mediated inhibition of HIV-1 in human tissue culture cells.

Materials

Reagents

- Human epitheloid cell line SW 480 (Leibovitz et al. 1976)
- Human T-lymphoid cell line MT-4 (Harada et al. 1985)
- Glass slides (10 mm diameter, Langenbrinck, Emmendingen, Germany)
- HIV-1 antigen ELISA (Organon Teknika, Durham, North Carolina)
- Infectious proviral HIV-1 DNA (pNL4–3, Adachi et al. 1986)
- T7-/T3-RNA polymerase (Boehringer Mannheim)
- NTPs: ATP, GTP, CTP, UTP (100 mM, Boehringer Mannheim)
- DNase I (RNase-free, Boehringer Mannheim)
- Sephadex G-50 medium (Pharmacia Biotech, Uppsala)
- Glass capillaries (GC 150F-10, 1.2 mm OD, 0.86 mm ID, with inner filament, Clarc Electromedical Instruments, UK)
- Geloader tips (1–10 μl, Eppendorf, Hamburg)

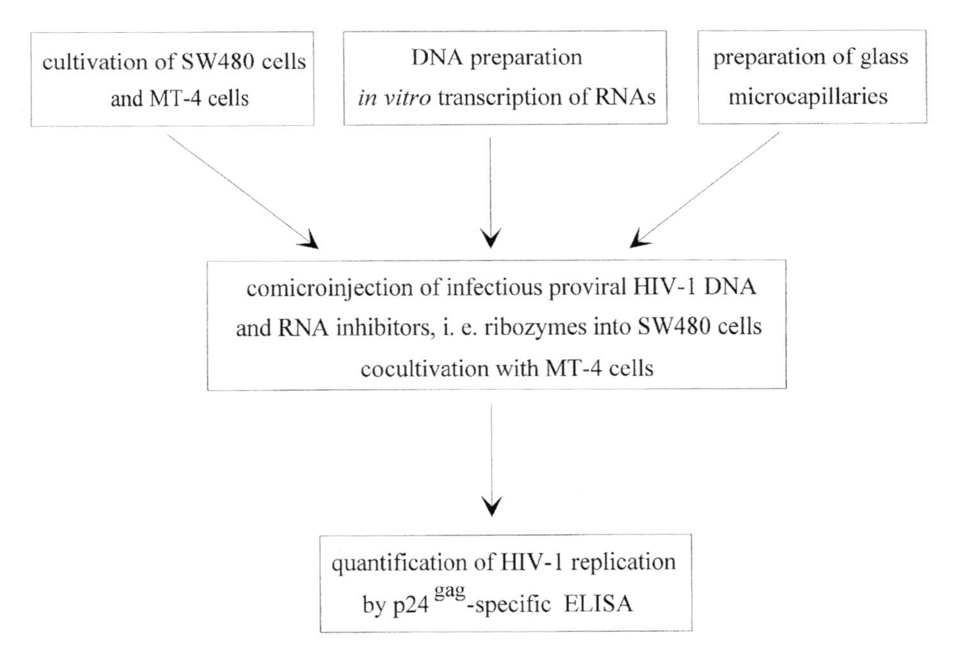

Fig. 7.1. Schematic flow sheet of the microinjection assay that was used to measure ribozyme-mediated inhibition of HIV-1 in human SW480 cells

Media, buffers and solutions

- DMEM: Dulbeccos minimal essential medium (Gibco BRL, Berlin, Germany)
- RPMI 1640: (Gibco BRL, Berlin, Germany)
- FCS (fetal calf serum, PAA, Linz, Austria)
- L-glutamine (200 mM, Eurobio, Les Ulis, France)
- Antibiotics mix (penicilline 10,000 U/ml, streptomycine 10,000 μg/ml, Eurobio, Les Ulis, France)
- Trypsine solution (0.125 % trypsine/0.125 % EDTA in H_2O)
- TE: 10 mM Tris/HCl pH 8, 1 mM EDTA
- 200 mM $MgSO_4$
- 3 M potassium acetate pH 5.2
- 5× in vitro transcription buffers:
  A: 90 mM Na_2HPO_4, 10 mM NaH_2PO_4, 100 mM DTT
  B: 40 mM $MgCl_2$, 20 mM spermidine
  C: 5 mM ATP, CTP, GTP, UTP, pH 7.5, fixed with NaOH
  D: 25 mM NaCl

Equipment

- Manual pipette puller (MPP1, Brindi, Germany)
- Microinjector 5242 (Eppendorf)
- Phase contrast inverted microscope (Olympus IMT2)
- Micromanipulator (Leitz, Germany)

7.1 Cell Lines and Preparation of Cells

Cell culture media (440 ml) were supplemented with 50 ml FCS (inactivated for 20 min at 56 °C), 5 ml 200 mM L-glutamine, and 5 ml antibiotics mix. Adherently growing SW480 cells were cultivated with supplemented DMEM (DMEM+) and the MT-4 cells that grow in suspension were cultivated with supplemented RPMI 1640 (RPMI+).

Procedure

Preparation of glass slides

1. Wash the glass slides for 5 min in acetone and subsequently 5 min in absolute ethanol.
2. Transfer a dozen glass coverslips in a 10 cm diameter plastic tissue culture dish and let them dry.

Preparation of SW480 cells

1. Suck off the medium of SW480 cells grown to 70 % confluence in a 75 cm^2 cell culture flask.
2. Treat the cells with 4 ml trypsine solution and detach cells with 1 ml trypsine solution for 5–10 min. **Note**: Detachment of the cells is increased at 37 °C (incubation time should be kept shorter than 5 min in order to avoid harming the cells).
3. Add 9 ml DMEM+ to the cell suspension to inactivate the trypsine by FCS and to dilute the SW480 cells.
4. Transfer 1 ml of the cell suspension in 10 ml DMEM+ in one tissue culture dish with the prepared glass slides.
5. Incubate cells at 37 °C until grown to semi-confluence (24 to 36 h) and 5 % CO_2.

Preparation of MT-4 cells

1. Propagate MT-4 cells at 37 °C and 5 % CO_2 in RPMI+.
2. Carefully suck off the old medium after 2–3 days, when color of the medium has changed to yellow (acidic pH value, density approximately $1-2\times10^6$ cells/ml).
3. Dilute the cells to 50 % with RPMI+ and incubate further at 37 °C and 5 % CO_2.

Note: One glass slide with microinjected SW480 cells will be cocultivated with $1-2.5\times10^5$ MT-4 cells.

7.2 Plasmids and Synthetic RNA

Procedure

Preparation of DNAs

1. The plasmids containing transcribable sequences and proviral HIV-1 DNA are purified by one CsCl gradient centrifugation.
2. CsCl contained in micronjected solutions can harm the cells, therefore the proviral HIV-1 DNA pNL4–3 is purified by gelfiltration as described in preparation of RNAs.

3. The DNA concentration is determined by UV-absorption (1 A_{260} ≅ 50 ng/µl dsDNA).

4. The plasmids containing the transcribable sequences between a T7- or T3-promoter and a restriction site were linearized. **Note:** Templates with 5'-overhanging ends attain better yields at the in vitro transcription.

Preparation of RNAs

In vitro transcription reaction

1. Mix 40 µl of each 5× in vitro transcription buffer A, B, C, and D.
2. Add 10 µg template DNA and 20 U T7- or T3-RNA polymerase in 40 µl H_2O to start the reaction. **Note:** Take precautions against RNases.
3. Incubate for 2 h at 37 °C.
4. Stop the reaction by adding 200 µl 20 mM $MgSO_4$ and 20 U DNaseI.
5. Incubate for further 30 min at 37 °C.
6. Remove proteins by two extractions with 400 µl phenol (equilibrated with TE) and extract once with 400 µl chloroform/isoamylalcohol (19:1/v:v).
7. Extract the aqueous phase and add 1/10 vol 3 M potassium acetate pH 5.2, and precipitate with 2.5 vol precooled ethanol for 1 h at −20 °C.
8. Cetrifuge 30 min at 15,000 rpm and 4 °C. Wash the pellet with 80 % ethanol in H_2O and dissolve the pellet in 200 µl TE. **Note:** Keep RNA solutions on ice or store at −20 °C.

Gel filtration

9. Purify the RNA by gelfiltration with Sephadex G-50, which is equilibrated in TE and contained within Pasteur pipettes. The small end of the pipettes is tamped with autoclaved glass wool.
10. Wash the column three times with 500 µl chilled TE.
11. Charge the column with the dissolved RNA and wash with 500 µl chilled TE.
12. Elute the RNA with chilled TE in four fractions of 120 µl each. Control the elution of the RNA on a 1 % agarose gel.

13. Pool the fractions containing RNA and measure RNA concentration by UV-absorption (1 A_{260} ≅40 ng/μl RNA). **Note:** A 200 μl in vitro transcription reaction with 10 μg DNA template and 20 U RNA polymerase yields approximately 50 μg RNA.
14. Prepare the mixtures of nucleic acids for microinjection experiments in TE. The mixtures contain 10 ng/μl pNL4-3 and 70 ng/μl test RNA or unspecific control RNA.

7.3 Preparation of Glass Capillaries for Microinjection

Procedure

A glass tube of 0.86 mm inner diameter, 1.5 mm outer diameter, and 10 cm length is clamped into the pipette puller to obtain a capillary with a tip of approximately 1 μm diameter. This is achieved by two separate strokes.

1. Introduce a constiction of about 10 mm in length and about 0.5 mm in diameter with the first stroke: elasticity 4.0, brake 35 mm, current 5.8 A.
2. Create the desired tip with the second stroke: elasticity 6.5, no brake, current 4.8 A.
3. Immediately remove the tip of the glass capillary out of the heated coil. **Note:** The listed parameters depend on the puller and the geometry of the heated platinum/iridium coil and must be optimized by varying the current (temperature of the heating wire) and elasticity (pulling forces at the carriage).
4. Load the capillary with 1-2 μl nucleic acids mixture through the rear opening by Geloader tips. The thin inner filament supports the transport of the test solution to the tip of the capillary. **Note:** To avoid clogging of the glass capillaries, centrifuge the test mixtures at least 30 min at 15,000 rpm at 4 °C and use supernatant for microinjection.

7.4
Microinjection of Nucleic Acid Mixes into Human Epitheloid Cells

Note: This experiment is performed under P2/S2 safety standard. Immediately after microinjection, cultured cells have to be transferred to a P3/S3 laboratory.

Procedure

1. Transfer two or three of the glass slides with tightly bound (adherently grown) SW480 cells into a plastic tissue culture dish of 6 cm diameter prefilled with 5 ml DMEM+ and fix the dish on the microscope table.
2. Clamp the prefilled capillary into the micromanipulator and put the tip under the surface of the medium in the dish. Handle capillaries very carefully to avoid breakage of the capillary tip. **Note:** Test solution contains proviral HIV-1 DNA! Prevent pricking your finger by handling extremely carefully and wearing two pairs of gloves (cotton gloves under powder free latex gloves).
3. Supply constant injection pressure (N_2) between 60 and 200 hPa by the microinjector. **Note:** Injecton pressure depends on the permeability of the tip of the capillary and the viscosity of the injected solution.
4. Focus the cell surface at the highest magnification (400×).
5. Center the capillary tip in the visuable field at the lowest magnification (40×). Repeat this with the next higher magnifications (100×, 200×, etc.) to bring the cells into focus and the tip of the capillary centered at the working magnification (400×).
6. Lower the capillary slowly until it reaches the cell surface (a white spot appears on the cell when the tip of the capillary touches the surface).
7. Inject the cell by further slow lowering of the capillary and observe the injected volume by the change in size and contrast of the cell compartments in relation to each other.
8. Avoid damaging the cellular and nuclear membranes by entering the injection capillary in an angle of 45° to 60° and by keeping the injection time between 0.1 and 1 s.

9. For microinjection of the nucleus, move the tip of the injection capillary downward to the top of the left third of the nucleus. Microinject the cytoplasm by entering the cell on the left side (when glass capillary is manipulated from the left side of the microscope table) below the nucleus (Fig.7. 2).

10. Use at least 20 cells per glass slide when microinjecting into nucleus and at least 50 cells when microinjecting into cytoplasm to obtain quantifiable amounts of HIV-1. Reproduce microinjection experiments at least five times to estimate the standard deviation of the experiments, which ideally is in the range of 15 to 30 % of the mean value.

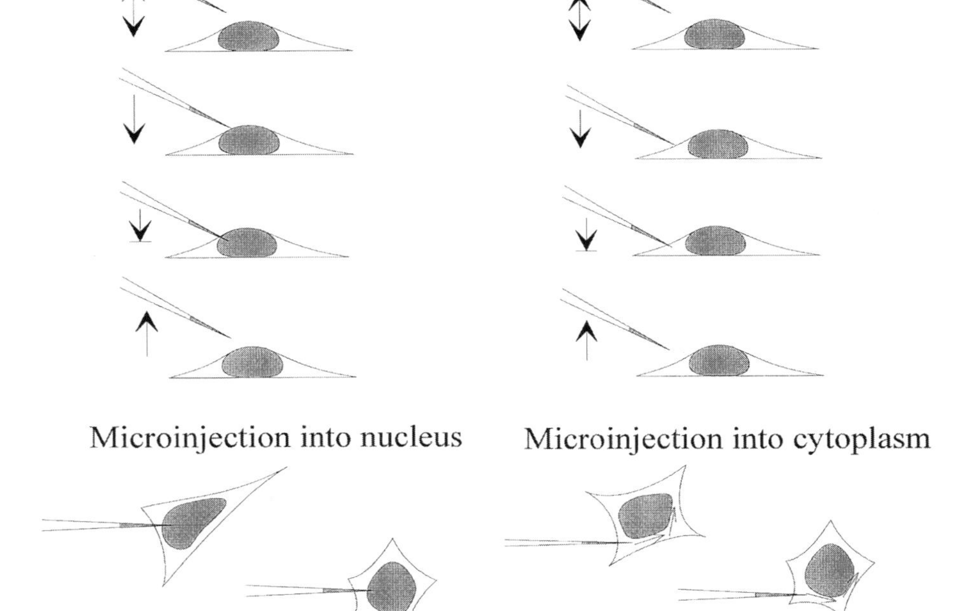

Fig. 7.2. Schematic drawing showing the movement of the glass capillary while microinjecting into nucleus and cytoplasm respectively. *Arrow* in the *upper panel* indicates the directions of movement of the injection capillary. *Arrows* in the *lower panel* (*right*) indicate the flow of the cytoplasmatically injected solution as is visible under the phase contrast microscope

11. Transfer microinjected SW480 cells on glass slide into a 48-well plate prefilled with 0.5 ml DMEM+.
12. Incubate for 24 h at 37 °C and 5 % CO_2. **Note:** Microinjection of pNL4–3 DNA induces HIV-1 replication and the release of infectious HIV-1 particles after approximately 6 h.
13. Amplify HIV-1 particles by cocultivation of microinjected SW480 cells with $1–2.5 \times 10^5$ MT-4 cells per well in 0.5 ml RPMI+1 day after microinjection. Quantify HIV-1 replication after 5 days.

7.5 Quantitative Measurement of HIV-1 Replication

Procedure

1. Determine replication of HIV-1 by quantification of HIV-1 matrix protein (p24) with a p24-specific antigen ELISA from cocultivated SW480/MT-4 cells 5 days after microinjection. **Note:** To estimate the amount of virus production, it is useful to pool supernatants of individual experiments with the same test RNAs, estimate the dilution which is next suited for the ELISA and subsequently dilute the supernatants before determinating virus production in each well.
2. Take 100 μl cell-free SW480/MT-4 coculture supernatants and incubate with disruption buffer, conjugate, and substrate as described in the instructions of the ELISA.
3. Relate the observed absorption data of different experiments to a microinjected control of pNL4–3 alone or a mixture of pNL4–3 with an unspecific RNA sample, e.g., $tRNA_{bulk}$, an HIV-1-derived sense RNA or an RNA derived from other unrelated sequences.

Results and Comments

- **Assay system**
  Microinjection of infectious proviral HIV-1 DNA into the nuclei of human fibroblasts leads to viral replication and release of infectious particles (Boyd et al. 1988). This observation is the basis for the presented transient assay to test and to compare the inhibitory effects of

antiviral agents including nucleic acids, proteins and low molecular weight compounds (Sczakiel et al. 1990; Rittner and Sczakiel 1991). The direct microinjection technique allows the delivery of any kind of water-soluble test compound together with cloned proviral HIV-1 DNA (pNL4–3; Adachi et al. 1986). The inhibitory effect of antiviral molecules can be tested in the nucleus as well as in the cytoplasm of adherently growing cells like SW480 cells. Due to the lack of CD4 antigen, the main receptor of HIV-1, SW480 cells cannot be reinfected by released HIV-1. Virus, initially produced in microinjected SW480 cells, therefore, is amplified in cocultivated $CD4^+$ MT-4 cells. Five days after microinjection, replication of HIV-1 is measured via HIV-1 p24 matrix protein in a monoclonal ELISA.

- **Inhibition of HIV-1 replication by hammerhead ribozymes**
  Single-stranded nucleic acids complementary to a given single-stranded target nucleic acid can act in vitro and in vivo as specific downregulators of gene expression and viral replication (for reviews see Hélène and Toulmé 1990; Weintraub 1990). These antisense nucleic acids were also examined as inhibitors of HIV-1 replication, whereby inclusion of the catalytic hammerhead ribozyme domain into an antisense RNA increases its inhibitory effect on HIV-1 replication (Homann et al. 1993). Hammerhead ribozymes are specifically structured RNAs which hydrolyze a phosphodiester bond of their target RNA in a sequence-specific manner. The target RNA is recognized by duplex formation via complementary sequences, whereby long duplex sequences seem to be more stable than short ones. The ability of ribozymes to cleave cognate RNA molecules (in trans) has made them of interest to molecular biology and medical research. The construction of symmetric and asymmetric hammerhead ribozymes is possible by changing the position of the catalytic domain in an antisense RNA (Tabler et al. 1994). Thus, the advantages of long antisense RNAs (stability, fast duplex formation) and of ribozymes (cleavage of the target sequence) can be combined.
  The described microinjection assay has been used to compare the efficacy of long-chain hammerhead ribozymes and parental antisense RNA against HIV-1. Table 7.1 summarizes data on the inhibition of HIV-1 replication with the long-chained hammerhead ribozymes αYRz195 (asymmetric ribozyme; Tabler et al. 1994) and 2as-Rz12 (symmetric ribozyme; Homann et al. 1993; Table 7.1). Inhibition data were related to an unspecific CAT RNA.

The comicroinjection of the set of ribozymes (70 ng/μl) into the nucleus of SW480 cells together with pNL4–3 (10 ng/μl) reveals only weak inhibition by the antisense effect (αYRz195inac) and increased efficacy by derivatives that contain an in vitro active hammerhead domain (αYRz195 and 2as-Rz12). However, none of these constructs shows significant inhibition when directly delivered into the cytoplasm of human SW480 cells (Table 7.1).
These findings indicate the importance of testing the inhibitory competence of antiviral agents in different cell compartments. The microinjection assay for HIV-1 replication as described here has been shown to be a useful alternative to other transient replication assays like cotransfection. When addressing the influence of the subcellular localization of the inhibitory RNA, this test seems to be the method of choice.

Troubleshooting

- **RNA preparation**
  Problems with low yields of RNA in the in vitro transcription reaction might be due to partially unlinearized template DNA, hydrolyzed NTPs, or RNase contamination. Thus, it is helpful to:
- Control restriction of template DNA on agarose gels
- Use freshly prepared NTPs

Table 7.1. Influence of subcellular localization of HIV-1-directed ribozymes on inhibition of HIV-1 replication in human cells
– indicates no inhibition; + indicates inhibition between 50 % and 80 %; ++ indicates inhibition greater than 90 %

RNA	Microinjection into nucleus	Microinjection into cytoplasm
CAT[c]	–	–
αYRz195[a]	++	–
αYRz195 inac[a]	+	–
2as-Rz12[b]	++	–

[a] Active (αYRz195) and inactive (αYRz195inac) HIV-1^{tat}-directed asymmetric hammerhead ribozyme with 220 nucleotides (Tabler *et al.* 1994).
[b] HIV-1^{gag}-directed symmetric hammerhead ribozyme with 420 nucleotides (Homann *et al.* 1993).
[c] CAT-directed antisense RNA with [850]nucleotides as a negative control.

- Add 10 U RNase inhibitor (RNasin, Boehringer Mannheim) to the in vitro transcription reaction
- Use autoclaved H_2O which was incubated for at least 2 h with diethyl-pyrocarbonate (DEPC, Sigma, Deisenhofen) to prepare buffers for the in vitro transcription reaction.

- **Microinjection**
  There are some problems with low virus production when microinjecting into cytoplasm. To obtain measurable amounds of virus:
- Test SW480 and MT-4 cells for contamination with mycoplasma by PCR (Mycoplasma PCR primer set, Stratagene, La Jolla, CA) and decontaminate cells with Mycoplasma removal agent (MRA, ICN Pharmaceuticals, Costa Mesa, CA)
- Use a higher concentration of pNL4-3 in the microinjected nucleic acids solutions.

References

Adachi A, Gendelman HE, König S, Folks T, Willey R, Rabson A, Martin MA (1986) Production of acquired immunodeficiency syndrome-associated retrovirus in human and nonhuman cells transfected with an infectious molecular clone. J Virol 59:284–291

Barre-Sinoussi F, Cherman JC, Rey R, Nugeryre MT, Chamaret S, Gruest J, Dauget C, Axler-Blin C, Vernizet-Brun F, Rouzioux C, Rosenbaum W, Montagnier L (1983) Isolation of a T-lymphotropic retrovirus from a patient at risk for acquired immunodeficiency syndrome (AIDS). Science 220:868–871

Boyd AL, Wood TG, Buckley A, Fischinger PJ, Gilden RV, Gonda MA (1988) Microinjection and expression of an infectious proviral clone and subgenomic envelope construct of a human immunodeficiency virus. AIDS Res Hum Retroviruses 4:31–41

Graessmann M, Graessmann A (1986) Microinjection of tissue culture cells using glass capillaries: methods. In: Celis JE, Graessmann A, Loyter A (eds) Microinjection and organelle transplantation techniques. Academic Press, London, pp 3–13

Harada S, Koyanagi Y, Yamamoto N (1985) Infection of HTLV-III/LAV in HTLV-I-carrying cells MT2 and MT-4 and aplication in a plaque assay. Science 229:563–566

Hélène C, Toulmé JJ (1990) Specific regulation of gene expression by antisense, sense and antigene nucleic acids. Biochim Biophys Acta 1049:99–125

Homann M, Tzortzakaki S, Rittner K, Sczakiel G, Tabler M (1993) Incorporation of the catalytic domain of a hammerhead ribozyme into antisense RNA enhances its inhibitory effect on the replication of human immunodeficiency virus type 1. Nucleic Acids Res 21:2809–2814

Leibovitz A, Stinson JC, McCombs III WB, McCoy CE, Mazur KC, Mabry ND (1976) Classification of human colorectal adenocarcinoma cell lines. Cancer Res 36:4562–4569

Popovic M, Sarngadharan MG, Read E, Gallo RC (1984) Detection, isolation and continuous production of cytopathic retroviruses (HTLV-III) from patients with AIDS and pre-Aids. Science 224:497–500

Rittner K, Sczakiel G (1991) Identification and analysis of antisense RNA target regions of the human immunodeficiency virus type 1. Nucleic Acids Res 19:1421–1426

Sczakiel G, Pawlita M, Kleinheinz A (1990) Specific inhibition of human immunodeficiency virus type 1 replication by RNA transcribed in sense and antisense orientation from the 5'-leader/gag region. Biochem Biophys Res Commun 169:643–651

Tabler M, Homann M Tzortzakaki S, Sczakiel G (1994) A three-nucleotide helix I is sufficient for full activity of a hammerhead ribozyme: advantages of an asymmetric design. Nucleic Acids Res 22:3958–3965

Weintraub HB (1990) Antisense RNA and DNA. Sci Am 262:34–40

Chapter 8

Cellular Microbiochemistry

RAINER PEPPERKOK[1*], OLAF ROSORIUS[2], AND JOCHEN SCHEEL[3]

Introduction

Capillary microinjection is efficiently used to introduce macromolecules into the nucleus or cytoplasm of living mammalian cells (Proctor 1992). Transfer can occur at well-defined stages of the cell cycle, and modifications of culture conditions are possible before, during, and after injection. The number of cells which can be injected per experiment is, however, limited. Therefore, in the past, biochemical analyses of microinjected cells has been possible, but difficult (Gautier-Rouviere et al. 1990; Lane et al. 1993).

Due to the computer automation of the technique (Ansorge and Pepperkok 1988; see also Ansorge and Saffrich, this Chap. 3) 500 to 1000 cells can now be injected within 30 min, allowing cellular microbiochemistry of microinjected cells on a routine basis (see Rosa et al. 1989; Pepperkok et al. 1993).

The following protocols are used in our laboratory to study secretory transport of the vesicular stomatitis virus glycoprotein (ts-O45-G, see Kreis 1986) between the endoplasmic reticulum and the Golgi complex biochemically. They can also be used as a basis to design experiments addressing different questions involving biochemical analyses of microinjected cells.

* corresponding author: phone: +44 171 269 3563; fax: +44 171 269 3585;
e-mail: Pepperko@icrf.icnet.uk

[1] Imperial Cancer Research Fund (ICRF), 44 Lincoln's Inn Fields, PO Box 123, London WC2A 3PX, UK

[2] Institut fur klinische und molekulare Virologie, Universität Erlangen, Schlossgarten 4, D-91054 Erlangen, Germany

[3] Department of Physical Biology, Max-Planck-Institute for Developmental Biology, Spemannstr. 35/I, D-72076 Tübingen, Germany

Outline

Short protocol

1. Grow 500 to 1000 cells in the center of a glass coverslip.
2. Label them with ^{35}S methionine.
3. Microinject all the cells on the coverslip.
4. Incubate for appropriate time and conditions.
5. Lyse cells and analyze proteins of interest.

Materials

Microinjection equipment

Raw glass capillaries were from Clark Electromedical Instruments (Reading, UK; cat. no. GC120TF-10). The pipette puller type P-87 from Sutter Instruments (Novato, USA). The Microinjector model 5242 for controlling the pressure applied to the micropipette, the microloaders for filling micropipettes (cat. no. 0030001.22) and Geloader tips (cat. no.: 0030001.222) were from Eppendorf (Hamburg, F.R.G.).

The automated microinjection system (AIS) including microscope, motorized scanning stage, and micromanipulator, and the control computer were from Zeiss (Jena, F.R.G.).

Chemicals and reagents

The following chemicals were obtained from Sigma (Buchs, Switzerland): cycloheximide (cat. no. C-7698), phenylmethylsulfonylfluoride (PMSF, cat. no. P7626), chymostatin (cat. no. C-7268), pepstatin (cat. no. P4265), aprotinin (cat. no. A1153), *trans*-epoxysuccinyl-L-leucylamido-(4-guanidino)butane (E64, cat. no. E3132), leupeptin-hemisulfate (cat. no. 62070) was obtained from Fluka (Buchs, Switzerland).

^{35}S methionine was obtained from NEN (Dupont, cat. no. NEG-009A). Endoglycosidase H (1 mU/μl, cat. no. 1088726), GTPγS (cat. no. 220647), brefeldin A (cat. no. 1347136) were from Boehringer (Mannheim, FRG). Cell culture petri dishes 35 mm diameter (cat. no. 153066) were from NUNC (Roskilde, DK). Glass coverslips were from Menzel (Braunschweig, FRG). MEM culture medium (cat. no. 21090–22), MEM culture medium without methionine (cat. no. 04101900M) and MEM culture medium containing low concentrations of carbonate (0.85 g/l, cat. no. 12565–024) were from Gibco Life Technologies (Switzerland). Penicillin/

streptomycin (10,000 U/ml, cat.no. A2213), L-glutamine (200 mM, cat. no. K0282), nonessential amino acids (100-fold concentrated solution, cat. no. K0293) were from Seromed (Biochrom KG, Berlin, Germany). Protein A Sepharose (cat. no. 17-0780-01) was obtained from Pharmacia Biotech (Switzerland) and prepared as a 50 % (v/v) slurry in PBS. The anti-ts-O45-G antibody αP4 (Kreis 1986) was provided by Dr. T. Kreis (University of Geneva, Switzerland).

Solutions

- Protease inhibitor mix1(PIM1)
  250 mM PMSF
  10 mg/ml chymostatin
  5 mg/ml pepstatin
  Dissolve in DMSO and store at −80 °C
- Protease inhibitor mix2 (PIM2)
  1 mg/ml aprotinin
  5 mg/ml leupeptin-hemisulfate
  0.7 mg/ml E64
  Dissolve in water and store at −80 °C
- Lysis buffer
  50 mM Tris pH 7.4
  100 mM NaCl
  1 % Triton X-100
  1 mM EDTA
- Microinjection buffer (pH 7.2)
  48 mM K_2HPO_4
  4.5 mM KH_2PO_4
  14 mM NaH_2PO_4
- Tissue culture medium for Vero cells (MEM^+)
  MEM culture medium
  100 U penicillin/streptomycin
  2 mM L-glutamine
  1× nonessential amino acids
  10 mM Hepes pH 7.4
  5 % fetal calf serum
- Tissue culture medium for microinjection (MEM^-)
  MEM culture medium containing low concentrations of carbonate (0.85 g/l)
  100 U penicillin/streptomycin
  2 mM L-glutamine
  1× nonessential amino acids
  30 mM Hepes pH 7.4

5 % fetal calf serum
- Labeling medium
  MEM culture medium without methionine
  100 U penicillin/streptomycin
  2 mM L-glutamine
  1× nonessential amino acids
  10 mM Hepes pH 7.4
  5 % fetal calf serum (should be dialyzed against PBS to deplete methionine)

Cell culture Stock cultures of monolayer Vero cells (African green monkey kidney cells, ATCC CCL81) were grown in tissue culture medium for Vero cells (MEM^+) at 37 °C and 5 % CO_2.

Procedure

Preparation of Cell Lysate

1. Grow Vero cells in ten petri dishes (15 cm diameter) to 70 % confluency ($\approx 10^8$ cells in total).
2. Place petri dishes on ice (work in the coldroom).
3. Remove medium from the first petri dish and add 10 ml of ice-cold lysis buffer.
4. Lyse cells for 1 min (rock the plate several times).
5. Transfer the lysate to the next plate.
6. Repeat steps **4** and **5** until cells in all dishes are lysed.
7. Rinse the plates subsequently with 5 ml lysis buffer and combine the two lysates.
8. Spin the lysate (15 ml in total) for 15 min in a minifuge (15,000 rpm) at 4 °C.
9. Remove the supernatant, aliquot it in 1 ml aliquotes, snap freeze them in liquid nitrogen, and store at −80 °C.

Plating Cells onto Glass Coverslips

1. Grow Vero cells to 80 % confluency in a 10 cm petri dish.
2. Typsinize cells and resuspended them in 10 ml complete culture medium (containing 5 % fetal calf serum).
3. Count the number of cells, using a hemocytometer.
4. Transfer the suspension into a 15 ml falcon tube and pellet the cells for 5 min at 200 *g*.
5. Resuspend the cell pellet in complete culture medium and adjust the cell density to 3.5×10^4 cells/ml.
6. Plate a droplet of 7 μl of this cell suspension in the center of a glass coverslip (10×10 mm ≈250 cells).
7. Place the coverslip into a humid chamber and incubate at 37 °C and 5 % CO_2 until cells have attached to the glass (this usually takes 6–8 h).
8. Transfer the coverslips into 3 cm petri dishes containing 2 ml of culture medium and let the cells grow for 2 days at 37 °C and 5 % CO_2. Usually between 500 and 1000 cells will be in the center of the coverslip after this time.

Microinjection, Metabolic Labeling, and Lysis of Cells

For experiments analyzing secretory transport of the temperature-sensitive vesicular stomatitis virus glycoprotein (ts-O45-G), cells are infected with vesicular stomatitis virus (ts-O45, see Kreis 1986; Pepperkok et al. 1993) before microinjection and metabolic labeling. Microinjection is carried out in MEM^- (see above for preparation) to keep pH constant during microinjection. Temperature during microinjection is controlled, using a home-made microscope stage temperature control (see Pepperkok et al. 1993). Incubations after injection are performed in MEM^+ in 5 % CO_2 at respective temperatures.

1. Microinject all the cells on the coverslip prepared as described in Section 8.4.2 (for details of the microinjection procedure see Chapts. 1 and 2, this Vol.).
2. Incubate cells for 15 min in labeling medium to deplete cellular stores of methionine.

3. Take the coverslip out of the culture dish, blot the edges dry with sterile filter paper, and place the coverslip into a separate new 3-cm petri dish without medium.
4. Immediately, pipette 40 µl of ^{35}S methionine (2 µCi/ µl in labeling medium) onto the coverslip. Transfer the petri dishes into a humid chamber and incubate for the appropriate time (typically 5 to 30 min) at 37 °C (39.5 °C for ts-O45 infected cells) and 5 % CO_2.
5. Wash cells three times with MEM^+ culture medium.
6. Chase cells for the appropriate time (depends on particular experiment) in 2 ml MEM^+ (containing 100 µg/ml cycloheximide) at respective temperatures and 5 % CO_2.
7. Take the coverslip out of the culture dish, blot the edges dry with sterile filter paper, and transfer it into 300 µl ice cold cell lysate (see Sect. 8.4.1 for preparation) in a 24-well plate.
8. Incubate the coverslip for 15 minutes at 4 °C; agitate gently.
9. Collect the 300 µl lysate and proceed with the biochemical analysis of lysed proteins. Alternatively, samples can be snap-frozen in liquid nitrogen and stored at −80 °C until further use.

Note: Microinjection (step **1**) can be performed after metabolic labelling (steps **2–5**).

Immunoprecipitation and Digestion with Endoglycosidase H (endoH)

1. Spin the lysate in a minifuge for 5 min at 4 °C and collect the supernatant.
2. Add antibody to the supernatant. The antibody should be in excess of the antigen to be precipitated.
3. Incubate the sample with gentle agitation overnight at 4 °C (use, e.g., a rotating wheel).
4. Add 15 µl protein A Sepharose (50 % slurry) to the sample and continue the incubation for 2 h at 4 °C.
5. Spin the sample for 1 min at 4 °C at 10,000 rpm in a minfuge to pellet the Sepharose.

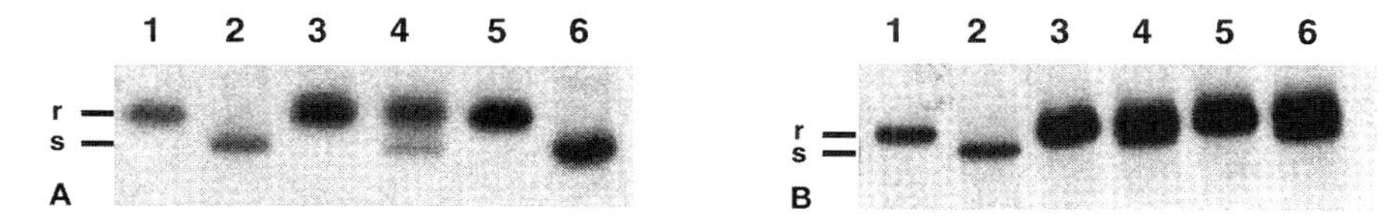

Fig. 8.1A, B. About 500 to 1000 Vero cells grown in the center of a glass coverslip (see Sect. 8.4.2) were infected with temperature-sensitive vesicular stomatitis virus, ts-O45, at 39.5 °C as described (Pepperkok et al. 1993). All cells on the coverslip were microinjected at 39.5 °C with either injection buffer alone (lanes *3, 4* in **A**) or injection buffer with 500 μM GTPγS (lanes *5, 6* in **A** and lanes *5, 6* in **B**). Cells were incubated for another 30 min at 39.5 °C, to let them recover from the injection, before metabolic labeling with ^{35}S methionine at 39.5 °C (ect. 8.4.3). **A** After labeling cells were chased in labeling medium (containing 100 μg/ml cycloheximide) for 1 h at 39.5 °C (lanes *1, 2*; not injected) or 31 °C (lanes *3–6*, injected), lysed, and vesicular stomatitis virus G-protein (ts-O45-G) was immunoprecipitated using a rabbit anti-ts-O45-G tail antibody (αP4; Kreis 1986). Immunoprecipitated ts-O45-G was digested with endoH and subsequently analyzed by gel electrophoresis and autoradiography. *Lane 1* 39.5 °C, noninjected, -endoH; *lane 2* as lane *1*, but +endoH; *lane 3* 1 h at 31 °C, control-injected, -endoH; *lane 4* as lane *3*, but + endoH; *lane 5* 1 h at 31 °C, GTPγS injected, -endoH; *lane 6* as lane *5*, but + endoH. **B** GTPγS injected (lanes *5, 6*) and noninjected cells (lanes *1–4*) were chased for 2.5 h at 39.5 °C in the absence (lanes *1, 2*) or presence of brefeldin A (lanes *3–6*). Then they were lysed and analyzed as described in **A**. *Lane 1* noninjected, -brefeldin A, -endoH; *lane 2* as lane *1*, but + endoH; *lane 3* noninjected, +brefeldin A, -endoH; *lane 4* as lane *3*, but + endoH; *lane 5* GTPγS injected, +brefeldin A, -endoH; *lane 6* as lane *5*, but + endoH

6. Remove the supernatant and add 0.5 ml lysis buffer to the beads.
7. Agitate gently for 1–5 min.
8. Repeat steps **5–8** three times.
9. After the last spin, remove the supernatant completely until beads are dry. Remaining buffer can be removed most conveniently by using Geloader tips.
10. Add 20 μl of 50 mM sodium citrate (pH 5.6) containing 0.4 % SDS (has to be added freshly to the citrate buffer) to the dry beads.
11. Incubate for 5 min at 95 °C.
12. Spin for 1 min to pellet the Sepharose.
13. Recover all the supernatant (you should obtain 20 μl; use Geloader tips).

14. Add 55 μl of citrate buffer containing PIM1 and PIM2 but no SDS.

15. Split the samples in two and add 2 mU endoH to one of them.

16. Incubate for 16 h at 37 °C.

17. Analyze the sample by gel electrophoresis and autoradiography.

In applications where digestion with endoH is not required, the following steps can be used after step **9**:

10a. Add 20 μl sample buffer for SDS-PAGE to the beads, mix gently, and boil the slurry for 3 min.
11a. Spin the sample as before and collect all supernantant (use Geloader tips).
12a. Analyze the sample by gel electrophoresis and autoradiography.

Results

About 500 to 1000 Vero cells grown in the center of a glass coverslip were infected at 39.5 °C with vesicular stomatitis virus ts-O45 (for details see Kreis 1986; Pepperkok et al. 1993). Cells were kept at 39.5 °C and either left uninjected as a control (Fig. 8.1A, lanes 1, 2), or microinjected with buffer alone (Fig. 8.1A, lanes 3, 4), or microinjected with GTPγS (500 μM final concentration; Fig. 8.1A, lanes 5, 6). After metabolic labeling with ^{35}S methionine at 39.5 °C cells were directly processed for analysis (Fig. 8.1A, lanes 1, 2) or shifted to 31 °C for 1 h (Fig. 8.1A, lanes 3–6). At the nonpermissive temperature the transmembrane glycoprotein ts-O45-G of ts-O45 is retained in the endoplasmic reticulum (ER). Since it lacks oligosaccharide side chain modifications by Golgi enzymes it is sensitive to cleavage by endoglycosidase H (endoH; Fig. 8.1A, lanes 1, 2). Incubation of cells at the permissive temperature, 31 °C, for 1 h allows transport of ts-O45-G to proceed through the Golgi complex, where its oligosaccharide side chains are modified to an endoH-resistant form (Fig. 8.1A, lanes 3,4). Anterograde transport of ts-O45-G to the Golgi complex at 31 °C is inhibited in cells microinjected with GTPγS and ts-O45-G remains endoH-sensitive (Fig. 8.1A, lanes 5, 6) to a similar degree as in cells kept at the nonpermissive temperature, 39.5 °C (Fig. 8.1A, lanes 1, 2).

When cells are treated with brefeldin A, Golgi resident enzymes are translocated back to the endoplasmic reticulum (see Klausner et al.

1992). Thus, whereas ts-O45-G remains sensitive to endoH at 39.5 °C in untreated cells (Fig. 8.1B, lanes 1, 2), ts-O45-G becomes resistant to endoH at 39.5 °C upon treatment of cells with brefeldin A (Fig. 8.1B, lanes 3, 4). Microinjection of GTPγS does not interfere with translocation of Golgi enzymes to the ER, since ts-O45-G becomes endoH resistant in GTPγS injected cells upon brefeldin A treatment (Fig. 8.1B, lanes 5, 6).

In conclusion, GTPγS inhibits anterograde transport from the ER to the Golgi complex but not brefeldin A-induced retrograde transport of Golgi enzymes to the ER.

Comments

- The protocols described above have also been used successfully to analyze the following processes in microinjected tissue culture cells: protein synthesis, post-translational protein modification, protein secretion (Pepperkok et al. 1993), protein degradation in the endocytic pathway (Olaf Rosorius, pers. comm.), protein phosphorylation.
- Total proteins from the lysate can be analyzed by gel electrophoresis after TCA precipitation of the lysate.
- Protein secretion is analyzed by transferring the coverslip containing labeled and injected cells into 300 μl of culture medium (MEM^+) in a 24-well plate. Secreted proteins can then be TCA precipitated from the culture supernatant and analyzed. If total proteins in the medium are to be analyzed by gel electrophoresis, the concentration of BSA in the chase medium should not be higher than 0.5 %.
- Exposure times of the autoradiograms vary according to the abundance and synthesis rate of proteins studied. Abundant viral proteins (e.g., ts-O45-G, as shown here) give a good signal on autoradiographs after only 2–3 days of exposure, whereas endogeneous cellular proteins (e.g., cathepsin D, see Pepperkok et al. 1993) need exposures of about 1 week or even longer.
- A different buffer for solubilization of proteins from labeled injected cells can be used. Buffers containing denaturing reagents need to be diluted or adjusted before immunoprecipitation to avoid denaturation of the antibody (see Harlow 1988). The presence of carrier proteins in the lysis buffer is, however, extremely important for the success of the experiment (e.g., immunoprecipitation).

- Affinity reagents other than antibodies (e.g., lectins, receptors) can be used to isolate specific proteins from the lysate.

Troubleshooting

In experiments where inhibition of cellular function is studied (see results in paragraph above) it is important to microinject all cells on the coverslip. The contribution of non-injected cells to the signals measured may be significant and distort the results. Microinjection of all cells is most conveniently achieved using computer-automated injection equipment and the protocol for plating 500 to 1000 cells in the center of a coverslip (see Procedure).

References

Ansorge W, Pepperkok R (1988) Performance of an automated system for capillary microinjection into living cells. J Biochem Biophys Methods 16:283–292

Gautier-Rouviere C, Fernandez A, Lamb NJC (1990) Ras-induced c-fos expression and proliferation in living rat fibroblasts involves C-kinase activation and the serum response element pathway. EMBO J 9:171–180

Harlow E, Lane D (1988) Antibodies: a laboratory manual. Cold Spring Harbor Laboratory, Cold Spring Harbor

Klausner RD, Donaldson JG, Lippincott-Schwartz J (1992) Brefeldin A: insights into the control of membrane traffic and organelle structure. J Cell Biol 116:1071–1080

Kreis TE (1986) Microinjected antibodies against the cytoplasmic domain of vesicular stomatitis virus glycoprotein block its transport to the cell surface. EMBO J 5:931–941

Lane HA, Fernandez A, Lamb NJC, Thomas G (1993) p70sk6 function is essential for G1progression. Nature 363:170–172

Pepperkok R, Scheel J, Horstman H, Hauri HP, Griffiths G, Kreis TE (1993a) β-COP is essential for biosynthetic membrane transport from the endoplasmic reticulum to the Golgi complex in vivo. Cell 74:71–82

Proctor GN (1992) Microinjection of DNA into mammalian cells in culture: theory and practise. Methods Mol Cell Biol 3:209–231

Rosa P, Weiss U, Pepperkok R, Ansorge W, Niehr, C, Stelzer EHK, Huttner WB (1989) An antibody against secretogranin I (chromogranin B) is packaged into secretory granules. J Cell Biol 109:17–34

Chapter 9

A New Strategy for Studying Microtubule Transport and Assembly During Axon Growth

PETER W. BAAS[1]

Introduction

There is widespread agreement that the net addition of new microtubule polymer to the axon is necessary for its growth, but there is controversy concerning the mechanisms by which this occurs. The earliest model held that preassembled microtubules are transported from the cell body of the neuron down the growing axon, while subsequent models held that new polymer is added at the distal region of the growing axon via local microtubule assembly. Since these early models were proposed, many workers have taken the view that microtubule transport and assembly events are mutually exclusive, and hence that evidence supporting one model refutes the other. We have taken a very different view, that microtubule transport and assembly are both important during axon growth. In our model, microtubule transport is required to increase the tubulin levels within the axon, and local assembly events are required to regulate the lengths of the microtubules (for review see Baas and Yu 1996).

The most controversial element of our model is that it hinges on the movement of assembled microtubules. Attempts to visualize microtubule transport down the axon using live-cell light microscopic methods have produced mixed and principally negative results (for review see Hirokawa 1993), leading some authors to conclude that all of the microtubules in the axon are stationary. These results have led to the speculation that tubulin may be actively transported down the axon not as polymer, but in another form such as free subunits or oligomers. It is also possible, however, that the movement of microtubules down the axon occurs, but is difficult to detect for technical reasons. For example, only a small

[1] Department of Anatomy, The University of Wisconsin Medical School, 1300 University Avenue, Madison, Wisconsin 53706, USA. phone: +01-608-2627307; fax: +01-608-2627306; e-mail: pwbaas@facstaff.wisc.edu

fraction of the microtubules may be moving at one time and the movement of these microtubules may be highly asynchronous. In addition, and entirely consistent with our model, the movement of these microtubules may be obscured by the fact that they are undergoing dynamic assembly events at the same time that they are moving.

These considerations suggest that higher resolution methods may be required to reveal microtubule transport in the axon and to test our hypothesis that microtubule assembly and transport events both occur during axon growth. We have recently developed a novel strategy for accomplishing this (Yu et al. 1996). In this strategy, biotinylated tubulin is microinjected into cultured neurons after the outgrowth of short axons. The axons are then permitted to grow longer, after which the cells are prepared for immunoelectron microscopic visualization of biotinylated-tubulin-containing polymer. We reasoned that any polymer that assembled after the introduction of the probe should label for biotin, while any polymer that was already assembled but did not undergo assembly or subunit turnover should not label. Therefore, the presence in the newly grown region of the axon of any unlabeled microtubule polymer indicates that this polymer was transported during axon growth. The presence of labeled polymer indicates that this polymer underwent assembly or subunit turnover during axon growth. Thus, using this technique, we are able to test our proposal that microtubule transport and assembly events both occur during axon growth.

Outline

Figure 9.1a summarizes the strategy of our approach, while Fig. 9.1b provides a flow chart of the steps that are required for its execution. Superior cervical ganglia are dissected from newborn rat pups, dissociated enzymatically with trituration, and plated onto a glass substratum that had been pretreated with polylysine and laminin. After waiting roughly 90 min for the neurons to attach to the substratum and for axons to grow roughly 75–100 microns in length, cultures are moved to the heated stage of an inverted microscope. Here, biotinylated tubulin is microinjected into the cell body of a small number of the neurons, after which the cultures are returned to the incubator. The axons are then permitted to grow longer over a period of 40 min postinjection. At this point, the cultures are extracted with a detergent in the presence of a microtubule-stabilizing buffer to remove free tubulin, and prepared for immunoelectron microscopic visualization of microtubule polymer that did or did

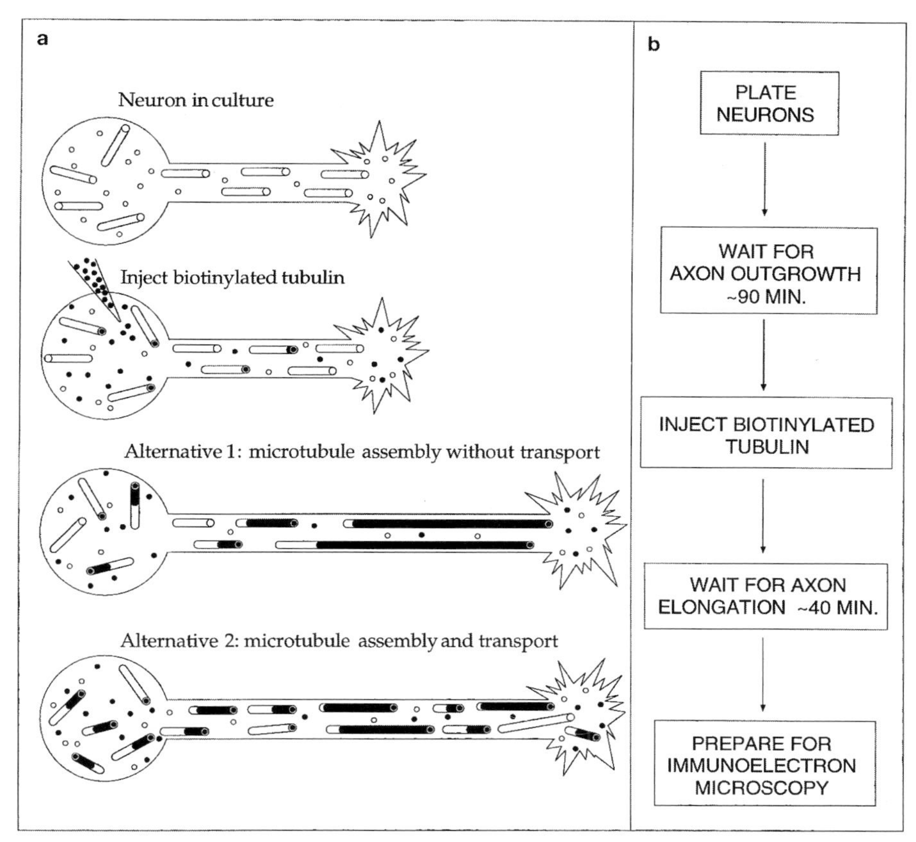

Fig. 9.1a, b. **a** Schematic illustration of our experimental strategy. Native microtubules and free tubulin are shown by *unblackened tubes* and *circles,* respectively. Biotinylated tubulin and regions of microtubules containing biotinylated tubulin are shown in *black*. **b** Flow chart of the steps required to execute this strategy. (**a** Yu et al. 1996)

not incorporate biotinylated tubulin. The presence in the newly grown region of the axon of any unlabeled microtubule polymer indicates that this polymer was transported during axon growth. The presence of labeled polymer indicates that this polymer underwent assembly or subunit turnover during axon growth. It should be noted that the levels of unlabeled polymer provide a minimal estimate of the contribution of microtubule transport, as it is possible that the labeled polymer also underwent transport as it turned over its subunits.

Materials

- Newborn Sprague-Dawley rat pups
- Dissection tools, dissecting microscope, laminar flow hood, incubator
- Enzyme solution containing 0.25 mg/ml collagenase (Worthington Biochemical Corporation, Freehold, NJ) and 0.25 mg/ml trypsin (Worthington Biochemical Corporation) in phosphate-buffered saline
- Tissue culture medium consisting of Leibovitz L15 (Sigma Chemical Company, St. Louis, IL) supplemented with 0.6 % glucose, 2 mM L-glutamine, 0.6 % methyl cellulose (Dow Chemical Company, Midland, MI), 100 U/ml penicillin, 100 μg/ml streptomycin, 10 % fetal bovine serum (Hyclone, Logan, UT), and 100 μg/ml 7S nerve growth factor (Upstate Biotechnology Incorporated, Lake Placid, NY)
- Special dishes prepared by drilling a 1-cm hole in the bottom of a 35 mm tissue culture dish, and adhering to the bottom of the dish a photoetched glass coverslip (Bellco, Vineland, NJ). Prior to cell culture, the special dishes must be pretreated first with a solution of 1 mg/ml polylysine (Sigma Chemical Company) in a borate buffer followed by a solution of 10 μg/ml laminin (Sigma Chemical Company) in serum-free L15 medium. Methods for the preparation of these dishes (Whitlon and Baas 1992), and polylysine/laminin treatment (Higgins et al. 1991) have been described.

Microinjection of biotinylated tubulin

- Aliquots of biotinylated bovine-brain tubulin, prepared by methods that have been described in detail elsewhere (Webster and Borisy 1989; Hyman et al. 1991). For our studies on peripheral neurons, the final product is suspended at 4 mg/ml in an injection buffer containing 50 mM potassium glutamate and 1 mM $MgCl_2$ at pH 6.8. For studies on central neurons, this injection buffer is toxic and should be replaced with a buffer consisting of 100 mM Pipes, pH 6.8
- An inverted microscope interfaced with an environmentally-controlled stage, and a pressure-regulated microinjection system. We use the Eppendorf system (Hamburg, Germany).

Immunoelectron microscopy

- Extraction solution consisting of microtubule stabilizing buffer (60 mM Pipes, 25 mM Hepes, 10 mM EGTA, 2 mM $MgCl_2$, pH 6.9) supplemented with 10 μM taxol and, containing 0.5 % Triton X-100 to remove unassembled tubulin
- Fixation solution consisting of the microtubule stabilizing buffer and 1 % glutaraldehyde

- Aldehyde-quenching solution consisting of 3 mg/ml sodium borohydride dissolved in half microtubule stabilizing buffer and half methanol
- Blocking solution consisting of 5 % normal goat serum and 2 % BSA in TBS-1 (10 mM Tris, 140 mM NaCl, pH 7.6)
- Primary antibody, a mouse monoclonal anti-biotin antibody conjugated directly to the fluorochrome Cy-3 (Jackson Immunoresearch, West Grove, PA), used at a concentration of 1:50 in TBS-1. Other antibiotin antibodies are available, but this one is particularly useful because it is conjugated directly to a fluorochrome
- TBS-2 (20 mM Tris, 140 mM NaCl, pH 8.2) containing 0.1 % BSA
- Goat anti-mouse second antibody conjugated to 5-nm colloidal gold particles (Amersham, Arlington Heights, IL), used at a concentration of 1:2 in TBS-2
- Second fixation solution consisting of 2 % glutaraldehyde in 0.1 M cacodylate also containing 0.2 % tannic acid
- Postfixaton solution consisting of 2 % osmium tetroxide in 0.1 M cacodylate
- For embeddment, one of the epon clones such as LX100 (Ladd Industries, Burlington, VA)
- Standard equipment and supplies for thin-sectioning and electron microscopy.

Procedures

Cell Culture

1. Superior cervical ganglia are dissected from newborn Sprague-Dawley rat pups under sterile conditions (see Higgins et al. 1991). The ganglia are collected into serum-free L15 medium, and then treated for 15 min at 37 °C with freshly prepared enzyme solution.

2. The ganglia are rinsed three times in serum-containing medium, and then dissociated into a single cell dispersion by triturating gently three to five times with a Pasteur pipette.

3. The cells are then plated at a density corresponding to roughly one ganglion per three 35-mm dishes into the polylysine/laminin treated special dishes. The cells are plated in the modified L15-based medium described by Bray (1991).

This medium has the advantage that it maintains pH in normal air, and hence there are no concerns about alterations in pH when moving from the incubator to the microscope stage. The L15-based medium is not good for culturing neurons over periods of time greater than 1–2 days, but is excellent for the short-term cultures utilized for the present studies. To ease the task of relocating specific cells after microinjection, we used special glass coverslips that are preetched with demarcating boxes (Belco, Vineland, NJ).

Microinjection Regime

1. After plating, the neuron cultures are placed in an incubator for roughly 90 min to permit the cells to adhere to the substratum and to extend axons.
2. At this point, after which most of the neurons had extended multiple axons that were 25–100 μm in length, cultures are placed on the prewarmed stage of an inverted microscope equipped with an environmental chamber that effectively maintains temperature at 37 °C. For microinjection, neurons are selected with axons that are not extensively branched and are clearly not fasciculated with the axons of neighboring cells. It is helpful to obtain a phase-contrast or differential-inference contrast (DIC) image of the entire neuron (we use a thermal videoprinter) either just before or just after microinjection.
3. Biotinylated tubulin is kept at 4 °C until the time of injection, and introduced into the neurons at a volume roughly, but not exceeding, 10 % of the volume of the cell. Neurons are notoriously difficult to inject because of their rounded shape and spongy consistency. Even with an automated microinjection system, each neuron must be injected manually and failed attempts are not uncommon. Successful introduction of the probe can usually be detected as a "rippling" effect through the neuron.
4. The culture is then returned to the incubator for 35 min to permit the axons to continue growing, after which a second set of phase-contrast or DIC images is obtained.
5. The two sets of images are subsequently used to assess the degree to which individual axons had grown during the 40-min period of time postinjection.

Immunoelectron Microscopy

1. Immediately after acquiring the second set of phase-contrast or DIC images, the cultures are rinsed briefly in the microtubule stabilizing buffer and then extracted for 7 min in the extraction solution.
2. The cultures are then fixed by adding an equal quantity of the fixation solution directly to the extraction solution. Neurons are typically not well adhered to their substratum and can lift off during fluid exchanges, especially after extraction. Adding the fixation solution directly to the extraction solution avoids the most dangerous fluid exchange.
3. After 10 min of fixation, the cultures are rinsed in microtubule stabilizing buffer and then incubated for 15 min in the sodium borohydride solution, rinsed in buffer, incubated for 30 min in the blocking solution, and then exposed to primary antibody overnight at 4 °C.
4. After incubation with the primary antibody, the cultures are rinsed six times for 10 min each with TBS-2 with BSA. Because the biotin antibody is directly conjugated to the fluorochrome Cy3, the microinjected cells can at this point be visualized and photographed using epifluorescence optics. This is helpful in confirming that the probe was effectively introduced into the neuron and that it effectively incorporated into the microtubule polymer.
5. The cultures are then incubated for 3 h at 37 °C with the gold-conjugated second antibody, rinsed six times in TBS-2, fixed in the second fixation solution for 10 min, rinsed in 0.1 M cacodylate, postfixed for 10 min, dehydrated in an ethanol series, and embedded in LX100 (Ladd Industries, Burlington, VA).
6. After curing of the resin, the glass coverslip is either removed mechanically or dissolved from the resin by a ten min incubation in hydrofluoric acid (see Whitlon and Baas 1992). Cells of interest are relocated using the DIC or phase-contrast images and the photoetched pattern transferred from the glass coverslip onto the resin, and thin-sectioned using an ultramicrotome.
7. The sections are picked up onto Formvar-coated slot grids, stained with uranyl acetate and lead citrate, and viewed with a standard transmission electron microscope.

Data Analysis

All sections through the injected neurons are viewed, and a typical middle section is used for quantification of labeled and unlabeled microtubule polymer. Distinguishing labeled and unlabeled polymer is difficult in some areas along the length of the axon because the tight spacing of the microtubules makes it impossible to know the correct microtubule with which many of the gold particles are associated. In other areas in which the microtubules are even more tightly bundled, there can be some question as to whether the polymer may not have labeled as a result of problems of accessibility of the gold particles.

To avoid misinterpretations, we use for our analyses only areas of the axons in which the microtubules had splayed apart sufficiently during extraction to minimize these potential problems. We uniformly select areas of the axon that were 4 μm in length, and score total lengths of labeled and unlabeled microtubule polymer within these regions.

Results

Figure 9.2a–c and d–f shows images of two different neurons immediately after injection of biotinylated tubulin (Fig. 9.2a, d), after 40 min of axon elongation (Fig. 9.2b, e), and after preparation for immunofluorescence visualization of biotinylated-tubulin-containing microtubules (Fig. 9.2c, f). In our recent study (Yu et al. 1996), a total of 19 axons grown from nine different neurons were analyzed. The average length of the axons at the time of injection was 60.2±12.9 μm, while the average amount of growth per axon was 17.5±6.1 μm. These rates of growth were generally similar to the rates at which the axons of uninjected neurons grew, typically 20–50 μm/h, indicating that the injection procedure did not markedly alter the growth properties of the axons. The immunofluorescence images in Fig. 9.2c and f show that the probe has incorporated into microtubules.

Figure 9.3a and b, respectively, are tracings obtained from the DIC videoprint images of an entire neuron immediately after injection and 40 min later. The remaining panels show immunoelectron micrographs from the cell body (Fig. 9.3c), and designated sites along the length of one of the axons, as indicated in Fig. 9.3b. This axon was 55 μm in length prior to injection, and grew an additional 11 μm after injection. The cell body of this neuron and all other neurons examined contained both labeled and unlabeled polymer, as did all regions of the axon proximal to

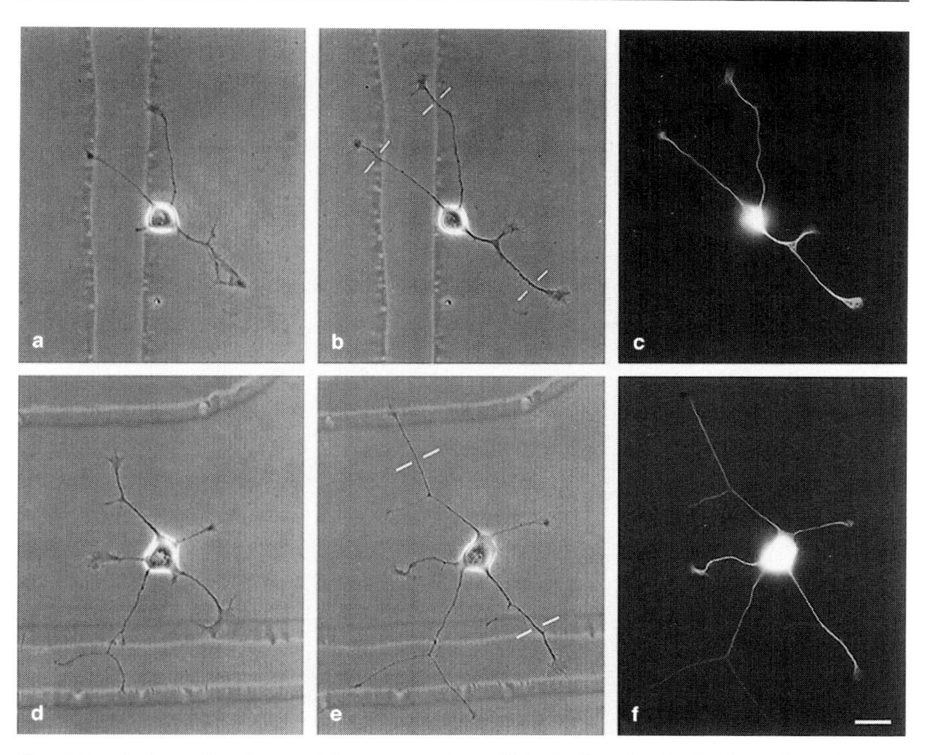

Fig. 9.2a–f. Introduction and incorporation of biotinylated tubulin into the microtubules of a cultured neuron. **a** Phase-contrast image of a typical neuron immediately after injection of biotinylated tubulin. **b** Axons elongated after 40 min. **c** Immunofluorescence image of the same neuron stained for biotin, showing incorporation of the probe into microtubules (note fibrous appearance of staining in some areas where the bundled microtubules splay apart). **d–f** Comparable images of another neuron with axons that grew somewhat more extensively over the 40-min period of time. *Bar* 20 μm (Yu et al. 1996)

the newly grown region. Figure 9.3d shows a site proximal to the newly grown region in which 33 % of the polymer was unlabeled. In the case of this axon, almost all of the polymer in the newly grown region was labeled. Figure 9.3e shows a fairly proximal site within the newly grown region in which a few unlabeled microtubule profiles intermingle with many labeled profiles. At this particular site, 10 % of the polymer was unlabeled. Examination of serial sections indicated that labeled and unlabeled polymer were continuous with one another, with the former elongating from the plus end of the latter. In no case did we observe unlabeled polymer extending from the plus end of labeled polymer. This

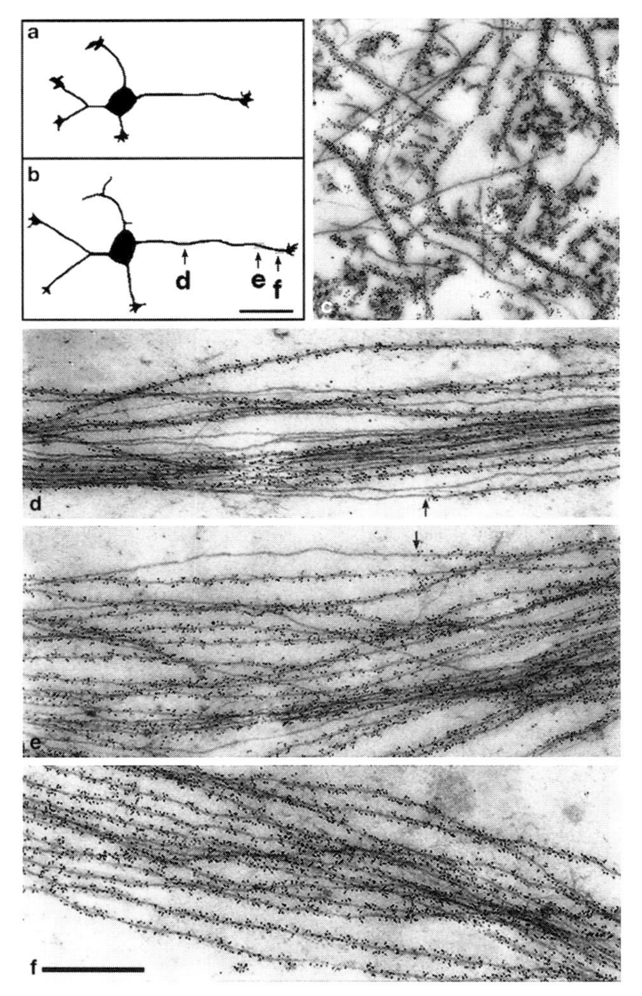

Fig. 9.3a–f. Immunoelectron microscopic visualization of microtubule polymer that did or did not incorporate biotinylated tubulin. **a** and **b**, respectively. Tracings obtained from the DIC videoprint images of an entire neuron immediately after injection and 40 min later. The remaining panels show immunoelectron micrographs from the cell body (**c**), and designated sites along the length of one of the axons as indicated in **b.** The cell body contains both labeled and unlabeled polymer, as do all regions of the axon proximal to the newly grown region. **d** Site proximal to the newly grown region in which 33 % of the polymer was unlabeled. **e** Fairly proximal site within the newly grown region in which a few unlabeled microtubule profiles intermingle with many labeled profiles. Ten percent of the polymer is unlabeled. *Arrows* in **d** and **e** indicate points where labeled and unlabeled polymer are continuous with one another, with the former elongating from the plus end of the latter. **f** Distal site near the growth cone in which all of the polymer was labeled. *Bar* **a**, **b** 20 μm, panels **c–f** 1.0 μm (Yu et al. 1996)

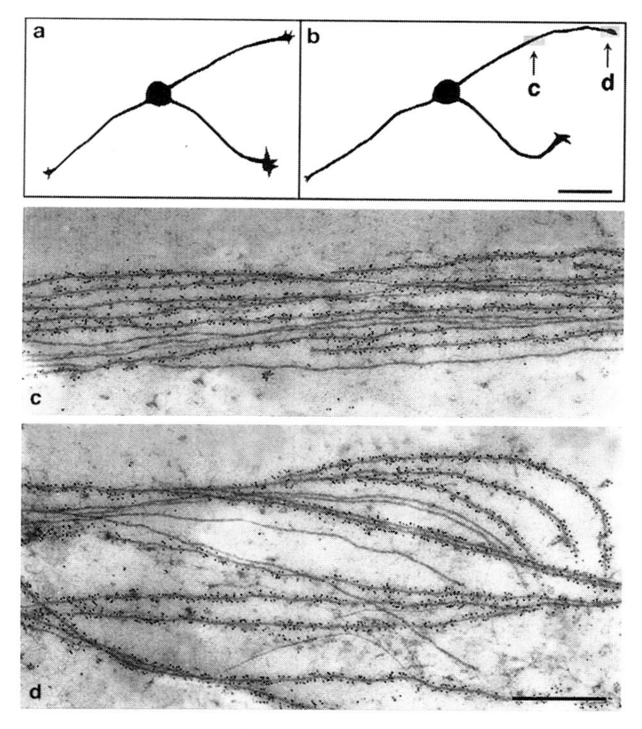

Fig. 9.4a–d. Immunoelectron microscopic visualization of microtubule polymer that did or did not incorporate biotinylated tubulin. **a** and **b**, respectively. Tracings obtained from the DIC videoprint images of a neuron immediately after injection and 40 min later. The remaining panels show immunoelectron micrographs from two different sites along one of the axons. **c** Site proximal to the newly grown region in which 35 % of the polymer was unlabeled. **d** Site within the newly grown region in which 27 % of the polymer was unlabeled. Notably, this site was located within the most distal region of the axon, directly behind the growth cone, a site which contained no unlabeled polymer in the axon shown in Fig. 9.3. *Bar* **a**, **b** 20 μm, **c**, **d** 0.5 μm (Yu et al. 1996)

was also apparent in many of the individual sections (see arrows in Fig. 9.3d, e). Beyond this proximal area of the newly grown region, all the microtubule profiles were labeled (Fig. 9.3f). The fact that we were able to detect unlabeled polymer, even at these relatively low levels, indicates that microtubules were transported into the newly grown region of the axon.

Figure 9.4a and b, respectively, are tracings obtained from the DIC videoprint images of another neuron immediately after injection and 40 min later. The remaining panels show immunoelectron micrographs

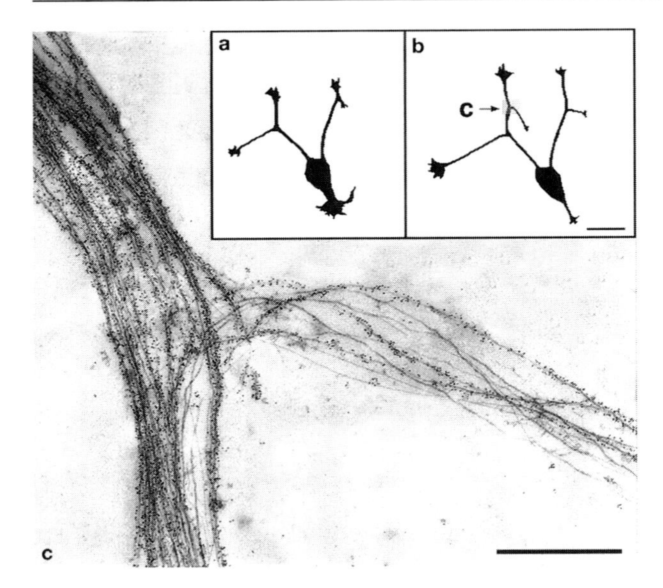

Fig. 9.5a–c. Immunoelectron microscopic visualization of microtubule polymer that did or did not incorporate biotinylated tubulin. **a** and **b**, respectively. Tracings obtained from the DIC videoprint images of a third neuron immediately after injection and 40 min later. During the 40-min time frame of the experiment, the growth cone at the tip of one of the axons underwent an asymmetric bifurcation. The larger branch grew straight, while the smaller branch curved off in another direction. **c** Immunoelectron micrograph showing the region of the parent axon contiguous with both branches. In the region of the parent axon just proximal to the branch point, 19 % of the polymer is unlabeled. In the larger branch, virtually all of the polymer is labeled. In regions along the length of the smaller branch, 40–50 % of the polymer is unlabeled. *Bar* **a**, **b** 20 μm, **c** 1.0 μm (Yu et al. 1996)

from two different sites along one of the axons. This axon was 62 μm in length prior to injection, and grew an additional 10 μm after injection. Figure 9.4c is a site proximal to the newly grown region in which 35 % of the polymer was unlabeled. Figure 9.4d is a site within the newly grown region in which 27 % of the polymer was unlabeled. Notably, this site was located within the most distal region of the axon, directly behind the growth cone, a site which contained no unlabeled polymer in the axon shown in Fig. 9.3. These data indicate that microtubules were transported into the newly grown region of this axon, and together with the data from Fig. 9.3, show that microtubule transport is more apparent in some axons than in others.

Figure 9.5a and b, respectively, are tracings obtained from the DIC videoprint images of a third neuron immediately after injection and 40 min later. During the 40-min time frame of the experiment, the growth cone at the tip of one of the axons underwent an asymmetric bifurcation. The larger branch grew straight and was essentially a continuation of the parent axon, while the smaller branch curved off in another direction. Figure 9.5c is an immunoelectron micrograph showing the region of the parent axon contiguous with both branches. There were notable differences between the two branches and the parent axon with regard to their content of unlabeled polymer. In the region of the parent axon just proximal to the branch point, 19 % of the polymer was unlabeled. In the larger branch, virtually all of the polymer was labeled. Notably, in regions along the length of the smaller branch, 40–50 % of the polymer was unlabeled. Thus, with this methodology, no microtubule transport was detectable in the larger branch, but significant microtubule transport was detectable in the smaller branch.

The starting lengths, amounts of axon growth, and percentages of unlabeled polymer scored in discrete regions at various points along the lengths of all 19 axons examined are shown schematically in Fig. 9.6. Each neuron is labeled with a Roman numeral, and individual axons from the same neuron are given separate letters. The axon shown in Fig. 9.3 is labeled IX-B in Fig. 9.5. The axon shown in Fig. 9.4 is labeled IV-A in Fig. 9.6. The axon shown in Fig. 9.5 is labeled III-B in Fig. 9.6. Axons IV-C and IV-D were already formed branches from a parent axon, as were axons 8-C and 8-D. Five of the 19 axons contained no detectable unlabeled polymer whatsoever within the sites we examined in their newly grown regions. An additional nine contained no unlabeled polymer directly behind their growth cones, but contained unlabeled polymer (6–50 %) at more proximal sites within their newly grown regions. The remaining five axons contained unlabeled polymer (1–28 %) throughout their newly grown regions, including sites directly behind their growth cones. Together, these data demonstrate that microtubule assembly and/or turnover is active in growing axons and also that preassembled microtubules are transported into newly grown regions of the axon. This transport is detectable in some, but not all, axons using our methodology.

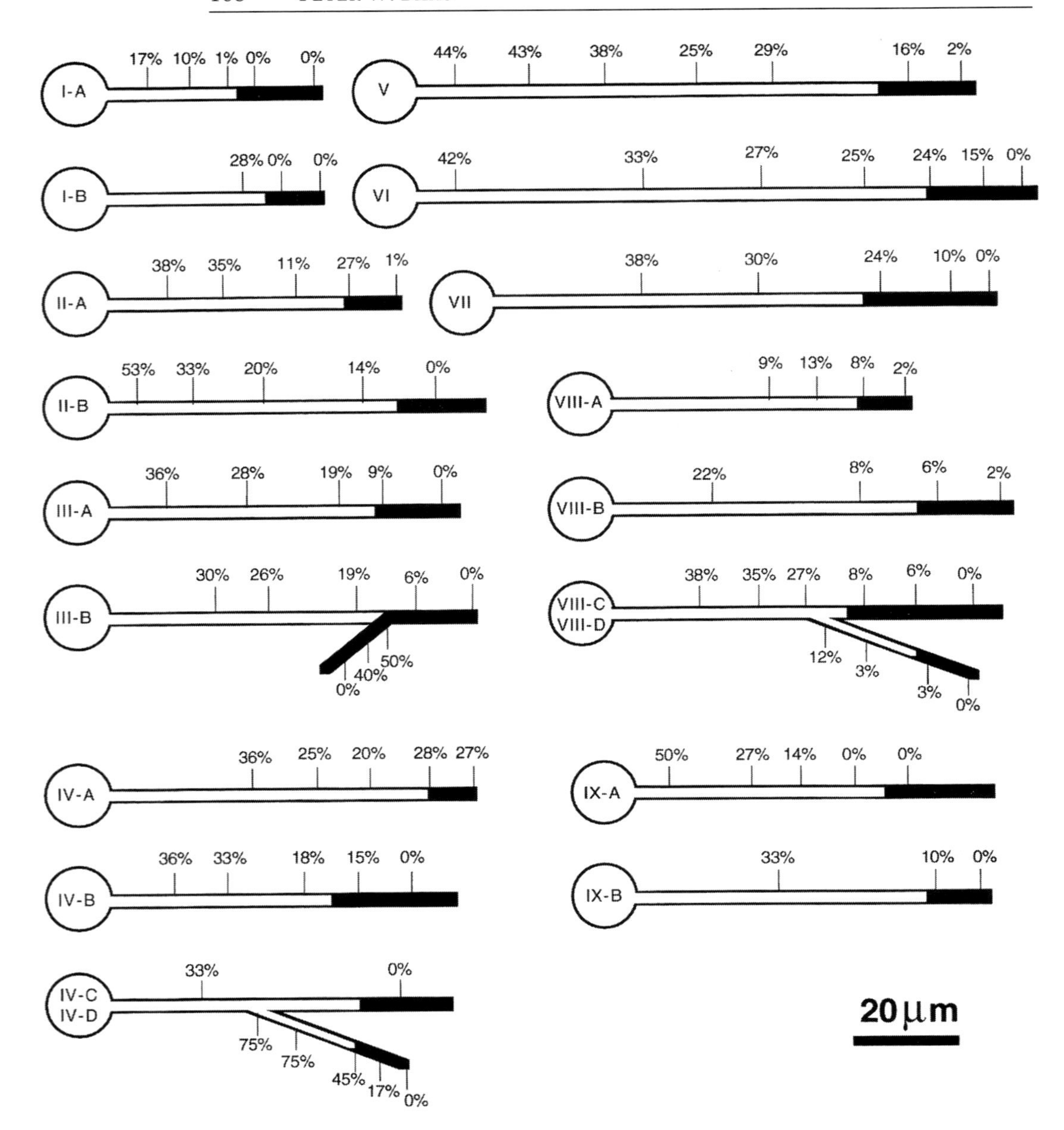

I-A
17% 10% 1% 0% 0%
I-B
28% 0% 0%
II-A
38% 35% 11% 27% 1%
II-B
53% 33% 20% 14% 0%
III-A
36% 28% 19% 9% 0%
III-B
30% 26% 19% 6% 0%
50% 40% 0%
IV-A
36% 25% 20% 28% 27%
IV-B
36% 33% 18% 15% 0%
IV-C
IV-D
33% 0%
75% 75% 45% 17% 0%
V
44% 43% 38% 25% 29% 16% 2%
VI
42% 33% 27% 25% 24% 15% 0%
VII
38% 30% 24% 10% 0%
VIII-A
9% 13% 8% 2%
VIII-B
22% 8% 6% 2%
VIII-C
VIII-D
38% 35% 27% 8% 6% 0%
12% 3% 3% 0%
IX-A
50% 27% 14% 0% 0%
IX-B
33% 10% 0%
20 μm

◀ **Fig. 9.6.** Summary of data from all axons examined. The starting lengths, amounts of axon growth, and percentages of unlabeled polymer scored at various points along the lengths of all 19 axons examined are shown schematically. Each neuron is labeled with a *Roman numeral*, and individual axons from the same neuron are given separate letters. Five of the 19 axons contained no detectable unlabeled polymer whatsoever in their newly grown regions. An additional nine contained no unlabeled polymer directly behind their growth cones but contained unlabeled polymer (6–50 %) at other sites within their newly grown regions. The remaining five axons contained unlabeled polymer (1–28 %) throughout their newly grown regions, including sites directly behind their growth cones. All percentages are indices calculated from measurements obtained from 4 μm regions of the axon. *Bar* 20 μm (Yu et al. 1996)

Troubleshooting

The protocol outlined here involves several different procedures, all of which require skill and practice. Nevertheless, with some effort and patience, all of these procedures can be accomplished. Specific problems (and remedies to these problems) associated with primary culture of neurons, preparation of conjugated tubulins, microinjection, and immunoelectron microscopy have been discussed elsewhere, and hence will not be discussed here.

Worth some discussion, however, are potential problems in adapting our strategy to other experimental systems. It is essential to keep in mind that microtubules are dynamic structures, constantly turning over their subunits. This occurs whether or not the microtubules are also undergoing transport. In our regime, one can conclude that an unlabeled microtubule in the newly grown region of the axon has been transported. However, one cannot conclude that a labeled microtubule has not undergone transport. Thus, the regime provides a minimal estimate of the contribution of microtubule transport, and does not provide a measure of the relative contributions of microtubule assembly and transport. In order to detect microtubule transport, one must select a system in which the axon is growing more rapidly than the rate at which all of the polymer in the distal region of the axon turns over its subunits. We suspect that longer axons might have the advantage of containing more slowly turning-over polymer, but longer axons present a problem of their own. The usefulness of the regime is contingent upon high levels of the probe reaching the axon tip before significant axon growth occurs, otherwise unlabeled polymer could be the result of assembly and not transport. For all of these reasons, a great deal of caution is recommended in adapting our experimental strategy to other systems, both in the selection of the system and in the interpretation of the results.

Acknowledgments. I would like to acknowledge the important contributions of Wenqian Yu, Matthew Schwei, and Fridoon Ahmad to the work presented here. My laboratory is funded by grants from the National Institutes of Health and the National Science Foundation.

References

Baas PW, Yu W (1996) A composite model for establishing the microtubule arrays of the neuron. Mol Neurobiol 12:145–161

Bray D (1991) Isolated chick neurons for the study of axonal growth. In: Banker G, Goslin K (eds) Culturing Nerve Cells. MIT Press, Cambridge, pp 119–135

Higgins D, Lein PJ, Osterhout DJ, Johnson MI (1991) Tissue culture of mammalian autonomic neurons. In: Banker G, Goslin K (eds) Culturing nerve cells. MIT Press, Cambridge, pp 177–205

Hirokawa N (1993) Axonal transport and the cytoskeleton. Curr Opin Neurobiol 3:724–731

Hyman A, Dreschsel D, Kellog D, Salser S, Sawin K, Steffan P, Wordeman L, Mitchison T (1991) Preparation of modified tubulins. Methods Enzymol 196:478–485

Webster DR, Borisy GG (1989) Microtubules are acetylated in domains that turn over slowly. J Cell Sci 92:57–65

Whitlon DS, Baas PW (1992) Improved methods for the use of glass coverslips in cell culture and electron microscopy. J Histochem Cytochem 40:875–877

Yu W, Schwei MJ, Baas PW (1996) Microtubule transport and assembly during axon growth. J Cell Biol 133:151–157

Chapter 10

The Use of Microinjection to Study Signal Transduction in Mammalian Cells

SERGE ROCHE[1]* AND SARA A. COURTNEIDGE[2]

Introduction

The field of signal transduction has boomed in recent years. Since the discovery of tyrosine phosphorylation 15 years ago, more than 25 different families of tyrosine kinases have been described. Furthermore, many signal transduction pathways, initiated by ligand binding to receptors, and leading to growth, survival, or differentiation, depending on cell type and signal context, have been elucidated. Many investigators have sought to determine the functions of these signaling pathways, using a combination of biochemistry and molecular biology. For example, receptor tyrosine kinase mutants lacking binding sites for particular signaling molecules have been introduced into cells lacking wild-type receptors (see, for example, Rönnstrand et al. 1992; Valius and Kazlauskas 1993). However, the conclusions from this approach could be compromised if the receptor does not function identically in a heterologous cell type. In an approach that does not require the use of heterologous cell types, so-called dominant negative forms of signaling proteins (that presumably compete with endogenous protein for substrates and other effector molecules) are introduced into cells. Here the problem is that it may not be possible to derive stable cell lines that express such inhibitory proteins. Several laboratories, including our own, have circumvented these difficulties by using microinjection to introduce dominant negative mutants (and inhibitory antibodies) into single cells (Mulcahy et al. 1985; Riabowol et al. 1988; Twamley-Stein et al. 1993; Sasaoka et al. 1994; Roche et al. 1994, 1995a, b, 1996). We will describe this approach here.

* corresponding author: phone: +33-67-543344; fax: +33-67-548039;
e-mail: roche@balard.pharma.univ-montp1.fr
[1] CNRS EP612 Faculté de Pharmacie, Ave Ch, Flahaut 34060 Montpellier, France
[2] SUGEN Inc., 515 Galveston Drive, Redwood City, California 94063, USA

Microinjection: an Alternative to Transient Transfection

There are many ways to express proteins transiently in mammalian cells in tissue culture. Most involve introducing DNA into cells by a variety of methods, including transfection, electroporation, or liposomal transfer. Protein can also be directly introduced into cells by scrape loading (McNeil et al. 1984) or other transient permeabilization methods. Although these techniques are extensively used, they suffer from several potential drawbacks. DNA transfection often leads to poor and/or highly variable expression efficiency. Furthermore, not all cell types can be efficiently transfected. Protein loading methods can compromise the long-term viability of the cell. An alternative technique, which has been in use for many years, but is just recently gaining in popularity, is microinjection (for further details, see Ansorge and Pepperkok 1988).

Advantages of Microinjection

- The procedure is well tolerated, and can be successfully applied to many cell types in culture.
- A comparison of the injected cells with the surrounding noninjected cells can help to rule out any artifactual effects due to culture conditions or variability of cell response.
- Proteins can be injected into either the cytoplasm or the nucleus (Roche et al. 1994).
- The time course of a response can be measured by, for example, injecting neutralizing antibody at various times after a stimulus, or at different stages of the cell cycle (Pagano et al. 1992; Twamley-Stein et al. 1993; Roche et al. 1994, 1995a).
- An initial determination of protein function can be made by this quick and simple assay, prior to spending weeks on the generation of stable transfectants. In addition, where the protein is growth-inhibitory, stable clones will not be obtained, and microinjection will be the method of choice for functional studies.

The Limitations of Microinjection

- The biggest drawback is the cost of the equipment required (microscope, micromanipulator, etc.).
- Some practice is required before one can successfully microinject 100 or more cells within a short time. However, automated microinjection systems have improved enormously in the last 5 years and some have been developed which provide rapid, precise, and reproducible microinjection: with some systems 100 cells can be injected in less then 5 min!
- Some cell types, for example nonadherent cells, will not be suitable candidates for microinjection.
- In most cases, it is only feasible to inject a few hundred cells on the coverslip, therefore videomicroscopic or immunofluorescence analyses will have to be used to determine the cells' response.
- Biochemical assays are for the most part excluded, i.e., changes in enzymatic activity or protein phosphorylation cannot be addressed (unless, for example, a phosphorylation site-specific antibody is available for the protein of interest).

Procedures

In our laboratories, we use microinjection to study the response of mammalian cells to growth factors, using DNA synthesis or cell division as a readout (Twamley-Stein et al. 1993; Roche et al. 1994, 1995a, b, 1996; Barone and Courtneidge 1995; Erpel et al. 1996). This approach to studying signal tranduction includes the followings steps (Fig. 10.1): purification of the protein (or DNA) for injection; microinjection into cells; growth factor stimulation; analysis of injected cells. Here we will describe these methods. For well-established protocols we will refer the reader to quoted reviews, or, where appropriate, to manufacturers' instructions. An example of this approach involving injection of purified protein is shown in Fig. 10.3

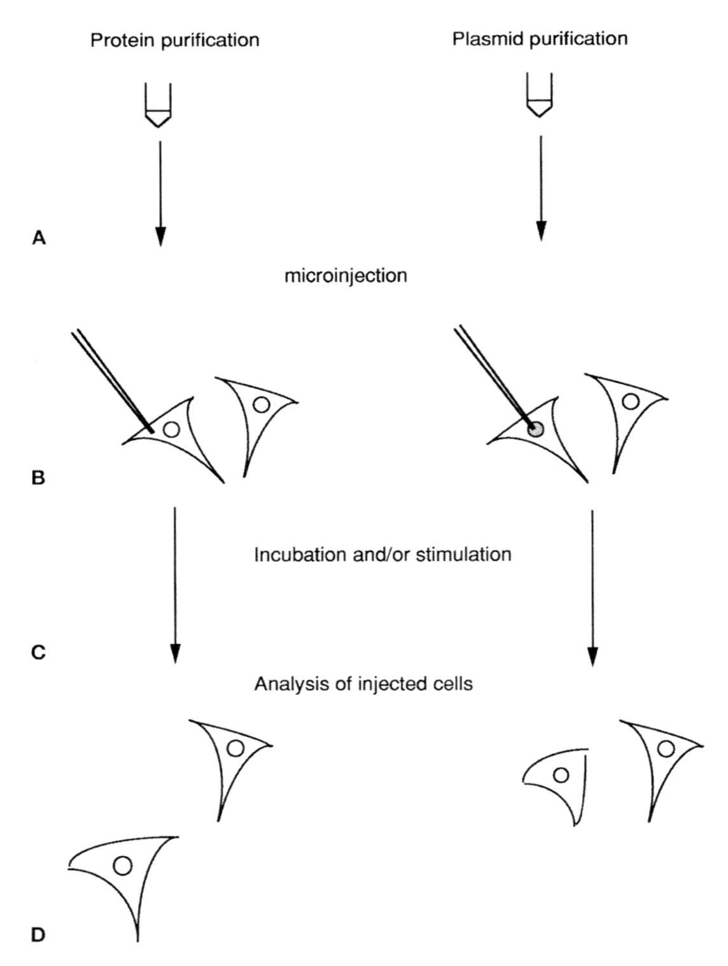

Fig. 10.1A–D. Using microinjection to study growth control in mammalian cells. Flow chart of the microinjection approach to study growth control in mammalian cells. **A** Purification of the protein to introduce (or DNA). **B** Microinjection into cells in the cytoplasm or in the nucleus. **C** Incubation or stimulation of the injected cells. **D** Analysis of the injected cells (DNA synthesis or cell division)

Purification of Proteins

Both purified antibodies and other proteins can be injected into cells.

Antibody purification

Two strategies are used: purification of the whole immunoglobulin fraction from serum (or from the cell supernatant in the case of a monoclonal antibody) and purification of the immunoglobulin specific to the antigen. The latter is much more specific, and highly recommended.

1. Since purified protein must be injected at high concentration (0.5–5 mg/ml), use at least 10 ml of serum as starting material (**Note:** the serum should not contain preservative).
2. To purify whole immunoglobulin from serum, ascites, or tissue culture supernatant, we use protein A or protein G Sepharose columns (Pharmacia). These are commercially available, and give quick and efficient purifications.
3. For affinity purification of antibody, the antigen must be coupled to a resin, such as Sepharose beads (Pharmacia). Several methods are available, and described in detail in the *Antibodies Handbook* (Harlow and Lane 1988).
4. Depending on the affinity of the particular antibody being used, and the abundance of the antigen in the cells, the antibody will need to be concentrated to between 0.5 and 5 mg/ml. Microfiltration with a minicon microconcentrator (Amicon) is one possibility. Alternatively, ammonium sulfate precipitation, followed by resuspension in a small volume of phosphate buffered saline and extensive dialysis, can be used.
5. Following purification, it is important to conduct biochemical tests to ensure that the antibody retains activity.
6. The antibody solution ready for injection should be stored in small volumes at −80 °C. It is important not to subject affinity purified antibodies to cycles of freezing and thawing.

Other protein purification

Introduction of purified protein into cells is a straightforward approach for functional studies (see, for example, Xiao et al. 1994; Roche et al. 1995a, 1996). There are now many ways in which proteins can be expressed to high levels in bacterial, yeast, or insect cells. Because of the ease with which one can express and purify the proteins, the use of gluthathione-S-transferase fusion proteins has gained in popularity. For an excellent description of the production and purification of GST fusion proteins, see Frangioni and Neel (1993). However, consider the following points:

1. It is important that the proteins be as pure as possible, that the final preparation contains no preservatives, and that the buffer in which the protein is stored is as near physiological as possible.
2. Check that after purification the protein is still active, i.e., it is properly folded.
3. Purified protein should be concentrated (to 0.5–5 mg/ml) before microinjection.
4. Store purified protein in small aliquots at −80 °C.

Purification of Plasmids

Microinjection of plasmid DNA is an excellent way to test the effects of overexpression of a wide variety of mutant proteins in cells (see, for example, Twamley-Stein et al. 1993; Roche et al. 1995b; Erpel et al. 1996). This approach is particularly appropriate when it is not possible to purify sufficient quantities of a protein for microinjection, or when the protein cannot be purified in the required form (for example, if the protein is only soluble in detergent, which cannot be microinjected into cells). Several mammalian expression vectors have been developed in which protein expression is under the control of a strong promoter. We often use plasmids that include a SV40 (pSG5, Invitrogen) or a CMV (pCDNA3, Invitrogen) promoter. The choice of vector for ectopic protein expression is very important. Questions to consider are:

- What level of overexpression do I wish to achieve? (for example, the CMV promoter is usually much stronger than SV40)
- Does a given promoter work in my cell type? (this can be tested with the promoter fused to a reporter construct such as lacZ)
- Is the activity of the promoter affected by the growth state of the cell? (many promoters function poorly in quiescent cells)

The purification state of plasmid to inject is very important for good efficiency of expression. Experience has taught us that only supercoiled plasmid DNA gives consistent expression. Therefore we recommend that all plasmid DNA is purified by banding on CsCl gradients twice (Sambrook et al. 1989). Purified plasmid is diluted in distilled water for microinjection and several DNA concentrations tested for in vivo protein expression (100 μg/ml is generally used as a starting DNA concentration). Note that the amount of protein expression finally achieved is in large part dependent on the concentration of DNA injected. Aim to determine the minimal concentration of DNA that gives a measurable effect. Extremely high level of protein expression from plasmid can nonspecifically inhibit growth factor responses.

Microinjection into Adherent Mammalian Cells

Preparation of Cells

Plate the cells on sterile coverslips in petri dishes. Use either commercially available cover slips preetched with a grid (for example CELLocate from Eppendorf), or (prior to plating the cells) a diamond-tipped pen to mark the coverslip, so that the area where the cells are microinjected can later be identified. For growth factor studies, quiesce the cells either before or just as they reach confluence, by placing them in medium containing low or no serum for the 24–48 h prior to microinjection (check that they are quiescent under these conditions before any assay). Just before microinjection, buffer the culture medium by adding 20–50 mM Hepes pH 7.5.

Preparation of the needles

The size of the needle (capillary) is critical for successful injection. Capillaries can be either bought or made with an automatic puller. Commercial needles are appropriate for most purposes (for example, Femtotips from Eppendorf). However, automatic pullers give capillaries with desired size, and this may be important for efficient and reproducible injections in some cases. For example, the nuclear volume is low compared to cytoplasm, and smaller volumes must therefore be delivered, which is best achieved by using a capillary with a narrower opening. Conversely, protein solution is rather viscous due to its high concentration, and for successful injection, capillaries with a wider bore are recommended. In addition, such capillaries provide a larger delivery (note that cytoplasm can increase up to 2 % of its volume without affecting cell viability). Two types of needle puller are commercially available, using either vertical and horizontal pulling. We routinely use the horizontal puller from Flaming Brown.

Microinjection procedure

Several microinjection systems are commercially available which offer flexibility and simplicity of use. Basically, each one is set up as described

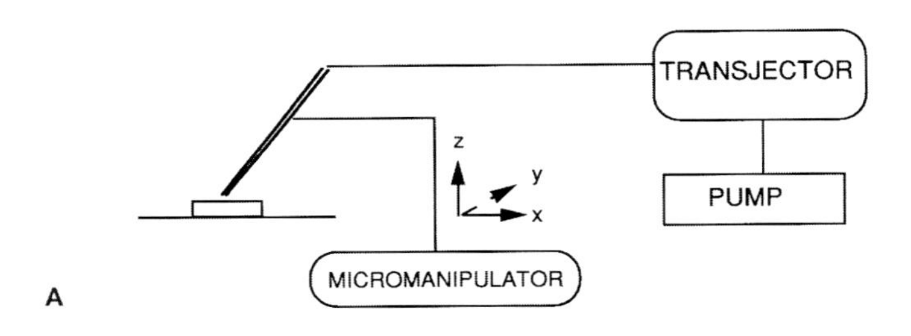

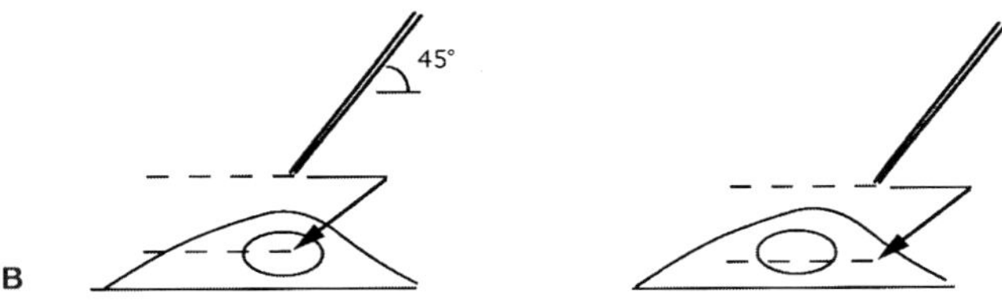

Fig. 10.2A, B. Description of the microinjection system. **A** Scheme of the microinjection system. The system is generally composed of three parts: a micromanipulator that controls movement of the needle in the x, y, and z directions; a transjector that controls microinjection into cells; a pump that controls pressure in the needle. **B** Description of cell microinjection performed by the transjector

in Fig. 10.2: a microscope with an associated micromanipulator, which controls movement of the capillary or the cell support (x, y, and z axis), and a transjector which controls inner needle pressure. Follow the recommendations of the manufacturer, but note the points below:

1. The sample must be centrifuged at high speed before injection to remove aggregates (for example, 12,000 rpm in an Eppendorf centrifuge for 10 min).
2. Since only picoliters of solution are delivered into cells, 1–2 μl of solution is enough to fill the capillary.
3. Before injection, check that the capillary is not blocked. To achieve that, a CLEAR function is included in most systems, that provides high pressure (<1000 bar) in the inner of the capillary.

4. A cell is successfully microinjected when the cytoplasm visually increases or the nuclear gains in contrast transiently.

5. Pump pressure is critical; too high and the cells will be damaged, too low and the capillary contents will not be delivered. The correct pressure depends on cell size, capillary shape and the solution being injected, and should be optimized for each assay. Generally, the pressure required for DNA injection is lower than that necessary for cytoplasmic protein injection.

Growth factor stimulation

After injection, replace the coverslips in the incubator, and allow the cells to recover for at least 30 min if they have been microinjected with protein. A period of 4–6 h is required after plasmid injection to allow for optimal protein expression from the introduced DNA. Then add growth factor, and if DNA synthesis is to be measured, bromodeoxyuridine, and continue incubation for the desired time (typically 16–18 h for fibroblasts).

Analysis of injected cells

The injected cells are identified by immunofluorescence, using an antibody specific to the introduced protein, followed by a secondary antibody conjugated to a dye (e.g., FITC or Texas red). In the case of antibody injection, only the secondary antibody step is needed. When the introduced protein is at too low a level for detection, coinjection with purified nonspecific immunoglobulin (0.5–1 mg/ml) or a plasmid expressing a reporter gene whose protein expression is easily detected (chloramphenicol acetyltransferase or β-galactosidase; Sun et al. 1994), will allow unambiguous detection. To measure the response to growth factor in the injected and noninjected cells, costaining with an anti-bromodeoxyuridine antibody is used (see Fig. 10.3). In very few cases, analysis has been achieved by a biochemical assay. For example, if all the cells on a coverslip are microinjected, then RNA transcripts can be detected by an RT-PCR approach (Barone and Courtneidge 1995), or protein synthesis measured by recording ^{35}S-methionine incorporation (Lane et al. 1993).

Troubleshootings

The three general areas in which problems may arise are: protein injections; the microinjection procedure; and the immunostaining analysis.

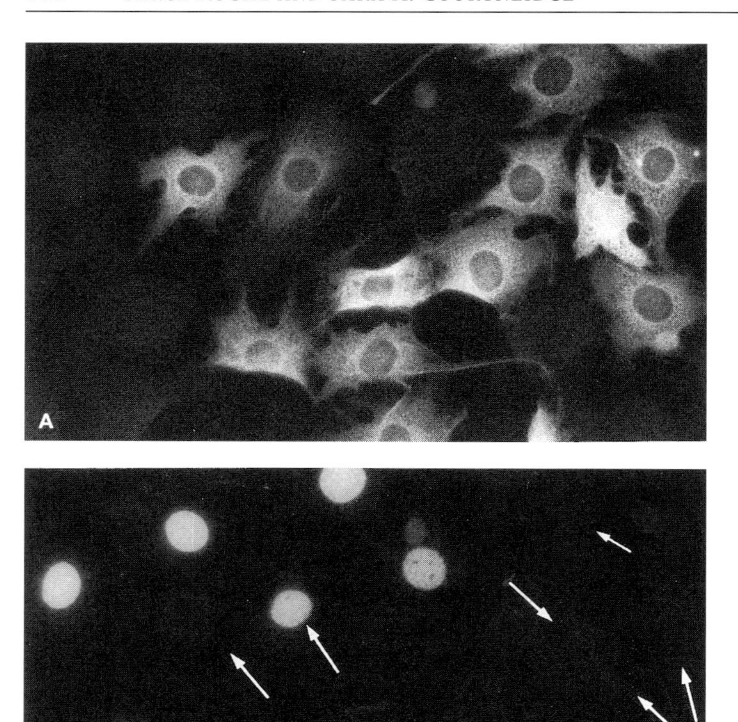

Fig. 10.3A, B. Example of microinjection experiment. Inhibition of PDGF-induced DNA synthesis by injection of the SH2 domain of the tyrosine phosphatase Shp2 in fibroblasts.
Quiescent fibroblasts were injected into the cytoplasm with a protein consisting of the SH2 domain of Shp2 fused to GST, and stimulated with the growth factor PDGF in the presence of BrdU. After 18 h of stimulation, cells were fixed and proccessed for immunostaining. For unambiguous detection of injected cells, purified protein was coinjected with nonimmune rabbit IgG. **A** Detection of injected cells by anti-rabbit coupled to FITC. **B** Detection of BrdU incorporation (as a marker of DNA synthesis) by monoclonal anti-BrdU followed by Texas red conjugated anti-mouse antibody. Injected cells are marked by an *arrow*. Note that cells containing the SH2 domain of Shp2 did not respond to the growth factor as they did not incorporate BrdU, whereas most of the surrounding, noninjected cells did (for further details, see Roche et al. 1996)

- **Protein purification**
  Difficulty may be experienced in affinity purifying the immunoglobulin. Common problems include too harsh elution conditions that denature the antibody, or high affinity antibodies that are not eluted from the affinity column. In our experience, elution from affinity columns with 1 M proprionic acid in the cold, followed by immediate neutralization, is suitable for most antibodies. However, the purification procedure may need to be customized for each antibody. Several suggestions are given in Harlow and Lane 1988. Purified protein may also be lost during the concentration step. Usually, this occurs when working with small amounts of purified protein, and the solution is to start with larger amounts of material.

- **Microinjection**
  The most common problem is blocked needles. This is generally due to protein aggregation or a glass fragment present inside the needle. The following steps avoid the problem:

- Use fresh protein solution whenever possible
- Do not use a sample that has been already thawed several times
- Centrifuge the sample before microinjection
- When sample is stored for a while on ice before injection, repeat the centrifugation just before use
- When the needle contains glass fragments, either use the CLEAR function on the microinjector to remove it, or use a new one
- Occasionally a cell fragment will stick to the capillary. It can be removed by increasing the working pressure
- Unsuccessful microinjection may be due to incorrect positioning of the capillary. Too high, and it will not contact the cell; too low, and it may go through the cell (usually manifest by the appearance of a white spot on the cell surface). In either case, reposition the capillary carefully. Note that the surface of a coverslip is not uniform, and cell volume is also somewhat variable, so the capillary position might need to be readjusted from one cell field to another.

- **Analysis of injected cells**
  Failure to detect injected cells can be for several reasons:

- The DNA solution might be too old or not pure enough. Use a fresh preparation of plasmid purified with cesium chloride
- Plasmid concentration may be too high or too low. Always titrate each batch of plasmid to optimize expression
- The injected protein may be too diluted. Coinject a high concentration of control immunoglobulin for unambiguous detection of the injected cells

- Antibody used for detection of ectopic protein does not work well in immunofluorescence assays. Express an epitope tagged version of the protein. Haemagglutinin, cMyc or FLAG epitopes are particularly appropriate and very good antibodies specific to these antigens are commercially available (12CA5 from Babco, 9E10 from ATCC, and anti-FLAG M2 from IBI respectively)
- The marker has leaked out during the fixation procedure. For example, dextran is frequently used as a marker, yet most is lost when methanol is used for cell fixation. Consider using dextran at 2 %, and paraformaldehyde as a fixative whenever possible.

Finally, when an effect is observed, it is very important to show that it is specific and not due to an artifact of the microinjection procedure. Suitable controls for antibody microinjections include the use of control immunoglobulin, or antibody preblocked with its antigen before microinjection. In the case of plasmid microinjections, several controls are also possible. For example, in the case of dominant negative mutants, the wild-type form of the protein can act as a control.

Remarks

Microinjection is widely used for the analysis of protein function. It has been particularly useful to identify the role of signaling molecules in the cell cycle. Microinjection systems have improved in recent years: they provide precise and rapid injections, strongly reducing killing effects due to the injection procedure. Furthermore, the recent development of fusion protein expression systems for high expression and quick purification makes protein microinjection an attractive approach.

References

Ansorge W, Pepperkok R (1988) Performance of an automated system for capillary microinjection into living cells. J Biochem Biophys Methods 16:283–292

Barone MV, Courtneidge SA (1995) Myc but not Fos rescue of PDGF signaling block caused by kinase inactive Src. Nature 378:509–512

Erpel T, Alonso G, Roche S, Courtneidge SA (1996) The Src SH3 domain is required for DNA synthesis induced by PDGF and EGF. J Biol Chem 271:16807–16812

Frangioni JV, Neel BG (1993) Solubilization and purification of enzymatically active gluthatione S-transferase (pGEX) fusion proteins. Anal Biochem 210:179–187

Harlow E, D (1988) Antibodies. A laboratory manual. Cold Spring Harbor Laboratory Press, Cold Spring Harbor

Lane HA, Fernandez A, Lamb NJC, Thomas G (1993) p70(S6k) function is essential for G1 progression. Nature 363:170–172

McNeil PL, Murphy RF, Lanni F, Taylor DL (1984) A method for incorporating macromolecules into adherent cells. J Cell Biol 98:1556–1564

Mulcahy LS, Smith MR, Stacey DW (1985) Requirement for ras proto-oncogene function during serum-stimulated growth of NIH 3T3 cells. Nature 313:241–243

Pagano M, Pepperkok R, Verde F, Ansorge W, Draetta G (1992) Cyclin A is required at two points in the human cell cycle. EMBO J 11:961–971

Riabowol KT, Vostka RJ, Ziff EB, Lamb NJ, Feramisco JR (1988) Microinjection of fos-specific antibodies blocks DNA synthesis in fibroblasts. Mol Cell Biol 8:1670–1676

Roche S, Koegl M, Courtneidge SA (1994) The phosphatidylinositol 3-kinaseα is required for DNA synthesis induced by some, but not all, growth factors. Proc Natl Acad Sci USA 91:9185–9189

Roche S, Fumagalli S, Courtneidge SA (1995a) Requirement for Src family protein tyrosine kinases in G2 for fibroblasts cell division. Science 269:1567–1569

Roche S, Koegl M, Barone VM, Roussel M, Courtneidge SA (1995b) DNA synthesis induced by some, but not all, growth factors requires Src family protein tyrosine kinases. Mol Cell Biol 15:1102–1109

Roche S, McGlade J, Kones M, Gish GD, Pawson T, Courtneidge SA (1996) Requirement of phospholipase Cγ, the tyrosine phosphatase Syp, the adaptor Shc and Nck for PDGF-induced DNA synthesis: evidence for the existence of Ras-dependent and Ras-independent pathways. EMBO J 15:4940–4948

Rönnstrand L, Mori S, Arridsson AK, Eriksson A, Wernstedt C, Hellman U, Claesson-Welsh L, Heldin CH (1992) Identification of two C-terminal autophosphorylation sites in the PDGFβ-receptor: involvement in the interaction with phospholipase Cγ. EMBO J 11:3911–3926

Sambrook J, Fritsch EF, Maniatis T (1989) Molecular cloning. A laboratory manual. 2nd edn. Cold Spring Harbor Laboratory Press, Cold Spring Harbor

Sasaoka T, Rose DW, Jhun BH, Saltiel AR, Draznin B, Olefsky JM (1994) Evidence for a functional role of Shc proteins in mitogenic signaling induced by insulin, insulin-like growth factor-1, and epidermal growth factor. J Biol Chem 269:13689–13694

Sun H, Tonks N, Bar-sagi D (1994) Inhibition of Ras-induced DNA synthesis by expression of the phosphatase MKP-1. Science 266:285–288

Twamley-Stein GM, Pepperkok R, Ansorge W, Courtneidge SA (1993) The Src family tyrosine kinases are required for platelet-derived growth factor-mediated signal transduction in NIH-3T3 cells. Proc Natl Acad Sci USA 90:7696–7700

Valius M, Kazlauskas A (1993) Phospholipase Cγ1 and phosphatidylinositol 3 kinase are the downstream mediators of the PDGF receptor's mitogenic signal. Cell 73:321–334

Xiao S, Rose DW, Sasaoka T, Maegawa H, Burke TRJ, Roller PP, Shoelson SE, Olefsky JM (1994) Syp (SH-PTP2) is a positive mediator of growth-factor-stimulated mitogenic signal transduction. J Biol Chem 269:21244–21248

Part II

Strategies and Protocols for Microinjection Experiments Using Embryos

Chapter 11

Supply and Husbandry of Mice for Transgenic Science

Jane Morrell*

Introduction

The use of healthy mice kept under good husbandry conditions is of paramount importance for reliable results in transgenic studies. The health status of mice and their husbandry conditions can have a significant impact on experimental results, and therefore must be considered initially when designing the experiment.

The aim of this chapter is to try to indicate which considerations of the supply and husbandry of mice are particularly important for transgenic studies and to suggest improvements which could be made in conventional animal units to try to maintain the health status of transgenic lines. Various people may be able to help with specific problems or advise on the local animal protection regulations etc. These are:

- Laboratory animal veterinarian
- Head of the animal house (may be the same person as above)
- Senior technician in the animal house
- Animal technician responsible for taking care of your mice

This chapter covers the following topics:

- Possible sources and health status of mice
- Housing and care of mice
- Disease surveillance
- Disease

* European Molecular Biology Laboratory, Meyerhofstrasse 1, 69117-Heidelberg, Germany; phone: +49-6221-387580; fax: +49-6221-387306;
e-mail: morrell@embl-heidelberg.de

- Personal safety in the animal house with respect to laboratory animal allergy.

Techniques for vasectomizing males, superovulating females, and embryo transfer have been included since they may be required for rederiving transgenic mice. These techniques are the same as those used in the creation of transgenic mice and will be familiar to many readers already.

General points

The following general points should be noted:

- Wherever possible, the mice should be virus antibody-free (VAF), if not specific pathogen-free (SPF)
- Experimental colonies should be maintained under high standards of hygiene
- A closed colony system is optimal
- Regular health screening should be carried out by a trained microbiologist in conjunction with the institute veterinarian
- Ideally, wild-type mice should be bred on site
- Strict quarantine procedures should apply to all incoming mice and cell lines or material to be injected.

Materials

Supply of Mice

Specific pathogen-free mice

Use of mice from an SPF breeding nucleus to establish the transgenic colony confers several advantages:

- Enables consistent reliable results to be obtained
- Scientific objectives can be achieved quickly
- Disease-related wastage of animals can be avoided
- Disease-related expenditure is minimized
- The size of experimental and breeding colonies can be minimized
- Reproductive efficiency can be maximized.

The disadvantage of using SPF mice is that they are highly susceptible to disease: a complete transgenic line can be lost by the accidental introduction of a pathogen.

Source of mice

Once the desired health status has been decided (SPF or non-SPF), a suitable source for the mice must be considered. The alternatives are to buy the mice or to breed them on site. Most mouse strains are readily available from local commercial suppliers but more obscure strains may be difficult to locate and will be expensive to purchase. Commercial breeders should be able to supply animals of a defined health status.

Advantages of in-house breeding

Breeding wild-type mice on site has several advantages:

- Provides the users with mice which are already acclimatized to the conditions in the animal house
- Allows control over factors such as health status and flexibility of supply, particularly of minority strains

There are also potential disadvantages:

- A breeding colony requires space, equipment, and a certain level of technical expertise
- Female mice will be required in larger numbers than males
- Demand may not be constant throughout the year, creating problems with disposal of unwanted animals.

These problems can be overcome by buying the mice from commercial breeders as required but with the inherent risks of bringing in disease or not finding a supplier of the strains or ages required when needed.

Housing of Experimental Mice

The mice may be housed either under barrier conditions (to maintain a high health status) or conventionally in a normal animal room.

Barrier unit

The "barrier" in a barrier unit is a physical one, consisting of the walls, floor, and ceiling of the unit and the maintenance of a differential air pressure. If the air inside is at a higher pressure than outside, pathogens will be kept out: all SPF units are run at positive pressure. Conversely, if the air inside is at a lower pressure than outside, pathogens will be sucked in and contained. Therefore, it is vital to maintain the unit at a higher pressure than the surroundings.

- Everything entering the unit must be sterilized in some way, either by fumigation, autoclaving or irradiation
- Personnel must not have been in contact with mice for a defined time (varying from 48 h to 1 week depending on local regulations), they must shower as they enter the unit, and put on sterilized clothes.

Undoubtedly, the barrier unit is the most secure way of housing mice in terms of maintaining their health status, but there may be some problems associated with working with transgenic mice in such a unit. Not all studies with transgenic mice are possible under such conditions:

- Access to the animals is limited
- Every person who enters the unit is a risk to the disease-free status
- Cross-breeding experiments with other transgenic lines of a different health status cannot be done in the barrier unit
- Rederivation of contaminated lines into the barrier unit is time-consuming, requiring approximately 3 months before adult animals are available for breeding, as shown in the following flow chart.

Embryo transfer
↓ ≈19 days
Babies
↓ ≈21 days
Weanlings
↓ ≈4–5 weeks
Mature adults
↓ ≈19 days
Babies

Total ≈ 11 weeks from embryo transfer to mature adults.

Conventional unit

Housing mice under conventional conditions interferes less with the conduct of experiments but cannot guarantee the health status of the mice or the validity of experimental results. Unfortunately, mouse pathogens are easily carried by people, in their respiratory tract or on their skin and clothing: therefore housing mice conventionally means that there is a considerable risk of disease entering the unit. Each experimenter is at the mercy of his colleagues, since the actions of one person working in the animal facility may affect all other scientists' results.

Tips

Since many scientists will not have access to a barrier unit, some practical tips which may help to reduce the possibility of contamination occurring in a conventional unit are given below:

- Restrict access to the animal facility (only authorized persons should enter)
- Wear clean protective clothing in the animal room, specific for that room
- Do not wear coats from the laboratory in the animal house
- Wear disposable masks, overshoes, and gloves; consider wearing disposable hats in addition
- Provide a footbath containing a suitable disinfectant, e.g., Virkon, by the entrance to each room
- Ensure that gloves are changed between rooms
- Test all material for potential pathogens before injecting it or keep injected mice in quarantine facilities
- Where possible, equipment should not be moved between animal rooms. In situations where such movement is unavoidable, suitable disinfectants should be used to clean the items thoroughly
- Screen the animals regularly for the presence of pathogens, e.g., according to FELASA recommendation

Containment Facilities

Mini-containment facilities

If access to an SPF facility is not possible but mice of high health status are required, various types of mini-containment facilities are available commercially which allow small numbers of animals to be kept under high standards of hygiene or to maintain SPF status. These include isolators, ventilated cabinets, and individually ventilated cage racks. The potential advantages and disadvantages of each are discussed below.

Fig. 11.1. Plastic film isolator containing a conventional rack for up to 50 mouse cages. Note the plastic sleeves and rubber gloves inserted into the plastic film, through which all manipulations are carried out

Plastic film isolators

Isolators (Fig. 11.1) consist of a plastic film stretched over a stainless steel frame, enclosing a space supplied with filtered air. Conventional racking and cages are contained within the isolator. The cages are serviced and work with the mice performed using thick rubber gloves and plastic sleeves built into the isolator wall. Movement by the operator is therefore somewhat restricted and there is quite an art to picking up mice while wearing thick rubber gloves. Mice can be housed safely and securely in isolators provided that proper procedures are followed for initial sterilization, moving in supplies, and taking out materials. However, all mice in one isolator share the same microenvironment and are exposed to each other's pathogens. Furthermore, mouse husbandry and experimental procedures are time-consuming and laborious. All supplies must be sterilized in some way, for example by irradiation, which increases the cost considerably.

Ventilated cabinets

Ventilated cabinets (Fig. 11.2) consist of a solid structure like a cabinet through which filtered air is pumped from bottom to top. Filter-top cages are placed on the shelves in the cabinet. Such ventilated cabinets effectively limit exposure of the mice to pathogens in the environment provided that the length of time the cabinet door remains open is mini-

Fig. 11.2. An example of a ventilated cabinet, the Scantainer. Note the filter-top cages inside the cabinet and the air-handling unit on the top which draws the air through the cabinet

mized and the filter tops on the cages are removed only in a laminar flow hood which has been sterilized (Fig. 11.3). Access to the mice for procedures is less restricted and husbandry less time-consuming than in an isolator, but is more limited than for conventionally housed animals. However, all the mice in the cabinet share the same microenvironment to some extent, depending on the efficiency of the filter tops.

IVC rack

Individually ventilated cage (IVC) racks (Fig. 11.4) resemble a conventional rack of cages, except that each cage has its own air inlet and outlet system through which filtered air is supplied and air from the cage removed. The advantage over the previous two systems is that each cage has a completely separate air supply and extract, therefore enabling mice of different health status to be kept in adjacent cages without cross-contamination occurring. Servicing the cages and working with the mice must be performed in a laminar flow hood, as for ventilated cabinets. The air flow through the cages is much better than in conventional cages, thus improving the living conditions of the mice.

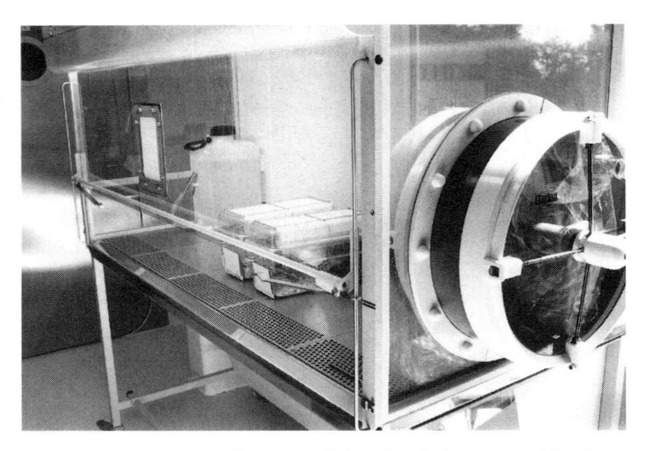

Fig. 11.3. Laminar flow hood (Holten) for use with the Scantainer. The front shield can be raised to the level indicated on the frame without compromising the sterility of the interior. Therefore the cages can be serviced and the mice examined without affecting their health status

Husbandry of Transgenic Mice

The principles of husbandry of transgenic mice are no different to those of nontransgenic mice, with one possible exception: extreme care must be taken in transferring all cage label information with the cage occutants. Transgenic mice of different lines may be phenotypically similar: therefore confusion can occur very easily if care is not taken when changing cages or weaning mice into single-sex groups. If in doubt of the origin of certain mice, the genotyping must be repeated.

Regulations for husbandry of experimental animals

Details of husbandry should comply with local and national regulations concerning the housing and care of animals used in experimental procedures. These regulations are laid out in several documents (see reference list): contact the head of the animal house to find out which regulations are pertinent to the institute. Animals should be looked after by properly trained animal technicians under the supervision or guidance of a laboratory animal veterinarian. The following environmental conditions are recommended for animal houses containing mice in Europe:

- Temperature: 20–24 °C
- Lighting: a defined light:dark cycle is important for reproduction. During the light cycle, the light intensity should not exceed, and should ideally be maintained at, around 350–400 lx (Clough 1982).

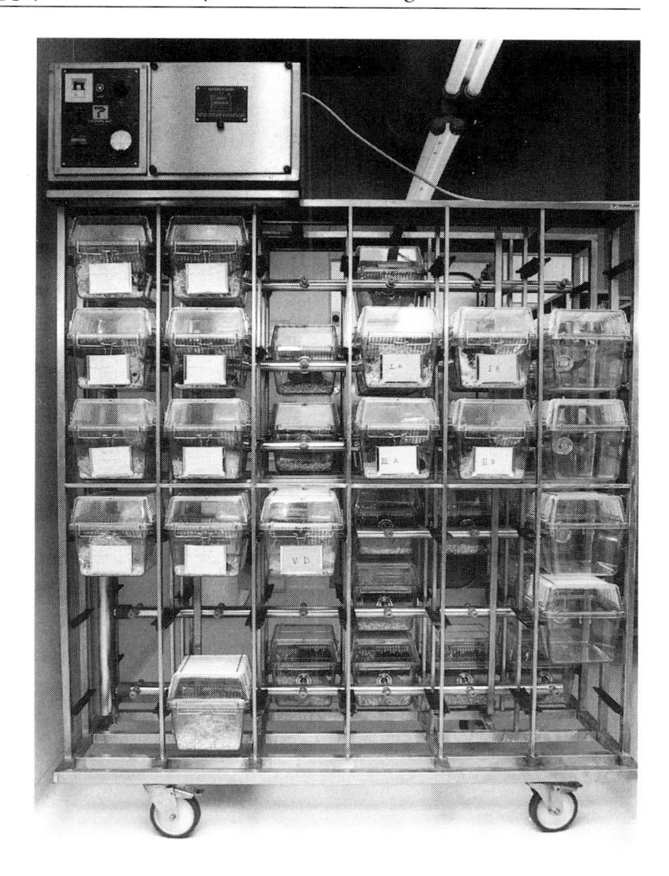

Fig. 11.4. An individually ventilated cage rack (Scanbur). The air-handling unit on top of the rack forces air into, and removes it from, each cage individually. The cages are maintained under positive pressure. Animals of different health status can be kept in adjacent cages without cross-contamination occurring

Albino mice may require protection from the light source. Fluorescent bulbs do not appear to have adverse effects on laboratory mice

- Humidity: 55±10 % is considered optimal
- Air changes: 15–20 air changes per hour are considered appropriate in most countries, unless the stocking density is very low, when fewer air changes may suffice
- Ventilation must be draught-free
- Noise should be kept to a minimum; beware of ultrasounds which are audible to mice (Clough 1982)

- Bedding and nesting material should be dust-free, clean, and prepared from nontreated wood. Products from wood which has been treated with insecticides, etc. or chemicals are unsuitable for use as animal bedding
- Feed should be free from chemical, physical, and microbiological contamination
- A commercially available rodent diet should be fed to ensure a well-balanced diet; stock animals can be given a maintenance quality diet, but breeding mice should be fed a specially formulated higher-protein diet
- Fresh water should be available at all times. It may be necessary to treat the drinking water of mice of a high health status, with either acid or chlorine, to reduce contamination by bacteria, yeasts, or fungi
- Exclusion of vermin and flies is an important feature of disease control

Quarantine facilities

Quarantine facilities should be provided for incoming animals. They serve a dual function: to protect animals already in the unit from disease brought in with the new animals, and to separate incoming animals from diseases which may be prevalent in the unit. Furthermore, quarantine serves to protect staff from zoonoses. The animal technician who takes care of the mice in quarantine should ideally not work with the mice in the main facility but, where it is not possible to have a separate member of staff, the quarantine facility should be serviced last.

Coping with Disease

Disease may interfere with experiments both directly and indirectly. Disease

- Reduces the number of animals which survive
- Causes others to be stunted
- Causes abnormal physiological development
- Reduces reproductive efficiency
- Causes immunological suppression.

Although some sick animals can be treated, they may represent a disease hazard to other mice. Furthermore, treatment may interfere with reproduction or with the experiment; for example, exposure to organophosphates, which are used as insecticides, e.g., for control of mites, may affect the results of embryo transfer (Morrell et al. 1995). Therefore strategies for disease eradication may be necessary to safeguard the health of the colony as a whole. Two possibilities may be considered:

- Rederivation of transgenic lines into SPF foster mothers in a clean facility, by embryo transfer rather than by hysterectomy, since some viruses may cross the placenta.
- The "fire-break" principle, i.e., the virus dies out if there are no new hosts to infect. The presence of susceptible animals is minimized by killing all babies born over a 6-week period, while prohibiting entry of other animals into the unit (Harkness and Wagner 1983).

Health screening

Since clinical and subclinical disease can have a profound effect on experimental results, it is vital to monitor the health status of the colony at regular intervals. One example of a disease causing havoc with transgenic studies is that of mouse hepatitis virus, which is found very frequently in transgenic colonies, and may interfere drastically with the immune and reproductive systems (Harkness and Wagner 1979; Kraft 1982). Other important mouse pathogens have similar drastic effects. If the presence of an infection is detected either serologically or through diagnosis from disease symptoms, the experimental findings should be interpreted accordingly.

FELASA guidelines

Health screening of experimental mouse colonies should ideally follow the guidelines for breeding colonies set out by the Federation of European Laboratory Animal Science Associations (FELASA) (FELASA working group 1994) as indicated below:

Viral, bacterial, mycoplasmal, and fungal infections should be monitored as follows:

- Screen at least 10 mice every 3 months (at least 2 weanlings, 4 young adults, and 4 mice older than 6 months)
- Where more than one mouse strain is present, the strains should be monitored successively
- Each strain should be monitored at least once per year.

Serological monitoring

Monitoring for the following viral infections should be carried out every 3 months using one or more of the techniques indicated. (**Note:** ELISA = enzyme linked immunosorbant assay; HI = hemagglutination inhibition; IFA = immunofluorescent assay)

• Minute virus of mice	ELISA, HI, IFA
• Mouse hepatitis virus	ELISA, IFA
• Pneumonia virus of mice	ELISA, HI, IFA
• Reovirus type 3	ELISA, IFA
• Sendai virus	ELISA, HI, IFA
• Theiler's murine encephalomyelitis virus	ELISA, HI, IFA

In addition, monitoring for the following rare infections should be carried out once per year or in rederived colonies:

• Ectromelia virus	ELISA, IFA
• Hantavirus	ELISA, HI, IFA
• Lactate dehydrogenase elevating virus	LDH plasma test
• Lymphocytic choriomeningitis virus	ELISA, IFA
• Mouse adenovirus	ELISA, IFA
• Mouse rotavirus	ELISA, IFA
• Mouse K virus	ELISA, HI
• Mouse polyoma virus	ELISA, HI, IFA
• Mouse thymic virus	ELISA, IFA
• Mouse cytomegalovirus	ELISA, IFA

ELISA or IFA kits can be purchased from a variety of sources (see list of suppliers) or commercial companies will perform the tests if sent either the live animal or serum.
Bacterial, mycoplasmal and fungal infections to be monitored by culture of the organism

- *Bordetella bronchiseptica*
- *Citrobacter freundii* (4280)
- *Corynebacterium kutscheri*
- *Pasteurella* spp.
- Salmonellae
- *Streptobacillus moniliformis*

- Streptococci-β-haemolytica
- *Streptococcus pneumoniae*

Bacterial and mycoplasmal infections to be monitored by serology

- *Leptospira*
- *Mycoplasma* spp. (followed by culture if positive)
- Tyzzer's disease (plus clinical signs, pathological lesions, histology)

Safety in the Animal House

Laboratory animal allergy

The incidence of allergies to laboratory animals appears to be increasing, particularly among personnel having only occasional contact with laboratory rodents. The allergens are contained in the animals' saliva, urine, and feces, and are spread on skin and hair by grooming activities. To protect yourself and your colleagues, always observe the following points.

- Wear disposable gloves at all times when handling animals or their cages
- Change gloves if torn
- Wear a face mask which protects against particles of an appropriate size (your institute safety officer should advise you)
- Wear only the protective clothing specifically provided for the animal house
- Handle animals correctly to avoid being bitten or scratched. The animal house veterinarian or senior animal technician will be glad to instruct you
- Do not take animals or cages outside the animal house unless contained within a suitable transport unit, e.g., a ventilated cabinet
- Do not wear your laboratory coat in the animal house
- Do not wear protective clothing from the animal house in the laboratory
- Wash your hands thoroughly after working with animals, particularly before touching your face or applying cosmetics

- Consider wearing a respirator as a precautionary measure
- Do not work alone in the animal house, including at night.

If you have mild symptoms of laboratory animal allergy, do not ignore them! Accidental exposure to allergens can trigger life-threatening anaphylaxis. Consider the following points:

- Be allergen-tested to discover the likely cause of symptoms
- Avoid working with laboratory animals as far as possible, particularly any to which you show a strong sensitivity in skin tests
- Ensure that your colleagues work safely, following the guidelines indicated above
- Wear a respirator if you must enter the animal house
- Inform the head of the animal house and the safety officer of your problem
- Ensure that there is a supply of adrenaline in the animal house and that there are at least two people present who know how to administer it.

Animals Needed for Transgenic Work

Wild-type mice

Mice of various ages will be needed for creating transgenic founders and for establishing a transgenic line. The strain or hybrid used depends to some extent on the wishes of individual scientists and on the nature of the problem to be studied by transgenic science. Whichever strain is chosen, several categories of mice will be needed as shown in the flow chart (Fig. 11.5).

These mice are:

- Juvenile mice, aged 21–28 days depending on strain (see note), for embryo donors
- Adult stud males
- Adult females to act as recipients for embryo transfer
- Vasectomized adult males to create pseudopregnant females
- Young adult wild-type males and females for mating with founders to establish the transgenic lines.

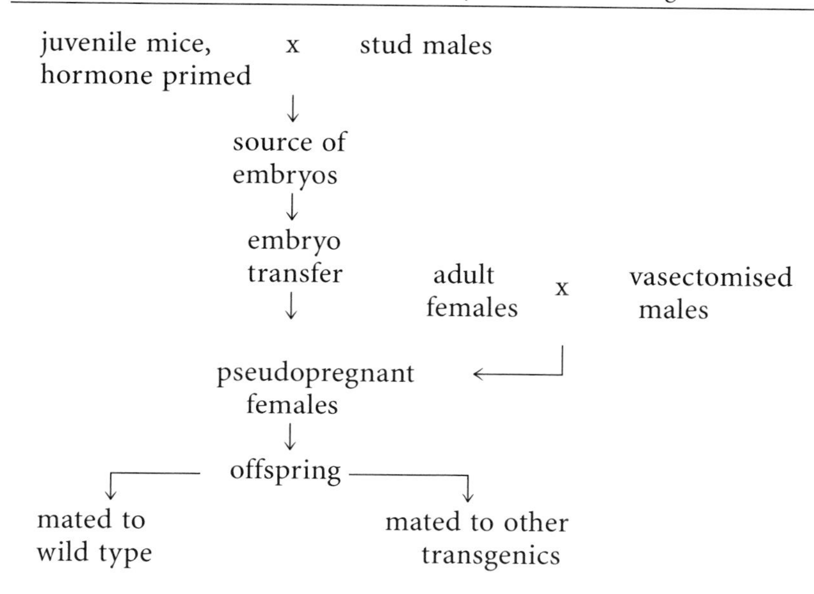

Fig. 11.5. Flow chart indicating which wild-type mice are required to make and breed transgenic mice

Note: The success of superovulation depends on the age of the juvenile mice, their strain and health status. While hybrid juvenile mice are usually capable of mating from about day 24, mice of some inbred strains may be too small to mate with adult males at this age without injury; therefore the age at which mice of a specified strain or hybrid are superovulated should be chosen according to the physical size and physiological development of that strain.
Note: Adult females can also be used as embryo donors but the number of embryos recovered will be less.

11.1 Preparation of Embryo Donors

Procedure

Juvenile mice are used, usually 21–28 days old according to strain. They should be acclimatized to the same light:dark cycle as the males. The following superovulation protocol works well for a 12-h light, 12-h dark cycle where the lights change at 07.00 and 19.00 h.

1. Administer 5 IU pregnant mare serum gonadotrophin (PMS) by intraperitoneal injection between 15.00 and 17.00 h.
2. Approximately 46 h later, administer 5 IU human chorionic gonadotrophin (hCG) by intraperitoneal injection.
3. Put to mate overnight with fertile males.
4. Check for the presence of vaginal plugs early the following day (use only those animals which have mated).

This superovulation protocol will produce approximately 30–60 embryos per juvenile female, depending on strain.

Adult females may be hormone-primed using the same protocol resulting in synchronization of ovulation rather than superovulation. Only mature follicles will ovulate rather than additional preovulatory follicles developing; therefore the recovery rate is usually 10–15 embryos per female, depending on strain.

Note: The timing of hCG administration should occur before the animal's endogenous luteinizing hormone (LH) surge occurs. The optimal timing may therefore be strain-dependent and should be tested in a preliminary study if working with an unusual strain or hybrid.

11.2 Preparation of Vasectomized Males

Males should be vasectomized well in advance of being required, so that test matings can be carried out to check for the effectiveness of the operation. Vasectomy can be carried out at about 6 weeks of age, but older males may be used provided that well-developed blood vessels are securely ligated during surgery. Consult the institute veterinarian for advice on aseptic technique, any authorization required to perform surgery, peri- and postoperative care, etc. The Universities Federation on Animal Welfare (UFAW) produces a set of guidelines on the care of laboratory animals and their use for scientific purposes, parts II and III of which deal with pain, analgesia and anesthesia, and surgical procedures, respectively.

Requirements

- Authorization from the official body responsible for regulating animal experiments
- Sterilized surgical instruments (round-ended forceps, scissors, needle holders)

- Resorbable suture material size 5/0 metric
- Warm lamp, warming chamber, or under-cage heating pad
- Anesthetics.

Anesthetics

Suitable anesthetics for use in mice include the following:

- Hypnorm and Midazolam (Hypnovel): mixture of 1 part hypnorm with 1 part water for injection and 1 part Midazolam with 1 part water for injection; give 7–10 ml/kg body weight (0.2–0.3 ml depending on weight of mouse) by intraperitoneal injection (Flecknell 1987). Anesthesia lasts 20–40 min.
- Tribromethanol (stock consisting of 1 g 2,2,2-tribromoethanol dissolved in 1 ml 2-methyl-2-butanol; add 0.25 ml stock solution to 9.75 ml phosphate buffered saline to prepare solution for injection. Administer 0.17 ml diluted solution per 10 g body weight (Hogan et al. 1986). Anesthesia lasts about 20 min.
- Methoxyfluorane administered by a face mask: anesthesia lasts only for the period of inhalation (Flecknell 1987).

Consult the institute veterinarian for advice on which anesthetics are available for use and prospective suppliers.

Procedure

Two Surgical Approaches are Possible for Vasectomies (Fig. 11.6)

Protocol 1: scrotal route

1. Weigh mouse.
2. Anesthetise the mouse.
3. Prepare the operation site by clipping away hair and cleansing the skin with a suitable disinfectant e.g., 70 % ethyl alcohol, chlorhexidine, etc. (UFAW guidelines on surgical procedures 1989b).
4. Apply gentle pressure to the lower abdomen to push the testes into the scrotal sac.
5. Make a midline incision in the scrotum.
6. Incise the membrane over the testis (tunica vaginalis).
7. Insert forceps medial to the testis and pull out a section of the vas deferens.

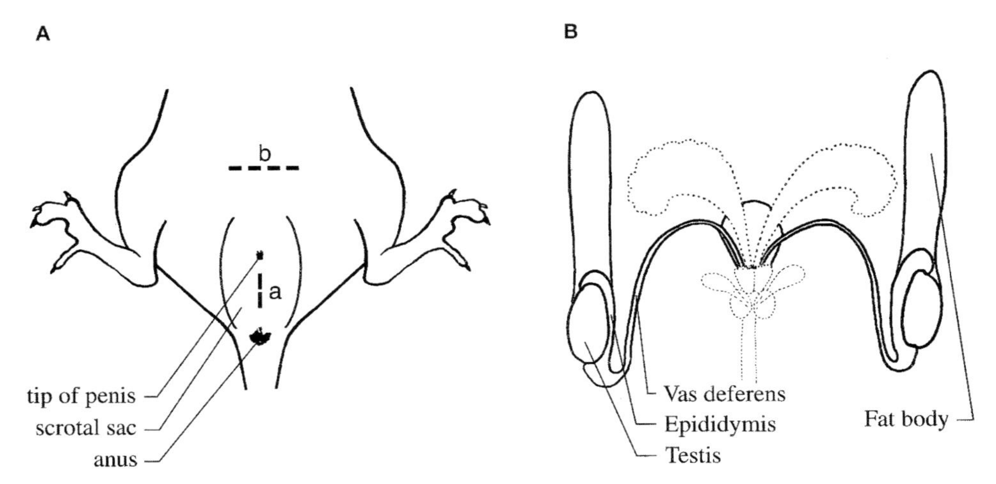

Fig. 11.6A, B. Diagram to show: **A** possible incision sites for vasectomy of mice, where a = scrotal route and b = abdominal route; and **B** the layout of the male reproductive tract showing the relationships of the various organs. Organs *outlined with a dotted line* are accessory sexual organs which are not relevant to the vasectomy procedure

8. Ligate well-developed blood vessels accompanying the vas.
9. Cut through the vas deferens at two sites, approximately 1–2 cm apart.
10. Remove section of vas.
11. Return any remaining tissue to its original place.
12. Incise second tunica vaginalis
13. Pull out vas deferens of second testis.
14. Repeat as for first vas.
15. Suture the incisions in the tunica vaginalis and scrotum.
16. Place mouse under a warm lamp or use a heating pad under the cage until consciousness has returned.
17. Check daily: the surgical wound should heal quickly with no indication of sepsis or herniation of scrotal contents.
18. Remove skin sutures or wound clip after approximately 7 days.

Protocol 2: abdominal route

1. Initial anesthesia and preparation as for protocol 1.
2. Make a transverse incision in the skin of the lower abdomen.
3. Incise the abdominal musculature and peritoneum.
4. Insert forceps to the right of mid line and pull up the fat pad which cushions the testis.
5. The vas deferens can be exposed by bringing the testis and associated tissue out of the peritoneum.
6. Ligate or cauterize well-developed blood vessels.
7. Cut the vas in two places and remove section of vas.
8. Return testis to the abdomen.
9. Insert forceps to the left of mid line and pull up second fat pad.
10. Repeat procedure on second vas deferens.
11. Appose peritoneum and muscle layers with sutures.
12. Use skin sutures or apply a wound clip.
13. Place mouse under a warm lamp or use a heating pad under the cage until recovery from anesthesia is complete.
14. Postoperative care as for protocol 1.

Note: The advantage of using the scrotal route is that potential tissue damage is less than for the abdominal route, resulting in faster wound healing. A potential disadvantage is trauma to the wound inflicted by cage mates: the perineal area is often the chosen site for such an attack.

11.3 Sterility Testing

It is important that vasectomized males should be sterile. Unintended pregnancies with wild-type embryos may result if males are mated too soon after vasectomy, since a few sperm may have survived in the remaining part of the vas. Rabbits and humans have been known to be fertile up to 21 days following vasectomy: a similar situation may occur in the mouse. Uncommonly, the sectioned vas may rejoin and be patent. The males can be shown to be sterile in test matings using one of several methods depending on the facilities and time available.

Procedure

Sterility test 1

1. Place a superovulated female with each male overnight.
2. Harvest the embryos from mated females and put in culture medium in the incubator.
3. Check for embryo development beyond the two cell stage.

Mating with sterile males should result in unfertilized eggs which do not develop into embryos.
Note: Occasionally, an egg may start to develop parthenogenetically: therefore one egg proceeding beyond the two-cell stage does not necessarily mean that the vasectomy was unsuccessful. The male should be test-mated again if there is any doubt concerning his sterility.

Sterility test 2

1. Put an adult female in with each male.
2. Check for plugs.
3. Separate female from male when plug found.
4. Record plugging date.
5. Examine females regularly approximately 10 days after plugging and look for signs of pregnancy.
6. The males which have produced pregnancies should be retested or culled.

Note: Procedure 1 provides a result in a much shorter time than procedure 2 (2 days versus 2 weeks), but depends on the presence of somebody with the technical expertise to harvest and culture embryos properly.

11.4 Preparation of Pseudopregnant Recipients for Embryo Transfer

Type of mouse for use as a recipient

The choice of strain for recipients varies according to the personal preference of the experimenter. Many people prefer to use the same strain of mouse for embryo donor and recipient to avoid problems of immune incompatibility between embryo and uterus, while others choose to use outbred mice which show good mothering characteristics, e.g., CD1. The consensus of opinion seems to be that embryo donors and recipients should have some similarity in their genetic background for maximum implantation efficiency.

Preference for a particular age of recipient also varies: some people prefer to use young adults around 8 weeks old, while others prefer to use females which have already successfully reared a litter. The former usually have less fat around the reproductive tract than the latter and less well-developed blood vessels, thus facilitating manipulation of the ovary and uterus and better visualization of the operation site.

Procedure

Pseudopregnant females

1. Calculate which day females are needed, depending on the developmental stage at which embryos will be transferred and the site of embryo transfer, e.g., day 0.5 for transfer of one, two, or more cell stage embryos into the oviducts, day 1.5 or 2.5 for uterine transfer of morulae or blastocysts, counting the day of plugging as day 0.5.
2. Synchronize females to be in estrous for the specified night, using male odor ("third night effect"), i.e., 2 days before females are put in with males they should be exposed to bedding from males' cages. This has the effect of inducing estrous on the third night.
3. Late in the afternoon put females in with vasectomized males.
4. Check for vaginal plugs early next morning.
5. Remove all females after plug checking, keeping plugged females separate.
6. Females which have not mated can be put in with a male again after 48 h.

Notes:

- Do not overuse vasectomized or stud males. Two, or at most three, opportunities to mate per week, preferably with a night's rest in between, is optimal
- Females should be acclimatized to the same light:dark cycle as the males
- Do not interrupt the dark cycle even for periods of a few seconds. Use a red light if the room must be entered during the dark phase

- Removal of a female which has not plugged in the morning tends to promote more male interest in a female placed in the cage later the same day.

11.5 Embryo Transfer

Most people working with transgenic mice are familiar with various techniques for embryo transfer. Although two techniques are described below, it is always better (and, in some countries, a legal requirement) to be taught by an experienced person before attempting them yourself.

Requirements

- Authorization from the official body responsible for regulating animal experiments
- Suitable training in aseptic surgical technique, anesthesia, pre-, peri- and post-operative care (Sect. 11.2)
- Pseudopregnant adult females at a suitable time after mating to a vasectomized male (see section 11.4)
- Sterilised surgical instruments (two pairs fine iris forceps, two pairs iridectomy scissors, needle holders)
- Transfer pipettes loaded with embryos (Hogan et al. 1986)
- Resorbable suture material size 5/0 metric
- Warm lamp, warming chamber or under-cage heating pad
- Anesthetics
- Cold light source
- Dissecting microscope (for transfer into oviducts).

Procedure

Embryo Transfer Into the Oviducts

This technique is suitable for early-stage embryos, from one- or two cell stage right up to early morulae.

1. Weigh mouse.
2. Anesthetize using a suitable anesthetic agent (see Sect. 11.2).
3. Place mouse in left lateral recumbency on a suitable sterile surface, e.g., a petri dish.

4. Remove hair and clean skin (Sect. 11.2) over left flank, midway between last rib and pelvis.
5. Incise skin in the center of prepared site.
6. Incise muscle layers and peritoneum.
7. Insert round-ended forceps into incision and pick up fat pad attached to ovary.
8. Gently draw out fat pad and ovary.
9. Attach a tissue clamp to the fat pad and place the fat pad flat on the mouse so that the ovary is uppermost.
10. Place animal under the microscope.
11. Either tear the bursal membrane covering the ovary using fine forceps or cut the membrane with small scissors.
12. Insert the tip of the loaded transfer pipette into the infundibulum and blow gently to release the contents into the oviduct.
13. Release fat pad from the tissue clamp and return ovary plus associated tissue to the peritoneal cavity.
14. Suture peritoneum and muscle layers.
15. Suture skin incision or apply wound clip.
16. Repeat procedure on right flank.
17. Place mouse under warm lamp to recover from anesthesia.

Transfer Into the Uterus

This technique is suitable for morulae and blastocysts.

1. Weigh mouse.
2. Anesthetize mouse using a suitable anesthetic agent (see Sect. 11.2).
3. Place mouse in left lateral recumbency on a suitable sterile surface, e.g., a petri dish.
4. Remove hair and clean skin over left flank, midway between last rib and pelvis.
5. Incise skin in the center of prepared site.

6. Incise muscle layers and peritoneum.
7. Insert round-ended forceps into incision and pick up fat pad attached to ovary.
8. Gently draw out fat pad and ovary.
9. Attach a tissue clamp to the fat pad and place the fat pad flat on the mouse so that the ovary is uppermost.
10. Grasp the uterine horn at the top near the junction with the oviduct.
11. Make a small hole in the uterine wall with a 25-g needle about one third of the way down the uterine horn.
12. Insert the tip of a transfer pipette into the hole and blow gently to release the contents.
13. Release fat pad from the tissue clamp and return ovary plus fat pad to the peritoneal cavity.
14. Suture peritoneum and muscle layers.
15. Suture skin incision or apply wound clip.
16. Repeat procedure on right flank if desired (see note).
17. Place mouse under warm lamp to recover from anaesthesia.

Note: Many people prefer to transfer the embryos into only one uterine horn, relying on transuterine migration to carry embryos into the other uterine horn for implantation. Others achieve a better implantation rate by making a transfer into each uterine horn.

Comments

Care of Pregnant Recipients

- Ideally, recipients should be caged in pairs or small groups, but this may only be possible if more than one female received embryos with the same genetic manipulation. The advantage is that the offspring are cared for by all foster mothers and may have a better chance of survival than if only one mother is present. The disadvantage is that it is not possible to trace which babies came from each recipient.

- Wound clips or skin sutures should be removed at an appropriate time after surgery: typically, after 7 days. Healing of the surgical incision is usually complete after 4 days in mice but many people question the wisdom of handling females at around the time of embryo implantation. Waiting until 7 days after transfer to remove remaining clips or sutures does not seem to cause any problems in terms of resorption of the fetuses.
- Nesting material should be provided at least in the last few days of pregnancy, but preferably for the entire gestation period. Disturbance of the nest and the embryo recipients themselves should be avoided. It is very tempting for scientists to want to check on the progress of the pregnancy, but too frequent handling of the mother may cause resorption in utero or rejection of the litter after birth. Housing the mice in transparent cages may facilitate examination of the mouse and offspring without removing them from the cage.
- With some strains of mice it is advisable not to clean out the cage for 1 week after the birth of the offspring. This factor should be considered when deciding how many embryo recipients can be housed together in cages of a given size.
- Wean the young mice at an appropriate age: siblings mate together (or with their foster mother) if sexually mature!

11.6 Breeding to Homozygosity

Once a transgenic mouse has been produced, it will usually be desirable to create a transgenic line if germ-line transmission occurs.

Procedure

Creation of the transgenic line can be done in the following manner:

1. Mate the transgenic mouse to a wild-type mouse of the desired genetic background.
2. Test the offspring to identify which, if any, are carrying the transgene.

3. Mate heterozygote offspring together, and test the offspring: 50 % will be heterozygous, 25 % wild-type, and 25 % homozygous unless the homozygous genotype is lethal.

4. Mate the homozygote either:
   a) with another homozygote to produce offspring which are all homozygous, or
   b) with a wild-type mouse to produce all heterozygous offspring for further expansion of the line.

References

Clough G (1982) Environmental effects on animals used in biomedical research. Biol Rev 57:487–523

FELASA Working Group (1994) Recommendations for the health monitoring of mouse, rat, hamster, guinea pig and rabbit breeding colonies. Lab Anim 28:1–12

Flecknell PA (1987) Laboratory anaesthesia. An introduction for research workers and technicians. Academic Press, London

Harkness JE, Wagner JE (1983) The biology and medicine of rabbits and rodents. 3rd edn. Lea & Febiger, London

Hogan B, Constantini F, Lacy E (1986) In: Manipulating the mouse embryo: a laboratory manual. Cold Spring Harbor

Kraft LM (1982) Viral diseases of the digestive system, chap 9. In: Foster HL, Small JD, Fox JG (eds) The mouse in biomedical research, vol II. Academic Press, London, pp 173–183

Morrell JM, Gore MA, Nayudu PL (1995) Retrospective observation of the effect of accidental exposure to organophosphates on the success of embryo transfer in mice. Lab Anim Scie 45:437–440

UFAW (1989a) UFAW guidelines on the care of laboratory animals and their use for scientific purposes. II Pain, analgesia and anaesthesia. UFAW, Potters Bar

UFAW (1989b) UFAW guidelines on the care of laboratory animals and their use for scientific purposes. III Surgical procedures. UFAW, Potters Bar

Home Office, 1986 Animals (Scientific Procedures) Act 1986, HMSO, United Kingdom.

Canadian Council on Animal Care (1984) Guide to the care and use of experimental animals: 2. Ottawa: Canadian Council on Animal Care.

Home Office (1989) Code of practice for the housing and care of animals used in scientific procedures. HMSO, UK.

Council of Europe (1986) European convention for the protection of vertebrate animals used for experimental and other scientific purposes. Strasbourg: Council of Europe.

US Animal Welfare Act (1970) 9 CER Part 3.

Glossary

Acclimatized: the animal has lived in the animal unit long enough to be accustomed to the conditions there
Animal protection regulations: regulations governing the husbandry, care, and use of vertebrate animals in scientific experiments
Barrier conditions: animal unit where every precaution is taken to prevent the entry of disease.
Closed colony: an animal population which is self-sustaining and where no animals from other colonies are allowed to enter
Contaminated lines: transgenic lines infected with a pathogen
Conventional facility: animal unit where few or no precautions are taken to prevent the entry of disease
Fire-break principle: removal of all susceptible animals from a population to prevent further spread of disease
Health screening: gross and microbiological examination of the animal to detect the presence of disease or potential pathogens
Herniation: protrusion of part of the intestine through an opening in the body wall
Human chorionic gonadotrophin (hCG): hormone derived from the urine of women in early pregnancy, which has luteinizing activity
Hysterectomy: surgical removal of fetuses from the uterus immediately prior to birth
Laboratory animal allergy: allergic reactions in humans to proteins secreted in the urine, faeces, saliva, etc. of laboratory animals
Laboratory animal veterinarian: a veterinarian specialized in laboratory animal science and the care of animals used in experimental procedures
Ligate: to tie off, e.g., blood vessels
Micro-environment: the immediate environment, i.e., in this context, the environment within the mouse cage
Ovarian bursa: fluid-filled sac containing the ovary and anterior portion of the oviduct (infundibulum)
Pathogens: disease-causing organisms
Pregnant mare serum gonadotrophin (PMS): hormone derived from the serum of pregnant mares which has follicle-stimulating activity
Pseudopregnancy: physiological state produced by mating to a sterile male (in mice)
Quarantine: period of time where new animals are kept away from the existing population until it can be shown that no disease problems are present

Rederivation: a process whereby embryos or fetuses are removed from one mother and transferred to a foster mother of a superior health status
Specific pathogen-free (SPF): status declared when no evidence of particular pathogens is found. These pathogens should be named
Transuterine migration: movement of embryos from one uterine horn to the other before implantation
Vas deferens: tube which carries sperm away from the epididymis
Virus antibody-free (VAF): no antibodies to viruses are detectable in the animal's serum
Zoonoses: disease transmitted from animals to man

Suppliers

- Anesthetics
  - Hypnorm: Janssen, Postfach 210440, 41470 Neuss, Germany
  - Midazolam, e.g., Hypnovel: Roche, Welwyn Garden City, UK
  - Methoxyfluorane Metofane Janssen (see above)
  - Tribromethanol and 2-methyl-2-butanol: Aldrich, Steinheim, Germany
- Animals
  - Biotechnology and Animal Breeding Division, BRL Biological Research, Wolferstrasse 4, CH-4414 Füllinsdorf, Switzerland
  - Charles River, e.g., Sandhofer Weg 7, D-97633 Sulzfeld, Germany
  - Harlan, e.g., Harlan UK Ltd., Shaw's Farm, Blackthorn, Bicester, OX6 0TP, UK
  - Jackson Laboratories (contact directly or through Charles River)
- Bedding and nesting material, e.g., aspen granules and wood wool from Tapvei Oi, 73620 Kortteinen, Finland
- Cold light source, e.g., Schott, obtainable from Buddeberg GmbH, Laborbedarf Optik, Markircherstr. 15–17, D-68229, Mannheim
- Diet, e.g., Special Diet Services, Boxmeer, Netherlands
- Disinfectants, e.g., Virkon, Antec International, Sudbury, Suffolk, UK
- Dissecting microscope, e.g., Carl Zeiss, Postfach 4041, D-37030 Goettingen, Germany

- Health monitoring (commercial laboratories performing health monitoring on mice in Europe), e.g., BRL, Charles River (see Animals), also Microbiological Laboratory Ltd., Northumberland Road, North Harrow, UK

- Hormones

- Pregnant mare serum gonadotrophin, e.g., Folligon: Intervet Laboratories Ltd., Science Park, Milton Road, Cambridge CB4 4FP, UK
- Human chorionic gonadotrophin, e.g., Chorulon: Intervet (see above)

- Laminar flow hood, e.g., Holten Industrial Group, Gydevang 17–19, DK-3450 Allerod, Denmark

- Minicontainment facilities

- Isolators, e.g., Harlan Isotec (see Harlan UK above)
- Ventilated cabinet, e.g., Scanbur, Gl. Lellingegard, Bakkeleddet 9, Lellinge, DK-4600 Koge, Denmark
- Individually ventilated cage rack, e.g., Scanbur (as above)

- Suppliers of kits for serological testing, e.g., Charles River, Harlan (see Animals)

- Surgical instruments, e.g., Holborn Surgical Ltd., Ramsgate Road, Margate Kent, UK

- Suture material, e.g., Vicryl: Ethicon GmbH & CO.KG, D-22851 Norderstedt, Germany

- Wound clips, applicator and remover, e.g., Clay Adams: Becton Dickinson and Company, 7 Loveton Circle, Sparks, MD 211252–0370. USA.

Chapter 12

Mouse Male Germ Cells in Culture: Toward a New Approach in Transgenesis?

MINOO RASSOULZADEGAN[1]*, JULIEN SAGE[1], AND VALÉRIE GRANDJEAN[1]

Introduction

To extend and diversify our methods for the manipulation of mammalian genomes, transfer of genes into germ cells might constitute a useful alternative to methods based on the use of fertilized zygotes, early embryos, or ES cells. These powerful techniques are time consuming and costly. Moreover, they are applicable only to a limited number of species. The search for alternative methods appears therefore as a worthwhile long term prospect. Since the function of germ cells is to transmit genes to successive generations, why not transfer DNA molecules directly into germ cells before fertilization? Very little information is available as to the ability of any genetic material, viral or cellular, to enter germ cells under natural conditions. If, during evolution, new genes may have been occasionally introduced into mammalian germ cells by retrovirus infection, this horizontal transfer has probably been a rare event. One experimental model is provided by a strain of laboratory mice (SWR/J), where efficient provirus acquisition in the female germline has been reported (Jenkins et al. 1985; Panthier et al. 1988).

DNA fragments can be introduce into a variety of cell types in culture. Two critical requirements have to be separately considered for any given cell type. Cell cultures must be established without the irreversible loss of differentiated properties that is often associated with immortalization. The transferred DNA fragment must be stably maintained and selectable genes expressed in order to amplify the clones of interest. In the case of germ cells, both are still largely problematic.

In this chapter, we will discuss the present state, the limitations and possibilities offered by the manipulation of mammalian germ cells, con-

* corresponding author: phone: +33 492 07 64 11; fax: +33 492 07 64 02; e-mail: Minoo@unice.fr
[1] Unité 470 de l'INSERM, Faculté des Sciences, Université de Nice, France

sidering only to the mouse, the species for which transgenic technology has been first established. We will also consider only the male germ line, which is clearly more suitable than the female for gene transfer studies. Studies in the female would be made more difficult by the facts that the first stages, up to the entry into meiosis, occur during embryonic life, that the number of germ cells is limited, and that completion of meiosis occurs upon fertilization.

Strategies should be aimed at introducing genes into diploid precusors. Haploid male cells undergo profound changes in the structure of their genomes, which ultimatly evolve into a transcriptionnally silent compacted form. It is not clear to what extent this is compatible with the integration of a transgene. It would, therefore, seem more adequate to introduce the constructs of interest into premeiotic precursors, to incubate the transfected cells under conditions that would allow meiosis to proceed, either in vivo or in culture, and to eventually recover the genetically transformed haploid cells in a state competent for fertilization.

Development of Male Germ Cells

To produce healthy spermatozoa in large amounts, a long developmental process is required, with a number of stages and successive rounds of cell amplification. The first amplification, from about 100 to 25,000 cells, occurs in utero during the migration of the earliest progenitors from the epiblast to the genital ridge (McLaren 1992, 1995). These primordial germ cells have been cultured (Buehr and McLaren 1993; De and McLaren 1983) and maintained in culture (Matsui et al. 1992). However, the resulting EG lines resemble ES cells and upon their injection into blastocytes behave just as a ES cells, giving rise to differentiated somatic cells of various lineage rather than being committed to evolve into germ cells. From day 12.5 on, cells recognized as male gonocytes are present in the seminiferous tubules of the fetal testis. The structure of the testis is then progressively formed by association with other cell types. Sertoli cells, present in the genital ridge at day 10.5, are the first to be associated with germ cells. Then the Leydig and peritubular cells migrate from the mesonephros to the fetal gonad (Buehr et al. 1993). From that stage until several days after birth, the male gonocytes will remain quiescent. Further studies are required to understand the process of the release from this block. Several days after birth, the gonocytes will multiply extensively and differentiate into spermatogonia. While some spermatogonia will remain as stem cells during all of adult life, their progeny will, after mitotic cycles, enter the meiotic differentiation pathway. The first and synchronous wave of meiosis starts from primary spermatocytes at about day 9 after birth, to produce the first haploid product, the round spermatid, at about day 18 after birth. These cells will, in turn, undergo spermiogenesis, generating the morphologically and functionally mature spermatozoa. In the mouse, the time from the first spermatogonial division to the release of a mature spermatozoon is 35 days, with roughly one-third of the time spent in spermatogonial mitosis, one-third in meiosis, and one third in spermiogenesis (reviewed in Russell et al. 1990).

Architecture of the Testis: Morphological and Functional Aspects

The sophisticated cytoarchitecture of the testis is a key feature of germinal differentiation and, at the same time, one of the main difficulties in devising experimental approaches. Its basic structure is the seminiferous tubule, which encloses the germ cells at meiotic and post-meiotic stages, tightly associated with a somatic component, the Sertoli cell. Sertoli cells form a continuous barrier, impenetrable to the blood and lymph, which closes around day 15 after birth. Peritubular cells form a monolayer around the basal membrane of the tubule, while the Leydig cells remain in interstitial space. Throughout spermatogenesis, differentiating germ cells form highly ordered, characteristic cellular associations with Sertoli cells. These characteristic complexes define the classical stages of the seminiferous epithelium cycle, designated I to XII in the mouse (Clermont and Perey 1957). The Sertoli cell exerts a variety of essential functions in supporting and directing the differentiation of the germinal component (Russell et al., 1990; Russell and Griswold 1993). Not only does it transfer nutrients into the adluminal compartment of the tubule, but it also receives a variety of endocrine and paracrine messages, and in turn sends signals to control the progress of germinal differentiation. Differentiation of germ cells thus appears to be critically dependent on functions of their somatic support (Griswold 1995). The problem of experimentally maintaining the germ cells in the proper environment for their differentiation may therefore be more difficult than for other cell types. It may eventually require complex coculture systems including at least Sertoli and germ cells.

Manipulation of Male Germ Cells

It is possible to obtain nearly pure populations of gonocytes from the embryo and the spermatogonia from very young males. In adult mice, only two cell types can be purified to a significant extent: early meiotic (pachytene) cells, and the haploid spermatid and sperm cells. Two techniques are available, elutriation centrifugation (Meistrich 1977) and zonal sedimentation (Romrell et al. 1976; Wolgemuth et al. 1985), both fractionating on the basis of the size and shape of the cells. None of these methods provides 100 % pure cell populations. Another possibility has been to prepare testicular cells from prepuberal animals (Russell et al. 1990). Until the start of the first wave of meiosis at day 8–9 after birth, the only germ cells present in the testis are spermatogonia. There are, however, two limitations: the first is the low number of cells that can be obtained from these very young mice, and the second, again a high risk of contamination by somatic cells. This technique has been more commonly used for the larger-sized rat, but the better-known genetics of the mouse makes the latter clearly a more interesting experimental animal. Another technique of potential interest is the microdissection of seminiferous tubules under transillumination developed by Parvinen and his colleagues (Parvinen 1993). It allows the preparation, in small amounts, of the defined assortment of premeiotic, meiotic, and postmeiotic germ cells, which, in a transverse section of the tubule, correspond to a given stage of the spermatogenic cycle.

Germinal Differentiation in Experimental Systems

Assuming that a foreign gene can be transferred into premeiotic or meiotic cells, it would then be necessary to provide conditions suitable for the next stages, at least up to the first haploid stages. Completion of spermiogenesis up to the spermatozoon stage will not be required, since fertilization of the oocyte was obtained after injection of round spermatid nuclei, and was followed by normal developement (Ogura et al. 1994; Kimura and Yanagimachi 1995). What is therefore required is a set of experimental conditions allowing transit through meiosis. The structural and functional complexity of the organ makes it difficult to imagine a simple cell culture system, such as an immortalized cell line, that would, by itself, progress through all stages of germinal differentiation.

Germ Cell Transplantation. Transplantation of testicular cells was recently shown to be possible. Germ cells from 5- to 28-day-old mice transplanted into the testis of an adult mouse after pharmacological depletion of the endogenous germ cells colonize the seminiferous tubules and initiate spermatogenesis. Genetic transmission of a transgene from the donor has been reported (Brinster and Avarbock 1994; Brinster and Zimmermann 1994), as well as complete gametogenesis from rat germ cell precursors grafted in the mouse testis (Clouthier et al. 1996).
Germ Cells in Culture. Lack of a cell culture system, which delayed the molecular analysis of the testis function, makes it difficult to develop gene transfer methods as it is today the case for ES cells. One possibility is to try to establish immortalized germinal cell lines, and several cases have been reported. However, these cells express only a limited range of germinal characteristics, and their use for the production of nuclei competent for fertilization has not been documented (Hofmann et al. 1994, 1995). The other possibility is to devise more complex cell culture systems that would reproduce part of the intercellular associations of the seminiferous epithelium. The intimate association of Sertoli cells with germ cells, and their important functions in supporting germinal differentiation make them a prime candidate for the role of a companion for germinal differentiation. It is possible to prepare primary cultures of Sertoli cells from immature males (Steinberger and Jakubowiak 1993). These cultures have been extensively used for physiological analysis, and provided an essential part of our present knowledge of the functions of the Sertoli cell. However, like all primary cultures, they present limitations in cell quantities and variability. Their short-term survival and nonclonal nature makes it impossible to derive stable variants with modified characteristics.

We, on the other hand, derived Sertoli cell lines from the testis of adult transgenic mice expressing the large T protein of polyoma virus (Paquis et al. 1993). The polyoma protein promotes the immortalization of cells in culture, but, unlike its SV40 homonym, it does not induce more advanced tumorigenic states that would not be compatible with the expression of transformed characteristics. As expected, the Sertoli lines established in this way conserve a number of the properties of the normal cell. In addition to the expression of characteristic marker genes (*Steel, WT1*), they are able to support male germ cell meiosis in a coculture system, starting either from pachytene cells fractionated from adult testis, or from premature 8-day-old baby testis. Exit from meiosis can be monitored by the appearance of haploid nuclei and by the expression of postmeiotic genes (Rassoulzadegan et al. 1993).

Gene Transfer into Germ Cells. Transfer and expression of DNA in germ cells is not as straightforward as it may have been in other cell types. Our recent results (H. J. S., V. G., M. R and F. Cuzin, unpubl.) would rather indicate that this may now constitute the main difficulty in devising a scheme for germ cell transgenesis. Whatever their mode of transfer (electroporation, calcium phosphate transfection, lipofection, retrovirus, or adenovirus infection), no significant expression was seen in germ cells for constructs carrying reporter genes (β-galactosidase, luciferase) under control of meiotic or postmeiotic promoters that are known to function faithfully in transgenic mice. Only a small fraction of the cells (10^{-6} to 10^{-5}) expressed the reporter genes. DNA appears to have entered the cell, as it could be detected by Southern blot analysis without gross alterations of structure or CpG methylation. Our preliminary results suggest that neither the survival of the germ cells nor their ability to attach to Sertoli cells is affected by gene transfer. Further investigations are needed to determine whether peculiar mechanisms operate in male germ cells to protect the germline from the adventitious entry of foreign DNA.

In conclusion, although recent significant progress has been made, prospects of generating transgenic animals by gene transfer into premeiotic precursors, and their subsequent differentiation into functional gametes, are still relatively remote. A better understanding of the molecular regulation of germinal differentiation will undoubtedly help to speed up this process.

Procedure

Procedure for Differentiation of Mouse Testicular Cells in Cocultures with 15P-1 Cells

Unless otherwise indicated the reader is referred to the report of Rassoulzadegan et al. (1993) for bibliographical references.

Preparation of germ cells

After removal of the tunica albuginea of the testis of one adult mouse (2 month old), seminiferous tubules are cut delicately into pieces and washed in PBS (phosphate-buffered saline) to remove mature sperm cells. The fragments are passed through a 100 μm pore size filter (Falcon) pushed by the piston of a 5-ml syringe. At this stage the suspension must be homogenous, it is centrifuged at 1500 rpm for 5 min, washed in PBS and resuspended in 20 ml DMEM medium supplemented with nonessential amino acids. The cell suspension is filtered on Cell Stainer 70 and 45 μm Nylon. Yields are in the range of 2×10^7 cells (from one animal), suspension at this stage contains diploid (premeiotic), tetraploid (pachytene), and haploid postmeiotic germ cells.

Coculture experiments can be performed using germ cells at this stage of preparation. These fractions contain preexisting haploid cells, but the total number of haploid cells increases to greater proportions between days 4 and 5, due to the accumulation of newly produced haploid meiotic products. Alternatively, purified fractions of germ cells can be obtained by either one of two well-established procedures: $1\times g$ sedimentation through preformed serum albumin gradients (Romrell et al. 1976; Wolgemuth et al. 1985) or elutriation centrifugation (Meistrich 1977). Both techniques allow efficient purification of a fraction of pachytene spermatocytes, the differentiation of which can be followed in cocultures. It is to be noted, however, that the efficiency of transit through meiosis, although still significant, is somewhat lower under these conditions (our unpublished results), thus revealing an activating effect of either postmeiotic spermatids or contaminating somatic cells in the initial suspension, the nature of which is currently under study. Another useful improvement would be the development of fractionation techniques that would allow the purification of premeiotic spermatogonia, for which even partial enrichment cannot be attained by any of the present methods.

Preparation of 15P-1 Sertoli cells and cocultures with germ cells

15P-1 cultures are propagated at 32 °C by plating, every 4th day, 10^5 cells per 10-cm petri plate in DMEM medium supplemented with 10% fetal calf serum. Figure 12.1a shows the morphology of growing 15P-1 cells. For coculture experiments, the cells are detached by trypsin treatment and resuspended in DMEM medium at a concentration of 2×10^4 cells/ml. 15P-1 and germ cells are seeded at the same time. 15P-1 and germ cells are mixed (50 germ cells per Sertoli cell), and incubated for 1 h at 32 °C to allow cell-to-cell recognition. Fetal calf serum (10%) is then added to the medium and 2 ml of suspension is seeded per culture chamber. The medium is changed every other day. Slides are prepared for the evaluation of the postmeiotic fractions, usually at days 1, 3, and 5. Upon more prolonged incubation periods, the phagocytic activity of the Sertoli cell line becomes predominant and germ cells are progressively taken up.

Culture substrate

Culture chambers on glass slides, such as the Labtek Chamber Slides (Nunc) offer suitable substrate for coculture experiments. This point is of importance, since we consistently observed a poorer conservation of germ cell morphology and less efficiency in differentiation on plastic

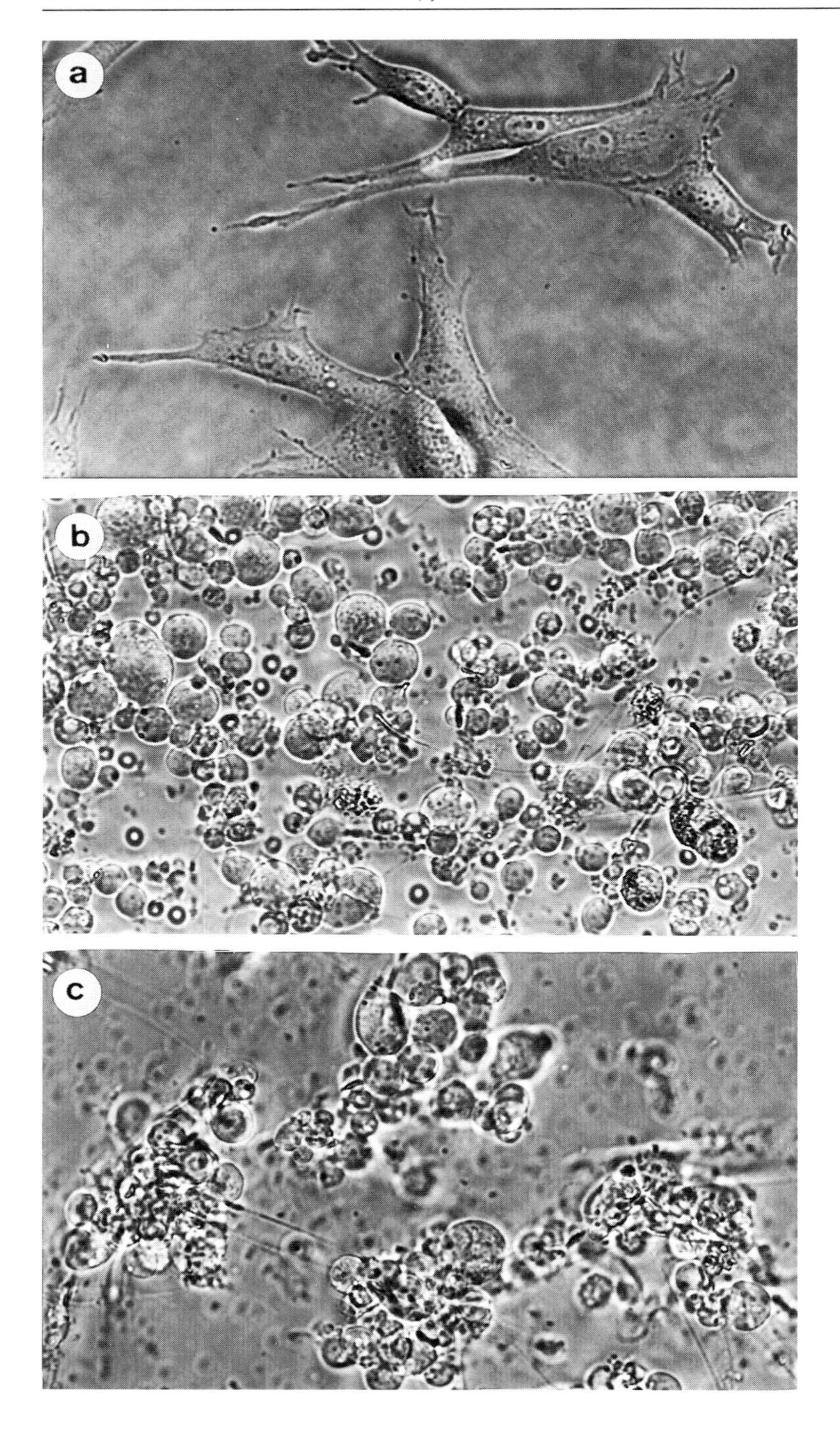
a
b
c

◀ **Fig. 12.1a–c.** Attachment of testicular germ cell to 15P-1 cells. Actively growing 15P-1 cells were prepared and 5×10^3 cells/cm^2 seeded in Labtek Chamber Slides (Nunc). The next day, 10^6 testicular germ cells were added to part of the cultures. Twenty four h later, the cells were photographed in phase contrast. **a** 15P1 cells alone. **b** Testicular germ cells. **c** Coculture of 15P-1 and germ cells (final magnification: 300×)

Petri plates. In addition, glass slides are useful since cells can be fixed and stained in various ways without damaging the substrate.

Visual observation

Specific aggregation of germ cells on top of 15P-1 cells can be observed within the first hours of the coculture (Fig. 12.1c). Formation of these complexes is a good index of the proper physiological state of the supporting cell line.

Identification of postmeiotic cells

Various assays can be developed to measure the fraction of post-meiotic cells in cultures. Two procedures have been used so far, based respectively on the activation of the promoter of the protamine *Prm1* gene, expressed only at the end of meiosis, and on the cytometric evaluation of nuclear DNA contents.

Germ cells with a genetic marker of postmeiotic differentiation are obtained from males of the transgenic a2 family, which express the *lacZ* gene of *Escherichia coli* under protamine *(Prm1)* control (Jasin and Zalamea, 1992). For in situ determination of β-galactosidase activity, cells on glass slides are fixed in 1 % buffered formaldehyde for 15 min, washed in PBS and stained in 1.0 mg/ml X-Gal, 2 mM $MgCl_2$, and 5 mM potassium ferricyanide, 5 mM potassium ferrocyanide in PBS. Enzyme activity in solution was measured by O-nitrophenyl-β-D-galactopyranoside hydrolysis (Miller 1972) on extracts prepared by three cycles of freezing and thawing followed by low speed centrifugation (1000 rpm for 5 min). Measurements of DNA contents can be performed by either image or flow cytometry. For image cytometry, cells on glass slides were fixed in buffered 10 % formaldehyde for 2 h, rinsed with PBS, and stained with Feulgen-Azur A, using the CAS100 Staining Kit (Becton-Dickinson 111025–5). Samples were analyzed on a CAS200 apparatus (Becton-Dickinson Cell Analysis System, Elmhurst, Illinois). Rat hepatocytes were used as an internal standard. For flow cytometry, detergent-isolated nuclei were stained with propidium iodide, and DNA histograms were generated using a FACScan apparatus (Becton-Dickinson).

References

Brinster RL, Avarbock MR (1994) Germline transmission of donor haplotype following spermatogonial transplantation (see comments). Proc Natl Acad Sci USA 91:11303–11307

Brinster RL, Zimmermann JW (1994) Spermatogenesis following male germ-cell transplantation. Proc. Natl Acad Sci USA 91:11298–302

Buehr M, McLaren A (1993) Isolation and culture of primordial germ cells. Methods Enzymol 58–77

Buehr M, Gu S, McLaren A (1993) Mesonephric contribution to testis differentiation in the fetal mouse. Development 117:273–281

Clermont Y, Perey B (1957) The stages of the cycle of the seminiferous epithelium of the rat: practical definitions in PA-Schiff-Hematoxylin and Hematoxylin-Eosin stained sections. Rev Can Biol 16:451–462

Clouthier DE, Avarbock MR, Maika SD, Hammer RE, Brinster RL (1996) Rat spermatogenesis in mouse testis. Nature 381:418–421

De FM, McLaren A (1983) In vitro culture of mouse primordial germ cells. Exp Cell Res 144:417–427

Griswold MD (1995) Interactions between germ cells and Sertoli cells in the testis. Biol Reprod 52:211–216

Hofmann M-C, Hess RA, Goldberg E, Millan JL (1994) Immortalized germ cells undergo meiosis in vitro. Proc Natl Acad Sci USA 91:5533–5537

Hofmann M-C, Abramian D, Millan JL (1995) A haploid and a diploid cell cycle coexist in an in vitro immortalized spermatogenic cell line. Dev Genet 16:119–127

Jasin M, Zalamea P (1992) Analysis of β-galactosidase expression in transgenic mice by flow cytometry of sperm. Proc Natl Acad Sci USA 89:10681–10685

Jenkins JR, Rudge K, Chumakov P, Currie GA (1985) The cellular oncogene p53 can be activated by mutagenesis. Nature 317:816–818

Kimura Y, Yanagimachi R (1995) Mouse oocytes injected with testicular spermatozoa and round spermatids can develop into normal offspring. Development 121:2397–2405

Matsui Y, Zsebo K, Hogan BL (1992) Derivation of pluripotential embryonic stem cells from murine primordial germ cells in culture. Cell 70:841–847

McLaren A (1992) Development of primordial germ cells in the mouse. Andrologia 24:243–427

McLaren A (1995) Germ cells and germ cell sex. Philos. Trans R Soc Lond, B Biol Sci 350:229–233

Meistrich ML (1977) Separation of spermatogenic cells from rodent testes. Methods Cell Biol 15:15–54

Ogura A, Matsuda J, Yanagimachi R (1994) Birth of normal young after electrofusion of mouse oocytes with round spermatids. Proc Natl Acad Sci USA 91:7460–7462

Panthier JJ, Condamine H, Jacob F (1988) Inoculation of newborn SWR/J females with an ecotropic murine leukemia virus can produce transgenic mice. Proc Natl Acad Sci USA 85:1156–1160

Paquis FV, Michiels JF, Vidal F, Alquier C, Pointis G, Bourdon V, Cuzin F, Rassoulzadegan M (1993) Expression in transgenic mice of the large T antigen of polyomavirus induces Sertoli cell tumours and allows the establishment of differentiated cell lines. Oncogene 8:2087–2094

Parvinen M (1993) Cyclic function of Sertoli cells. In: Russell LD, Griswold MD (eds) The Sertoli cell. Cache River Press, Clearwater, pp 349–364

Rassoulzadegan M, Paquis FV, Bertino B, Sage J, Jasin M, Miyagawa K, van Heyningen V, Besmer P, Cuzin F (1993) Transmeiotic differentiation of male germ cells in culture. Cell 75:997–1006

Romrell LJ, Bellve AR, Fawcett D (1976) Separation of mouse spermatogenic cells by sedimentation velocity. Dev Biol 49:119–131

Russell LD, Griswold MD (1993) The Sertoli cell. Cache River Press, Clearwater

Russell LD, Ettlin RA, Sinha Hikim AP, Clegg ED (1990) Histological and histopathological evaluation of the testis. Cache River Press, Clearwater

Steinberger A, Jakubowiak A (1993) Sertoli cell culture: historical perspective and review of methods. In: Russell LD, Griswold MD (eds) The Sertoli Cell. Cache River Press, Clearwater, pp 155–180

Wolgemuth DJ, Gizang-Ginsberg E, Engelmeyer E, Gavin BJ, Ponzetto C (1985) Separation of mouse testis cells on a Celsep apparatus and their usefulness as a source of high molecular weight DNA or RNA. Gamete Res 12:1–10

Chapter 13

Sperm-Mediated Gene Transfer

MARIALUISA LAVITRANO[1], VALENTINA LULLI[2], BARBARA MAIONE[2], SABINA SPERANDIO[1], AND CORRADO SPADAFORA[2*]

Introduction

Spermatozoa are highly specialized cells which play a central role in the process of fertilization, carrying the male genome into the oocyte. Although widely studied, the biological process of fertilization remains still obscure in many aspects, and the molecular steps leading to the formation of the diploid zygote after the interaction of the gametes are not yet entirely clear.

A feature of sperm cells that had long been unknown is their spontaneous ability to take up exogenous DNA and transfer it into eggs during fertilization. This feature was first observed in 1971 by Brackett et al. (1971), yet remained ignored for many years by the scientific community. In 1989 it was rediscovered in our laboratory and was further pursued when we realized that the interaction between sperm cells and foreign DNA could be exploited to generate transgenic mice (Lavitrano et al. 1989), as schematically diagrammed in Fig. 13.1. The finding that spermatozoa can spontaneously bind DNA of foreign origin has deeply modified our views on the role and function of spermatozoa: these cells are not only highly specialized vectors of their own genome, but also act as potential vectors of any exogenous DNA released in their environment with which they may come in close contact. Such spontaneous ability of sperm cells offers a potentially powerful tool in the fields of animal transgenesis and biotechnology, and may also provide a key to disclose hidden pathways in the evolutionary processes.

If fully developed, sperm-mediated gene transfer may in principle be successfully applied to all animal species whose reproduction is medi-

* corresponding author: phone: +39-6-49902972; fax: +39-6-49387120; e-mail: Lorenz@net.iss.it

[1] Department of Experimental Medicine, University La Sapienza, Rome, Italy

[2] Institute of Biomedical Technology, CNR, Via G.B. Morgagni 30/E 00161 Rome, Italy

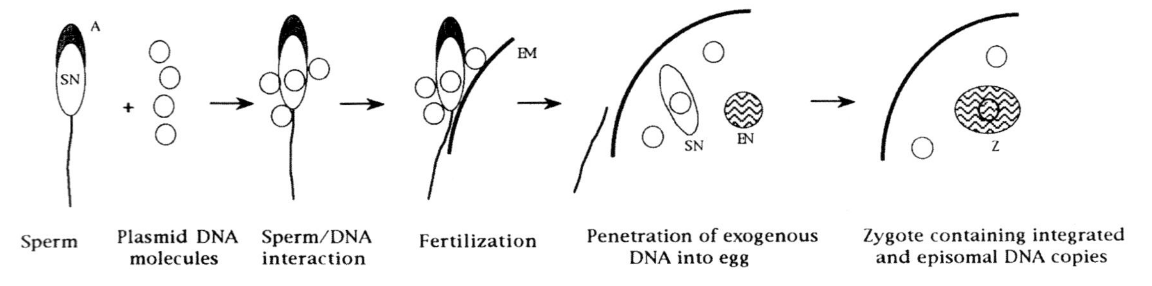

Fig. 13.1. Schematic representation of the sperm-mediated gene transfer process. The exogenous DNA is represented by *circles*. *A* Acrosome; *SN* sperm nucleus; *EM* egg plasma membrane; *EN* egg nucleus; Z zygote

ated by gametes. In our and other laboratories experience, the ability of sperm cells to bind foreign DNA is not restricted to one or few species but represents instead a widespread feature of spermatozoa from all species, ranging from echinoids to mammals (see below). However, it is not possible to establish a unique sperm-mediated gene transfer protocol suitable for all species, because of the wide differences in the environmental requirements for fertilization in different species, such as, for example, water or seawater for amphibians and fish and the female genital tract for mammals. Therefore, differential protocols adapted to different experimental conditions must be developed for each species, though animals from closely related species may share similar, if not identical, conditions.

The sperm-mediated gene transfer method potentially offers obvious advantages compared to the more widely used microinjection: it does not require sophisticated and expensive equipment such as micromanipulators, can be easily performed in field experiments as it only requires very basic equipment, does not need any particular skill, and can be successfully applied to those animal species, relevant both to research and to biotechnology, which have proved to be refractory to microinjection, such as fish, swine, and bovine.

The major obstacle currently interfering with a wide application of the technique is the lack of information concerning the molecular events occurring in gametes, particularly in sperm cells, during the initial steps of fertilization. This was found to be particularly important with mice, because the ignorance of the parameters and factors involved both in the mechanism of DNA binding to sperm cells and its transfer to ova makes it difficult to exert an experimental control on sperm-mediated gene transfer (see below); however, in other mammalian species, such as bovine and swine, for example, the method gives reliable and reproducible results. In the past few years most of our efforts have been devoted to the dissection of the sperm-mediated gene transfer process in order to clarify its underlying molecular mechanisms: we have now acquired relevant informations on the mechanism of interaction between sperm cells and the DNA. Furthermore, the sperm-mediated gene transfer process has been successfully applied in several animal species in our and in other laboratories.

Problems and Achievements

Shortly after the publication of our first article (Lavitrano et al. 1989), Brinster et al. (1989) reported unsuccessful attempts to reproduce our results. The authors pointed out a problem of which we were unaware, as they showed that the sperm-mediated gene transfer protocol was not as straightforward and reproducible as it had originally proved in our experiments.

The absence of a simple explanation for the discrepancy between our and other group's results, and the absence of preexisting studies in this field which could orient our search for answers, prompted us to undertake a systematic study by dissecting the sperm-mediated gene transfer process in different steps. These studies have yielded crucial information and disclosed a novel area of research. We now know that each mouse epididymal sperm cell can take up $1.5-4\times10^6$ DNA molecules (on average 5 kb) and that the binding occurs via ionic interactions, is reversible and takes place in the subacrosomal segment of sperm heads (Fig. 13.2); 90–95 % of epididymal sperm cells take up the foreign DNA straightforwardly, whereas ejaculated spermatozoa only bind the DNA after extensive washes and complete removal of the seminal fluid (Fig. 13.3; Lavitrano et al. 1992). The seminal fluid contains, in fact, a factor which exerts a powerful inhibitory effect on the binding between sperm cells and DNA (Fig. 13.4).

Sperm-bound DNA is rapidly internalized into the nuclei in 60–65 % of sperm cells and is found embedded in the sperm chromatin (Francolini et al. 1993). The binding of exogenous DNA to sperm cells is mediated by a group of positively charged proteins of apparent molecular weight ranging between 30 and 35 kDa, and is sequence-independent. These proteins are located within or beneath the plasma membrane, are electrophoretically conserved in various species and behave as DNA-binding proteins in band-shift assays (Zani et al. 1995).

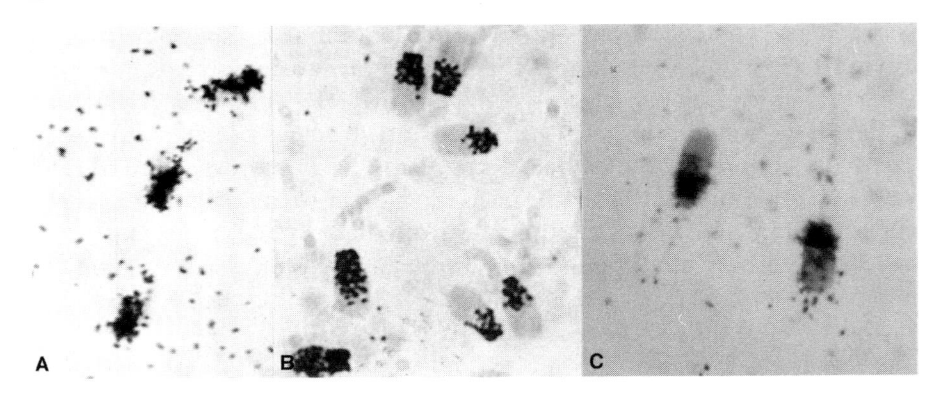

Fig. 13.2A–C. Autoradiography of mouse epididymal sperm cells (**A**), and of bovine (**B**) and swine (**C**) ejaculated and washed sperm cells associated to ^3H end-labeled pSV_2CAT DNA. Sperm cells were incubated with ^3H end-labeled DNA for 30 min, washed with FM (mice) or TALP (bovine and swine), smeared on glass slides and autoradiographed as described

Results from other laboratories have provided further evidence that the ability to bind foreign DNA is a general feature of spermatozoa from all species (Atkinson et al. 1991; Bachiller et al. 1991; Castro et al. 1991; Horan et al. 1991).

In the past few years, we have carried out experiments of sperm-mediated gene transfer in mice, using epididymal spermatozoa, as well as in swine and bovine, using ejaculated washed semen. These studies have revealed striking differences between the murine, on one hand, and the swine and bovine systems on the other hand. Genetically transformed mice have been obtained with a very high efficiency (30–80 %) in clusters of positive experiments. A variable number of totally negative experiments separate two series of successful experiments; the parameters determining the positive or negative outcome of the experiments with mice are currently under study (unpublished results). In sharp contrast with such all-or-nothing distribution in mice is the response of swine and bovine species to sperm-mediated gene transfer experiments. Indeed, transformed individuals can be obtained in the latter species with a lower efficiency than in mice, i.e., about 20 % bovine and 6 % swine respectively, yet positive individuals are obtained practically in every experiment (Sperandio et al. 1996). A homogeneous versus a polarized production of transformed individuals may reflect intrinsic differences between the species; however, in our opinion, it is most probably due to the use of epididymal (mice) versus ejaculated (bovine and

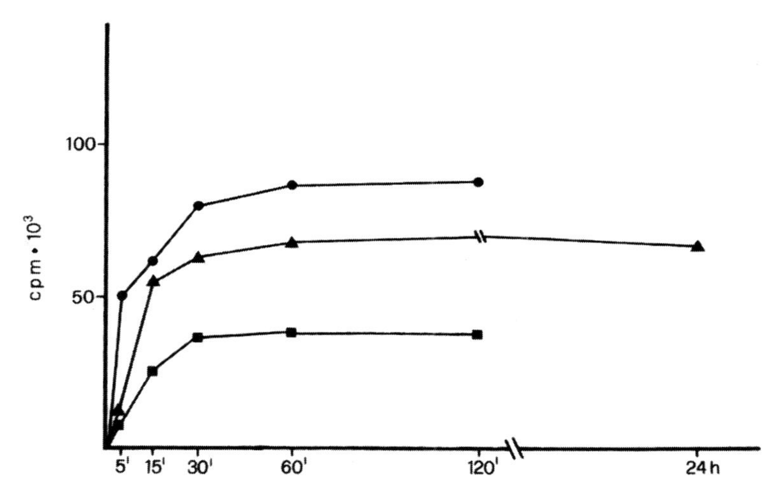

Fig. 13.3. Time course association of end-labeled pSV_2CAT DNA molecules with mouse epididymal sperm cells (●), and with ejaculated and washed sperm cells from boar (▲) and bull (■). DNA molecules from one preparation of end-labeled pSV_2CAT were incubated with spermatozoa of all three species. Samples containing 10^6 sperm cells were withdrawn at the indicated times and washed thoroughly with FM (mouse) or TALP (bovine and swine) medium. Cell pellets were dissolved and counted as described in the text

swine) sperm cells. Seminal fluid may induce deep modifications in spermatozoa, both at the structural and at the metabolic level, and therefore the fate of the internalized foreign DNA into epidydimal sperm cells may be very different from that of DNA internalized into ejaculated cells.

The generation of genetically transformed embryos (Arezzo 1989; Hochi et al. 1990; Perez et al. 1991; Nakanishi and Iritani 1993;) and born individuals (Rottman et al. 1991; Khoo et al. 1992; Sin et al. 1993; Schellander et al. 1995) has been reported by several laboratories, suggesting again that sperm-mediated gene transfer reflects a general biological process which can be performed in virtually all animal species under a variety of experimental conditions.

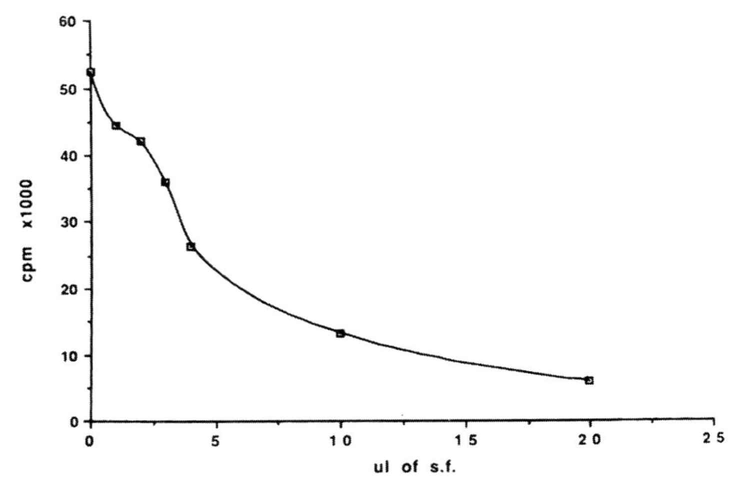

Fig. 13.4. Inhibition of the DNA uptake by mouse sperm cells in the presence of increasing amounts of human seminal fluid. Epidydimal mouse sperm cells (5×10^6/ml concentration) were incubated with the indicated volumes of human cell-free seminal fluid for 30 min, at 37 °C in 7 % CO_2 in air; 100 ng of end-labeled plasmid DNA were added to each sample for 30 min; sperm cells were then washed and counted

13.1 Sperm-Mediated Gene Transfer in Mice

Short Protocol

1. Squeeze cauda epididymis in FM medium supplemented with BSA.
2. Allow the sperm cells to disperse for about 30 min.
3. Count sperm cells.
4. Withdraw the required amount of sperm cells.
5. Add exogenous DNA to the sperm cell suspension (0.01 – 1 μg/10^6 sperms) and incubate for 30 min.
6. While incubation is proceeding, remove oviducts from superovulated females and squeeze the cumulus out (FM supplemented with BSA).
7. Mix eggs from two oviducts with DNA-loaded spermatozoa ($1-2 \times 10^6$ sperms/dish).

8. After 5 h transfer nondegenerated eggs into Ham's F-10 medium.
9. Allow the embryos to grow to the two-cell stage (overnight)
10. Transfer the two-cell stage embryos in prewarmed PBS supplemented with 1 mg/ml BSA.
11. Batches of 15–20 two-cell stage embryos are surgically implanted into mouse oviducts.

Materials

Equipment

Basic equipment for a cell biology laboratory is required, including: lamina flow hood, CO_2 incubator, inverted microscope and stereomicroscope equipped with heating stage, fiber optic lighting system, heating block, hemocytometric counter.

In vitro fertilization is very sensitive to environmental conditions. Controlled temperature and sterility are two key parameters for achieving successful experiments and therefore must be accurately set in the working room.

Reagents and culture media

All media are filtered through 0.22 μm filters and stored at 4 °C under sterile conditions for no more than 4 weeks. The medium used for in vitro fertilization experiments (FM) is based on Whittingham's Tyrode solution (Whittingam 1971), from which sodium lactate, penicillin, and streptomycin are omitted; NaH_2PO_4 is replaced by 0.15 mM Na_2HPO_4 and NaCl is increased to 120 mM.

– Fertilization medium (FM) 1×

Powder	g/l
NaCl	6.970
$NaHCO_3$	2.106
Glucose	1.000
KCl	0.201
$Na_2HPO_4.2H_20$	0.027
$CaCl_2.2H_2O$	0.264
$MgCl_2.6H_20$	0.102
Na Pyruvate	0.055
Osmolarity	275–290

The medium is supplemented with 4 mg/ml bovine serum albumin (BSA Fraction V, Miles or Sigma).

- Ham's F10 medium is hypoxanthine-free, supplemented with $NaHCO_3$ (2.106 g/l) and 4 mg/ml BSA. Osmolarity should also be between 275–290
- Falcon 3653 dishes are routinely used for squeezing epididymis, collecting eggs, fertilizations, and embryo cultures
- Silicon oil (Aldrich, Cat. no. 14615–3) is routinely used to cover embryo cultures
- All procedures (squeezing of epididymis, fertilization, embryo cultures) are carried out in the inner wells of Falcon 3653 dishes containing 1 ml of appropriate medium and overlaid with autoclaved silicon oil. Four ml of BSA-free PBS are placed in the outer wells. This arrangement ensures optimal conditions for both cell and embryo survival. It is recommended to prepare two sets of dishes respectively containing FM and Ham's F10, as described above, 15–24 h before starting the experiment and place them in a 37 °C incubator under 7 % CO_2 in air to equilibrate both the temperature and the pH. To accurately evaluate the pH of different media (optimum pH is 7–7.5), 3-ml aliquots of each medium are dispensed into conical tubes and incubated in parallel with the dishes.
- The source of water for the media preparation is a very critical factor for the efficiency of fertilization. We have found that most commercially available purified H_20 lots are not suitable for fertilization experiments. In our laboratory, water is routinely purified through a Millipore system, repurified through a Barnstead Easypure system containing a Pretreatment cartridge R/O and an Easypure high purity/low TOC Cartridge, and finally quartz-distilled in a Barnstead Mega-Pure water still.

Animals Mouse CD1 and BDF1 strains are purchased from Charles River. CD1 males are routinely used as sperm donors. Vasectomized CD1, tested for sterility, are mated with CD1 recipients 8–12 h before implanting the embryos. CD1 recipients are retired females. BDF1 females 3–10 weeks old are routinely used as egg donors.

Superovulation is induced following intraperitoneal injection of BDF1 mice with 5 IU of PMSG (Folligon, Intervet) at 8.30–9.30 pm; 48 h later, 5 additional IU of hCG (Corulon, Intervet) are injected intraperitoneally.

Anesthetic Concentrated 40× avertine anesthetic stock is prepared as follows: 10 g tribromoethilen are dissolved in 10 ml of tert-Amyl alcohol and stored at 4 °C. The working solution is obtained by diluting the stock to 1:40 with physiological solution. Mice, intraperitoneally injected with 15 μl of anesthetic/g of bodyweight, are anaesthetized for about 20 min.

Procedure

Preparation of Gametes

1. Mice are maintained in cycles of 13 h light and 11 h dark.
2. Eggs are collected from 3–10-week-old B6D2F1 mice. Females are induced to superovulate; 13 h after hCG injection, females are sacrificed by cervical dislocation, oviducts are removed and placed in a PBS-containing dish.

Two oviducts are then placed in the outer well of each dish and the cumulus is squeezed out using a 1-ml syringe.

The cumulus is transferred in the inner well containing 1 ml of fertilization medium (FM) supplemented with BSA (4 mg/ml, preequilibrated overnight in the incubator) and overlaid with silicon oil.

Spermatozoa for in vitro fertilization are obtained from CD1 strain mice. Sperm cells are collected from the cauda epididymis of proven males that had abstained for at least 3 days but no longer than 1 week. The sperm suspension is prepared by squeezing the terminal part of the vas deferens and puncturing the middle part of the epididymis in 1 ml of pre-equilibrated FM supplemented with 4 mg/ml BSA overlaid with silicone oil.

The sperm suspension is allowed to disperse by incubating the drop for 30 min at 37 °C in 7 % CO_2 in air. It is reccomended to pool sperm cells squeezed from the epididymis of two males.

DNA Uptake

1. After dispersal, sperm cells are counted as follows: 10 μl of the sperm suspension are diluted with 990 μl of HCl 0.1 N. 10 μl are spotted in a hemocytometric chamber and counted. The concentration of sperm cells is usually $25–35 \times 10^6$/ml. The required amount of sperm cells is withdrawn and mixed with plasmid DNA at a concentration ranging from 0.01–1 μg/10^6 sperms. Plasmid DNA can be used either in the linear or in the circular form; plasmids are usually linearized using enzymes that restrict at the cloning sites and separate the gene from the vector sequences.
2. Spermatozoa and DNA are routinely incubated for 30 min at 37 °C in 7 % CO_2 in air. The incubation time can be reduced to 10 min without loss of efficiency. After incubation, aliquots of spermatozoa are withdrawn and used for IVF.

In Vitro Fertilization (IVF)

1. Aliquots of $1-2\times10^6$ spermatozoa are withdrawn and added to the egg-containing dishes – usually 30–70 eggs collected from two oviducts are pooled in one dish. Dishes containing both sperm cells and eggs are placed in the incubator for 5 h.
2. After incubation, non degenerated eggs are transferred into dishes containing 1 ml of Ham's F10 and incubated overnight. Transfer of fertilized eggs from FM to Ham's F10 takes place outside the incubator, therefore it is recommended to operate as quickly as possible to minimize embryo damage. The fertilization rate is assessed 26–28 h after mixing the gametes by measuring the percentage of two-cell embryos. Trial fertilization experiments with control epididymal sperms routinely give an efficiency included between 60 and 100 %; under optimal fertilization conditions, the use of DNA-loaded sperm cells does not substantially modify this efficiency. In contrast, the efficiency is significantly reduced in those batches of eggs whose fertilization efficiency by control sperms was below 60 %: thus, the experiments are not carried on any further below 60 %.

Implantation of Embryos into Foster Mothers

1. Two-cell stage embryos (15–20) are transferred to a 37 °C prewarmed dish containing 50 μl of PBS supplemented with 1 mg/ml BSA and overlaid with silicone oil. Under these conditions the dish can temporarily be kept at 37 °C on a heating block outside the incubator.
2. Groups of 15–20 embryos from one PBS droplet are surgically implanted into the oviduct of plugged retired CD1 females that had been mated to CD1 vasectomized males 8–12 h before. During surgical implantation of the embryos, the foster mothers are anesthesized with avertin.

Results

Screening of the mouse offspring is routinely performed by Southern blot analysis of the DNA extracted from the tails of the animals. Radiolabeled DNA probes, corresponding to the transgene and lacking all vector sequences, are used to identify positive individuals in several hybridization rounds. The results obtained with mice by sperm-mediated gene transfer are generally characterized by three essential features: (1) The production of genetically transformed animals is not scattered among different experiments, but tends to concentrate in series of clustered positive experiments. Positive experiments usually yield a high percentage (30–80 %) of genetically transformed animals. Series of negative experiments of variable length of time separate two series of positive experiments. (2) The hybridization patterns of transgenes in positive animals indicate rearrangement events that have occurred in the foreign DNA during the sperm-mediated transfer process. (3) By restriction analysis, the transgene rearrangements appear to be very similar, if not identical, in positive individuals generated within the same experiment. This suggests that the exogenous DNA molecules are exposed to the same rearrangement event within each batch of transformed sperms, and that the transgene integration into the host genome takes place in one or few sites.

Troubleshooting

- Sperm cell aggregation:
  - Indicates compromised health conditions and/or heavy stress of the animals
- Low efficiency or no fertilization at all:
  - Sterilize the working area. If the problem persists, prepare new media
- Southern blot analysis shows faint signals
  - Technical problems can be ruled out by introducing a single-copy gene probe in the hybridization mixture. If the single-copy gene is visible while other bands are weaker, you may be observing a DNA population below one copy per genome in mosaic animals.

13.2
Sperm-Mediated Gene Transfer in Swine

Short Protocol

1. Collect semen from boar donor in a thermostatic container.
2. Dilute 5-ml aliquots to 50 ml with TALP medium supplemented with BSA and prewarmed at 37 °C (from now on keep TALP at room temperature).
3. Centrifuge at 800 *g* for 8–10 min at 25 °C.
4. Discard supernatants without perturbing the pellets.
5. Add Talp medium to 50 ml.
6. Gently invert the tubes a few times to resuspend the pellets.
7. Centrifuge at 800 *g* for 8–10 min at 18 °C.
8. Discard supernatants leaving 1–2 ml of medium.
9. Resuspend carefully the sperm pellets in the residual TALP using a wide-tip pipette and combine pellets in one tube.
10. Count sperm cells.
11. Transfer $0.7-1.5\times10^9$ sperm cells in:
    a) 15-ml Falcon tube and dilute to 12 ml with TALP preequilibrated at 18 °C (suggested for surgical intrauterine insemination);
    b) 150-ml Ehrlenmayer flask and dilute to 100–120 ml with 18 °C preequilibrated TALP (suggested for artificial insemination).
12. Add plasmid DNA (0.4–0.5 μg/10^6 sperm cells).
13. Incubate at 18 °C for 2–4 h or longer.
14. DNA-loaded sperm cells are now ready for insemination.

Surgical intrauterine insemination:

15. Centrifuge as above.
16. Discard supernatants leaving 1–2 ml TALP.
17. Suspend sperm cells in the residual TALP medium and use for surgical insemination.

Artificial insemination

15. Sperm/DNA incubation mixtures of 100–120 ml can be used immediately for insemination. 12-ml incubation mixtures must be diluted to 120 ml with fresh TALP medium before insemination.

Materials

Equipment

- Water bath at 37 °C
- Thermostat at 18 °C
- Refrigerated bench centrifuge
- Microscope equipped with heating stage
- Hemocytometric counter chamber

Reagents and culture media

All media are filtered through 0.22 μm membranes and stored at –20 °C for no more than 2 months under sterile conditions.

- Calcium-free TALP medium (1×) (Ball et al. 1983)

Powder	g/l
NaCl	5.4
KCl	0.23
Na_2HPO_4	0.04
$MgCl_2.6H_2O$	0.31
$NaHCO_3$	2.10
HEPES	2.38
Na pyruvate	0.11
Na lactate 60 %	3.7 ml

Osmolarity 278–290

- TALP can either be freshly prepared every time or a stock solution (10×) can be prepared and stored at –20 °C in 10–50-ml aliquots. TALP medium (1×) is supplemented with 6 mg/ml BSA, fraction V (Miles or Sigma) before use. Kanamycin (7.5 mg/ml) is added only when sperm cells are to be incubated with the DNA for a long time (12 h).

TALP-IVF medium	(1×)
Powder	g/l
NaCl	5.4
KCl	0.23
Na_2HPO_4	0.04
$MgCl_2.6H_2O$	0.1
$NaHCO_3$	2.1
HEPES	2.38
Na pyruvate	0.055

$CaCl_2$	0.39
Na lactate 60 %	2 ml

Osmolarity 280, pH 7.6

- TALP-IVF medium is supplemented with 6 mg/ml BSA (SIGMA, Cat. A8806) and with kanamycin 7.5 mg/ml.
- Tyrode acid solution (1×)

Powder	g/l
NaCl	8
KCl	0.2
$CaCl_2.2H_2O$	0.2
$MgCl_2.6H_2O$	0.1
Glucose	1
Polyvinyl alcohol (PVA)	1

Adjust pH to 2.5 with HCl

PBS-PVA: PBS supplemented with 1 mg/ml PVA

- Solutions for nuclei preparation

DTT buffer:	DTT 100 mM, Tris 50 mM, pH 7.5
CTAB solution:	CTAB 10 %, DTT 10 mM
lysis buffer:	Tris 20 mM pH7.5, EDTA 1 mM, NaCl 10 mM, SDS 1 %

- Solutions for lysis and neutralization of embryos for PCR analysis

lysis solution:	50 mM DTT, 200 mM KOH
neutralizing buffer	900 mM Tris-HCl (pH 8.5), 200 mM HCl, 300 mM KCl

- DL-dithiothreitol (DTT) Sigma
- Cetyltrimethylammonoum bromide (CTAB) Aldrich
- TRIS (hydroxymethyl) aminomethane Sigma
- Sodium dodecyl sulfate (SDS) Sigma
- L(+)-lactic acid hemicalcium salt: hydrate Sigma.

Anesthetic

- 65–75 kg gilts are anesthetized as follows:
  Preanesthesia: 6 ml (240 mg) Azaperone (Stresmil, Jansen) and 2 mg atropine are injected in each animal.
  Anesthesia: 1–1.5 g thiopental sodium (sodium pentothal) are injected into each animal.

Procedure

Selection of Donor Boars

1. Sperm donors are first selected on the basis of microscopic inspection of the semen on a prewarmed slide: vitality of the sperm cells must be at least 90 % and motility not below 65–70 % after the washing procedures.
2. The second step of selection is a test of the ability of sperm cells to take up foreign DNA after removal of seminal fluid (see below). DNA uptake is assessed either
   a) by checking the association of radiolabelled DNA to sperm cells by scintillation counting or autoradiography or
   b) by direct Southern blot analysis of the foreign DNA internalized into nuclei.

Interaction of radioactive DNA with sperm cells

3.a Washed sperm cells (see Preparation of Sperm Cells below) are resuspended at a concentration of $5-10\times10^6$ cells/ml, mixed with linearized, end-labeled double-stranded plasmid DNA (0.5 μg/ml) and incubated at 18 °C.

4.a Aliquots containing 10^6 cells are withdrawn from the incubation mixture at specific times, diluted in Eppendorf tubes containing 1 ml of TALP medium, and washed by centrifuging ar 4000 rpm for 5 min in a microfuge.

5.a Spermatozoa are washed a second time by suspending in 0.5 ml TALP and centrifuging as above.

6.a Pellets are then disolved in 100 μl NaOH 1 M for at least 1 h at 37 °C, neutralized with an equal volume of 1 M HCl, diluted in toluene scintillation cocktail, and counted. The results of the uptake assay should be comparable to those reported in Fig. 13.3.

Autoradiographic experiments are performed by incubating aliquots of sperm cells with H^3 end-labeled double-stranded plasmid DNA, washed and treated according to published procedures (Lavitrano et al. 1992; Zani et al. 1995). Binding of foreign DNA to the sperm cells of the different species should be similar to that shown in Fig. 13.2.

Southern blot analysis of nuclear internalized DNA

3b. Washed sperm cells are resuspended as above and incubated at 18 °C with linear or circular nonradioactive plasmid DNA at a concentration of 0.5 μg/10^6sperm cells.

4b. Aliquots containing at least 3×10^6 sperm cells are withdrawn and thoroughly washed with TALP medium.

5b. Nuclei are prepared essentially as described by Balhorn et al. (1977): briefly, sperm cells are suspended in DTT buffer (10^6 sperm cells in 26 μl) and incubated 30 min in ice; 1/9 of the volume of CTAB buffer is added (1 % CTAB final) and further incubated 45–60 min, in ice.

6b. Nuclei are pelleted at 9000 rpm in a microfuge for 5 min, at room temperature and pellets are washed with 500 μl of 50 mM Tris pH. 7.5–8 by centrifuging as above.

7b. Nuclear pellets are disolved in 300–500 μl of lysis buffer and incubated overnight with 0.5 mg/ml of Proteinase K (Boehringer) at 37 °C. DNA is extracted, purified, restricted with appropriate enzymes, fractionated through 1 % agarose gel, blotted, and the filter is hybridized with a radioactive probe.

8b. For a quick quantitative evaluation of the foreign DNA present in the sperm nucleus, a dot blot analysis of the extracted DNA is sufficient.

Preparation of Sperm Cells

1. Semen is collected from the selected donor that had abstained for 4–5 days. Semen is collected in a sterile plastic bag put in a thermostatic container so as to avoid temperature shocks. It is convenient to collect only the initial 30–40 % of the ejaculate because this fraction contains most of the sperm cells and a low proportion of the seminal fluid, which is well known to antagonize the binding of the DNA to the sperm cells (see Zani et al. 1995).
2. Five-ml aliquots of semen are transferred in 15-ml Falcon tubes and mixed with an equal volume of TALP medium supplemented with 6 mg/ml BSA prewarmed at 37 °C (leave the TALP medium from now on at room temperature).
3. Incubate for 5 min, then transfer to 50-ml Falcon tubes and fill the tubes with TALP/BSA to 50 ml. Centrifuge at 800 *g* for 8–10 min at 25 °C.

4. Discard as much of the supernatant as possible without perturbing the soft sperm pellets and leave pellet phases of 3–5 ml.
5. Fill the tubes again with TALP/BSA. Repeat centrifugation at 800 *g* for 8–10 min at 18 °C and discard supernatants. The final dilution of swine semen is 1:100.
6. Carefully resuspend each pellet in a few ml of TALP and combine the suspensions in one tube. Count the cells by diluting 10 μl of the suspension in 990 μl of 0.1 M HCl solution and spot a few microliters in a hemocytometric counter chamber.

DNA Uptake

DNA uptake experiments can be performed either with (a) concentrated ($70-100\times10^6$/ml) or (b) diluted ($5-12\times10^6$/ml) sperm suspensions, leaving the DNA/sperm cells ratio unaltered (we routinely use 0.4–0.5 μg/10^6 spermatozoa), regardless of the total volume of sperm suspension.

1. Transfer aliquots containing $0.7-1.5\times10^9$ sperm cells into 15-ml conical Falcon tubes and add TALP medium to 12 ml (a) or suspend the same amount of sperm cells in 100–120 ml of TALP in a Erlenmeyer flask (b).
2. Add plasmid DNA at the desired concentration and incubate at 18 °C for 2–4 h, or until required to achieve a complete interaction between sperms and the DNA. Very gently invert the tubes every 15–20 min to prevent sedimentation of the sperm cells.
3. During the last 20 min of incubation, withdraw the incubation mixtures from the 18 °C thermostat and keep them at room temperature; just before insemination, heat them at 37 °C for 1 min.

Artificial Insemination

Artificial inseminations can be performed in normally cycling or in artificially induced sows. In the latter case, prepuberal gilts (65–75 kg) are injected with 1250–1500 IU eCG (Folligon, Intervet, Holland) and 60 h later with 750 IU hCG (Corulon, Intervet, Holland). Ovulation occurs 40 h after hCG injection. Inseminations are performed 43 h after hCG injection using $1.5-2\times10^9$ DNA-treated sperm cells per animal.

Results can be improved by surgically depositing DNA-loaded sperm cells directly in the uterus horns. Sows are anesthetized and the ends of each horn, proximal to the tubes, are reached through a midventral incision and punctured with a blunted 19-gauge needle; 1–2 ml TALP containing $0.5-1 \times 10^9$ sperm cells are then introduced into each uterus horn using a sterile Pasteur pipette. This method yields better results, possibly because surgical fertilization prevents the competition between DNA-loaded and normal sperms which may otherwise occur in the female genital tract and disfavor the DNA-loaded sperm cells.

Analysis of Offspring

The screening of the offspring is performed by Southern blot analysis of the DNA extracted from both ear and tail tissues from each animal. Before sampling, the tails and ears of each animal are carefully cleaned first with a detergent and subsequently with alcohol to minimize contamination by feces or bacteria. After purification, DNA samples are restricted and analyzed by Southern blot. It is recommended to avoid probes containing vector DNA sequences (pBR322) and only use sequences corresponding to the relevant gene.

Results

In contrast with the highly variable results obtained with mice, sperm-mediated gene transfer experiments in swine give a homogeneous production of transformed animals, consistently included between 2 and 20 %. Southern blot analysis is the major selection procedure for the identification of transgenic animals and therefore it is very important to control as rigorously as possible all potential sources of artifact. In our experience, when tail and ear DNAs are probed using the entire plasmid DNA (i.e., vector + insert), the results obtained with each DNA source can differ. In particular, analysis of tail DNA samples reveals a high percentage of positive individuals (above 50 %), characterized by heterogeneous hybrydization patterns. In contrast, analysis of ear DNA samples usually yields a lower percentage of positivity (up to 20 %) and shows simplified patterns of hybrydization. On the other hand, when probing with the purified gene sequences and no vector DNA, the tail DNA pattern is more similar to that of ear DNA, though still showing a higher positivity and a somewhat higher complexity. Only rarely did we find identical patterns in both ear and tail DNA. This observation may either reflect an artifact due to bacterial contamination of the tail but not of the ear tissues; or an alternative possibility is that the complex hybrydization signals in tail DNA reflect a mixture of integrated and episomal foreign DNA sequences in tissues of different embryological origin which eventually contribute to the adult tail tissue. Final selection of the positive animals is generally based on several rounds of screening of ear DNA.

It is worth mentioning that restriction patterns of transgenes, both from ear and tail DNA, are usually altered compared to that of the original plasmid, suggesting that the original DNA was rearranged during the process of sperm-mediated gene transfer.

Troubleshooting

- Sperm cells are aggregated
  The presence of sperm cell aggregates of various sizes (from three to four cells to hundreds) after the washing procedures suggests a non-healthy condition of the donor due to bacterial/viral infections or to heavy stress
- Low or no DNA binding to sperm cells
  Sperm cells were not thoroughly washed and trace amounts of seminal fluid are still present
  Check viability of sperm cells as well as pH and ionic strength of media
- Low fertilization efficiency
  Washing conditions may be too hard and cause damage to sperm cells: reduce speed and/or time of centrifugation
  Overloading sperm cells with exogenous DNA may cause poisoning effects: challenge sperm cells with reduced amounts of DNA
  The animal is unhealthy as above
  Check estrus of sows
- Southern blot analysis shows faint signals
  Technical problems can be ruled out by introducing a single copy gene probe in your hybridization mixture. If the single copy gene is visible while other bands are weaker, you may be observing a DNA population below one copy per genome in mosaic animals

- Low number of offspring delivery by sows
  When fertilizing with DNA-loaded sperm cells, we have sometimes observed that sows delivered 4–5, rather than the expected 8–12, piglets. This may be due to embryo lethality possibly induced by the foreign DNA (both in qualitative or/and quantitative terms). Experimental conditions should be modified either by reducing the ratio of DNA/sperm cells in the transformation step, or by testing different DNA constructs – when disruption of a vital locus can be suspected in the embryos
  Artificially induced estrus may cause a reduced number of offsprings due to egg degeneration. A higher efficiency is obtained with naturally cycling animals.

13.3 Sperm-Mediated Gene Transfer in Bovine

The sperm-mediated gene transfer protocol described for swine can be applied to bovines in essentially the same way. Only minor modifications have been introduced in the bovine protocols. Essential steps are reported below.

Procedure

Selection of Donor Bulls

The same criteria described for the selection of swine sperm donors are adopted for the selection of bovine donors. Slight differences in the washing procedures are reported in the next paragraph.

Preparation of Sperm Cells and DNA Uptake

1. 0.5 ml of the ejaculated semen are diluted in 9.5 ml of calcium-free TALP medium prewarmed at 37 °C and supplemented with 6 mg/ml BSA. Mix by inverting the tubes two to three times and dilute to 50 ml with TALP (1:100 dilution).
2. Diluted semen is distributed in 50-ml Falcon conical tubes. Cells are spinned down at 800 *g* for 8–10 min at 25 °C and supernatants are carefully discarded, leaving 2–4-ml pellet phases in each tube.

3. Fill the tubes again with TALP to 50 ml (final dilution is about 1:1000) and centrifuge at 18 °C.
4. Sperm cells are counted using a hemocytrometic counter chamber and diluted in TALP-IVF to 10×10^6 sperm/ml.
5. The interaction between sperm cells and DNA is carried out by adding 0.4–0.5 $\mu g/10^6$ sperm cells of plasmid DNA to the sperm suspension and incubating for 2 h at 18 °C. Sperm cells are then ready for IVF or artificial insemination.

IVF Production of Cattle Embryos

Cattle embryos are produced in vitro as described by Parrish et al. (1986) with minor modifications (Sperandio et al. 1996).

1. Embryos are cultured in vitro to the morula/blastocyst stage; the zona pellucida is then removed by treatment with Tyrode acid solution.
2. Embryos are washed three times with calcium/magnesium-free PBS and either immediately processed for PCR amplification, or stored in liquid nitrogen in 0.5-ml tubes containing microliter amounts of PBS-PVA (polyvinyl alcohol).

PCR Analysis of Embryos

1. Embryos are individually lysed by heating at 65 °C in 2.5 ml of lysis buffer for 15 min.
2. 2.5 μl of neutralizing buffer are then added.
3. Lysed embryos are directly used in a 100-μl PCR amplification reaction essentially as described by Chou et al. (1992). Conditions are adjusted according to the composition of the oligonucleotide primers and are optimized for each primer pair so as to detect 10–50 DNA molecules per reaction. Under these conditions, positive embryos are detectable using 34–45 cycles. In all our PCR analysis, one negative control (i.e., embryos developed from oocytes fertilized with sperm cells that had not been incubated with the foreign DNA) and one no-DNA sample are routinely amplified every fifth sample.

Results

Results obtained by PCR amplification of blastocysts show that sperm-mediated gene transfer occurs with a high frequency in bovine, giving large percentages of genetically transformed embryos. We have obtained 22 % of transformed blastocysts when sperm cells were challanged with pSV_2CAT plasmid DNA. Two unrelated plasmids, pRSV-LT and pALU, yielded only 2.1 and 0.4 % positive embryos respectively, suggesting that the DNA sequence carried by the sperm does significantly affect the overall efficiency of embryo transformation.

Troubleshooting

- False negatives
  Check lysis of the embryos by performing amplification of an endogenous gene
- False positives
  Adopt general precautions for PCR analysis: autoclave and aliquot every solution as well as the reaction tubes. carry out PCR amplifications in a laboratory area completely free of plasmid contamination; if necessary, work in an area equipped with a UV ray source. Use DNA-free pipettes and aerosol-resistant tips.

13.4 Sperm-Mediated DNA Transfer in Other Animal Species

In order to provide the readers with a complete and exhaustive information, we report below further applications of the sperm-mediated gene transfer method which appeared in the literature. In this section we will briefly examine the results of experiments of genetic transformation of embryos and offsprings in a variety of species using sperm cells as vectors of the foreign DNA. We will report only the essential results and refer to the original paper for protocol details.

Procedure

Fish

Transgenic zebrafish were obtained by Khoo et al. (1992) using the pSV$_2$CAT plasmid; 23 and 37 % of transgenic individuals were generated after the spontaneous interaction of sperm cells with plasmid DNA in the circular or in the linear form, respectively. F_1 transgenic offspring as well as transgenic F_2 individuals were also obtained from founders. However, expression of the reporter gene was not detected. Transgenic chinoock salmon fry were obtained by Sin et al. (1993) by electroporating sperm cells in the presence of plasmid DNA. The transformation efficiency reported by these authors ranged between 5 and 10 %. The same group reported that conditions which improve sperm electroporation in the presence of plasmid DNA surprisingly result in a reduced efficiency of embryo transformation (Symonds et al. 1994).

Amphibia

Sperm-mediated gene transfer was also successfully performed in Amphibia: Habrovà et al. (1996) obtained transgenic *Xenopus laevis* individuals by fertilizing eggs with sperm cells which carried the entire *Rous sarcoma* virus DNA. Viral sequences were detected by Southern blot analysis of *Xenopus* DNA, and expression was determined by Northern assays in different tissues. Furthermore, transgenic animals showed an altered myogenic differentiation which correlated with expression of the viral transgene.

Chicken

Lipofection was used by Rottmann et al. (1991) to improve internalization of the foreign DNA into chicken sperm cells; after fertilization, 26 % of transgenic chicken were obtained. Interestingly, analysis of the DNA showed that under these experimental conditions the transgene was not integrated into the host genome but persisted as episome. Transfer of foreign DNA into chicken oocyte by sperm cells was also obtained by Nakanishi and Iritani (1993). The authors confirmed that lipofection, but not electroporation of sperm cells, improved the efficiency of DNA transfer.

Comments

Genetically transformed animals of a variety of species can be generated by sperm-mediated gene transfer. Different species respond quite differently to transgenesis mediated by sperm cells: this may reflect different requirements in the experimental conditions, or intrinsic differences in the species as well as in their sperm cells. We have reported in this chapter significant differences concerning both the efficiency and the reproducibility of the protocol in the murine and swine systems; it is worth recalling that epididymal sperm cells have to be used for mice, while ejaculated sperms are used for larger mammals after extensive washes. We do not presently know if the different type and degree of maturation of spermatozoa is responsible for the observed differences between the two species; we do, however, know that seminal fluid exerts an important role in the interaction between sperm cells and exogenous DNA (Lavitrano et al. 1992; Zani et al. 1995). It is plausible that factors present in the seminal fluid contribute to the capacitation of spermatozoa by altering the membrane structure, thereby causing modifications in the sperm permeability to foreign molecules. Sperm-mediated gene transfer remains as yet poorly understood in most of its molecular and biochemical aspects. We need to learn more about its molecular basis in order to establish reproducible protocols in all species. Significant progress in the identification of the basal factors involved in the interaction of sperm cells with DNA have started to reveal certain important features, such as the existence of DNA "receptors" on sperm cells and of antagonizing factor(s) in the seminal fluid of mammals. The precise biochemical nature of these molecules remains an open question and a challenge worth pursuing, both for its fundamental implications in our understanding of sperm cells and for the purpose of generating transgenic animals.

Acknowledgments. We thank Dr. R. Lorenzini for providing laboratory space to C. S. and B. M. The work described in this chapter was supported by grants from the Italian Ministry of Agriculture, D.M. 15262, in the framework of the RAIZ project Improvement of Farm Animal Reproduction and by CNR (Consiglio Nazionale delle Ricerche) Progetto Finalizzato Ingegneria Genetica.

References

Arezzo F (1989) Sea urchin sperm as a vector of foreign genetic information. Cell Biol Int Rep 13:391–404

Atkinson PW, Hines ER, Beaton S, Matthaei KI, Reed KC, Bradley MP (1991) Association of exogenous DNA with cattle and insect spermatozoa in vitro. Mol Reprod Dev 29:1–5

Bachiller D, Schellander K, Pell J, Ruether U (1991) Liposomes-mediated DNA uptake by sperm cells. Mol Reprod Dev 30:194–200

Balhorn R, Gladhill BL, Wyrobek AJ (1977) Mouse sperm chromatin proteins: quantitative isolation and partial characterization. Biochemistry 16:4074–4080

Ball GB, Leibfired ML, Lenz RW, Ax RL, Bavister BD, First NL (1983) Factors affecting successful in vitro fertilization of bovine follicular oocytes. Biol Reprod 28:717–725

Brackett BG, Boranska W, Sawicki W, Koprowski H (1971) Uptake of heterolougous genome by mammalian spermatozoa and its transfer to ova through fertilization. Proc Natl Acad Sci USA 68:353–357

Brinster RN, Sandgren EP, Behoursinger RR, Palmiter RD (1989) No simple solution for making transgenic mice. Cell 59:239–241

Castro FO, Hernandez O, Uliver C. Solano R, Milanes C, Aguilar A, Perez A, de Armas R, Herrera C, de la Fuente J (1991) Introduction of foreign DNA into the spermatozoa of farm animals.Theriogenology 34:1099–1110

Chou Q, Russel M, Birch DE, Raymond J, Bloch W (1992) prevention of the pre-PCR mis-priming and primer dimerization improves low-copy number amplification Nucleic Acids Res 20:1717–1723

Francolini M, Lavitrano M, Lora Lamia C, French D, Frati L, Cotelli F, Spadafora C (1993) Evidence for nuclear internalization of exogenous DNA into mammalian sperm cells. Mol Reprod Dev 34:133–139

Habrovà V, Takàc M, Navratil J, Macha J, Ceskova N, Jonàk J (1996) Association of Rous sarcoma virus DNA with *X. laevis* spermatozoa and its transfer to ova through fertilization. Mol Reprod Dev 44:332–342

Hochi S. Minomiya T, Mizuno A, Homma M, Yuchi A (1990) Fate of exogenous DNA carried into mouse eggs by spermatozoa. Anim Biotechnol 1:25–30

Horan R, Powell R, Mc Quaid S. Gannon F, Houghton JA (1991) The association of foreign DNA with porcine spermatozoa. Arch Androl 26:83–92

Khoo HW, Ang LH, Lim HB, Wong KY (1992) Sperm cells as vectors for introducing foreign DNA into zebrafish. Aquaculture 107:1–19

Lavitrano M, Camaioni A, Fazio VM, Dolci S, Farace MG, Spadafora C (1989) Sperm cells as vectors for introducing foreign DNA into eggs: genetic transformation of mice. Cell 57:717–723

Lavitrano M, French D, Zani M, Frati L, Spadafora C (1992) The interaction between exogenous DNA and sperm cells. Mol Reprod Dev 31:161–169

Nakanishi A, Iritani A (1993) Gene transfer in the chicken by sperm mediated methods. Mol Reprod Dev 36:258–261

Parrish JJ, Susko-Parrish JL, Liebfried-Rutledge ML (1986) Bovine in vitro fertilization with frozen-thawed semen. Theriogenology 25:591–600

Perez A, Solano, R, Castro FO, Lleonart R, de Armas R, Martinez R, Aguilar A. Herrera L, de la Fuente J (1991) Sperm cells mediated gene transfer in cattle. Biotecnol Apl 8:90–94

Rottmann OJ, Antes R, Hoefer P, Maierhofer G (1991) Liposomes mediated gene transfer via spermatozoa into avian eggs cells. J Anim Breed Genet 109:64–70

Schellander K., Peli J, Schmall F, Brem G (1995) Artificial insemination in cattle with DNA-treated sperm. Anim Biotechnol 6:41–50

Sin FYT, Bartley AL, Walker SP, Sin IL, Symonds JE, Hawke L, Hopkins CL (1993) Gene transfer in Chinook salmon (*Oncorhynchus* tshawytscha) by electroporating sperm in the presence of Prsv-Lacz DNA. Aquaculture 117:57–69

Sperandio S, Lulli V, Bacci ML, Forni M, Maione B, Spadafora C, Lavitrano M (1996) Sperm-mediated DNA transfer in bovine and swine species. Anim Biotechnol 7:59–68

Symonds TE, Walker SP, Sin FYT, Sin I (1994) Development of mass gene transfer method in chinook salmon: optimization of gene transfer by electroporated sperm. Mol Mar Biol Biotech 3:104–11

Whittingham DG (1971) Culture of mouse ova. J Reprod Fertil Suppl 14:7–21

Zani M, Lavitrano M, French D, Lulli V, Maione B, Sperandio S, Spadafora C (1995) The mechanism of binding of exogenous DNA to sperm cells: factors controlling the DNA uptake. Exp Cell Res 217:57–64

Chapter 14

Application of Ligand-Dependent Site-Specific Recombination in ES Cells

Pierre-Olivier Angrand*[1], Catherine P. Woodroofe[1], and A. Francis Stewart[1]

Introduction

Mammalian genetics has benefited greatly from the development of methods for culturing and genetically modifying mouse embryonic stem (ES) cells. ES cells are totipotent cells derived from the inner cell mass of 3.5-day blastocysts. They have two main characteristics:

- They can be modified in culture and reintroduced back into blastocysts, giving rise to manipulated chimeric animals (Thompson et al. 1989)
- They can spontaneously differentiate to various cell types under certain growth conditions.

As such, ES cells provide a fundamental model to explore the function of a gene in differentiation and development processes (Wiles 1993). A variety of genetic tools for ES cell manipulations are now available including homologous recombination, site-specific recombination, repressible promoters, and gene trap vectors (Friedrich and Soriano 1993; Bronson and Smithies, 1994; Ramírez-Solis and Bradley 1994). These different tools can be used individually or in combination in order to achieve many kinds of genetic alterations in ES cells. This chapter focuses on site-specific recombination since it represents one of the most promising ways to manipulate ES cells (Kilby et al. 1993; Sauer 1993, 1994; Rossant and Nagy 1995).

* corresponding author: phone: +49-6221-387260; fax: +49-6221-387518; e-mail: angrand@embl-heidelberg.de
[1] European Molecular Biology Laboratory, Gene Expression Program, Meyerhofstrasse 1, D-69117 Heidelberg, Germany

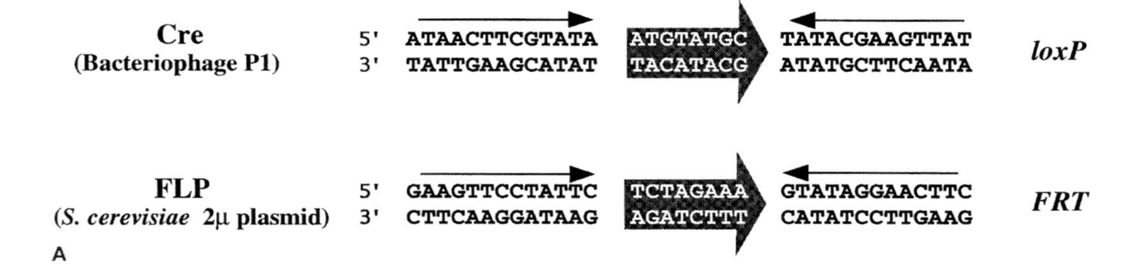

Fig. 14.1A, B. **A** Sequence and structure of the target sites, the *loxP* site, target for Cre, and the *FRT* site, target for FLP. SSRs bind to the 13-bp inverted repeat elements that flank the central core region. The core sequence is not involved in SSR binding, but is the region of strand exchange. The asymmetry of the core region gives directionality to the target sites. **B** Examples of SSR-mediated reactions. Depending on target site orientation and localization on DNA molecules, SSRs can induce DNA inversion, excision, integration, or translocation. Recombinase target sites are indicated as *triangles*

Site-Specific Recombinases

Site-specific recombinases (SSRs) are enzymes from bacterial and yeast elements that cleave and ligate DNA at specific targets inducing recombination. The simplest SSRs that have been well characterized are two members of the integrase family: the Cre recombinase from the bacteriophage P1, which recognizes a 34-bp sequence called *loxP* (Sternberg and Hamilton 1981), and the FLP recombinase from the *Saccharomyces cerevisiae* 2 μ circle, which recognizes the 34-bp *FRT* site (Falco et al. 1982). Each recognition target site consists of two 13-bp inverted repeats flanking an 8 bp core region (Fig. 14.1A). With an appropriate positioning of target sites, SSRs should allow the experimenter to invert, delete, insert, or translocate DNA molecules (Figure 14.1B):

a) Recombination between two inverted sites on the same DNA molecule inverts the DNA between them.

b) SSR-mediated recombination between two directly repeated sites causes excision of the intervening DNA (and one target site) as a circular molecule, leaving one target site resident in the parental DNA.

c) The converse of (b) permits the integration of circular DNA into another circular or linear molecule.

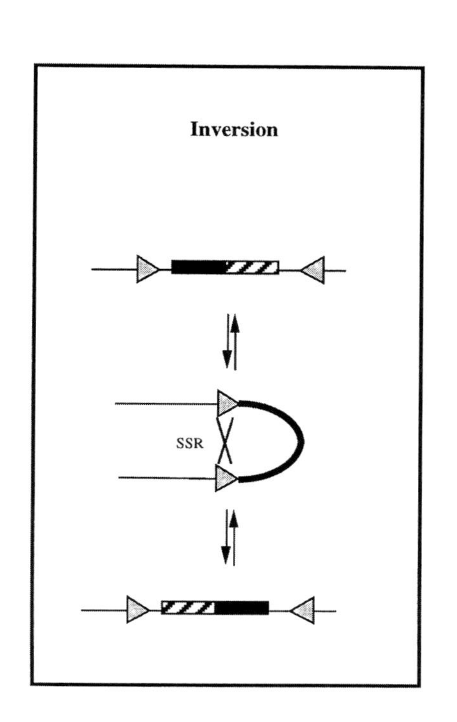
Inversion
SSR

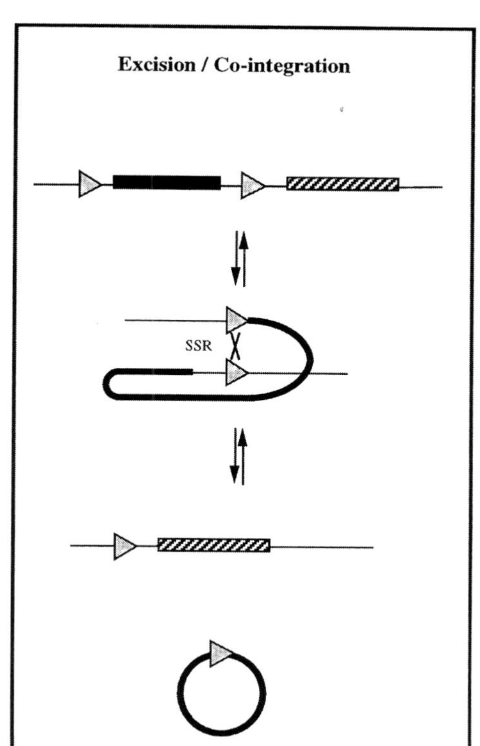
Excision / Co-integration
SSR

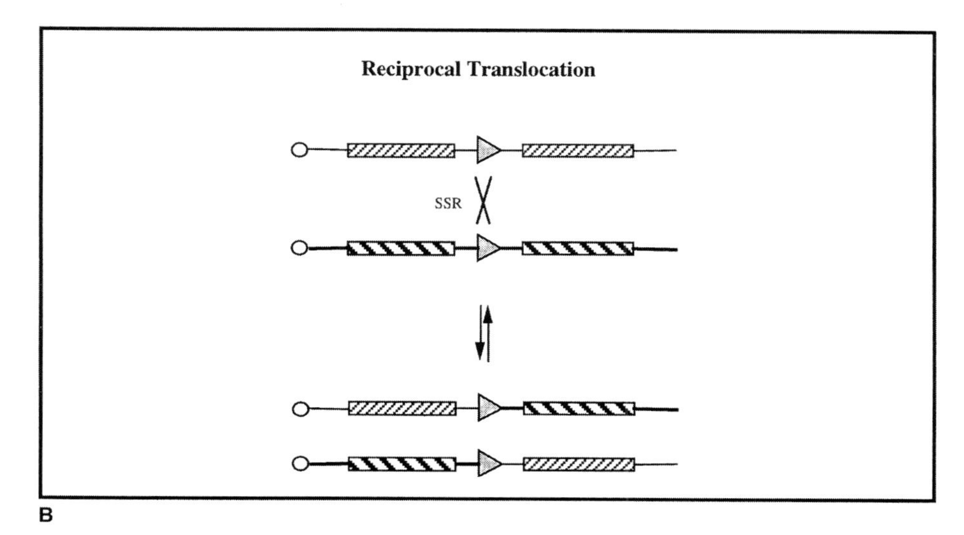
Reciprocal Translocation
SSR
B

Fig. 14.1B.

d) If target sites are present on separate linear molecules, recombination produces a mutual exchange of the distal regions.

Extensive biochemical analysis has shown that neither Cre nor FLP require additional proteins or cofactors for recombination activity. Thus, both Cre and FLP can be used in bacteria, yeasts, plants, flies, mammalian cells and mice (Cox 1983; Sauer 1987; Sauer and Henderson 1988; Golic and Lindquist 1989; Odell et al. 1990; Dale and Ow 1991; O'Gorman et al. 1991; Lakso et al.1992; Orban et al.1992). In particular, the potential of SSRs for DNA manipulation in mammals, as well as for studying differentiation processes, can be fully realized by their use in murine ES cells.

SSR Activity in ES Cells

Both Cre and FLP have been shown to induce recombination in ES cells in culture. In combination with conventional homologous recombination, the removing of a selectable marker flanked by two recognition target sites has been performed using both FLP/*FRT* and Cre/*loxP* systems. The J_H-E_μ switch region of an immunoglobulin H gene was flanked with *loxP* sites using homologous recombination, and subsequently deleted together with a *tk* selectable marker by transient expression of Cre (Gu et al. 1993). Similarly, FLP recombinase has also been used to delete an *FRT*-flanked *neo* selectable cassette (Jung et al. 1993). The Cre/*loxP* system has also been used to introduce large modifications such as 3–4-cM deletions, in the genome of ES cells (Ramírez-Solis et al. 1995).

Applications of site-specific recombination in ES cells is not limited to excisive recombination or intramolecular events. Recently, by placing *loxP* sites on different chromosomes, Cre-mediated chromosomal translocations between mouse chromosomes 2 and 13 and mouse chromosomes 12 and 15 have been described (Smith et al. 1995; Van Deursen et al. 1995).

However, in all cases reported so far, the source of SSR activity is provided by transient expression, and transfection efficiencies of ES cells probably account for the low efficiencies of SSR-mediated recombination achieved in these studies.

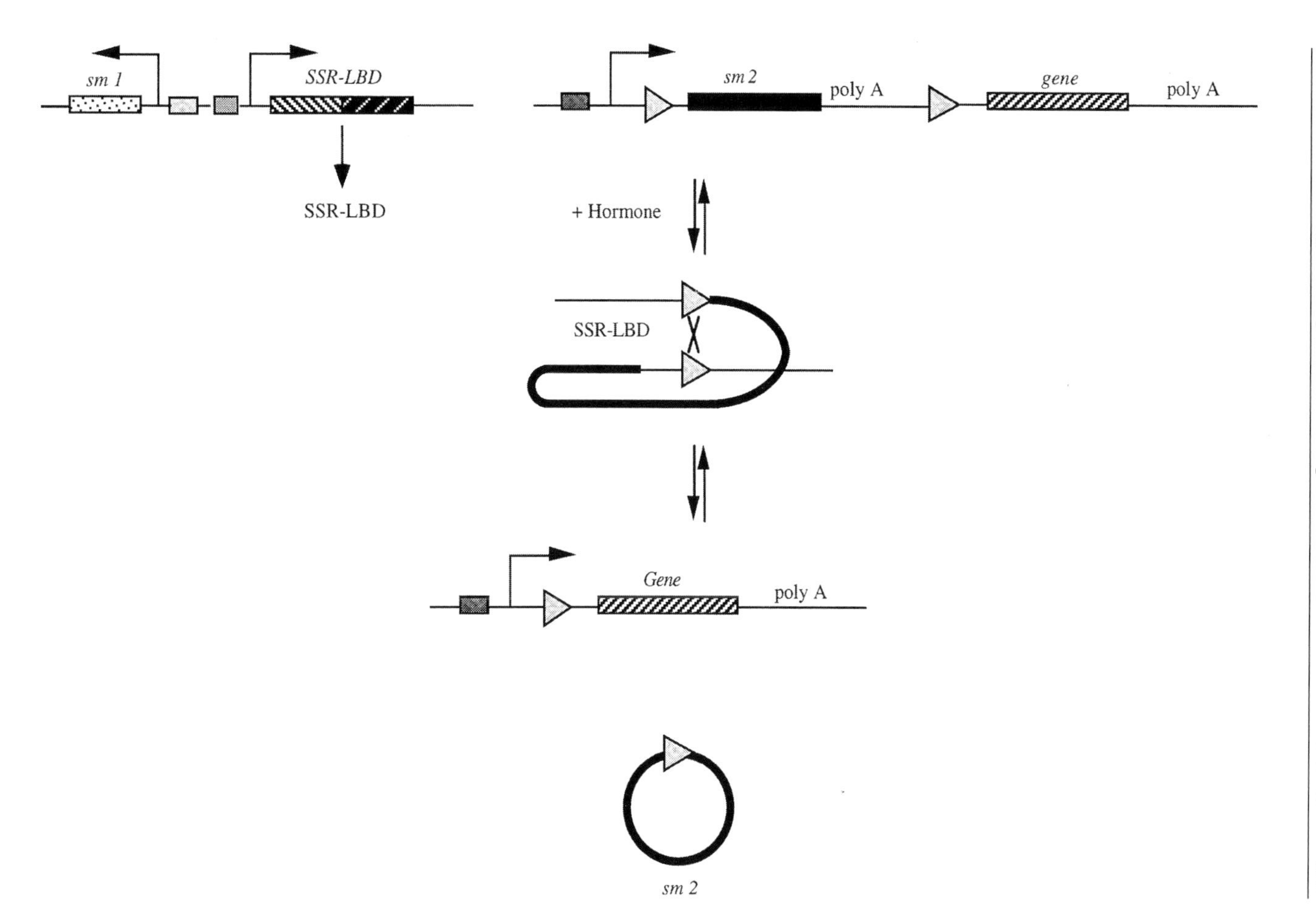

Fig. 14.2. Strategy used to set up the SSR-LBD-mediated inducible expression system.Two vectors are involved in the SSR-LBD inducible expression system; the SSR-LBD expression plasmid, at the *left*, is integrated into the genome using the first selectable marker (*sm 1*). The target-inducible expression vector is shown at the *right*. A promoter drives the expression of the second selectable marker gene (*sm 2*), used to integrate the inducible construct into the genome. This selectable marker gene is flanked by two recombinase target sites (*triangles*). Addition of the inducible ligand will activate the SSR-LBD, that will excise the selectable marker gene 2, and the inducible gene will be activated. *Arrows* indicate the location of the transcription start points

Ligand-Regulated SSR Activity

For many applications, regulation or control in time of the activity of SSRs may be necessary. Regulation of the activity of an SSR stably integrated in cells is also an alternative to transient expression of the protein. Because of the absence of satisfying regulated expression systems in ES cells, we recently developed an alternative strategy for regulating SSR activity. This strategy relies on fusing a ligand-binding domain (LBD) from steroid receptors onto the SSRs, thereby creating recombinases that require a ligand for activity.

The fusion of the LBD of steroid receptors to a variety of heterologous proteins, including transcription factors, oncoproteins, an RNA-binding protein, serine/threonine and tyrosine kinases results in a ligand-dependent control of their activity (Picard 1993; Mattioni et al. 1994). We and others have shown that LBDs can regulate FLP and Cre recombinases in different cell lines such as 293, CV-1, F9, as well as in ES cells (Logie and Stewart 1995; Metzger et al. 1995; Kellendonk et al. 1996; Zhang et al. 1996).

We report here the use of the SSR-LBD-regulated system as an inducible expression system.

SSR-LBD Inducible Expression System. Strategy and Vectors

The use of SSR-LBD to induce gene expression is based on a two step strategy that requires two components stably integrated into the genome of ES cells, using two different selectable marker (sm, Fig. 14.2). In this strategy, the first component is a SSR-LBD expression vector, and the second is the inducible gene expression plasmid. This construct contains a promoter driving the expression of a selectable marker gene cassette (sm2), flanked by two SSR recognition targets in a direct orientation. The downstream inducible gene is not expressed until SSR-LBD-mediated recombination is activated by the corresponding ligand. Recombination excises the selectable marker gene, whose presence has been blocking expression of the inducible gene.

Figure 14.3A shows an SSR-LBD expression vector. To select for integration of the SSR-LBD plasmids, a selectable marker gene cassette is included in the construct. A modified polyoma enhancer / HSV-Tk promoter (from pMC1 – Stratagene) drives the expression of the *hpt* (hygromycin phosphotransgerase) gene that confers resistance to hygromycin B. The *SSR-LBD* gene is under the control of the *phosphoglycerate kinase*

1 gene promoter (PGK), driving expression in ES cells. A number of different LBDs from steroid receptors can regulate SSRs, but rather than using the wild-type LBDs, mutant LBDs that are insensitive to endogenous ligands were chosen. In pHPK-FlpER1, the FLP recombinase is fused to the mutant human estrogen receptor, G521R (homologous to the murine mutant G525R; Danielian et al. 1993; Littlewood et al. 1995). This mutation confers to the FLP-EBD^{521R} fusion protein an activity dependent on the presence of synthetic antagonists such as 4-hydroxytamoxifen or raloxifene, but refractory to 17β-estradiol (C. Logie, M. D. Nichols and A. F. Stewart, submitted for publication). Therefore, phenol-red medium can be used and stripping of steroid hormones from the fetal calf serum is not necessary.

Figure 14.3B shows inducible gene expression plasmid for FLP (pPGKpaZ22). The puromycin resistance gene (*pac* – puromycin acetyltransferase; de la Luna and Ortín 1992) is placed between two *FRTs*, and under the control of the PGK promoter. The SV40 early polyadenylation signal (EpA) included before the second SSR target site, prevents expression of the downstream *lacZ* reporter gene. In this vector, the *pac* selectable gene was chosen for the following reasons:

1. PAC expression does not interfere with hygromycin B or G418 selections. Therefore, it is possible to perform double-selection experiments, and to transform cell lines that already express one of the resistant gene.
2. Selection with puromycin is fast because sensitive cells die after about 24 h of treatment. This property can be used as a screening procedure in a puromycin killing assay (see below).
3. In contrast to the *neo* gene, which has been shown to decrease the expression of co-transforming genes, *pac* does not influence adjacent promoters (Artelt et al. 1991).
4. The *pac* gene can be used both as a selectable marker and as a reporter gene. Therefore, it is possible to predict the relative expression of the inducible gene by determination of PAC activity before recombination.
5. *pac*-expressing ES cell can be grown in presence of puromycin for at least three passages and are still able to give chimeric mice with germ line transmission (P.-O. Angrand, A. R. Plück, and A. F. Stewart, unpubl.).
6. The cost of puromycin required for the selection is about 50-fold less expensive than that for either neomycin or hygromycin B.

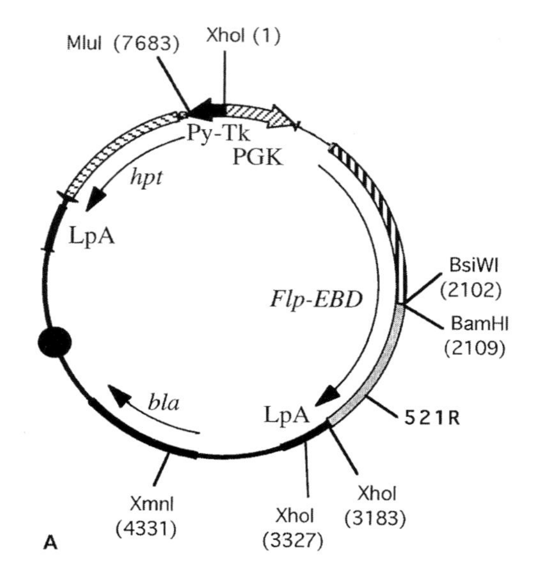
MluI (7683)
XhoI (1)
Py-Tk
PGK
hpt
LpA
BsiWI (2102)
Flp-EBD
BamHI (2109)
bla
LpA
521R
XmnI (4331)
XhoI (3327)
XhoI (3183)
A

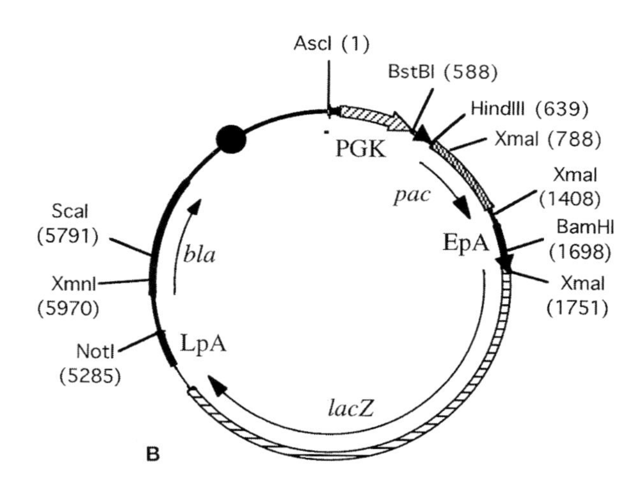
AscI (1)
BstBI (588)
HindIII (639)
XmaI (788)
PGK
pac
XmaI (1408)
ScaI (5791)
bla
EpA
BamHI (1698)
XmnI (5970)
XmaI (1751)
NotI (5285)
LpA
lacZ
B

◀ **Fig. 14.3A, B.** Vectors of the SSR-LBD inducible expression system. **A** SSR-LBD expression vector. In pHPK-FlpER1, the PGK promoter drives the expression of FLP-EBD[521R], and a modified polyoma enhancer / HSV-Tk promoter drives the expression of the *hpt* selectable gene. Position of the mutation of the estrogen binding domain that confer ability to bind synthetic antiestrogens, but not the physiological ligand is indicated as *521R*. **B** Target-inducible expression vector. In pPGKpaZ22, the PGK promoter drives the expression of the *pac* selectable gene. The *pac* gene is flanked by two *FRTs* (*triangles*) in direct orientation. The *lacZ* reporter gene is placed in 3' of the downstream *FRT. PGK* Phosphoglycerate kinase 1 gene promoter; *Py-Tk* F441 polyoma enhancer / HSV-Tk promoter; *EpA* SV40 early polyadenylation signal; *LpA* SV40 late polyadenylation signal; *bla* beta-lactamase gene; *pac* puromycin acetyltransferase gene; *hpt* hygromycin phosphotransferase gene. colE1 replication origins are indicated by *black circles*. Restriction site positions are indicated in base pairs

14.1 SSR-Mediated Inducible lacZ Expression

The setup of an inducible expression system based on SSR-LBD requires two rounds of tranfection / selection. In the first step, the target-inducible plasmid is introduced into ES cells, and transfectant cells are isolated in a puromycin-containing medium. These clones are subsequently subjected to a second transformation step to introduce the SSR-LBD expression vector, and secondary clones are isolated using hygromycin B selection.

Among the different methods for introducing DNA into ES cells, electroporation is the most efficient. SSR-mediated inducible *lacZ* expression is followed by in situ X-Gal staining of cells grown in absence and in presence of the inducing ligand.

Materials

- ES cells exponentially growing
- ES medium:
  DMEM with high glucose (4.5 g/l)
  15 % fetal calf serum
  Note: It is very important to use a batch of serum that gives optimal growth, with no observable cell differentiation
  100 units/ml Penicillin – 100 μg/ml Streptomycin (from a 100× stock, Seromed no. A 2213)

100 μM nonessential amino acids (from a 100× stock, Seromed no. K 0293)
1 mM sodium pyruvate (from a 100×stock, Gibco BRL no. 11360–039)
1 μM β-mercaptoethanol (from 100 μM stock)
Note: To make 100 μM β-mercaptoethanol stock: 70 μl of β-mercaptoethanol (Sigma no. M-6250) in 100 ml water, store aliquots at −20 °C.
2 mM L-Glutamine (from a 100× stock, Gibco BRL no. 25030–024)
500 u/ml LIF ESGRO (Gibco BRL no. 13275–029)

- Trypsin / EDTA (0.05/0.02 %, Seromed no. L 2143)
- 0.1 % gelatine (Sigma swine skin type II no. G-2500)
- Phosphate buffered saline (PBS):
  137 mM NaCl
  2.7 mM KCl
  6.4 mM Na_2HPO_4
  1.4 mM KH_2PO_4
- 1 μg/ml Puromycin. Prepare a 1 mg/ml stock solution: dissolve 10 mg of puromycin (Sigma no. P-7255) in 10 ml water, filtrate on sterile filters (0.22 mm), store 1-ml aliquots at −20 °C
- 400 μg/ml Hygromycin B (from a 2500× stock solution, stored at 4 °C, Calbiochem no. 400051)
  Note: Resistance of ES cells to various components can vary from one cell line to another. It is necessary to test the toxicity of drugs beforehand to find the lowest dose giving 100 % killing cells within 6 days.
- Electroporation apparatus (e.g. Bio-Rad Gene Pulser no. 165–207) equipped with a capacitance extender (Bio-Rad no. 165–2078)
- 0.4-cm electrode gap electroporation cuvettes (Bio-Rad no. 165–2088).

Procedure

1. Begin with enough exponentially growing ES cells. One 10-cm dish contains about $2-5\times10^7$ cells. Per electroporation, approximatively one 10-cm dish is needed. Wash plates with 3 ml of trypsin/EDTA, add 3 ml of fresh trypsin/EDTA, leave in the incubator for 10 min. Pipette up and down, check for single cells under microscope. If necessary, let incubate in trypsin longer. It is very important to have single cells for electroporation.

2. Add 7 ml of ES media, pipette up and down several times, then preplate in incubator for 15 min to let the feeder cells settle out.

3. In the meantime, gelatinize 10-cm dishes by adding 5 ml of 0.1 % gelatine onto the plate, take off, then aspirate excess liquid with a Pasteur pipette. Leave plates in the hood to dry a little.
4. After 2 min, gently collect the cells by washing over the plate and transfer into a 50-ml Falcon tube. Spin cells at 1000 rpm for 5 min, aspirate media, wash once in PBS, and resuspend cells at a density of 7×10^7 cells/ml in PBS.
5. Prepare electroporation cuvettes, add 40 µl of linearized DNA (0.5 µg/µl), add 800 µl cell suspension, mix gently, and cap tube. Set Bio-Rad electroporator at 240 V, 500 µF. Switch to time constant and pulse the cells. The pulse time should be between 5 and 6 ms. Leave cell/DNA suspension at room temperature for 5–10 min. Prepare tube with 5 to 10 ml of ES media per cuvette, add cells to media, rinse cuvettes, then plate cells into two to four 10-cm dishes, place dishes in the incubator at 37 °C, in a humidity-saturated 5 % CO_2 atmosphere.
6. The medium should be changed the next day to remove debris. ES cells should form small colonies, 10–20 % subconfluent.
7. 48 h after electroporation the medium is changed to ES cell selection medium.
8. Change media every 2 days and pick colonies soon after they appear (around 10 days after electroporation).

Note: After thawing, cells should be passaged twice before being used for electroporation experiments. It is important to maintain the starting cell population as well as their engineered derivatives under optimal growth conditions in culture. Therefore, to preserve totipotency of the modified ES cell lines, it is recommended to grow the ES cell line as well as the derivatives (when not grown under selection presure) on feeder layers.

14.2 In Situ X-Gal Staining of Cells

Materials

- 200 mg/ml X-Gal stock solution. 5-bromo-4-chloro-3-indolyl-β-D-galactopyranoside (X-Gal – Biomol no. 02249) is dissolved in dimethyl formamide (Merck). Strore solution at −20 °C

- 50 mM potassium ferricyanide stock solution. To make solution, dissolve 0.33 g potassium ferricyanide (Sigma no. P-8131) in 20 ml PBS. Store solution in the dark at 4 °C
- 50 mM potassium ferrocyanide stock solution. To make solution, dissolve 0.42 g potassium ferrocyanide (Sigma no. P-9387) in 20 ml PBS. Store solution in the dark at 4 °C
- Phosphate buffered saline (PBS):
  137 mM NaCl
  2.7 mM KCl
  6.4 mM Na_2HPO_4
  1.4 mM KH_2PO_4
- 25 % glutaraldehyde (Sigma no. G-6257)
- 37 % formaldehyde (Merck)
- 1 M $MgCl_2$

Procedure

1. Rinse cells with PBS (volume used is around 5 ml for a 10-cm dish, 3 ml for a 6-cm dish, 500 μl for a 24-well), and remove.
2. Add fixative solution. Fixative solution must be made fresh before use as follows:

For 10 ml, mix together:
9.42 ml PBS
541 μl formaldehyde (37 %)
40 μl glutardaldehyde (25 %)

3. Leave 2 min. The length of time the cells are exposed to fixative is important. Two min results in good presentation of cell structure, but fixation for longer reduces the intensity of the histochemical reaction product and renders the assay more difficult to interpret.
4. Wash plates with 3× PBS. The first wash must be as short as possible.
5. Add staining solution. Staining solution must be made fresh before use as follows:
   For 10 ml, mix together:
   7.9 ml PBS
   20 μl $MgCl_2$ (1 M)
   1 ml potassium ferricyanide (50 mM)
   1 ml potassium ferrocyanide (50 mM)
   50 μl X-Gal (200 mg/ml)

Note: Heat the staining solution at 50 °C before adding the X-Gal, in order to avoid its precipitation.

6. Incubate plates at 37 °C. A blue reaction product starts to be visible within 2 h, but the intensity of coloration continues to increase over 12 to 24 h.

Results and Comments

FLP-EBD^{521R}-Mediated Inducible Expression

R1 cells (Nagy et al. 1993) were first electroporated with the ScaI-linearized pPGKpaZ22 target plasmid. After selection with puromycin (1 μg/ml final concentration) for 10 days, puromycin-resistant colonies were picked randomly for expansion. Of these clones, none expressed detectable levels of β-galactosidase activity before exposure to FLP. After exposure to FLP, most clones began expressing *lacZ*, indicating that most clones are competent for FLP-mediated recombination. Some primary transfectants failed to respond to FLP. Since the frequency of spontaneous resistance to puromycin is low (about 10^{-7}), it is likely that in some cases the pPGKpaZ22 construct was rearranged during integration so that the rearranged construct is unable to give *lacZ* expression after FLP-mediated recombination.

Several independent clones were chosen for electroporation with the XmnI-linearized pHPK-FlpER1 SSR-LBD expression vector. Electroporated cells were seeded onto three plates:

a) In the first dish, cells were grown in presence of the double selection presure (puromycin 1 μg/ml plus hygromycin B 400 μg/ml). The effect of double selection pressure is
   - hygromycin selects for integration of the FLP-EBD^{521R} expression construct and
   - puromycin selects against recombination. Therefore, this plate served as the source of the secondary transfectant clones

b) In the second dish, cells were grown in presence of hygromycin B (400 μg/ml).

c) In the third dish, cells were grown in presence of hygromycin B (400 μg/ml) and the inducing ligand (100 nM 4-hydroxytamoxifen; OH-Tx).

eZ2::pHPK-FlpER1

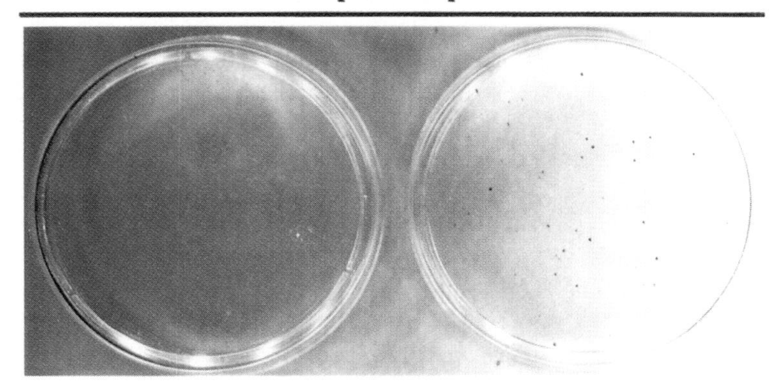

A

Hygromycin

Hygromycin
4 OH-Tamoxifen (100 nM)

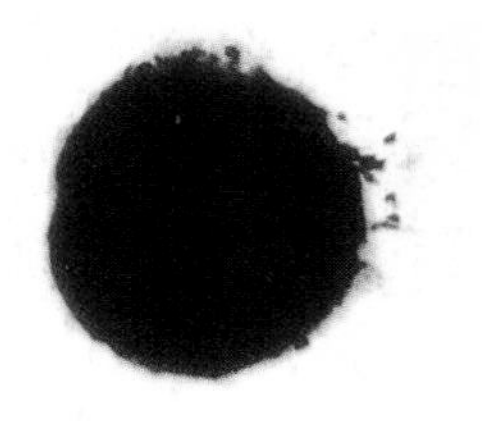

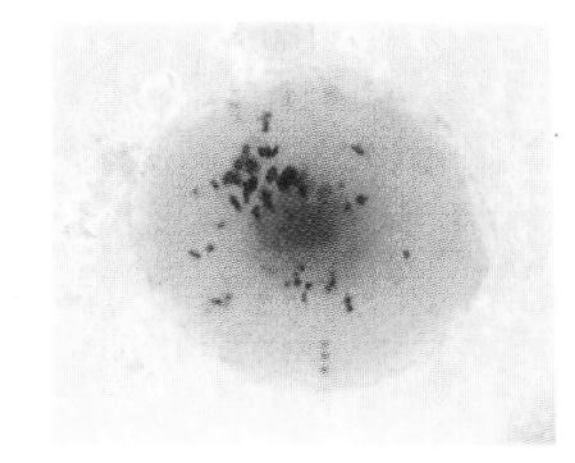

B

eZ2::pHPK-FlpER1

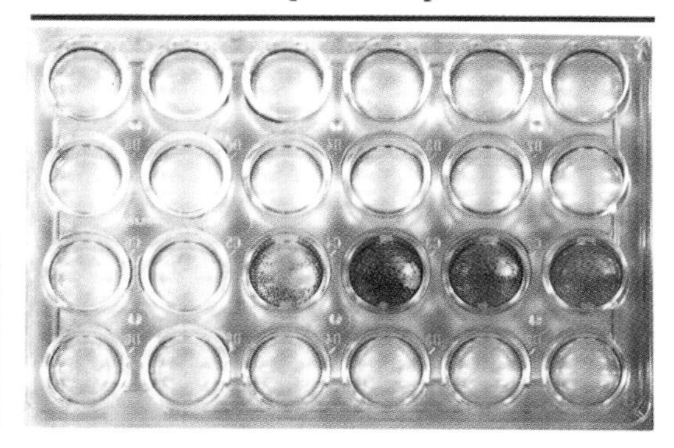

4 OH-Tamoxifen (100 nM)

C

◀ **Fig. 14.4A–C.** In situ *X-Gal* staining of eZ2 cells electroporated with the pHPK-FlpER1 construct. **A** A primary transfectant clone stably expressing the pPGKpaZ22 construct (eZ2) has been electroporated with the XmnI-linearized pHPK-flpER1 construct. After electroporation, cells are grown in an hygromycin B-containing medium without (*left*) or with (*right*) 4-hydroxytamoxifen (100 nM) for 10 days. The colonies are then fixed and stained for the β-galactosidase activity. **B** Detail at a higher magnification of a [lacZ$^+$] colony (*left*) and a mosaic colony (*right*). **C** Twelve individual secondary transfected clones were cultured in a 24-well dish, without (*two top rows*), or with 4-hydroxytamoxifen (*two bottom rows*) for 3 days. The cells were then fixed and stained for the β-galactosidase activity. As can be seen, the clones in wells 3 to 6 express *lacZ* upon 4-hydroxytamoxifen treatment. Very little or no *lacZ* expression in the absence of ligand can be seen

As soon as hygromycin-resistant colonies are visible, these two dishes (b) and (c) are stained for β-galactosidase activity. This test indicates the level of inducible, ligand-mediated recombination. The purpose of the X-Gal staining test was to identify the frequency of responding transfectants, so that clones could be picked from the corresponding hygromycin- plus puromycin-containing dish (a).

An example of plates b) and c), stained for the β-galactosidase activity is shown in Fig. 14.4A. Typically, the frequency of [lacZ$^+$] colonies (Fig. 14.4B) is about 1/20, and the frequency of mosaic colonies (Fig. 14.4B) is around 1/5. In absence of ligand, no [lacZ$^+$] colonies, and occasionally a small number of mosaic colonies were observed, indicating that the FLP-EBD^{521R}-mediated inducible expression system is tightly regulated.

Picked clones from the dish (a) were tested for FLP-EBD^{521R}-mediated *lacZ* induction by OH-Tx individually in 24-well dishes (Fig. 14.4C). Typically, about 1/4 of the individual secondary transfectants give *lacZ* expression upon activation of FLP-EBD^{521R} by OH-Tx, and 80 % of them do not show [lacZ$^+$] cells in absence of ligand.

14.3 SSR-LBD-Mediated cDNA-Inducible Expression

SSR-LBD-inducible recombination can be applied to activate expression of cDNAs as well as antisense RNAs.

Strategy and vectors. In order to induce expression of cDNAs, the *lacZ* gene has been replaced by a multiple cloning site in the inducible gene expression plasmid. This multiple cloning site can be used to insert cDNAs, and the experimental procedure is as described in Section 14.3. However, the in-situ X-Gal staining of cells cannot be used as a screening procedure. This time, the secondary transfectants are screened by a puromycin killing assay.

Puromycin Killing Assay and Methylene Blue Staining

The principle of the puromycin killing assay is that the *pac* gene is:

1. The resistance gene marker used to select for stable integration of the inducible gene expression construct.
2. It is it self-surrounded by the SSR target sites.

If the SSR-LBD fusion protein is induced by the corresponding ligand, the *pac* gene will be excised. Therefore, the positive clones will die in the presence of both the inducing ligand and puromycin. Since puromycin kills sensitive cells very quickly, this screen can be easily performed in 24-well dishes, enabling the test of reasonably large numbers of colonies. Simple visual examination of the wells after about 3 days of treatment with the inducing ligand and puromycin permits rapid identification of inducible clones.

To preserve the result of the puromycin killing assay, a methylene blue staining of cells can be done as follows.

Materials

- Phosphate buffered saline (PBS)
  137 mM NaCl
  2.7 mM KCl
  6.4 mM Na_2HPO_4
  1.4 mM KH_2PO_4
- 99 % ethanol
- 1 % methylene blue solution (Sigma no. M-9140)

Procedure

1. Rinse cells with PBS (volume used is around 500 μl for a 24-well).
2. Add ethanol to fix the cells.
3. Leave for 5 min and discard ethanol.
4. Add 1 % methylene blue solution.
5. Leave for 5–10 min.
6. Wash two or three times with water.
7. Let the dishes dry in the air.

The Use of a cDNA-Reporter Dicistronic Strategy

Alternatively, in order to report the SSR-LBD-mediated cDNA inducible expression, we have developed a strategy based on a dicistronic mRNA. The multiple cloning site for cDNA expression is cloned upstream of the encephalomycarditis virus (EMCV) internal ribosome-entry site (IRES; Jackson et al. 1994) driving translation of the *lacZ-neo* fusion gene β*geo* (IRES-β*geo*; Mountford et al. 1994). The IRES-β*geo* cassette functions efficiently in undifferentiated ES cells (Mountford et al. 1994). Therefore, activation of the SSR-LBD fusion protein with the inducing ligand will not only excise the *pac* selectable marker gene and turn on expression of the cDNA, it will also give expression of the β*geo* selectable/reporter cassette, permiting selection for or screening by X-Gal staining for the inducible responsive clones.

Applications

Here, we describe an inducible expression system based on regulated activity of the SSR-LBD fusion proteins. In contrast to many other regulated expression systems that are not fully satisfying for application in ES cells, the SSR-LBD-mediated expression system is tight and efficient in these cells. One interesting corollary to this strategy is that very toxic genes (such as the diphtheria toxin gene; P.-O. Angrand and A. F. Stewart, unpubl.) can be tolerated in an inactive state, and then activated upon ligand administration.

In addition to the FLP-based experiments described here, we have recently shown that the Cre recombinase can similarly be used in an inducible expression system in ES cells in culture (Kellendonk et al. 1996). However, in constrast to FLP, all secondary transfectants (ES cells stably transfected with both a Cre-LBD expression plasmid and *loxP*-containing, inducible vector) showed variable levels of [lacZ$^+$] cells in absence of induction, indicative of a background level of ligand-independent recombination. This difference could reflect inherent differences in FLP and Cre activities (F. Buchholz, L. Ringrose, P.-O. Angrand, F. Rossi and A. F. Stewart, Nucleic Acid Res., in press).

Applications of SSR-LBD regulation are not restricted to the simple inducible expression systems. Because an SSR-LBD can be stably present in cells in an inactive form, and can be activated by the inducing ligand at any time, this system can, for instance, be used to remove a selectable marker in a conventional homologous recombination strategy. In this case, we predict that the efficiency of excision of a recognition target site-flanked cassette will be significantly superior to what is obtained by transient expression of the SSR. Moreover, the SSR-LBD system could be useful to generate time-controlled gene knockouts in mice. In this context, it is worth mentioning that the murine *PGK1* promoter drives widespread expression in transgenic mice (McBurney et al. 1994). The use of the G521R mutated version of the human estrogen receptor in the SSR-LBD fusion protein relies on induction mediated by synthetic steroids. Thus, interference by endogenous mechanisms is limited. Similarly, we have shown that the Cre recombinase fused to mutated version of the human progesterone receptor hPR891 (Vegeto et al. 1992) containing a 42-amino acid C-terminal deletion is active in response to the synthetic steroid RU486 at a concentration 100- to 1000-fold lower than required for abortion or antinidation, but not in response to the physiological progesterone (Kellendonk et al. 1996). We and others are testing the viability of the SSR-LBD system in animals.

Acknowledgments. Many thanks to the members of our laboratory for discussion and critical support. P-O A is a recipient of an EU Human Capital and Mobility Fellowship.

More detailed information about the vectors described here can be obtained by e-mail to angrand@embl-heidelberg.de.

References

Artelt P, Grannemann R, Stocking C, Frield J, Bartsch J, Hauser H (1991) The prokaryotic neomycin-resistance-encoding gene acts as a transcriptional silencer in eukaryotic cells. Gene 99: 249–254

Bronson SK, Smithies O (1994) Altering the mice by homologous recombination using embryonic stem cells. J Biol Chem 269:27155–27158

Cox MM (1983) The FLP protein of the yeast 2-microns plasmid: expression of a eukaryotic genetic recombination system in *Escherichia coli*. Proc Natl Acad Sci USA 80:4223–4227

Dale EC, Ow DW (1991) Gene transfer with subsequent removal of the selection gene from the host genome. Proc Natl Acad Sci USA 88:10558–10562

Danielian PS, White R, Hoare SA, Fawell SE, Parker MG (1993) Identification of residues in the estrogen receptor that confer differential sensitivity to estrogen and hydroxytamoxifen. Mol Endocrinol 7:232–240.

de la Luna S, Ortín J (1992) *pac* gene as efficient dominant marker and reporter gene in mammalian cells Methods Enzymol 216:376–385

Falco SC, Li Y, Broach JR, Botstein D (1982) Genetic properties of chromosomally integrated 2 mu plasmid DNA in yeast. Cell 29:573–584

Friedrich G, Soriano P (1993) Insertional mutagenesis by retroviruses and promoter traps in embryonic stem cells. Methods Enzymol 225 681–701

Golic KG, Lindquist S (1989) The FLP recombinase of yeast catalyzes site-specific recombination in the *Drosophila* genome. Cell 59:499–509

Gu H, Zou YR, Rajewsky K (1993) Independent control of immunoglobulin switch recombination at individual switch regions evidenced through Cre-loxP-mediated gene targeting. Cell 73:1155–64

Jackson RJ, Hunt SL, Gibbs CL, Kaminski A (1994) Internal initiation of translation of picornavirus RNAs. Mol Biol Rep 19:147–159

Jung S, Rajewsky K, Radbruch A (1993) Shutdown of class switch recombination by deletion of a switch region control element. Science 259:984–987

Kellendonk C, Tronche F, Monaghan A-P, Angrand, P-O, Stewart AF, Schütz G (1996) Regulation of Cre recombinase activity by the synthetic steroid RU 486. Nucleic Acids Res 24:1404–1411

Kilby NJ, Snaith MR, Murray JAH (1993) Site-specific recombinases – tools for genome engineering. Trends Genet 9:413–421

Lakso M, Sauer B, Mosinger Jr B, Lee EJ, Manning RW, Yu SH, Mulder KL, Westphal H (1992) Targeted oncogene activation by site-specific recombination in transgenic mice. Proc Natl Acad Sci USA 89:6232–6236

Littlewood TD, Hancock DC, Danielian PS, Parker MG, Evan GI (1995) A modified oestrogen receptor ligand-binding domain as an improved switch for the regulation of heterologous proteins. Nucleic Acids Res 23:1686–1690

Logie C, Stewart AF (1995) Ligand-regulated site-specific recombination. Proc Natl Acad Sci USA 92:5940–5944

Mattioni T, Louvion J-F, Picard D (1994) Regulation of protein activities by fusion to steroid binding domains. Methods Cell Biol 43:335–352

McBurney MW, Staines WA, Boekelheide K, Parry D, Jardine K, Pickavance L (1994) Murine PGK-1 promoter drives widespread but not uniform expression in transgenic mice. Dev Dyn 200:278–293

Metzger D, Clifford J, Chiba H, Chambon P (1995) Conditional site-specific recombi-

nation in mammalian cells using a ligand-dependent chimeric Cre recombinase. Proc Natl Acad Sci USA 92:6991–6995
Mountford P, Zevnik B, Duwel A, Nichols J, Li M, Dani C, Robertson M, Chambers I, Smith A (1994) Dicistronic targeting constructs: reporters and modifiers of mammalian gene expression. Proc Natl Acad Sci USA 91:4303–4307
Nagy A, Rossant J, Nagy R, Abramow-Newerly W, Roder JC (1993) Derivation of completely cell culture-derived mice from early-passage embryonic stem cells. Proc Natl Acad Sci USA 90:8424–8428
Odell J, Caimi P, Sauer B, Russell S (1990) Site-directed recombination in the genome of transgenic tobacco. Mol Gen Genet 223:369–378
O'Gorman S, Fox DT, Wahl GM (1991) Recombinase-mediated gene activation and site-specific integration in mammalian cells. Science 251:1351–1355
Orban P C, Chui D, Marth JD (1992) Tissue- and site-specific DNA recombination in transgenic mice. Proc Natl Acad Sci USA 89:6861–6865
Picard D (1993) Steroid-binding domains for regulating the fuctions of heterologous proteins in cis. Trends Cell Biol 3:278–280
Ramírez-Solis R, Bradley A (1994) Advances in the use of embryonic stem cell technology. Curr Opin Biotech 5:528–533
Ramírez-Solis R, Liu P, Bradley A (1995) Chromosome engineering in mice. Nature 378:720–724
Rossant J, Nagy A (1995) Genome engineering: the new mouse genetics. Nat Med 1:592–594
Sauer B (1987) Functional expression of the cre-lox site-specific recombination system in the yeast *Saccharomyces cerevisiae*. Mol Cell Biol 7:2087–2096
Sauer B (1993) Manipulation of transgenes by site-specific recombination: use of Cre recombinase. Methods Enzymol 225:890–900
Sauer B (1994) Site-specific recombination: developents and applications. Curr Opin Biotech 5:521–527
Sauer B, Henderson N (1988) Site-specific DNA recombination in mammalian cells by the Cre recombinase of bacteriophage P1. Proc Natl Acad Sci USA 85:5166–5170
Smith AJH, De Sousa MA, Kwabi-Addo B, Heppell-Parton A, Impey H, Rabbitts P (1995) A site-directed chromosomal translocation induced in embryonic stem cells by Cre-loxP recombination. Nat Genet 9:376–385
Sternberg N, Hamilton D (1981) Bacteriophage P1 site-specific recombination. I. Recombination between loxP sites. J Mol Biol 150:467–486
Thompson S, Clarke AR, Pow AM, Hooper ML, Melton DW (1989) Germ line transmission and expression of a corrected HPRT gene produced by gene targeting in embryonic stem cells. Cell 56:313–321
Van Deursen J, Fornerod M, Van Rees B, Grosveld G (1995) Cre-mediated site-specific translocation between nonhomologous mouse chromosomes. Proc Natl Acad Sci USA 92:7376–7380
Vegeto E, Allan GF, Schrader WT, Tsai MJ, McDonnell DP, O'Malley BW (1992) The mechanism of RU486 antagonism is dependent on the conformation of the carboxy-terminal tail of the human progesterone receptor. Cell 69:703–713
Wiles MV (1993) Embryonic stem cell differentiation in vitro. Methods Enzymol 225:900–918
Zhang Y, Riesterer C, Ayrall A-M, Sablitzky F, Littlewood TD, Reth M (1996) Inducible site-specific recombination in mouse embryonic stem cells. Nucleic Acids Res 24:543–548

Chapter 15

A Modified Protocol for the Generation of Transgenic Mice and Freezing of Mouse Embryos

CAROL MURPHY[1]*, LUIS MARTIN-PARRAS[2], AND ULRICH RÜTHER[3]

Introduction

The generation of transgenic mice has become routine in many laboratories, and the procedures for microinjection of fertilized mouse eggs are well described in Hogan et al. (1986). Therefore, here we will only mention steps which deviate from those widely described. We have, over the years, simplified several steps, and consider it worthwhile to document this procedure.

Another point of interest is an economical and fast method to store transgenic mouse lines, avoiding the use of expensive equipment. We present two methods which we have used, the second is rather more convenient and successful than the first, and has been developed by Shaw et al. (1991a).

15.1 Preparation of DNA for Microinjection

Plasmid or cosmid DNA is purified on Qiagen columns and digested to separate vector sequences. Cesium chloride gradients of the plasmid DNA are unnecessary, and cruder methods to isolate the plasmid DNA can also be used.

The digested DNA is separated by agarose gel electrophoresis. The use of low melting point agarose is unnecessary. The fragment is cut from

* corresponding author: phone: +30-651-28388; fax: +30-651-33442; e-mail: cmurphy@cc.uoi.gr

[1] Laboratory of Biological chemistry, Medical School, University of Ioannina, 45110 Ioannina, Greece

[2] Institut de Génétique et de Biologie Moléculaire et Cellulaire (IGBmc), 1 rue Laurent Fries, 67400 Illkirch - C.U. de Strasbourg, France

[3] Institut für Molekularbiologie, Medizinische Hochschule, Hanover, Germany

the gel, avoiding the use of UV light (see below), purified using glass milk (Bio 101) and resuspended in microinjection buffer at a concentration of >100 ng/μl (10 mM Tris pH 7.6, 0.25 mM EDTA, filtered through a 0.22 μm filter) and stored at −20 °C.

The method we use to excise the agarose gel piece without exposure to UV light is as follows:

Procedure

1. The agarose gel is placed onto a sheet of plastic and, using a scalpel, a small fraction of the lane containing the digested DNA for microinjection is cut away from the gel. Only this small piece is visualized by UV light and the position of the fragment of interest is marked with the scalpel. The gel pieces are then aligned and the band of interest is cut out.
2. Following excision of the DNA fragment the gel is checked by UV light to ensure that the band of interest if no longer present. The fragment can also be electroluted and does not adversely effect the number of positive transgenic mice generated.
3. The DNA solution is then centrifuged 30 min in a microfuge at top speed, and diluted to a concentration of 1 ng/μl. This is then used for microinjection and is centrifuged for 30 min prior to injection.

15.2 Isolation of Fertilized Mouse Eggs

Materials

Media

- Modified M2 medium

10× Hanks	5 ml
40× Hepes	1.25 ml
300× Na-pyruvate	165 μl
Pen/strep	0.5 ml
ddH_2O	43 ml
1 M NAOH	0.25 ml
BSA	0.2 g

- Modified M16 medium

10× Earles	1 ml
37.5× $NaHCO_3$	270 μl

300× Na-pyruvate	33 μl
Pen/strep	100 μl
ddH_2O	8.6 ml
BSA	40 mg

Both solutions should be made up in a sterile flow hood and are stable when stored at 4 °C for 1 week.

Suppliers

– 10× Earles	Gibco	042–4050 H
– 10× Hanks	Gibco	042–4060 H
– 100 mM Na-pyruvate	Gibco	043–1360 H
– 1 M Hepes	Gibco	043–5630 H
– 7.5 % $NaHCO_3$	Gibco	043–5080 H
– BSA	Sigma	A–9647

- Mineral oil: Mineral oil purchased from Sigma is already tested for embryo toxicity (M8410).

These are the media we routinely use for incubation of eggs for microinjection. They are not suitable for culturing the eggs. We use these media as an alternative to the M2 and M16 described by Hogan et al. (1986), due to the ease of preparation. Alternatively, one can also now buy the media from Sigma.

Our procedure for microinjection is as follows;

Procedure

1. Isolation of fertilized eggs in modified-M2 medium.

2. Hyaluronidase treatment in mod-M2.

3. Washing away hyaluronidase in mod-M2.

4. Microinjection immediately in mod-M2, overlaid with mineral oil.

5. Transfer into the infundibulum of a pseudopregnant female.

The entire procedure takes approximately 3–4 h. If the eggs are isolated and microinjection is not carried out immediately, then the eggs are incubated in mod-M16 medium overlaid with mineral oil at 37 °C, 5 % CO_2.

15.3
Preparation of Slides for Microinjection

Materials

- Histowax: Jung Histowax, Reichert Jung. Cat. no. 037 408585, 2.5 kg
- Coverslips 24×60 mm

Procedure

The method we use is shown schematically in Fig. 15.1. Plastic slides are prepared to fit the microscope stage. Each contains a circular hole in the center. The plastic slide is placed on a heating block at approximately 70–80 °C. Histowax is melted and several small drops placed on the plastic slide around the central hole. A glass coverslip is placed on top of the melted wax, over the hole, and then when the slide/coverslip is placed on the bench at RT, the wax solidifies and the chamber is ready for use. Following use, the plastic slide/coverslip is again heated to remove the coverslip as the wax melts, the coverslip is discarded, and the plastic slide is washed and recycled.

15.4
Detection of Transgenic Mice

Materials

- Tail buffer
  50 mM Tris-HCl, pH8.0
  100 mM EDTA
  100 mM NaCl
  1 % SDS
- Proteinase K: Merck, cat. no. 24568, 100 mg. Stock solution of 10 mg/ml. Resuspend in 10 ml distilled H_2O, aliquot and store at −20 °C.
- 10× PCR buffer
  500 mM KCL
  200 mM Tris-HCl pH 8.4
  25 mM $MgCl_2$
  1 mg/ml BSA (Boehringer, Mannheim cat. no. 711454, 20 mg/ml).

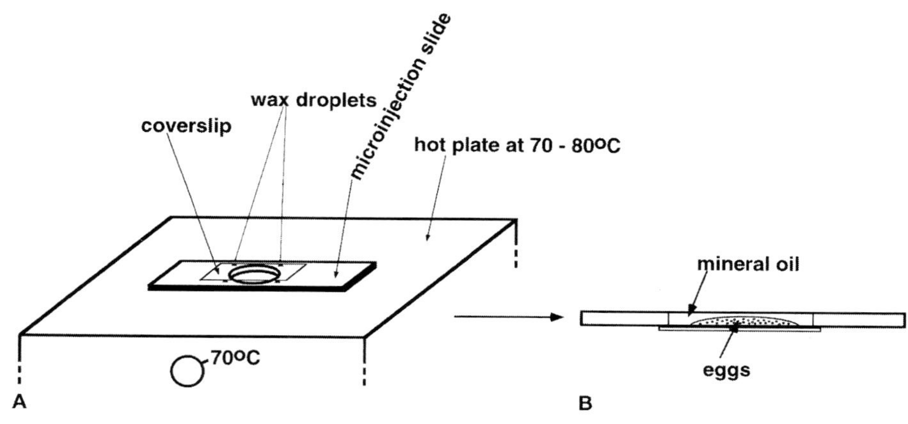

Fig. 15.1A,B. Construction of microinjection slides. **A** The slide is heated to 70–80 °C on a hot plate and several drops of melted wax are placed onto the heated slide. A glass coverslip is then placed on the wax. When the wax spreads out and surrounds the central hole in the slide, the slide is then placed on a cold surface until the wax solidifies. **B** A drop of either M2 or M16 modified medium is placed on the coverslip and overlaid with light mineral oil. Using a glass capillary, the eggs are then placed into the medium

- dNTP mix: 2 mM stock
- Source of dNTPS: Pharmacia 100 mM stocks:
  dATP, cat. no.: 27-2050-01
  dGTP, cat. no.: 27-2070-01
  dTTP, cat. no.: 27-2080-01
  dCTP, cat. no.: 27-2060-01
  The nucleotides are diluted to 2 mM in distilled H_20, and stored at −20 °C.
- Taq-polymerase: Eurotaq, cat. no. ME 0060 05. Eurogentec, 4102 Seraing, Belgium

PCR program

The program is, of course, dependent on the annealing temperature of the oligonucleotides and on the size of the DNA fragment to be amplified. We design primers to amplify a DNA fragment of approximately 500–600 bp, and use a quick cycling program.

94° 30 s, 50° 30 s, 72° 30 s, 30 cycles; 25 µl are loaded on an agarose gel for analysis.

A quick and reliable method for the extraction of genomic DNA to detect transgenic mice is presented. It is based on the salting-out approach described by Miller et al. (1988). The analysis of founder transgenic animals is carried out by digestion of the genomic DNA, selecting a restriction pattern to ensure that no major rearrangement of the transgene has occurred, blotting the DNA and hybridization. Furthermore, Southern blot analysis of founder mouse tail DNA will indicate whether the transgene has integrated in more than one site, and will allow an accurate determination of transgene copy number. Once the founder animals are identified, we use polymerase chain reaction (PCR) analysis to detect the positive offspring. Tails can be cut between 5–7 days after birth.

Procedure

Tail DNA preparation

1. Cut tail and place in 750 μl tail buffer plus 40 μl proteinase K (400 μg) incubate overnight at 55 °C. This time can be reduced depending on the age of the animal.
2. Shake for 2 min on an Eppendorf mixer.
3. Add 250 μl 6 M NaCl.
4. Mix 2 min on an Eppendorf mixer.
5. Spin at maximum speed in a microfuge at RT for 5–10 min.
6. Take 750 μl supernatant, avoiding the pellet and lipid floating on top of the supernatant.
7. Add 500 μl isopropanol.
8. Mix for 2 min on an Eppendorf mixer.
9. Spin for 1 min at maximum speed in a microfuge at RT.
10. Discard supernatant, wash pellet with 1 ml 70 % ethanol. Do not dry the pellet.
11. Resuspend the pellet in 10 mM Tris/1 mM EDTA pH 8.0, 250–500 μl depending on size of tail (3 weeks old, 1-cm cut, resuspend in 500 μl; 5 days old, 0.5–1-cm, cut resuspend in 250 μl). Alternate between Eppendorf mixer and 37 °C heating block for at least 2 h to resuspend the DNA. There is no need to phenol-extract the DNA. Vortex before taking 5 μl for PCR or 50 μl for genomic Southern analysis.

For genomic Southern analysis, the DNA can be digested with most enzymes, Bam HI, Eco RI, Bgl II, etc.; however, some enzymes which are more sensitive to DNA purity should be avoided (e.g., Sac I).

Digestion

1. 50 μl tail DNA.
2. Add 6 μl appropriate 10× enzyme buffer and 30–40 units enzyme.
3. Mix on Eppendorf mixer for 1 min and incubate overnight at 37 °C in a 37 °C oven.

For Southern blot analysis, 10 μl of loading dye is added to the 60 μl digestion and 50 μl loaded directly onto an agarose gel; there is no need to precipitate the DNA. Following electrophoresis, the gel is blotted onto a membrane and prehybridization and hybridization carried out according to Church and Gilbert (1984).

PCR

1. To 5 μl tail DNA add 50 μl distilled H_2O in a PCR tube. Heat at 95 °C for 5 min in the PCR machine.
2. Spin down (optional) and add:
   10 μl 10× PCR buffer
   2 μl dNTP mix
   40 pmol each primer
   1.5 units Taq
   Adjust volume to 100 μl with H_2O
   Overlay with oil and cycle.

The Taq polymerase used is important. We have tested the enzyme from various companies and found that Stratagene and Eurogentec Taq polymerases are the best for use with the above method. Eurogentec is by far the cheapest enzyme and therefore we use it routinely for DNA tail analyses.

15.5 Mouse Embryo Freezing

Materials

- Freezing medium 1: EFS
  40 % ethyleneglycol (Sigma E9129) in1× PBS, 1 % FCS, 30 % Ficoll 70 (Pharmacia 17–0410–01), 0.5 M sucrose
  Slowly add to 70 ml of 1× PBS (stir on a magnetic plate, heat to approx. 40 °C), 30 g Ficoll, and 17 g sucrose. After complete dissolution add 1 ml of FCS. Take 64 ml of this solution and add 36 ml of ethyleneglycol. This solution of EFS can be stored at RT and used without any sterilization

- D-lactose was purchased from Sigma, L1768
- Straws for freezing embryos:
  Mini-pailletten 100 per pack, transparent: cat. no. A-201/90265 available from: IMV, Veteriner-medizinische Erzeugnisse, A. Albrecht, 88326 Aulndorf, Postfach 1351, Germany.
- Freezing medium 2:
  0.25 M sucrose, 4 mg/ml BSA (as used above in modified M2 medium) and 4.5 M dimethyl sulfoxide (DMSO; tissue culture tested from Sigma, cat. no. D-2650) in modified M2 medium. Lower concentrations of DMSO may cause chromosomal damage (Shaw et al. 1991b).

Procedure

Method 1: One to Four-Cell Stage

1. Embryos are isolated at the one-, two-, or four-cell stage. This method is not suitable for blastocyst freezing. Following hyaluronidase treatment, embryos are washed in mod-M2 medium and placed into 3 ml 1×PBS (phosphate buffered saline)/1 % fetal calf serum (FCS) and incubated at room temperature (RT) for 10 min.
2. Transfer between 10–15 eggs into a small volume into 3 ml of EFS (see above).
3. Wash the eggs in this 3 ml by sucking once into a capillary.
4. Suck all eggs into a freezing straw (see below) and seal it at one end with a hot forceps. There is no need to sterilize the straws.
5. After about 3 min place the straw straight into liquid nitrogen (LN).
6. Repeat the procedure with the next 10–15 eggs. Freeze in total at least ten straws of one batch of eggs.
7. Store them in a liquid nitrogen (LN) tank. Before you can be sure about a successful freezing, implant the eggs from one straw into a pseudopregnant female and analyze the mice born.

Defrosting of eggs

1. Take the straw out of LN and warm it up at 37 °C for a few seconds.
2. Expel the cryoprotectant containing the embryos into 3 ml 0.5 M lactose in PBS/1 % FCS and transfer immediately into 3 ml of fresh lactose/PBS/FCS-solution.

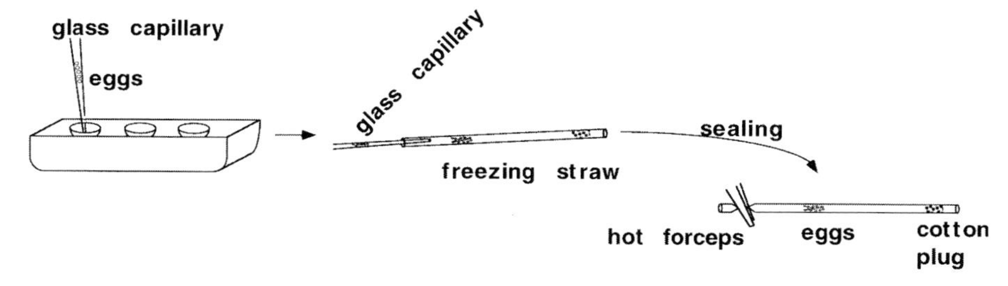

Fig. 15.2 Freezing of fertilized eggs. The eggs are transferred into freezing medium and inserted into the freezing straw using a glass capillary. The straw is then sealed at one end using a hot forceps. Following incubation on ice the straw is then frozen in liquid nitrogen

3. Incubate for 2 min at RT and transfer the eggs into PBS/1 % FCS.
4. Incubate for 20 min at RT and implant the intact eggs into pseudo-pregnant foster mothers.

Method 2: Mouse Embryo Freezing

1. The method is shown schematically in Fig. 15.2, and has been developed by Shaw et al. (1991a). Isolate either one-cell stage fertilized eggs or later stage embryos. Following hyaluronidase treatment and washing in modified-M2 medium, the embryos are transferred into the freezing medium 2 in a minimal volume of modified-M2. The embryos are placed into fresh freezing medium and 50 μl containing approximately 20 embryos are transferred into a precooled freezing straw.
2. The straw is heat-sealed at one end as described in freezing method 1 above and after 20–40 min at 0 °C the straws are frozen in LN.

Defrosting of eggs

Straws were thawed in a water bath at 37 °C for approximately 5–10 s, the end cut, and the embryos expelled into modified-M2 medium containing 0.25 M sucrose at RT for 10 min. Following this period the sucrose is removed by washing in fresh modified-M2 medium and embryos are transferred into pseudopregnant foster mothers.

Survival Rate of Embryos Frozen Using Methods 1 and 2

Method 1 above has been used to freeze mostly fertilized eggs and in our experience the survival rate is approximately 80 % before transfer into a pseudopregnant female, and 50 % following transfer. Shaw et al. (1991a) have systematically checked the survival of embryos frozen using method 2 at various stages of development. The highest survival rate (88 %) was achieved when eight-cell stage embryos were frozen following a 40-min incubation in cryoprotectant at 0 °C, using method 2 above. The values for fertilized eggs, two-cell stage, morulae, early to mid blastocysts and hatching blastocysts were 54, 69, 62, 65, and 33 %, respectively, all were incubated for 40 min at 0 °C with the exception of hatching blastocysts, which were incubated for 20 min The survival rate using method 2 is very good; the procedure is easy and reliable.

References

Church GM, Gilbert W (1984) Genomic sequencing. Proc Natl Acad Sci USA 81:1991–1995

Hogan B, Constantini F, Lacy E (1986) Manipulating the mouse embryo, a laboratory manual. Cold Spring Harbour Laboratory, Cold Spring Harbour

Miller SA, Dykes DD, Polesky HF (1988) A simple salting out procedure for extracting DNA from human nucleated cells. Nucl Acids Res 16:1215

Shaw JM, Diotallevi L, Trounson, AO (1991a) A simple rapid 4.5 M dimethylsulfoxide freezing technique for the cryopreservation of one-cell to blastocyst stage preimplantation embryos. Reprod Fertil Dev 3:621–626

Shaw JM, Kola I, MacFarlane DR, Trounson AO (1991b) An association between chromosomal abnormalities in rapidly frozen two-cell mouse embryos and the ice-forming properties of the cryoprotective solution. J Reprod Fertil 91:9–18

Chapter 16

The Use of Morula Aggregation to Generate Germline Chimeras from Genetically Modified Embryonic Stem Cells

Miguel Torres*

Introduction

Embryonic stem cells (ES) are undifferentiated totipotent cells. They are derived from the inner cell mass of mouse blastocysts and can be maintained indefinitely in culture (see for review Robertson 1987). When incorporated into preimplantation stage embryos, ES cells are able to form chimeras. Such embryos develop into adult mice in which ES cells contribute to all tissues, including the germline (reviewed by Robertson 1986). These unique characteristics have allowed the development of strategies based on the introduction of foreign DNA into ES cells, screening for rare integration events, and subsequent generation of mouse lines derived from the in vitro-modified ES cells. This strategy has been successfully employed in gene targeting experiments by homologous recombination, allowing the introduction of designed mutations into selected genes, and the generation of mouse lines carrying the mutated allele (see for review Capecchi 1989). ES cell technology has also allowed the use of gene trap strategies to isolate and mutate genes relevant in mouse development (Gossler et al. 1989). Methods for derivation, culture, and genetic modification of ES cells, either by gene targeting or gene trap, have been described elsewhere (Gossler and Zachgo 1993; Wurst and Joyner 1993).

Two alternative methods have been described for the introduction of ES cells into embryos:

- Microinjection of ES cells into the blastocoelic cavity of blastocyst stage preimplantation embryos
- Aggregation of ES cells with morula stage embryos (Bradley 1987).

* Departamento de Inmunología y Oncología, Centro Nacional de Biotecnología, Universidad Autónoma, Madrid 28049, Spain. phone: +34-1-5854530; fax: +34-1-3720493; e-mail: mtorres@samba.cnb.uam.es

Early experiments, however, showed that while both methods produce chimeric animals, only injection chimeras showed acceptable rates of germline contribution. For this reason, blastocyst microinjection has been the method of choice for most researchers to introduce genetically modified ES cells into embryos during the last decade. Only recently, non-injection methods have been improved in various laboratories, and it is becoming an interesting alternative to injection methods (reviewed by Wood et al. 1993). The availability of ES cell lines with specially good performance in aggregation experiments, the use of appropriate wells to hold the aggregates, and some modifications of previous culture media (Nagy et al. 1989, 1993) have made aggregation an efficient and versatile method for the generation of germline chimeras.

The use of aggregation to produce chimeras has several advantages with respect to the blastocyst injection method:

- No need for micromanipulation equipment
- Short time required to learn the technique
- Processing of larger numbers of embryos per experiment
- No need to use inbred mouse strains to obtain germline chimeras
- Possibility of generating complete ES cell-derived embryos and animals.

The protocols to perform morula aggregation described in this chapter have been used successfully during the past 4 years for the generation of germline chimeras from genetically modified ES cells our laboratory (Subramanian et al. 1995; Torres et al. 1995; Yamada et al. 1995; Mansouri et al. 1996; Torres et al., unpubl. results). These protocols were established for the R1 cell line and represent a modification of those described previously for this cell line (Nagy et al. 1993). However, such procedures have been shown not to work for a variety of other commonly used ES cell lines. Therefore, the use of the R1 cell line is strongly recommended. Morula aggregation is described here in detail, while references are quoted for those methods common to microinjection which have been profusely documented.

Outline

Figure 16.1 shows the schedule of a typical morula aggregation experiment.

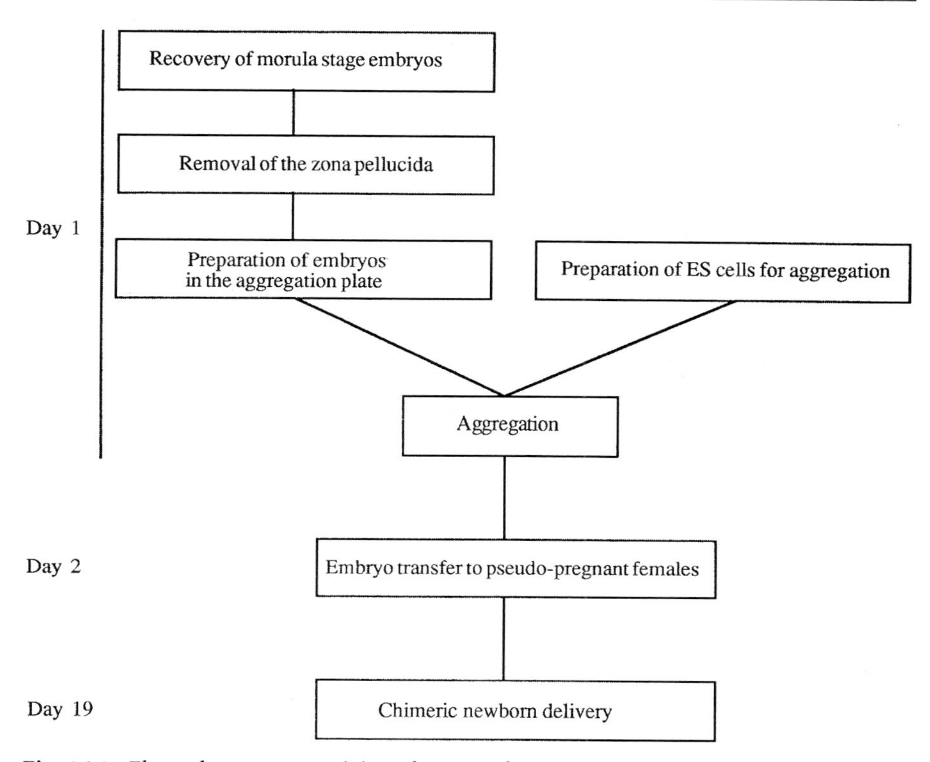

Fig. 16.1. Flow chart summarizing the morula aggregation procedure in temporal sequence

Materials

- Mice strains suitable for the experiments described here are available from Charles River
- R1 ES cells (Nagy et al. 1993) were obtained from Andras Nagy (Samuel Lunenfeld Research Institute, Mount Sinai Hospital, 600 University Av., Toronto, Canada M5G 1X5)
- Mouth-controlled micropipette: use a glass Pasteur pipette, heat the narrow end on the flame, pull, and break it to obtain a smooth end of approximately twice the diameter of an embryo. Connect the wide end of the pipette to a flexible rubber tube ending in a mouth adapter
- Conical-pointed darning needle. In principle, any needle with a smooth, conical point is adequate. If there are problems to find the proper needle, Biochemical Laboratory Service Ltd. (Budapest, Fax: 361 407-2602) provides suitable ones. Sterilize the needle with 70 % ethanol, never flame it or the point of the needle might be damaged.

- 3.5 cm diameter tissue culture plates (Falcon, Becton Dickinson)
- 30-gauge syringe needles (Becton Dickinson)
- 1-ml syringes
- Surgical instruments: watchmaker's forceps, blunt-ended forceps, scissors

Equipment

- 37 °C warming plate
- Dissection stereomicroscope with 10×–60× magnification and transmitted light illumination
- Humidified incubator, 37 °C and 5 % CO_2

Media and buffers

All reagents should be handled in sterile disposable plastic material in all steps, avoiding the use of detergents or any organic contaminants. Use double-distilled or Milli-Q water to prepare all buffers and media.

Acid Tyrode's solution, M2 and M16 media were prepared and handled as described previously (Hogan et al. 1994).

- Mineral oil Sigma M-8410
- Phosphate-buffered saline (PBS) without Ca and Mg
  8.0 g NaCl
  0.2 g KCl
  1.15 g $Na_2HPO_4.2H_2O$
  0.2 g KH_2PO_4

Adjust pH to 7.2 and bring to a total volume of 1 l. Sterilize by autoclaving and store at room temperature.

- ES cell culture medium

Component	Reference
1× DMEM (highglucose+Na-pyruvate+ glutamine)	Gibco 41966–029
1 ml/100 ml nonessential amino acids	Gibco 11140–035
10^{-4} β-mercaptoethanol	Sigma M7522
500 units LIF (Esgro)	Gibco 13275–029
20 % fetal calf serum	Tested batches

Keep at 4 °C

- Trypsin solution
  8.0 g NaCl
  0.40 g KCl
  0.10 g $Na_2HPO_4.2H_2O$
  1.0 g glucose
  3.0 g Trizma base

0.01 g phenol red
2.50 g trypsin (Difco 1:250)

Adjust pH to 7.6 and bring to a total volume of 1 liter, filter sterilize and store in 20 ml aliquots at −20 °C. This stock is diluted 1:4 in saline/EDTA for use as 0.05 % Trypsin/EDTA solution. Store at −20 °C in aliquots of 5–10 ml.

- Saline/EDTA for dilution of trypsin
  0.2 g EDTA (disodium salt)
  8.0 g NaCl
  0.2 g KCl
  1.15 g $Na_2HPO_4 \cdot 2H_2O$
  0.2 g KH_2PO_4

Adjust to pH 7.2 and bring to a total volume of 1 l, filter sterilize or autoclave. Store at room temperature.

Procedure

Recovery of Morula Stage Embryos

1. Open the abdominal cavities of 2.5-days postcoitum (dpc) pregnant females. Superovulated females from inbred (C57BL/6) and outbred (NMRI or CD1) strains are equally suitable for this procedure. Grasp the uterus with the blunt forceps at approximately 0.3 mm from the oviduct (see Fig. 16.2A), separate mesometrium by tearing with the watchmaker's forceps and, without releasing the uterus, cut with the scissors, first in the uterus (cut 1: in Fig. 16.2A) and then, between the oviduct and the ovary (cut 2 in Fig. 16.2A) without damaging the oviduct. The oviduct together with the upper part of the uterus, should be now free to be removed.

2. Proceed in the same manner with all mice and transfer the dissected pieces to a petri dish in M2 medium at 37 °C. Minimize blood carryover; it may be also necessary to clean the dissected pieces with Wipes.

Flushing embryos

3. Transfer the oviducts one by one to an embryological watch dish in a minimal amount of M2 medium. Fill a 1-ml syringe holding a 30-gauge needle with prewarmed M2 medium. Use the watchmaker's forceps and the tip of the needle to identify the opening of the oviduct (infundibulum) (Fig. 16.2B).Grasping the infundibulum with the for-

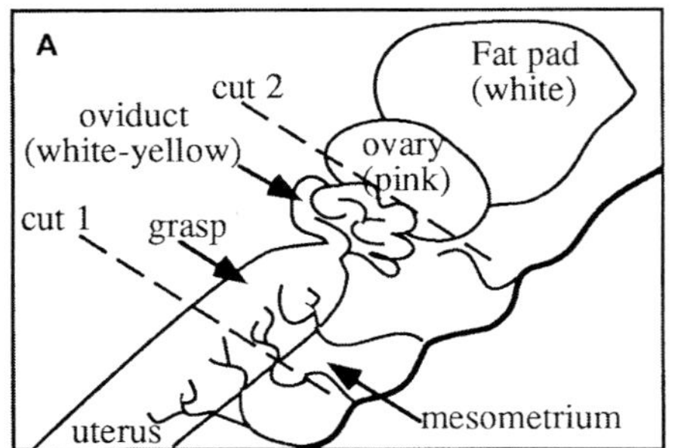

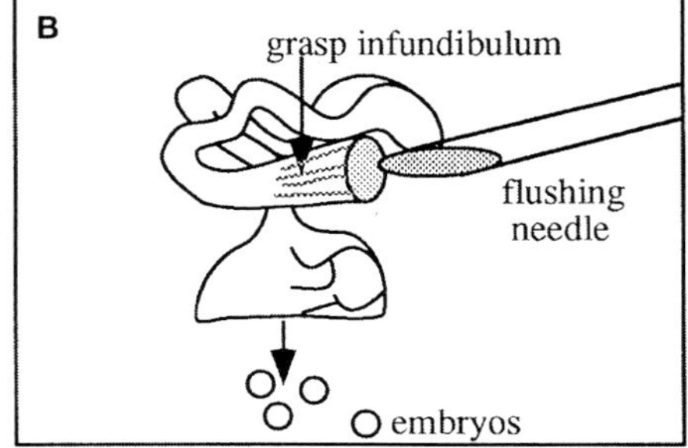

Fig. 16.2A, B. Eight-cell stage embryo recovery by oviduct flushing. **A** Disposition of the relevant components within the abdominal cavity and the dissection strategy. **B** Dissected oviduct and flushing strategy

ceps, introduce the tip of the needle through its lumen and flush about 0.3 ml of medium. The embryos should flow out through the cut end of the uterus. Discard the flushed oviduct and proceed the same with the rest of dissected uteri, all in the same dish.

4. Collect all embryos using the mouth pipette; select eight-cell stage embryos and discard underdeveloped embryos. Compacted and uncompacted embryos are equally suitable for aggregation. Transfer the selected morulas to an M2 drop in the zona removal dish.

Removal of the Zona Pellucida

1. Preparation of the zona removal dish. Prepare a 3.5 cm tissue culture dish with several 10-μl drops of acid Tyrode's (AT) solution and six M2 drops. Cover drops with mineral oil and place the dish on the warming plate at 37 °C.
2. Transfer embryos in successive groups of approximately 20 (beginners should start with lower numbers) into an AT drop, minimizing M2 carryover.
3. Monitor the dissapearance of the zona by constantly observing under the binocular microscope. This will take a variable time, from a few seconds to a minute, depending on the volume of M2 medium carried over together with the embryos. Use a fresh of AT drop for each group of embryos.

4. As soon as the zona disappears, transfer the embryos to the M2 (otherwise they will lyse), wash them by successive changes through five different M2 drops. From this moment on, the embryos are sticky, and care should be taken not to leave them in contact with each other because they will fuse.
5. Transfer embryos to a 20-µl drop of M16 in the aggregation dish (see next protocol).

Preparation of the Aggregation Dish

1. Prepare M16 medium supplemented with 2 % ES cell medium if using C57BL/6 embryos and 4 % if outbred embryos are used. This small proportion of ES cell medium does not compromise embryonic development and helps to favor a good contribution of ES cells to the embryo and good germline transmission (M. Torres, unpubl. observations).

Aggregation dish

2. Place two drops of 20 µl and ten drops of 10 µl of M16 supplemented with ES cells medium and cover them with oil.
3. Make six conical depressions on the surface of the dish underneath each small drop by pressing vertically with a darning needle while performing a slight circular movement (see Fig. 16.3).
4. Place one embryo in each conical depression and return the dish back to the incubator until the ES cells are ready for aggregation.

Preparation of ES Cells and Aggregation

Following the schedule outlined in Fig. 16.2, this is usually done late in the morning or in the early afternoon. The following steps are performed under the hood in the cell culture room.

1. ES cells for aggregation are split into a 3.5-cm plate, 2 days before aggregation is to be performed. Cell density should be kept low enough so that cells will have formed medium-sized colonies by the time of aggregation, but will not have reached confluence. When starting the experiment from a frozen stock, cells should be split at least twice before aggregation.

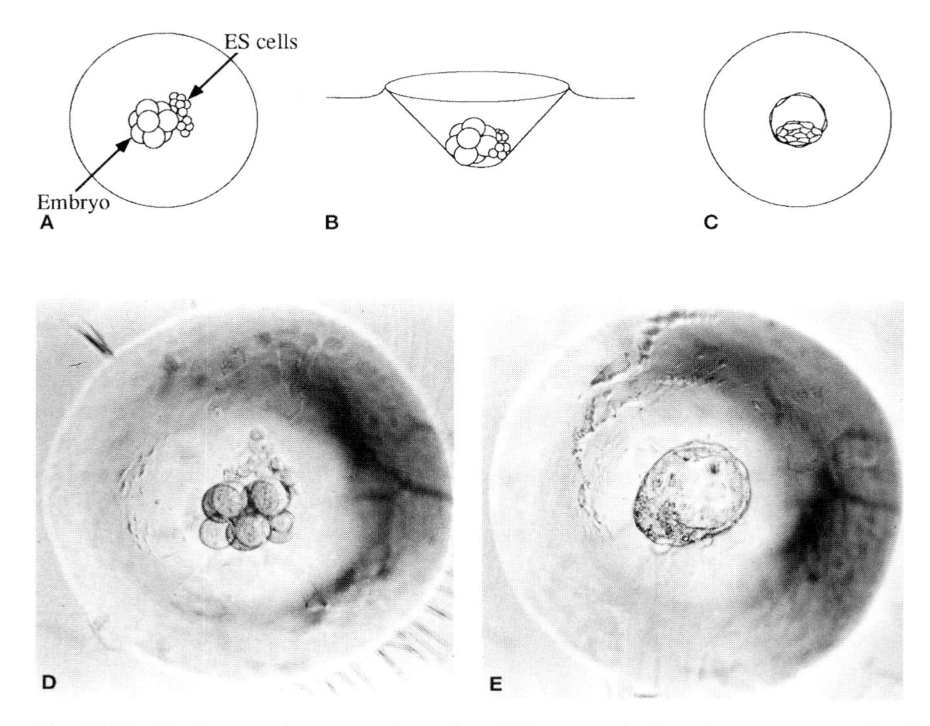

Fig. 16.3A–D. Aggregation procedure. **A** and **B**, respectively. Schematic top and side views of the conical depressions containing a clump of ES cells aggregated to an eight-cell stage embryo. **C** Top view of a blastocyst developing from the aggregate after overnight culture in the conical depression. **D** Eight-cell stage embryo aggregated with a clump of ES cells in the bottom of the depression. **E** A ready-to-transfer blastocyst, developing after overnight culture from an aggregate such as the one shown in **D**

Trypsinizing ES cells

2. Wash ES cells twice with PBS, add 0.5 ml of trypsin solution and incubate at 37 °C for 5 min.

3. Add 2 ml of ES cell culture medium and pipette up and down slowly several times using a 2 ml pipette. Check the cells under the microscope periodically and continue pipetting until cell clumps of the appropriate size are obtained. The number of cells that aggregate with each embryo is crucial for the success of the experiment. Aggregation with too many cells will kill the embryo, while too few cells results in a poor yield of chimeras. We have found that the number of ES cells which should aggregate to each embryo for an optimal result is in the

range of 10 to 15 for embryos from outbred strains and 6 to 8 for the C57BL/6 strain.

4. Transfer cell suspension to a sterile 15-ml plastic tube (Falcon or similar), dilute up to 4–5 ml with ES cell medium, cap tube, and bring it to the animal house. The cells should be used for aggregation within the next 30 min, otherwise cell clumps will aggregate to each other.

5. Pipette 2 ml of the ES cell clump suspension into a 3.5-cm plate and, under the binocular, load the pipette with an excess number of cell clumps of approximately the appropriate size (see step **3**). If counting the number of cells in each cell clump is difficult under the binocular, it is advisable to determine the number of cells in clumps of different sizes under phase contrast using an inverted microscope. Once the number of cells is determined, observe the clump size under the binocular and select clumps similar in size to the ones containing the right number of cells. Transfer the loaded clumps to the 20-μl drops in the aggregation plate.

Assembling aggregates

6. Further select individual clumps with the appropriate number of cells from the 20-μl drops, choose clumps containing cells of a similar size and showing a round shape. After this final selection, transfer the clumps to the aggregation drops and place one in each depression containing an embryo. The conical shape of the depression will force contact between ES cell clumps and embryos (see Fig. 16.3). Once aggregation has been completed in all wells, transfer the dish to the incubator (37 °C, 5 % CO_2) until the next day.

7. By the afternoon of the following day, most embryos will have incorporated all the ES cells and will have developed to the blastocyst stage (no more than 5 % should be underdeveloped; Fig. 16.3). Choose those embryos that have developed to the blastocyst stage and transfer six to eight of them into each uterus of 2.5-day pseudo-pregnant females as described (Hogan et al. 1994).

Chimeras

8. Birth will take place 16 to 17 days after embryos were transferred. Since ES cells carry the wild-type allele for the *agouti* locus, chimeric animals can be easily identified at birth by their skin and eye pigmentation, if an albino outbred mouse strain was used as recipient, or after 5 or 6 days by the hair color, if black recipients such as the C57BL/6 strain are used.

9. Once chimeric animals are identified, adjust litter size to six to eight pups by removing excess nonchimeric animals or by adding pups from a different litter of the same age.

Mating chimeras

10. Once chimeric animals reach sexual maturity (approximately 6 weeks), cross them back to animals of the strain from which embryos were obtained for aggregation. Germline transmitters are identified by the presence of *agouti* progeny.

11. Genotype agouti progeny at 3 weeks of age to identify those animals carrying the genetic modification(s) introduced into the ES cells. The most reliable method for genotyping is genomic Southern blot, but this requires considerable time and effort when handling large mouse colonies. PCR is a good alternative to Southern blot and minimizes genotyping effort but care should be taken to avoid sample contamination, which may lead to false positives.

Results

Using these protocols for gene targeting in the R1 ES cell line, we have obtained germline transmissions of 80 % of the targeted clones in different series of gene targeting experiments (Subramanian et al. 1995; Torres et al. 1995; Yamada et al. 1995; Mansouri et al. 1996; Torres et al. unpubl. results). On average, 4 % of the aggregated embryos will yield a germline transmitter; however, there is considerable variability in the efficiency of the different clones to produce germline chimeras. We recommend aggregating at least three independent clones for each targeting experiment and transfering of around 100 embryos for each clone. Considering that one person can aggregate and transfer 60 embryos each day, a gene-targeting experiment will take on average 6 days of one person's work.

The use of morula aggregation is a specially adequate method for large-scale analysis of gene trap ES cell clones, in which chimeras from a large number of clones need to be routinely generated. We have analyzed the expression pattern of more than 400 genes trapped in ES cells using either transient or germline chimeras (M. Torres, P. Bonaldo, G. Chalepakis, K. Chowdhury, S. Gajovic, C. Kioussi, A. Mansouri, A. Stoykova, T. Thomas, A. Voss, and P. Gruss, unpubl.). An example of a typical gene trap experiment is summarized in Table 16.1. This experiment was designed to minimize the effort required for each single clone, while still keeping a good proportion of clones producing germline chimeras. To

Table 16.1. Efficiency in the generation of germline chimeras by morula aggregation of gene trap clones

	Lines aggregated	Lines producing chimeras	Lines producing germline chimeras
Each week (average)	8	6	5
Total	140	112(82 %)	84 (60 %)

obtain these results, two persons were aggregating ES cells to embryos 4 days a week and each person aggregated one clone per day, transferring on average 40 embryos/day to pseudopregnant female mice.

Troubleshooting

- If the embryos do not develop properly to the blastocyst stage, check for possible sources of contamination. Remember that all material directly or indirectly in contact with the embryos should be maintained free of possible contaminants (especially any organic substances). Detergents should be avoided at all steps. When establishing embryo culture for the first time in a laboratory, it is advisable to compare different batches of oil and sources of water.
- If, after overnight culture during aggregation, ES cells do not incorporate into the blastocysts and are found spread on the bottom of the well, the amount of ES cell medium in the aggregation drops should be increased to determine empirically the optimal percentage.
- If the number of chimeric animals is consistently low, but there is good recovery of nonchimeric animals, the number of ES cells in each clump or the percentage of ES cell medium in the M16 should be increased to determine empirically the optimal.
- If the number of chimeric animals is consistently low and the number of nonchimeric animals is also low (e.g., less that three per foster mother), the number of ES cells in each clump or the amount of ES cell medium in the M16 medium should be decreased.

Acknowledgments. The author would like to thank Prof. Peter Gruss for constant support during the time these protocols were being developed in his laboratory, Sharif Mashur, Ronald Scholz, Mirjam Breuer, and Jens Krull for technical assistance, and Cathy Mark for critical reading of the manuscript. During the establishment of these procedures the author was supported by EMBO, European Community, and a grant from the AMGEN company.

References

Bradley A (1987) Production and analysis of chimeric mice. In: Robertson EJ (ed) Teratocarcinomas and embryonic stem cells, a practical approach. IRL Press, Oxford, pp 113–151

Capecchi MR (1989) The new mouse genetics: altering the genome by gene targeting. Trends Genet 5:70–76

Gossler A, Zachgo J (1993) Gene and enhancer trap screens in ES cell chimeras. In Joyner AL (ed) Gene targeting, a practical approach. Oxford University Press, Oxford, pp 181–228

Gossler A, Joyner AL, Rossant J, Skarnes WC (1989). Mouse embryonic stem cells and reporter constructs to detect developmentally regulated genes. Science 244:463–465

Hogan B, Beddington R, Constantini F, Lacy E. (1994) Manipulating the mouse embryo 2nd edn. Cold Spring Harbor Laboratory Press, Cold Spring Harbor

Mansouri A, Stoykova A, Torres M, Gruss P (1996) Dysgenesis of cephalic neural crest derivatives in *Pax-7 -/-* mutant mice. Development 122:831–838

Nagy A, Markkula M, Sass M (1989) Systematic non-uniform distribution of parthenogenetic cells in adult mouse chimeras. Development 106:321–324

Nagy A, Rossant J, Nagy R, Abramow-Newerly W, Roder JC (1993) Derivation of completely cell culture-derived mice from early-passage embryonic stem cells. Proc Natl Acad Sci USA. 90:8424–8428

Robertson EJ (1986) Pluripotent stem cell lines as a route into the mouse germline. Trends Genet. 2:9–13

Robertson EJ (1987) Embryo-derived stem cell lines. In: Robertson EJ (ed) Teratocarcinomas and embryonic stem cells, a practical approach. IRL Press, Oxford, pp 71–112

Subramanian V, Meyer BI, Gruss P (1995) Disruption of the murine homeobox gene Cdx1 affects axial skeletal identities by altering the mesodermal expression domains of Hox genes. Cell 83:641–653

Torres M, Gómez-Pardo EG, Dressler GR, Gruss P (1995) *Pax-2* controls multiple steps of urogenital development. Development 121:4057–4065

Wood SA, Allen ND, Rossant J, Auerbach A, Nagy A (1993) Non-injection methods for the production of embryonic stem cell-embryo chimaeras. Nature 365:87–89

Wurst W, Joyner AL (1993) Production of targeted embryonic stem cell clones. In: Joyner AL (ed) Gene targeting, a practical approach. Oxford University Press, Oxford, pp 33–62

Yamada G, Mansouri M, Torres M, Blum M, Stuart ET, SchultzM, de Robertis E, Gruss P (1995) Targeted mutation of the mouse *Goosecoid* gene leads to neonatal death and craniofacial defects in mice. Development 121:2917–2922

Chapter 17

The Generation of Transgenic Mice with Yeast Artificial Chromosomes

HOLGER HIEMISCH[1], THORSTEN UMLAND[1], LLUÍS MONTOLIU[2], AND GÜNTHER SCHÜTZ[1]*

Introduction

Since the original descriptions of methods on how to generate transgenic mice, scientists have been confronted with the limited cloning capacity of the standard plasmid vectors available. Constructs transferred to the germline often lack important *cis*-regulatory elements and, therefore, fail to precisely reproduce the expression pattern of the endogenous counterpart. When integrated into the genome, the constructs might be influenced by nearby regulatory elements, leading to silenced, enhanced, or altered expression of the transgene. Only rarely have transgenes been found to express in a way which is not dependent on their integration site. In order to overcome the limitations mentioned above, yeast artificial chromosomes (YACs) have been introduced as a new vector system. YACs are linear molecules with a cloning capacity of over one megabase (Mb). Cloned DNA is flanked by two vector arms containing all the necessary elements for stable maintenance of the artificial chromosome in yeast cells. Telomeric sequences guarantee chromosomal stability. The long vector arm harbors a centromer and an autonomously replicating sequence required for chromosomal segregation and replication, respectively. Making use of yeast metabolic marker genes in both arms, one can select for the presence of the YAC. The highly active homologous recombination system in yeast offers the possibility to introduce any desired mutations into the YAC.

Recently, a number of groups succeeded in transferring YACs into the germline of the mouse (for reviews see Lamb and Gearhart 1995;

* corresponding author: phone: +49-6221-423411; fax: +49-6221-423470; e-mail: g.schuetz@dkfz-heidelberg.de
[1] Division Molecular Biology of the Cell I, German Cancer Research Center, Im Neuenheimer Feld 280, D-69120 Heidelberg, Germany
[2] Department of Biochemistry and Molecular Biology, Faculty of Veterinary, Autonomous University of Barcelona, 08193 Bellaterra (Barcelona), Spain

Umland et al. 1997). Three different methods have been applied: spheroplast fusion, lipofection, and pronuclear microinjection. The two former methods represent different versions of ES cell transfections. In order to be able to select for ES cell colonies with integrated YACs, a marker gene cassette has to be introduced into the YAC first. Following selection in cell culture, individual ES cell clones are injected into blastocysts to produce chimeric mice. For lipofection, the YAC DNA is purified prior to transfection, therefore no contaminating yeast genomic DNA will be cotransferred. Using this method, most of the ES cells will receive only fragmented YACs, therefore many clones have to be analyzed to identify the few clones containing intact copies. Spheroplast fusion does not require YAC DNA purification. Instead, entire yeast cells whose cell walls have been digested are fused with mouse ES cells. This will ultimately lead to the transfer of all or parts of the yeast genome into the mouse germline with unpredictable phenotypic consequences. In contrast to lipofection, spheroplast fusion seems to be a more robust technique since a high percentage of ES cell clones carry intact YACs and, moreover, comparably large YACs have ben successfully transferred. However, the most rapid procedure to produce YAC transgenic mice is the direct microinjection of YAC DNA into fertilized mouse oocytes. There is no need for modifications of the YAC or time-consuming ES cell work, nor any potential uncertainties due to cotransferred yeast DNA. As a prerequisite, the YAC DNA has to be purified from the yeast genomic DNA and maintained in a special buffer to prevent breakage in solution or shearing upon passage through the injection needle. The protocol is designed such that sufficiently high YAC DNA concentrations are obtained for microinjection. This can be achieved either by an ultracentrifugation step or, alternatively, using a second dimension gel electrophoresis (as described below). Although molecules approaching 1 Mb can be microinjected intact, a certain percentage of transgenic mice will contain fragmented YACs.

Using the following protocol, our laboratory has generated several YAC transgenic mice with efficiencies of 10–20 %, which compares well to the 10–30 % obtained by microinjections of standard plasmid derived constructs. Figure 17.1 shows a summary flow chart of the purification protocol.

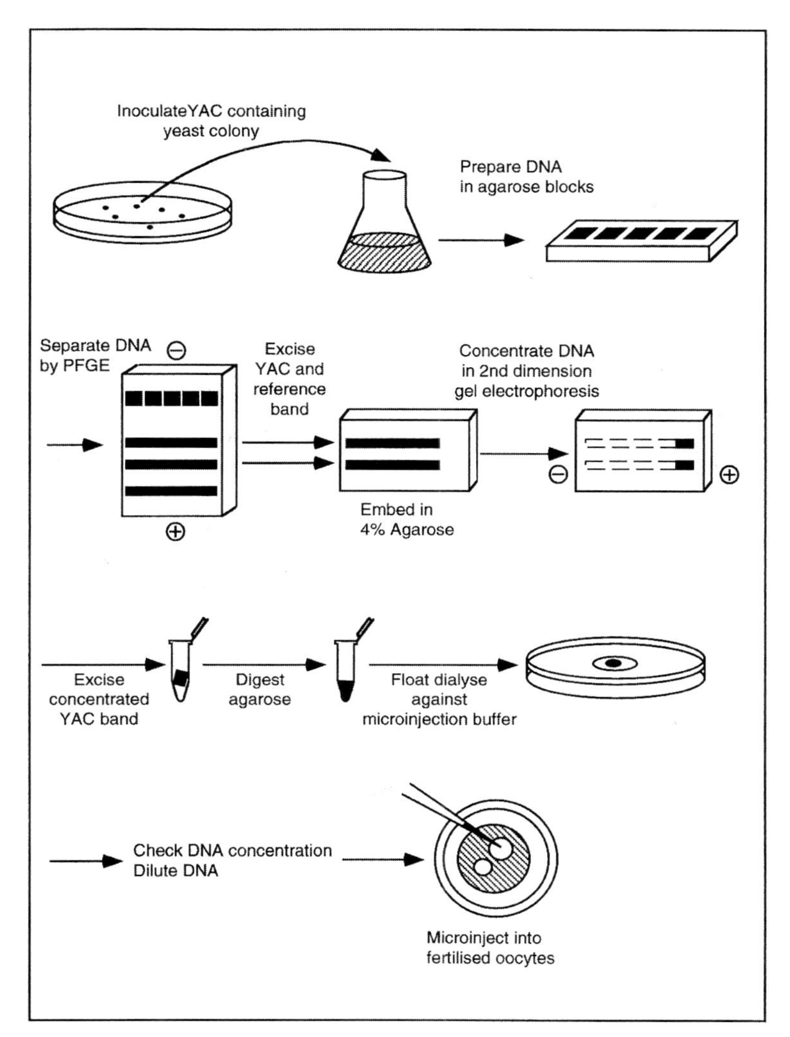

Fig. 17.1. Schematic flow chart summarizing the purification protocol of YAC DNA for microinjection into fertilized oocytes

Materials

- Plug molds (e.g., from Pharmacia, Uppsala, Sweden)
- Pulsed field gel electrophoresis (PFGE) apparatus (e.g., CHEF DRIII system from BioRad Laboratories, Richmond, CA)
- Dialysis membrane filter type VM, pore size 0.05 μm (Millipore, Bedford, MA)
- Dodecyl lithium sulfate (Sigma)
- Zymolyase-20T (ICN Biomedicals Inc., Costa Mesa, CA)
- Seaplaque GTG LMP agarose (FMC, Rockland, ME)
- Nusieve GTG LMP agarose (FMC, Rockland, ME)
- Gelase (Epicentre Technologies, Madison, WI)
- Spermine, Spermidine (Sigma).

17.1 Preparation of High-Density Agarose Plugs

The main problem when preparing high molecular weight DNA is that it is easily sheared in solution. Therefore, YAC DNA is generally prepared in agarose plugs. Try to obtain plugs of the highest possible concentration since this will help to meet the required YAC DNA concentration for microinjection. The protocol described here is based on Huxley et al. (1991).

Materials

Solutions

- Solution I
  1 M sorbitol
  20 mM EDTA, pH 8.0
  14 mM β-mercaptoethanol, always add fresh
- Enzyme solution
  6 mg/ml zymolyase-20T (ICN) in solution I
- Agarose solution
  1 M sorbitol
  20 mM EDTA, pH 8
  2 % Sea plaque GTG agarose (FMC)

  Melt in microwave, cool to 40 °C, then add β-mercaptoethanol to 14 mM final concentration and keep at 40 °C. **Note:** Always prepare fresh.

- Solution II
  1 M sorbitol
  20 mM EDTA, pH 8
  10 mM Tris/HCl, pH 7.5
  14 mM β-mercaptoethanol
  2 mg/ml zymolyase-20T (ICN)
  Always prepare fresh.
- 100 % NDS
  350 ml H_2O
  93 g EDTA
  0.6 g Trizma-Base

  Adjust to pH > 8 with NaOH pellets, dissolve 5 g N-laurylsarcosine in 50 ml H_2O and add, adjust to pH 9.5 with concentrated NaOH, bring volume to 500 ml with H_2O, store at room temperature.

- LDS
  100 mM EDTA, pH 8
  10 mM Tris/HCl, pH 8
  1 % dodecyl lithium sulfate

Procedure

Plug preparation procedure

1. Inoculate 100 ml of suitable yeast medium (e.g., AHC for YACs carrying *URA3* and *TRP1* markers) with 0.2–1 ml of an overgrowth culture of your yeast clone. Grow in 1-l flasks for 1–2 days at 30 °C with shaking (200–250 rpm) until saturation. Count with a hemocytometer, the culture should have at least 5×10^7 cells/ml.
2. Prepare solution I, the enzyme solution and the agarose solution. Keep agarose solution at 40 °C. Tape the bottom of the plug molds and place them on ice.
3. Spin down cells at 1000 *g* for 5 min at room temperature. Discard the supernatant.
4. Resuspend cells in 40 ml 50 mM EDTA (pH 8) and transfer to a preweighed 50-ml Falcon tube. Spin down as in step **3.**
5. Take up cells in 40 ml solution I, spin down cells as above. Carefully remove remaining liquid from the inside of the tube.
6. Weigh the pellet of cells and calculate the volume assuming that 1 mg = 1 μl.

7. Briefly warm the tube to 37 °C, add half of the volume of enzyme solution, resuspend, put back to 37 °C.
8. Quickly add an equal volume of agarose solution (equilibrated to 40 °C). Resuspend with a cut-off blue tip to obtain a homogeneous suspension.
9. Quickly aliquot into the precooled plug molds using a cut-off yellow tip. Leave on ice for 10–15 min to solidify.
10. Transfer plugs into a 50-ml Falcon tube with 5 ml of solution II per plug. Incubate for 2 h at 37 °C with gentle agitation.
11. Transfer plugs to 30 ml LDS, incubate for 1 h and, after changing solution, leave overnight at 37 °C with gentle agitation.
12. Wash plugs in 30 ml 20 % NDS for 2 h at room temperature with gentle agitation.
13. Repeat wash.
14. Wash plugs three times 30 min in TE buffer pH 8. Plugs can be stored at 4 °C.

Comments

- A saturated yeast culture should have a concentration of about $0.5-1\times10^8$ cells/ml. However, certain selective media might not allow the cells to grow to such density. For plug preparation either increase culture volume or use rich medium if possible.
- Try to obtain plugs with evenly distributed cells and without any trapped air bubbles to ensure high quality DNA preparation.
- Although the plugs should be of the highest concentration, refrain from using less agarose since this will result in fragile plugs that easily fall apart.
- Yeast plugs can be stably stored in TE buffer pH 8 or in 0.5 M EDTA pH 8 for up to 1 year at 4 °C.
- Zymolyase can be replaced by yeast lytic enzyme (ICN), a crude enzyme preparation which is cheaper and works almost equally well.

17.2 Purification of YAC DNA for Microinjection

YAC DNA is separated from the endogeneous yeast chromosomes by preparative pulsed field gel electrophoresis (PFGE) of the agarose plugs. DNA from the cut-out YAC band is subjected to a second-dimension gel electrophoresis into a high percentage agarose thereby concentrating the DNA. After excision of the YAC DNA in a minimal volume agarose is removed by gelase digestion and the YAC DNA is dialyzed against microinjection buffer (Schedl et al. 1993a).

Materials

- Equilibration buffer
  1× TAE
  100 mM NaCl
  30 μM spermine
  70 μM spermidine
  Note: Always prepare fresh!
- Microinjection buffer
  10 mM Tris/HCl, pH 7.5
  0.1 mM EDTA, pH 8
  100 mM NaCl
  30 μM spermine
  70 μM spermidine
  Note: Always prepare fresh!
- 50× TAE
  2 M Tris
  250 mM sodium acetate
  50 mM EDTA

  Adjust to pH 7.8 with acetic acid.

Procedure

1. Cast a 0.5× TAE, 1% PFGE gel using Seaplaque GTG agarose. Tape 5–8 cm of the comb to obtain a large preparative slot in the middle of the gel.

2. Equilibrate the DNA plugs in 0.5x TAE for 3×30 min under gentle agitation.
3. Load the agarose plugs into the preparative slot. Line them up closely, filling to the height of the gel. Include marker lanes in flanking slots. Seal the slots with 1 % Sea plaque agarose (0.5x TAE) and leave it to solidify.
4. Start the PFGE run using conditions for optimal resolution of the YAC DNA band.
5. After the run, cut off the outer parts of the gel containing the marker lanes plus about 0.5 cm of the preparative lane at either side. Stain them in 0.5x TAE buffer supplemented with 0.5 μg/ml ethidium bromide for 1 h.
6. Under UV light, mark the exact position of the YAC band at the inner edges of the gel slices with incisions using a scalpel. Also mark an endogenous chromosome, it will serve as a reference lane in the second gel electrophoresis.
7. Reassemble the gel (no UV!) and precisely excise the preparative YAC lane guided by the marked positions from the stained parts. Cut out the reference lane in the same way.
8. Equilibrate the gel slices in 1x TAE 3×20 min.
9. Place the gel slices into a minigel tray parallel to the running direction (i.e., perpendicular to the previous PFGE run). Leave approximately 2 cm to the lower end of the gel tray. Cast a 4 % Nusieve GTG LMP agarose gel in 1× TAE around the slices. Do not cover them with agarose.
10. Run the gel for approximately 8 h at 4 V/cm with circulating buffer. For long slices the running time has to be increased.
11. Cut off the half of the gel containing the marker lane and stain it with 0.5 μg/ml ethidium bromide. Mark the position of the stained DNA and excise the YAC DNA from the unstained part in the smallest possible volume.
12. Equilibrate the gel slice in equilibration buffer for 3×30 min under gentle agitation.
13. Weigh a 1.5-ml Eppendorf tube. Remove all liquid from the gel slice and transfer it to the tube. Weigh again to calculate the volume.

14. Melt the agarose for 10 min at 65 °C. Bring down all agarose by a very short centrifugation.
15. Place in a 42 °C water bath. Add 4 U of gelase per 100 mg of gel. Let the gelase prewarm in the tip of the pipette before adding. Mix gently by swirling with the pipette. Incubate for 2 h at 42 °C.
16. Place the tube on ice for 10 min and check for complete digestion. If the digestion is incomplete, repeat steps **14** and **15.**
17. Pour 40 ml of microinjection buffer into a sterile petri dish. Centrifuge the DNA solution for 30 s and transfer it carefully onto a floating dialysis membrane (pore size 0.05 μm) on the surface of the microinjection buffer. Dialyze for 3–4 h.
18. Carefully remove the DNA from the membrane and transfer to an Eppendorf tube. To determine the DNA concentration, compare 1–2 μl aliquots with DNA of known concentration on a 0.5× TAE 0.6 % agarose minigel with 0.5 μg/ml ethidium bromide. The obtained concentration should be at least 2 ng/μl.
19. YAC DNA can be stored at 4 °C. **Note:** Never freeze YAC DNA because it will break.

Comments

- YAC DNA concentration can be increased by targeting the YAC with a retrofitting vector that introduces a conditional centromer (Schedl et al. 1993a; Smith et al. 1993). A flanking inducible promotor overrides the centromere function leading to accumulation of 10–100 YAC copies per yeast cell.
- For maximal separation of the YAC band from endogeneous chromosomes it is recommended to use single pulse conditions rather than a time ramp in the PFGE run. Test the conditions on analytical gels before wasting your high-density plug preparation.
- In case your YAC comigrates with an endogeneous chromosome, you can transfer the YAC by a very simple and efficient method to a so-called window strain, a set of strains with chromosome-free areas (windows) in their electrophoretic karyotypes (Hamer et al. 1995).

- After the excision of the DNA from the preparative gel lanes you can stain the remaining gel pieces to control whether you have precisely cut out the bands.
- Stabilization of the fragile YAC molecules in solution is absolutely essential. Although a 100-mM NaCl solution helps to stabilize YACs, only the addition of 100 μM polyamines (spermine plus spermidine) allows the formation of compact and highly condensed spherical units as a prerequisite for minimal shearing during manipulation and microinjection (Montoliu et al. 1995). Using these conditions, YACs as large as 850 kb have been transferred intact into the mouse germline (L. Montoliu, T. Umland and G. Schütz, unpubl.).
- Polyamines are delivered as hygroscopic powders and should immediately be dissolved upon opening. We prepare a 1000× stock solution of spermine and spermidine, filter sterilize it, and store aliquots at −20 °C.
- It is crucial to completely dissolve the agarose because YAC DNA will stick to the agarose and this will ultimately lead to low DNA concentration in the batch. Furthermore, agarose pieces might block the injection needle or even exert toxic effects on the oocytes.
- There is no filtration step included in the YAC DNA purification protocol because of the shearing forces involved. If there are undigested gel pieces remaining, the DNA solution should be centrifuged for 5 min. In the presence of salt and polyamines the YAC DNA will remain stable in the supernatant.
- An aliquot of the YAC DNA preparation can be checked for integrity on an analytical PFGE alongside the original plug. Pour an agarose plug of the purified YAC DNA to be able to efficiently load and run it on the gel. Purified DNA might exhibit an altered migration rate when compared to plugs containing all yeast chromosomes.

17.3
Microinjection into Fertilized Oocytes

Procedure

Microinjections into fertilized oocytes are carried out by standard techniques as described in Hogan et al. (1994) and are not outlined in detail here.

Comments

- Store YAC DNA in small aliquots at 4 °C, do not freeze.
- When you have prepared a new batch of DNA, first inject control oocytes and incubate them overnight until the two-cell stage to check the DNA quality. The percentage of surviving embryos should not differ from standard plasmid microinjections. If in doubt prepare new DNA, paying particular attention to completely dissolve the agarose.
- The YAC DNA obtained after the purification procedure should be diluted at least 1:2 in microinjection buffer. Dilutions will substantially increase survival rates after microinjection. Routine microinjections have been carried out at concentrations of 1–2 ng/μl of YAC DNA.

17.4
Analysis of Transgenic Offspring

Transgenic founder mice are identified among the offspring by Southern blotting or PCR analysis of DNA from tail biopsies. This can be done by exploiting sequence differences between the transgene and its endogeneous counterpart of the mouse strain used. When this is not possible, components of the YAC vector arms can be used to detect the trangene. We routinely obtain between 10 and 20 % of transgenics from the surviving offspring. If there are too few transgenic founders or no pups born at all, recheck your DNA batch and repeat the YAC DNA purification.

To verify the integrity of the transgene we check the DNA in Southern blots using probes to both vector arms. Alternatively, RecA assisted restriction endonuclease cleavage (RARE) can be applied to cut out the entire YAC from the integration site, which is then separated on a PFGE and visualised by Southern blotting (Gnirke et al. 1993). Using internal probes spanning different portions of the YAC, one can test for possible rearrangements.

The exact integration site can be mapped using fluorescence in-situ hybridization (FISH) on metaphase spreads from cultured cells of the transgenic animal.

The use of YACs does not per se guarantee correct expression of the transgene. Depending on the representation of the gene locus in the YAC, remote regulatory elements might be missing. Nevertheless, all the lines with intact YAC transgenes tested in our lab express faithfully, i.e., show copy number-dependent and position-independent expression of the transgene (Schedl et al. 1993b; Hiemisch et al., in prep.). Where analyzed, the transgene exactly recapitulated the expression pattern of the endogenous counterpart.

References

Gnirke A, Huxley C, Peterson K, Olson MV (1993) Microinjection of intact 200- to 500-kb fragments of YAC DNA into mammalian cells. Genomics 15:659–667

Hamer L, Johnston M, Green ED (1995) Isolation of yeast artificial chromosomes free of endogenous yeast chromosomes: construction of alternate hosts with defined karyotypic alterations. Proc Natl Acad Sci USA 92:11706–11710

Hogan B, Beddington R, Constantini F, Lacy E (1994) Manipulating the mouse embryo. Cold Spring Harbor Laboratory Press, Cold Spring Harbor

Huxley C, Hagino Y, Schlessinger D, Olson MV (1991) The human HPRT gene on a yeast artificial chromosome is functional when transferred to mouse cells by cell fusion. Genomics 9:742–750

Lamb BT, Gearhart JD (1995) YAC transgenics and the study of genetics and human disease. Curr Opin Genet Dev 5:342–348

Montoliu L, Bock CT, Schütz G, Zentgraf H (1995) Visualization of large DNA molecules by electron microscopy with polyamines: application to the analysis of yeast endogenous and artificial chromosomes. J Mol Biol 246: 486–492

Schedl A, Larin Z, Montoliu L, Thies E, Kelsey G, Lehrach H, Schütz G (1993a) A method for the generation of YAC transgenic mice by pronuclear microinjection. Nucleic Acids Res 21:4783–4787

Schedl A, Montoliu L, Kelsey G, Schütz G (1993b) A yeast artificial chromosome covering the tyrosinase gene confers copy number-dependent expression in transgenic mice. Nature 362:258–261

Smith DR, Smyth AP, Strauss WM, Moir DT (1993) Incorporation of copy-number control elements into yeast artificial chromosomes by targeted homologous recombination. Mamm Genome 4: 141–147

Umland T, Montoliu L, Schütz G (1997) The use of yeast artificial chromosomes for transgenesis. In: Houdebine LM (ed) Transgenic animals – generation and use. Harwood Academic Publ, Amsterdam (in press)

Chapter 18

Following Up Microinjected mRNAs: Analysis of mRNA Poly(A) Tail Length During Mouse Oocyte Maturation

FATIMA GEBAUER*

Introduction

The mRNA poly(A) tail regulates gene expression by participating in processes related to mRNA metabolism including mRNA export, stability, and translation. The role of the poly(A) tail as a modulator of translation is especially critical during oocyte maturation and early embryogenesis of many animals, where polyadenylation of maternal messages in the cytoplasm triggers their recruitment onto polysomes and their subsequent translation, whereas deadenylation causes the release of these messages from polysomes. Recently, the relevance of cytoplasmic polyadenylation for meiosis and pattern formation has become apparent (Gebauer et al. 1994; Sallés et al. 1994; Sheets et al. 1995; for recent reviews on cytoplasmic polyadenylation see Vassalli and Stutz 1995; Richter 1996).

Several methods have been used to measure the poly(A) tail length of both endogenous and microinjected mRNAs. Traditionally, the method of choice for endogenous mRNA studies has been the Northern blot assay. To determine the size of the poly(A) tail using this method, part of the RNA sample is hybridized to oligo(dT) and digested with RNase H, resulting in nonadenylated RNA that is used as a size marker (Mercer and Wake 1985). This technique, however, has two major drawbacks: first, relatively large amounts of RNA are required; second, the resolution of the agarose gels usually employed for Northern blot analysis is low and precludes detection of small poly(A) tails. To overcome this second limitation, investigators have reduced the size of the RNA under study by cleaving off the 3' UTR of the mRNA using antisense oligonucleotides and subsequent digestion with RNase H (Brewer and Ross 1988).

* European Molecular Biology Laboratory, Meyerhofstrasse 1, D-69117-Heidelberg, Germany. phone: +49-6221-387502; fax: +49-6221-387518;
e-mail: gebauer@embl-heidelberg.de

Another method to analyze the poly(A) tail length consists of the thermal elution of the mRNA (Palatnik et al. 1979; Jacobson 1987). This procedure is based on the retention of poly(A)+ RNA in poly(U)-Sepharose columns: the longer the poly(A) tail, the tighter the retention. RNAs with short poly(A) tails elute from the column at low temperatures while higher temperatures are required to elute RNAs with longer poly(A) tails. After elution, RNAs are analyzed by Northern blot or amplified by PCR. Because this method does not allow direct visualization of the poly(A) tail, it requires a previous assessment of the elution profile of RNAs with known poly(A) tail length. Furthermore, this is not the procedure of choice when a precise quantitation of the poly(A) tail size is needed, because RNAs with a range of different poly(A) tail lengths will elute at any given temperature.

A PCR-based method has been recently developed that allows quantitation of the poly(A) tail length of endogenous and injected mRNAs using a small starting sample. This procedure is termed the poly(A) test (PAT) assay and has been described in detail elsewhere (Sallés and Strickland 1995). Briefly, several molecules of oligo(dT) are hybridized along the complete length of the poly(A) tail, and are subsequently ligated together. The resulting poly(dT) is also ligated to an anchor sequence, yielding a poly(dT)-anchor that is used to prime reverse transcription of the mRNA. The poly(A) tail size is determined by PCR amplification using as primers the anchor sequence and a message-specific oligodeoxynucleotide. Although extreme care should be taken to avoid amplification of spurious bands, this method is powerful because of its sensitivity and simplicity.

A more direct test of the poly(A) tail length is accomplished by injection into oocytes of radioactively labeled RNA, allowing detection of poly(A) tail size changes after injection. This method, however, involves working with considerable amounts of radioactivity at the microinjection stage. A sensitive technique that overcomes this problem is the RNase protection assay: the injected RNA is not labeled and is analyzed by hybridization to an antisense labeled RNA. This procedure allows detection of as little as 5 pg of RNA. Ideally, if 10 pl of a 0.5 μg/μl RNA sample is injected into mouse oocytes, only one oocyte would be required for detection. Practically, however, ten oocytes should be injected to assess detection of the poly(A) tail by the RNase protection assay, because the size of adenylated RNAs is not uniform and, therefore, the signal is usually weak. This chapter describes this method in detail.

General Considerations

A schematic representation of the RNase protection assay is shown in Fig. 18.1. The sequence of interest must be cloned immediately upstream of a poly(dA) stretch in a vector such as Bluescript (Stratagene) that allows synthesis of RNA from two opposite promoters. One promoter is used to synthesize sense, nonadenylated RNA for injection while the opposite promoter is used to synthesize antisense labeled RNA containing a poly(U) stretch. Vectors containing poly(dA) have been described by Munroe and Jacobson (1990) and Gebauer et al. (1994).

Sense, nonadenylated RNA is microinjected into the cytoplasm of GV stage mouse oocytes (i.e., oocytes that contain a germinal vesicle and therefore have not matured) cultured in the presence of 150 μM isobutylmethylxanthine (IBMX), which prevents spontaneous maturation. After injection, the oocytes are washed several times with medium without IBMX and are incubated for 16 h at 37 °C to allow for maturation, which is denoted by the emission of the first polar body. Total RNA is then extracted from the injected oocytes and is annealed to the antisense poly(U)-containing probe. The RNA hybrids are digested with RNase A, which cleaves single-stranded RNA 3' of pyrimidines, and the protected fragments are resolved by denaturing polyacrylamide gel electrophoresis. The segment of the probe that is complementary to the cDNA should not be longer than 100 nt to facilitate detection of small size changes due to polyadenylation. The maximum size of the poly(A) tail detected depends on the size of the poly(U) in the probe.

The restriction site located between the cDNA and the poly(dA) (that of enzyme 2 in Fig. 18.1) is an important issue to consider here. This site should contain only adenosine residues 3' of the nucleotide cleaved by the enzyme (Fig. 18.2). This requirement is necessary to avoid the formation of an unpaired region between the sense RNA and the probe during the RNase protection assay, which would preclude detection of the poly(A) tail (Fig. 18.2). Given that endogenous mRNAs are unlikely to fulfil this condition, the RNase protection assay is only applicable to the analysis of the polyadenylation status of injected RNAs. This, in its turn, allows to distinguish between endogenous and injected RNAs.

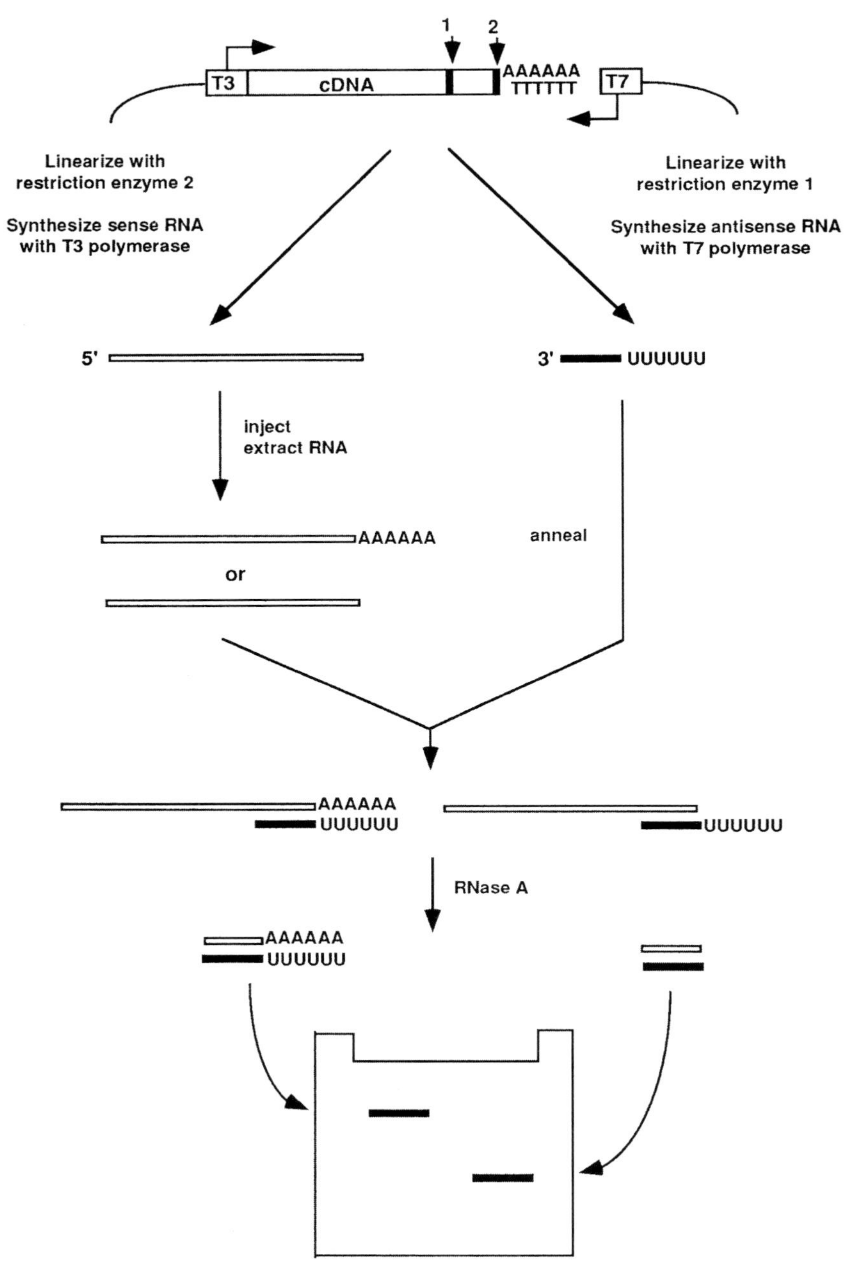

Fig. 18.1. Schematic diagram of the RNase rotection procedure used to analyze the polyadenylation status of microinjected RNAs (Gebauer et al. 1994)

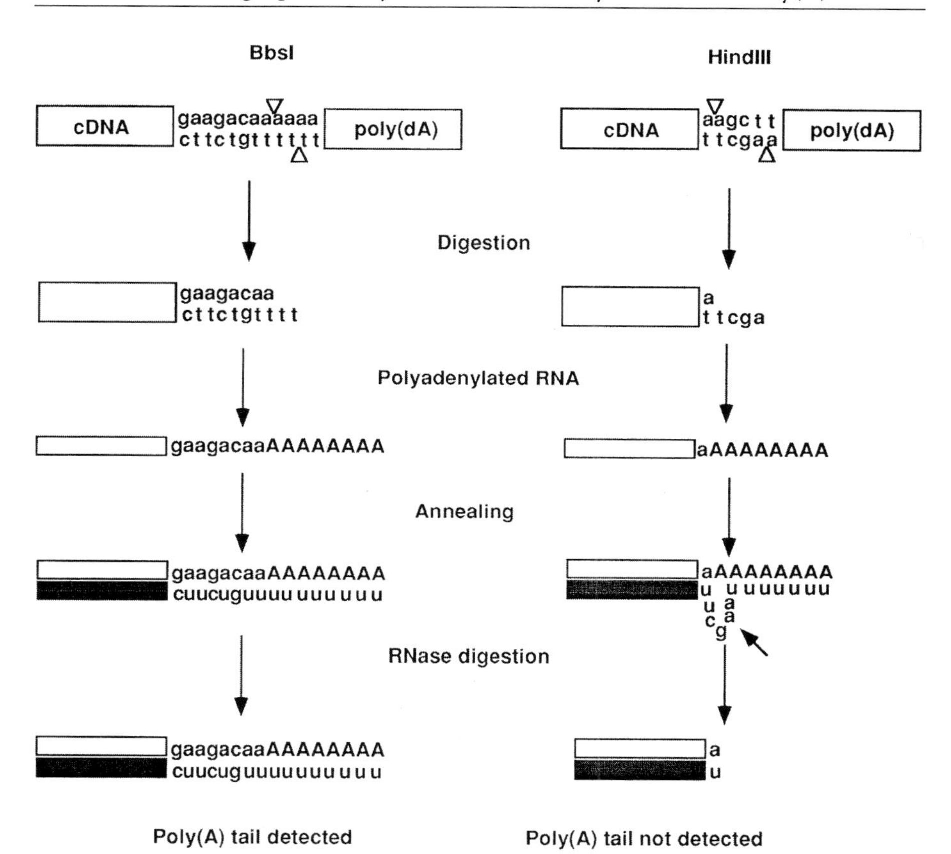

Fig. 18.2. Characteristics of the restriction site located between the cDNA and the poly(dA). A condition to obtain a successful RNase protection assay is that this site contains only adenosine residues 3' of the nucleotide cleaved by the enzyme. Examples of restriction sites that do (BbsI) or do not (HindIII) meet this condition are shown. The cleavage sites are represented by *arrows*. When a HindIII site is present, the probe contains extra residues that are not complementary to the sense RNA. As a consequence, a gap is formed during the annealing step. This gap is recognized by the RNase A during the digestion step, precluding detection of the poly(A) tail. When a BbsI site is present, however, no gap is formed, allowing detection of the poly(A) tail

18.1
Synthesis of RNA for Microinjection

Procedure

The RNA synthesized should be capped to avoid rapid degradation upon oocyte microinjection. The Message Machine kits from Ambion (Austin, Texas, USA) are recommended to obtain high amounts of quality capped RNA. After synthesis following the recommendations of the manufacturer, the DNA template must be removed by DNaseI digestion. The sample is then treated once with phenol/chloroform, once with chloroform, and precipitated with 0.5 M final amonium acetate and 2.5 vol of absolute ethanol. The RNA pellet is washed twice with 70 % ethanol to remove residual salts that are toxic to oocytes, dried in a Speed-vac concentrator, and resuspended in deionized-sterile water to a final concentration of 0.5–2 μg/μl.

18.2
RNA Extraction from Mouse Oocytes

Total RNA is extracted from mouse oocytes following the method described by Chomczynski and Sacchi (1987) with some modifications.

Materials

- Solution D:
  4 M guanidine thiocyanate
  25 mM sodium citrate, pH 7
  0.5 % sarcosyl
  Autoclave
  Add mercaptoethanol to 0.1 M final
  Protect from light
  Store at room temperature for a month
- 2 M sodium acetate, pH 5
- tRNA (Sigma)
- isopropanol
- phenol/chloroform (1:1), equilibrated in 50 mM TRIS-HCl pH 7.5
- 70 % ethanol

Procedure

1. Homogenize the oocytes in 200 µl of solution D by vigorous vortexing.
2. Add 5 µg of tRNA, 20 µl of 2 M sodium acetate and 200 µl of phenol/chloroform. Mix by vortexing and incubate on ice for 10 min.
3. Spin for 10 min at 4 °C.
4. Recover aqueous phase and add 260 µl of isopropanol. Precipitate the RNA by incubating for 1 h at −20 °C.
5. Spin for 15 min at 4 °C and discard the supernatant.
6. Resuspend the pellet in 60 µl of solution D.
7. Add 7 µl of 2 M sodium acetate and 70 µl of isopropanol. Precipitate the RNA by incubating for 1 h at −20 °C.
8. Spin for 15 min at 4 °C and discard the supernatant.
9. Wash the pellet with 70 % ethanol, dry and store at −70 °C until use.

18.3 Preparation of the RNA Probe

Synthesizing a high quality probe is one of the most important factors to achieve a clean and sensitive RNase protection assay. A protocol to prepare this labeled antisense RNA is described below.

Materials

- RNasin, RNA polymerase (Promega)
- 5× transcription buffer (200 mM TRIS-HCl pH 7.5, 30 mM $MgCl_2$, 10 mM spermidine, 50 mM NaCl; supplied by Promega together with the enzyme)
- 0.1 M DTT
- $^{32}P\alpha CTP$ (3000 Ci/mmol)
- AUG mix (2.5 mM ATP, UTP, and GTP)
- 150 µM CTP
- 1 µg/µl linearized template DNA

- Formamide dye:
  90 % formamide
  1 mM EDTA
  0.1 % xylene cyanole
  0.1 % bromophenol blue
  Store at −20 °C
- Probe extraction buffer:
  0.1 % SDS
  0.5 M ammonium acetate
  10 mM magnesium acetate
  Store at room temperature
- tRNA (Sigma)
- Phenol/chloroform (1:1), equilibrated in 50 mM TRIS-HCl, pH 7.5
- Chloroform
- Absolute ethanol
- 3 M sodium acetate, pH 5

Procedure

1. Mix the following reagents in the indicated order at room temperature:
   4 μl 5× transcription buffer
   2 μl 0.1 M DTT
   0.8 μl 40 U/μl RNasin
   4 μl AUG mix
   2 μl 150 μM CTP
   1 μl 1 μg/μl template DNA
   5 μl ^{32}PαCTP
   1 μl RNA polymerase
   The label should be fresh and **should not** be ^{32}PαUTP because the probe will contain a poly(U) stretch and limiting the amount of UTP in the reaction (a condition necessary to label with high specific activity) will prevent the synthesis of a full length probe. The reagents should be mixed at room temperature to avoid precipitation of the DNA template due to the spermidine contained in the reaction buffer (Promega Protocols and Applications Guide 1991).
2. Incubate at 37 °C for 1 h.

Gel purification of the probe

Gel-purifying the probe is necessary to remove the DNA template and other minor RNA products of unwanted size, such as template-size RNA that arises when template linearization is not complete. To do this, follow the directions detailed below.

3. Add 10 μl of formamide dye to the reaction mix from step 2, and boil for 5 min to denature the sample.
4. Transfer to ice and spin briefly to remove the condensation at the top of the tube. Load in a 5 % polyacrylamide-8 M urea gel.
5. After the run is complete remove only one of the glass plates, wrap the gel, and expose it for 1 min to visualize the probe band. Cut the band in small pieces and transfer them to an Eppendorf tube. Reexpose the gel to make sure that the band has been cut properly.
6. Add 500 μl of probe extraction buffer and incubate in a rocking platform at 37 °C for 20 min.
7. Spin 5 min at room temperature and transfer the supernatant to a new tube. Approximately 50 % of the counts should be recovered.
8. Add 1 μg of tRNA and 40 μl of 3 M sodium acetate.
9. Extract once with phenol/chloroform and once with chloroform.
10. Add 2.5 vol of absolute ethanol and precipitate at −20 °C for 20 min.
11. Spin for 15 min at 4 °C, discard the supernatant, and dry the pellet. Resuspend the pellet in water to 6×10^4 cpm/μl (the specific activity of the probe should be about 1.2×10^8 cpm/μg).

18.4 RNase Protection

Materials

- Hybridization buffer:
  80 % deionized formamide
  0.4 M NaCl
  40 mM PIPES, pH 6.4
  5 mM EDTA
  Make small aliquots and store at −80 °C. Use each aliquot once only.
- Mineral oil (Sigma)

- Digestion buffer:
  10 mM TRIS, pH 7.5
  5 mM EDTA
  300 mM NaCl
  10 μg/ml RNase A
  Make fresh
- Proteinase K (5 mg/ml in 2 mM Cl_2Ca)
- 10 % SDS
- 3 M sodium acetate, pH 5
- Absolute ethanol
- Formamide dye (see Sect. 18.3 for recipe)
- Phenol/chloroform (1:1), equilibrated in 50 mM TRIS-HCl, pH 7.5
- Chloroform

Procedure

Hybridization

1. Coprecipitate the probe and the RNA sample by mixing the following:
   1 μl probe (6×10^4 cpm, approx. 500 pg)
   X μl RNA sample
   1 μl 3 M sodium acetate
   H_2O to a final volume of 10 μl
   25 μl absolute ethanol
   The probe is added in excess to target RNA in order to increase sensitivity and accelerate the hybridization kinetics. Incubate for 15 min at −20 °C.
2. Spin for 15 min at 4 °C, discard the supernatant and dry the pellet.
3. Resuspend the pellet in 20 μl of hybridization buffer. The use of this small volume also favors the kinetics of hybridization. Overlay 15 μl of mineral oil to avoid evaporation during the subsequent incubations.
4. Denature the sample by incubating for 5 min at 85 °C, and cool slowly to 45 °C for optimal annealing of the probe with the target RNA.
5. Incubate overnight at 45 °C to stabilize hybrids. Due to the harsh hybridization conditions (80 % formamide at 45 °C) the probe normally hybridizes very specifically.

Digestion

6. Add 280 μl of digestion buffer, mix by vortexing and incubate for 1 h at 30 °C.

7. Add 20 μl of 10 % SDS and 10 μl of 5 mg/ml proteinase K. Incubate for 30 min at 37 °C to inactivate the RNase A present in the digestion buffer.
8. Add 2 μg of carrier tRNA. Extract once with phenol/chloroform and once with chloroform.
9. Precipitate by adding 33 μl of 3 M sodium acetate and 2.5 vol of absolute ethanol, and incubating for 15 min at −20 °C.
10. Spin for 15 min at 4 °C, discard the supernatant, and dry the pellet. Resuspend the samples in 10 μl of water and add 10 μl of formamide dye. Boil the samples for 5 min, transfer them to ice, and spin them briefly to remove the condensation at the top of the tubes. Resolve the protected products by denaturing polyacrylamide-gel electrophoresis.

Controls

The following controls should be included in the RNase protection assay:

- To address possible background coming from incomplete digestion of the probe, the probe alone (without target RNA) should be included in the assay as a negative control.
- To confirm the sensitivity of the assay, uninjected RNA should be included as a positive control; a curve with known amounts of uninjected RNA allows an estimation of the concentration of the RNA recovered from microinjected oocytes.
- If available, in-vitro synthesized polyadenylated RNA can be used as a size marker and as an indicator of the resolution of the gel.
- It is also advisable to load an aliquot of the undigested probe in the gel to check whether it remained intact.

Troubleshooting

A troubleshooting guide for the RNase protection assay is shown in Table 18.1.

Results

The c-*mos* proto-oncogene product, Mos, is a serine-threonine kinase whose expression is required at specific times during oocyte maturation. Failure of mouse oocytes to express Mos results in their parthenogenetic activation (O'Keefe et al. 1989; Hashimoto et al. 1994; Colledge et al.

Table 18.1. Troubleshooting guide

Problem	Cause	Solution
Background in all lanes, including the negative control	Digestion step did not work	Use a new RNase A aliquot Use additional RNase T1 Increase digestion temperature to 37°C
	Probe hybridizes with itself	Decrease probe size to eliminate self-hybridizing segments
Background in the lanes except in the negative control	Probe hybridizes nonspecifically	Increase hybridization temperature Decrease probe size Use fresh deionized formamide in the hybridization buffer
Signal is weak, including the positive control	RNA used for injection, or probe are degraded Probe is weakly labeled	Make new probe with fresh label Make new RNA for injection
Signal is weak, except in the positive control	RNA has been degraded after injection	Use fresh 7-methylguanosine cap during the transcription reaction Use fresh solutions for RNA extraction

1994), indicating a role for Mos in metaphase II (MII) arrest. De novo synthesis of Mos is regulated by cytoplasmic polyadenylation (Gebauer et al. 1994). This polyadenylation event must occur before MII because Mos is required at this stage. The RNase protection assay was used to analyze the timing of c-*mos* mRNA polyadenylation and determine whether it correlated with any of the landmarks of oocyte maturation before MII: germinal vesicle breakdown (GVBD) and first polar body emission (PB), which occur at 2 and 10–12 h of maturation, respectively. The cDNA for c-*mos* 3' UTR, containing the signals required for polyadenylation, was cloned upstream of a poly(dA) of 82 residues as depicted in Fig. 18.1. Sense RNA was synthesized from the T3 promoter and was microinjected into the cytoplasm of GV stage oocytes that were subsequently allowed to mature. At different times after injection, oocytes were collected, their RNA extracted, and the polyadenylation status of the microinjected RNA analyzed by hybridization to a labeled antisense RNA synthesized from the T7 promoter. The results (Fig. 18.3) indicated

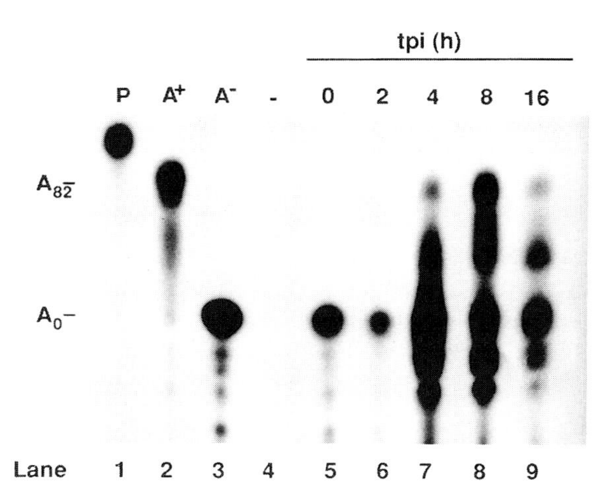

Fig. 18.3. Time course of c-*mos* mRNA 3' UTR polyadenylation. GV stage oocytes were injected with c-*mos* mRNA 3' UTR (0.7 pg/pl) and analyzed for polyadenylation at several times thereafter. *Lane 1* Integrity of the probe used in the assay (244 nt); *lane 2* 50 pg of in vitro synthesized polyadenylated RNA; *lane 3* 50 pg of the nonadenylated RNA used for injection (the size of the protected fragment is 113 nt); *lane 4* digested probe alone; *lanes 5–9* RNA from oocytes collected at 0, 2, 4, 8, and 16 h after injection (10, 24, 26, 24, and 26 oocytes were collected at each time point respectively) (Gebauer et al. 1994)

that polyadenylation started between 2–4 h of maturation and reached a maximum at approximately 8 h. Therefore, polyadenylation of c-*mos* 3' UTR occurs after GVBD and increases dramatically before PB, at a time when accumulation of Mos is required in preparation for MII arrest.

Acknowledgments. I thank Dr. Juan Valcárcel for useful comments on this manuscript, and Dr. Joel D. Richter for his continuous encouragement. I was supported by fellowships from NATO and the Spanish Ministry of Education and Science.

References

Brewer G, Ross J (1988) Poly(A) shortening and degradation of the 3' A+U-rich sequences of human c-*myc* mRNA in a cell-free system. Mol Cell Biol 8:1697–1708
Chomczynski P, Sacchi N (1987) Single-step method of RNA isolation by acid guanidinium thiocyanate-phenol-chloroform extraction. Anal Biochem 162:156–159
Colledge WH, Carlton MBL, Udy GB, Evans MJ (1994) Disruption of c-*mos* causes parthenogenetic development of unfertilized mouse eggs. Nature 370:65–68

Gebauer F, Xu W, Cooper GM, Richter JD (1994) Translational control by cytoplasmic polyadenylation of c-*mos* mRNA is necessary for oocyte maturation in the mouse. EMBO J 13:5712–5720

Hashimoto N, Watanabe N, Furuta Y, Tamemoto H, Sagata N, Yokoyama M, Okazaki K, Nagayoshi M, Takeda N, Ikawa Y, Aizawa S (1994) Parthenogenetic activation of oocytes in c-*mos*-deficient mice. Nature 370:68–71

Jacobson A (1987) Purification and fractionation of poly(A)+ RNA. In: Berger SL, Kimmel AR (eds) Guide to molecular cloning techniques. Methods Enzymol, vol 152. Academic Press, London, pp 254–261

Mercer JFB, Wake SA (1985) An analysis of the rate of metallothionein mRNA poly(A) shortening using RNA blot hybridization. Nucleic Acids Res 13:7929–7943

Munroe D, Jacobson A (1990) mRNA poly(A) tail, a 3' enhancer of translational initiation. Mol Cell Biol 10:3441–3455

O'Keefe SJ, Wolfes H, Kiessling AA, Cooper GM (1989) Microinjection of c-mos antisense oligonucleotides prevents meiosis II in the maturing mouse egg. Proc Natl Acad Sci USA 86:7038–7042

Palatnik CM, Storti RV, Jacobson A (1979) Fractionation and functional analysis of newly synthesized and decaying messenger RNAs from vegetative cells of *Dictyostelium discoideum*. J Mol Biol 128:371–395

Richter JD (1996) Dynamics of poly(A) addition and removal during development. In: Hershey JWB, Mathews MB, Sonenberg N (eds)Translational control. Cold Spring Harbor Laboratory Press, Cold Spring Harbor, pp 481–503

Sallés FJ, Strickland S (1995) Rapid and sensitive analysis of mRNA polyadenylation states by PCR. PCR Methods Appl 4:317–321

Sallés FJ, Lieberfarb ME, Wreden C, Gergen P, Strickland S (1994) Coordinate initiation of *Drosophila* development by regulated polyadenylation of maternal messenger RNAs. Science 266:1996–1999

Sheets MD, Wu M, Wickens M (1995) Polyadenylation of c-*mos* mRNA as a control point in *Xenopus* meiotic maturation. Nature 374:511–516.

Titus E (ed) (1991) Promega Protocols and Applications Guide. Promega Corporation, Madison, Wisconsin

Vassalli JD, Stutz A (1995) Awakening dormant mRNAs. Curr Biol 5:476–479

Chapter 19

Mouse Preimplantation Embryos and Oocytes as an In Vivo System to Study Transcriptional Enhancers

SADHAN MAJUMDER*

Introduction

Multicellular organisms, as opposed to their unicellular counterparts, face a unique problem in carrying out life-sustaining functions. Whereas, in unicellular organisms the same cell performs all the necessary functions, in multicellular organisms there is a division of labor: specific cell types carry out specific functions in a spatial and temporal manner. For example, as a fertilized mouse one-cell embryo divides, differentiates, and develops into a complete animal, it is crucial to express the right gene at the right time and by the right cell type. One of the important mechanisms by which multicellular organisms achieve such a goal is to regulate transcription of RNA polymerase II promoters through enhancers. Our present knowledge of the principles that regulate mammalian transcription, including the enhancer function, mainly stems from studies involving cell-free in vitro systems, or in vivo systems comprised of tissue culture cells or animal viruses. In fact, although a wealth of knowledge on the mechanism of RNA polymerase II trancription that occurs at the promoter site has been gained from in vitro sytems consisting of purified transcription factors, most of these systems do not exhibit enhancer function, unless the template DNA is reconstituted into chromatin. Similarly, it has been relatively unknown how these principles apply in physiological processes that regulate, for example, the development of a fertilized one-cell embryo into an animal. This inability is primarily due to the limited availability of sufficient embryos to carry out biochemical studies. It is only recently that technologies have been developed allowing the study of transcription and replication

* Departments of Neurooncology and Molecular Genetics, University of Texas MD Anderson Cancer Center, 1515 Holcombe Boulevard, Box 100, Houston, TX 77030, USA. phone: +01-713-7928920; fax: +01-713-7926054; e-mail: majumder@utm-dacc.mda.uth.tmc.edu

in mammalian embryos as early as a fertilized one-cell embryo, and in as few as a single embryo, using microinjection techniques. It provides an unprecedented opportunity to study these regulatory processes in the context of a living animal.

What is a transcriptional enhancer? Transcription by RNA polymerase II is thought to be primarily controlled by two DNA elements: promoters and enhancers. Promoters determine the site of transcriptional initiation and function close to but upstream from this site. On the other hand enhancers stimulate weak promoters in a tissue-specific manner and function distal to the start site from either upstream or downstream orientation. Whereas the primary function of promoters is generally believed to facilitate assembly of an active initiation complex, there is controversy as to what the primary function of an enhancer is. There are two competing models:

- There is no basic difference between the role of an enhancer and that of a promoter; they both promote formation of an active complex. Enhancers simply provide promoters with additional transcription factor activity and thus stimulate transcription (Carey et al. 1990). This became a favored model when it was observed that the same transcription factors that can constitute a promoter could also constitute an enhancer.
- There is, indeed, a difference between the primary role of an enhancer and that of a promoter; whereas a promoter facilitates formation of an active complex, the primary role of an enhancer is to relieve repression of weak promoters, brought in most likely by chromatin structure (Felsenfeld 1992; Studitsky et al. 1995; Struhl 1996; Wolffe and Pruss 1996). This might result in an increase in the probability of transcription (Walters et al. 1995).

Recent analysis of gene expression in mouse preimplantation embryos (see below), as well as cell-free transcription systems that are capable of constituting chromatin structure, support the latter model (Laybourn and Kadonaga 1991, 1992; Paranjape et al. 1994). Although most of the data from cell-free transcription systems suggest that histone H1 is the major transcription inhibitory component of the chromatin structure, another publication suggested that histone H1 may exert only a minor role in chromatin-mediated repression under certain circumstances (Kamakaka et al. 1993; Pazin et al. 1994).

As described below, gene expression studies carried out in mouse preimplantation embryo and oocyte system revealed that the major enhancer function appears first in two-cell embryos. The lack of enhancer function in the paternal pronuclei of S-phase-arrested one-cell embryos, at least in part, depends on the absence of promoter repression. Thus, the paternal pronuclei not only provide us with a system where the role of chromatin in enhancer function can be determined by introducing in vitro reconstituted chromatin templates, rather than naked plasmid DNA, but the system can also help clarify some of the contradictory phenomena observed in cell-free systems. Whether the environment of the paternal pronuclei would destabilize such injected chromatin templates is yet to be seen. However, even in that case, the role of chromatin destabilization and reorganiztion in enhancer function can be studied in this system. Oocytes, on the other hand, do show promoter repression, but the lack of enhancer function in this system appears to be due to the absence of a critical coactivator activity required for enhancer function. Thus, the oocyte system can be utilized to identify critical components of enhancer function, and study how these regulators control gene expression at the beginning of animal development.

I will first describe the mouse system employing oocytes and early embryos to study enhancer function in vivo, and then the theoretical and experimental observations indicating the validity of microinjection as a method for studying transcriptional regulation in this system. This will be followed by experimental protocols to study enhancer function utilizing mouse oocytes and preimplantation embryos as an in vivo system.

The Mouse System

Fertilization of a mammalian egg by a sperm triggers a complex developmental program leading to the formation of an animal. The initial stages of this pathway provides us with a unique system to understand mammalian transcriptional mechanisms that set into motion the onset and further control of such a developmental program. A schematic representation of early mouse development is shown in Fig. 19.1. In sexually mature females, oocytes undergo the first meiotic reductive division to become mature unfertilized eggs. The latter, when fertilized by sperm, go through the second meiotic division to produce a one-cell embryo containing a paternal and a maternal haploid pronucleus. Each pronucleus undergoes DNA replication and then they fuse during the first mitosis, to generate a two-cell embryo containing one zygotic diploid nucleus per cell. Although translation of maternally inherited mRNAs occurs continuously in mature eggs and one-cell embryos, a process regulated

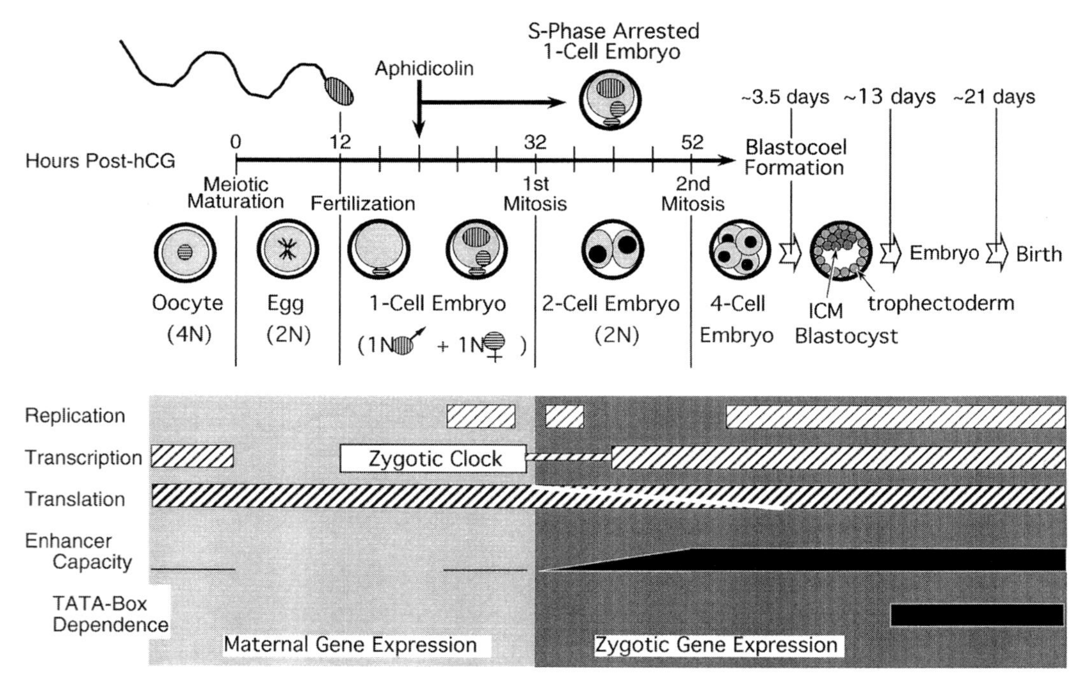

Fig. 19.1. Schematic representation of events at the beginning of mouse development. *Upper panel:* Morphological and cell cycle events are indicated as a function of time after injection of human chorionic gonadotropin (post-hCG), a hormone that stimulates ovulation. The paternal pronucleus is indicated by *vertical striations*, oocyte and maternal pronucleus are denoted by *horizontal striations*, and the zygotic nuclei are represented by *filled circles*. Addition of aphidicolin to one-cell embryos prior to the appearance of pronuclei insures their morphological arrest as they enter S-phase; *lower panel*: events in gene expression are divided into two phases, maternal and zygotic. Periods of DNA replication, transcription and mRNA translation are indicated by *bars*. The ability of oocytes and embryos to utilize enhancers encoded in plasmid DNA injected into cell nuclei is indicated as (a) the general capacity of enhancers to stimulate promoter activity, and (b) the dependence of enhancer stimulation on the TATA-box of the promoter. The resumption of transcription in embryos is delayed by a time-dependent biological clock mechanism (zygotic clock)

mostly through posttranscriptional modification (Richter 1991; Bachvarova 1992; Manley and Proudfoot 1994), expression of zygotic genes begins at the two-cell stage (Schultz and Heyner 1992; Schultz 1993), and is characterized by transcription-dependent (α-amanitin sensitive) protein synthesis. Transcription of zygotic genes begins prior to S-phase in two-cell embryos with the synthesis of a small number of proteins, but after S-phase, both the number and overall rate of protein synthesis increases (Conover et al. 1991; Latham et al. 1992). Concurrent with these events is a rapid degradation of maternal mRNAs, reflecting a switch from maternal to zygotic control of the embryonic development (Telford et al. 1990).

If one-cell embryos are arrested in S-phase, using inhibitors of DNA replication, zygotic gene expression still occurs at the same time post-fertilization that it normally would occur in developing embryos (two-cell stage), despite the fact that the arrested embryos remain morphologically one-cell embryos (Martinez-Salas et al. 1989; Conover et al. 1991; Wiekowski et al. 1991). Therefore, initiation of zygotic gene expression is governed by a biological clock (zygotic clock) that initiates transcription at a specified time after fertilization. The zygotic clock is also observed during early development in other mammals, although the developmental stage in which transcription begins is species-specific. Thus, zygotic gene expression begins in two-cell mouse and hamster embryo, in four-cell pig embryos, in four- to eight-cell human embryos, and in 8- to 16-cell sheep, rabbit, and cow embryos (Telford et al. 1990; Seshagiri et al. 1992; Schultz 1993). The purpose of the zygotic clock may be to prevent spurious transcription while paternal and maternal chromosomes are undergoing remodeling from a postmeiotic state to one that is competent for regulating gene expression (Majumder and DePamphilis 1995). The existence of this clock allows the requirements for transcription and replication in mouse oocytes and one-cell embryos to be compared to those in two-cell stage to blastocyst stage embryos. It was this comparison that revealed the sudden appearance of enhancer function in two-cell embryos, and the lack of a general repression mechanism in the paternal pronuclei of one-cell embryos corresponding to their inability to utilize enhancer function. This comparison also revealed the presence of repression mechanisms in both oocytes and two-cell embryos, but the presence of enhancer function in two-cell embryos only, and not in oocytes. This indicated the possibility of a missing coactivator function in oocytes that is required for the optimal activity of an enhancer.

The Microinjection Technique

Microinjection of DNA as a method for studying transcriptional regulation in early mammalian embryos: the fact that mammalian embryos are available in limited quantities places a serious roadblock in front of any effort in identifying *cis*-acting sequences and *trans*-acting factors that are required for DNA transcription or replication at the beginning of mammalian development. One solution to this problem has been to inject plasmid DNA into the germinal vesicles of oocytes (precursor to the maternal pronucleus), paternal or maternal pronuclei of one-cell embryos or the zygotic nuclei of two-cell embryos and then identify sequences that are required to either replicate the plasmid or to express an encoded reporter gene. These transient assays, like those used during transfection of cultured cell lines, reveal DNA replication or transcriptional environment: the capacity of cells to replicate or express genes, their ability to utilize specific *trans*-acting factors, and their ability to respond to specific *cis*-acting sequences. This information can be applied to understanding events at the beginning of mammalian development, because injected DNA responds to the same cellular signals that regulate endogenous DNA replication and gene expression.

Injected DNA undergoes replication and transcription

- Only when unique eukaryotic regulatory sequences are present
- Only in cells that are competent for that function (Schultz 1993; Majumder and DePamphilis 1994a, 1995).

For example, because mouse oocytes are arrested in prophase of their first meiosis, they cannot replicate DNA. Accordingly, plasmid DNA does not replicate when injected into mouse oocytes, even if the injected DNA contains a viral origin and is provided with the appropriate viral proteins (Martínez-Salas et al. 1988; DePamphilis et al. 1988). However, plasmid DNA will replicate when injected into either one-cell or two-cell embryos, but only if it contains a viral origin, and only in the presence of the cognate origin recognition protein. In the absence of a valid replication origin, plasmid DNA does not replicate in mouse embryos.

Although mouse oocytes cannot replicate DNA, they express some of their genes. Likewise, injected plasmids are also immediately expressed. The same sequence that provides oocyte specific expression of zona pellucida protein-3 (ZP3) when integrated into the chromosomes of transgenic animals (Lira et al. 1990; Schickler et al. 1992) also provides oocyte-specific expression when present on injected plasmid DNA (Millar et al. 1991). On the other hand, promoters injected into the transcriptionally inactive pronuclei of S-phase-arrested one-cell embryos remain inactive until the zygotic clock initiates expression of the endogenous genes (Martinez-Salas et al. 1989; Wiekowski et al. 1991). Promoters injected into either S-phase arrested or developing two-cell embryos are active immediately. Therefore, expression of injected genes is governed by the same mechanisms that regulate expression of zygotic genes. Moreover, the same promoter elements that are required for expression of genes injected into S-phase-arrested one-cell embryos are also required when DNA is injected into two-cell embryos. The only significant difference observed so far is that enhancers are required for efficient promoter and origin activity in two-cell to 16-cell embryos, whereas enhancers exhibit little or no activity in one-cell embryos or oocytes. These sequence requirements are the same whether or not two-cell and four-cell embryos are arrested in S-phase or allowed to continue development. Similar regulation of gene expression from microinjected plasmids is also observed in rabbits (Delouis et al. 1992; Christians et al. 1994). Therefore, we conclude that expression observed from microinjected plasmids is not an artifact of the experimental protocol, but accurately reflects the cell's capacity and requirements for carrying out transcription in vivo. Furthermore, microinjection has been used to introduce a wide variety of materials such as DNA, RNA, proteins, nucleoprotein complexes, and even cytoplasmic contents into mammalian oocytes and preimplantation embryos (DePamphilis et al. 1988; Clarke et al. 1992; Simerly and Schatten 1993). Therefore, microinjection techniques provide us with a system to manipulate cellular components of these embryos and oocytes, and thus study their role in various biological processes.

Using the mouse embryonic system and the microinjection techniques described above, experiments done so far revealed a surprising observation that the components of enhancer function that are required ubiquitously in cultured cells and cell extracts are actually acquired sequentially during the course of mouse embryonic development (for review see Majumder and DePamphilis 1994a, 1995):

- The primary role of enhancers first appears at the two-cell stage in mouse development, coincident with the onset of zygotic gene expression; this role is to relieve repression of promoters. Promoter repression is most likely brought in by chromatin structure (Majumder et al. 1993).
- Enhancers stimulate promoters by a TATA-box-independent mode in undifferentiated embryonic cell types, and later change to a TATA-box-dependent mode as cell differentiation becomes evident (Majumder and DePamphilis 1994b).
- Enhancers appear to require a putative coactivator(s) for their optimal activity. The activity of this coactivator(s) is not available until zygotic gene expression begins at the two-cell stage (Majumder et al. 1997).

19.1
Isolation and Culture of Mouse Oocytes and Embryos

Isolation and culture of mouse oocytes and embryos in vitro are described in detail elsewhere (Hogan et al. 1986; Wassarman and DePamphilis 1993), and a brief version is described here that we follow in our laboratory. Several strains of mice have been successfully used for this purpose. Among them, the most frequently used are the C57BL/6J inbred strains and the CD-1 outbred strains. Although CD-1 mice mate with a frequency of about 50–80%, and yield about 15–25 eggs per superovulated mouse, as compared to C57BL/6J mice (80–90%, and 20–30 respectively), the former are considerably less expensive than the latter. The development of CD-1 embryos is a little slower than that of C57BL/6J embryos. Male mice are raised one per cage, while the females can be raised together, about five to ten per cage, depending on the cage size. Mice colonies are maintained in a 14-h light/dark cycle (for example, lights off at 19.00 h and lights on at 05.00 h). Mice are usually sacrificed by cervical dislocation.

Procedure

1. For isolation of oocytes, use 14- to 15-day-old females.
2. Dissect out ovaries in about 3 ml of oocyte culture medium + HEPES (OCM+HEPES, Table 19.1: MEM with 0.03 mg/ml sodium pyruvate; 4 mg/ml bovine serum albumin; 0.5 mg/ml penicillin/streptomycin; 100 mg/ml dibutyryl cAMP to prevent germinal vesicle breakdown) containing 25 mM HEPES pH 7.4 (to maintain pH outside the CO_2 incubator), and overlayed with 3 ml mineral oil (Sigma) in a 60-mm tissue culture dish. Instead of ECM+HEPES, commercially available M2 medium (Specialty Media, Inc., NJ, USA) could also be used for this purpose. However, we find the ECM+HEPES medium to be more suitable for culture of embryos and their capacity for expression of microinjected genes.
3. Observe the ovaries at 20–50-fold magnification under a stereo dissecting microscope (Wild M5A), and poke several times with a 23-gauge needle attached to an empty syringe to dislodge oocytes into the medium.

Table 19.1. Oocyte culture medium

Water used to prepare all solutions should be double distilled and endotoxin-tested (Sigma W1503)	
1.1× MEM:	
Weigh an empty 250-ml Corning plastic flask, add about 25 ml water; add:	
MEM powder (Sigma M0769)	940 mg
200 mM glutamine (Sigma G7513	500 μl
0.5 mg/ml penicillin (Sigma P3032)	500 μl
0.5 mg/ml streptomycin (Sigma S9137)	500 μl
Fetal bovine serum (Hyclone)	5 ml
Water (excluding weight of the empty flask) to	95 g
Mix well; freeze aliquots at −20 °C for several months	
100× dibutyryl cAMP (Sigma D0627):	
10 mg/ml solution; freeze aliquots at −20 °C for several months	
10× HEPES (Sigma H9136):	
2.5 g HEPES in about 40 ml water; adjust pH to 7.4 with 1 M KOH; (adjust volume to 50 ml; store at 4 °C for several months)	
1× OCM (for oocyte culture at 37 °C, 5 % CO_2 incubator); 2 ml:	
1.1× MEM	1.8 ml
110 mg/ml sodium bicarbonate (Sigma S5761) made fresh in water	200 μl
100× dibutyryl cAMP	20 μl
1× OCM + HEPES (for oocyte manipulation at room temperature); 10 ml:	
1.1× MEM	9 ml
110 mg/ml sodium bicarbonate (Sigma S5761) made fresh in water	166 μl
10× HEPES	1 ml
Dibuyryl cAMP	100 μl

4. Rinse the freed oocytes twice in fresh medium using a Pasteur pipette drawn out to an inner diameter slightly larger than an oocyte or egg, and attached to a mouthpiece by a flexible tubing that passes through a trap containing a 0.22-μM Millipore filter trap to prevent contamination.

5. Transfer rinsed oocytes into drops of OCM (without HEPES; Table 19.1) overlayed with oil in a 60-mm dish that has been preincubated in a chamber at 37 °C, 5 % CO_2 for at least 1 h, and then return them to the incubation chamber for further culture. Oil overlay prevents evaporation of the medium and reduces the change of pH and the rate of heat loss. Generally, 20 to 70-μl drops with a concentration of 1 μl/oocyte or embryo are used for in vitro culture. Excess volume of medium per embryo inhibits the development of embryos presumably

Table 19.2 Embryo culture medium (ECM)

Water used to prepare all solutions should be double-distilled and endotoxin tested (Sigma W1503); All chemicals below are from Sigma. Catalog numbers are given in parenthesis.

10× salts (100 ml):	
NaCl (S5886)	4.003 g
KCl (P5405)	0.358 g
KH_2PO_4 (P5655)	0.150 g
$MgSO_4$ (M2643)	0.271 g
Glucose (G6152)	0.989 g
Dissolve in about 90 ml water; adjust pH to 7 with 2N NaOH; readjust the volume to 100 ml. Aliquots can be sotred at −20 °C for several months	
10× sodium lactate (50 ml):	1.84 g
(L7900)	
Aliquots can be stored at 4 °C for several months	
10× calcium lactate (50 ml):	0.2455 g
(L4388)	
Aliquots can be stored at −20 °C for several months	
10× bovine serum albumin (10 ml):	0.40 g
(A4161)	
Aliquots can be stored at −20 °C for several weeks	
100× antibiotics (10 ml):	
Penicillin (P3032)	0.060 g
Streptomycin (S9137)	0.050 g
Aliquots can be stored at −20 °C for several months	
100× EDTA (100 ml):	
(E5134)	0.150 g
Adjust pH to 8; stored at 4 °C for several months	
100× phenol red (10 ml):	
(P5530)	0.010 g
Stored at 4 °C for several months	
100× pyruvate (10 ml):	
(P5280)	0.050 g
Stored at 4 °C for about 1 week	
10× HEPES (50 ml):	
(H9136)	2.5 g
Adjust pH to 7.4; can be stored at 4 °C for several months	
10× $NaHCO_3$ (5 ml):	
(H9136)	0.105 g
Prepare fresh	
2× stock ECM (20 ml):	
Water	2.4 ml
10× salts	4 ml
10× sodium lactate	4 ml
10× calcium lactate	4 ml

Table 19.2 continued

10× bovine serum albumin	4	ml
100× antibiotics	0.4	
100× EDTA	0.4	ml
100× phenol red	0.4	ml
100× pyruvate	0.4	ml
Mix well; store at –20 °C for several months		
1× ECM (for embryo culture at 37 °C, 5 % CO_2 incubator); 2 ml:		
Water	0.8	ml
2× stock ECM	1	ml
10× $NaHCO_3$	0.2	ml
Mix well; filter through 0.22-m filter (Millipore; SLGV025LS)		
1×ECM+HEPES (for embryo manipulation at room temperature); 10 ml		
Water	3.83	ml
2× stock ECM	5	ml
10× $NaHCO_3$	0.166	ml
10× HEPES	1	ml
Mix well; filter through 0.22-m filter (Millipore; SLGV025LS)		

because of the dilution of useful growth factors secreted by these embryos into the medium.

6. To obtain greater yield of one- or two-cell embryos, superovulate 7- to 8-week-old females by intraperitoneal injection of 10 U of pregnant mare's serum (PMS; Sigma G-4877) at about 18.00 h, followed 48 h later by 10 U of human chorionic gonadotropin (hCG; Sigma CG-5). Stock solution for both PMS and hCG are made at 100 U/ml, and 0.1 ml is injected per mouse.
7. Place each of the injected females with a single male that is older than 10 weeks. Injection of hCG triggers ovulation 11–13 h later, and mating takes place at about midnight.
8. Next morning, the presence of a white copulation plug in the vagina indicates successful mating.
9. To isolate fertilized eggs (one-cell embryos), both oviducts are removed from pregnant females 17 h after hCG, placed in embryo culture medium + HEPES (ECM+HEPES, Table 19.2: 102 mM NaCl; 2.7 mM KCl; 8 mM Na_2HPO_4; 1.42 mM KH_2PO_4; 0.93 mM $CaCl_2$; 0.48 mM $MgCl_2$; 13.2 mM sodium pyruvate; 1 mg/ml glucose; 4 mg/ml bovine serum albumin; 10 μg/ml phenol red; 0.5 mg/ml penicillin/streptomycin; supplemented with 25 mM HEPES, pH 7.4) taken in a

60-mm culture dish, and observed under Wild M5A microscope as described above. Instead of ECM+HEPES, commercially available M2 medium (Specialty Media, Inc.; NJ, USA) could also be used for this purpose. However, we find the ECM+HEPES medium to be more suitable for culture of embryos and their capacity for expression of microinjected genes.

10. A portion of the oviduct is usually swollen (ampulla), that contains one-cell embryos embedded in cumulus cells, and can be easily seen under the microscope. Using a no. 5 forceps and a 30-gauge needle, tear the ampulla, releasing the contents into the medium.

11. Transfer the cumulus masses into ECM+HEPES medium containg hyaluronidase (120 μg/ml; Sigma H-6254) in a 60-mm-dish using a transfer pipette as described above for oocytes.

12. Place them on a heating block at 37 °C, or at room temperature, and observe often under the microscope.

13. As soon as the cumulus cells are dissociated from the embryos (usually a few min), wash the latter in fresh medium, transfer them into drops of ECM (without Hepes, Table 19.2; the presence of HEPES in the culture medium appears to hinder development of embryos) overlayed with oil, and place them in the incubation chamber at 37 °C, 5 % CO_2. M16 medium (Specialty Media, Inc.) can also be used in place of ECM, although it is less suitable for gene expression assays used in these studies. As described for oocytes, the incubation medium is preincubated in this chamber at least for 1 h prior to embryo culture.

14. To obtain two-cell embryos, remove oviducts from pregnant mice at 40 to 42 h post-hCG, place in ECM+HEPES medium, and flush the embryos out of the oviduct by injecting medium through infundibulum (the end of the oviduct closest to the ovary) using a 1-ml syringe attached to a 30-gauge blunt-ended needle.

15. These embryos are then washed, and incubated in ECM at 37 °C, 5 % CO_2 as described above for one-cell embryos.

16. Where necessary, isolate and culture embryos in the presence of DNA-replication inhibitor, aphidicolin (4 μg/ml, dissolved in dimethyl sulfoxide and stored in aliquots at −20 °C; Aphidicolin from Boehringer-Mannheim 724-548), to arrest their development at the beginning of S-phase. Since the first S-phase had not yet begun at the time of isolation of one-cell embryos, aphidicolin causes them to retain their two pronuclei, male and female, throughout the experiment. Since two-cell embryos are isolated after they had undergone DNA replication, aphidicolin would cause them usually to cleave into four cells. In the absence of aphidicolin, injected one-cell and two-cell embryos developed up to the morula stage.

19.2 Preparation of DNA

Various plasmid DNAs have been used to study gene expression in the mouse system (Majumder and DePamphilis 1995). Plasmid DNAs described in this chapter are: pluc, ptkluc, pF101tkluc, and are described below. Preparation of good-quality plasmid DNA is very important for gene expression studies utilizing the mouse system. Plasmid DNA prepared by two CsCl-ethidium bromide equilibrium density gradients, as compared to procedures utilizing commercially available kits (like Qiagen) was consistently found to be a much better template for transcriptional studies in the mouse preimplantation embryo system. However, plasmid DNA prepared by the latter methods was found to be of sufficiently good quality for the purpose of constructing transgenic mice. Extraction of DNA grown in bacteria can be achieved by any of the several methods routinely used in various laboratories (Sambrook et al. 1989). However, we find that the following Brij/sodium doecyl sulfate lysis method (Miranda et al. 1993) consistently yields DNA of excellent quality, as compared to the alkali lysis method (Sambrook et al. 1989). Although it generates a much lower yield, it is a gentle procedure, and avoids nicking of plasmid DNA, specially when the plasmids are large (>10 kb).

Procedure

1. Harvest overnight a 500 ml culture of bacteria harboring the plasmid at 4229 *g* for 5 min (5000 rpm in a GS3 or similar rotor), and resuspend the bacterial pellet in ice cold 12 ml ST buffer (25 % sucrose; 50 mM Tris.HCl, pH 8).
2. To this suspension, add 1 ml of freshly prepared lysozyme in 50 mM Tris, pH 8 (5 mg/ml), incubate for 10 min on ice, followed by 1.25 ml

ice-cold 0.5 M EDTA, pH 8, another 10 min on ice, followed by 5 ml Brij lysis solution (1% Brij 58; 0.4% SDS; 60 mM EDTA, pH 8; 50 mM Tris, pH 8), and incubate a further 15 min on ice. It is advisable not to incubate the last step longer, for otherwise the plasmid DNA may be difficult to separate from the cellular DNA.

3. Centrifuge the resulting mixture at about 100,000 *g* (30,000 rpm in 45Ti or 32,000 rpm in 70 Ti, or 23,000 rpm in SW28 rotor) to obtain a clear lysate.
4. Treat the lysate (with pancreatic RNase (10 μg/ml, 37 °C, 1 h) followed by proteinase K (100 μg/ml in the presence of 0.5% SDS, 37 °C, 1 h) to get rid of contaminating RNA and proteins.
5. Dissolve solid CsCl to the DNA mixture at 1 g/ml.
6. Add ethidium bromide (10 mg/ml stock solution) to give a final concentration 750 μg/ml.
7. Centrifuge the mixture at about 7700 *g* (8000 rpm in SS34 rotor) for 5 min, and discard the resultant crust of ethidium bromide-bacterial protein complexes at the top.
8. Transfer the underlying clear red solution into an ultracentrifuge tube (Beckman Quick-seal or equivalent) using a disposable syringe fitted with a large-gauge needle.
9. Fill the rest of the tube with either paraffin oil or CsCl/ethidium bromide solution (prepared separately with CsCl concentration of 1 g/ml), seal and centrifuge at 20 °C using a Beckman vertical VTi65 rotor (45,000 rpm for 16 h), or any of the other angle rotors (Ti50: 45,000 rpm for 48 h; Ti65: 60,000 rpm for 24 h; Ti 70.1: 60,000 rpm for 24 h).
10. Isolate the plasmid DNA band (form I; Sambrook et al. 1989), and centrifuge the DNA a second time in CsCl-EtBr gradient centrifugation as described above.
11. Collect the plasmid DNA band, and remove ethidium bromide by extracting with isobutyl alcohol saturated with water.
12. Dialyze the DNA extensively against TE (10 mM Tris, pH 8; 1 mM EDTA) buffer. Plasmid DNA dissolved in TE buffer can be stored at 4 °C for 1 month or at −20 °C for prolonged period of time.

13. Prior to microinjection, precipitate the DNA with 2.5 M ammonium acetate or 0.1 M NaCl plus 2.5 vol of 90 % ethanol, and centrifuge at full speed in an Eppendorf tabletop centrifuge.
14. Rinse the DNA pellet with 75 % ethanol, centrifuged again, and semidry at room temperature (DNA pellet may be difficult to resuspend if overdried).
15. Resuspend the DNA pellet in injection buffer. Two injection buffers have been routinely used in our laboratory: PIB (48 mM K_2HPO_4; 4.5 mM KH_2PO_4; 14 mM NaH_2PO_4; pH 7.2) or TIB (10 mM Tris.HCl, pH 7.6; 0.25 mM EDTA). Injection buffers are filtered through 0.22 μ Millipore filter prior to storage at 4 °C.

19.3 Microinjection of Plasmid DNA into Mouse Oocytes and Embryos

Microinjection procedure is basically carried out as described elsewhere (Wassarman and DePamphilis 1993), and in previous chapters on generating transgenics in this book. The holding pipette described there is the same that we use for our gene expression studies. However, we use a different kind of injection pipette, and the procedure is described elsewhere (Miranda and DePamphilis 1993). For gene expression studies, as compared to transgenic studies, considerably higher concentration of plasmid DNA is used for microinjection. This causes high viscosity of the DNA solution that is difficult to inject, and also causes nuclear material to stick to the regular injecting needle, making it easier to clog. For example, on the average, one pipette that is generally used in generating transgenic mice can only inject about ten to twenty embryos or oocytes, before they become clogged. To circumvent these problems, we use siliconized borosilicate glass pipettes that have beveled open tips of about 1 μm in inner diameter, for microinjection purposes (Miranda and DePamphilis 1993). Approximately 100 embryos and oocytes can be injected with DNA using one such injection pipette. However, personal choice might dictate the use of several of the former type of injection needle for one given experimental condition, because these needles are easier to make.

Procedure

1. Place between 50 to 100 oocytes or embryos per DNA sample in the culture medium + HEPES (Tables 19.1 and 19.2) in a microinjection dish.
2. Inject plasmid DNA into the germinal vesicles of oocytes (within 6 h of isolation), one of the pronuclei of S-phase arrested one-cell embryos (between 24 and 28 h post-hCG), or one of the zygotic nuclei of S-phase arrested two-cell embryos (between 44 and 48 h post-hCG), and transfer them back into the incubation chamber. About 100 injections can be performed per hour.
3. Re-examine the injected embryos and oocytes under a dissection microscope approximately 30 min later, and the remove the surviving ones into a new drop of equilibrated medium (~1 μl of medium per embryo or oocyte). These are assayed after the required incubation period (see above).

19.4 Luciferase assay

Procedure

1. Assay individual oocytes and embryos that survived injection for luciferase (~80 % of two-cell embryos, ~65 % of one-cell embryos, ~90 % of oocytes). Oocytes are assayed 24 h after injection and embryos are assayed 44 h after injection, as previously described (Miranda et al. 1993).
2. Working under a dissection microscope, transfer each embryo or oocyte into an Eppendorf tube containing 50 μl luciferase reaction mix (LRM: 25 mM glycylglycine, pH 7.8; 10 mM magnesium acetate; 0.5 mM ATP, pH 7; 100 mg/ml bovine serum albumin; 1 mM dithiothreitol; can be stored at 4 °C for a month) containing freshly added 0.1 % Triton X-100, and freeze in a dry ice/ethanol bath. Frozen embryos can be stored at −70 °C until assayed.
3. Thaw samples at 37 °C and subject to short centrifugation in an Eppendorf centrifuge at 10,000 rpm, to collect the embryo extract at the bottom of the tube.
4. Prewarm all solutions to ambient temperature for further steps.

5. Mix the embryo extract with 300 μl of LRM in a polystyrene cuvettes, and measure luciferase activity using a Monolight 2010 luminometer (Analytical Luminiscence) that dispenses 100 μl of 1 mM luciferin solution in water (Analytical Luminiscence; stock solution can be stored at 4 °C for 1 month) plus 1 mg/ml coenzyme A (grade II, Boehringer Mannheim; stock solution of 100 mg/ml can be stored at −70 °C for months), and integrates the emitted light over a period of 10 s. Other commercially available luminometers can also be used for the purpose.

6. Represent the mean value of 40 to 150 oocytes or embryos as each data point, as shown in Fig. 19.1, and express the variation among individual embryos as ± standard error of the mean. While the range of luciferase activities among individual embryos could vary as much as 1000-fold (Miranda et al. 1993), the mean value obtained from several independent experiments was reproducible to within 13 to 25 %. Moreover, the relative activity between different types of embryos and different promoters was always reproducible, even when DNA injection was performed by different workers.

Results

- Expression of promoters and enhancers in one-cell and two-cell embryos

To determine the relative transcriptional activity of the paternal pronuclei of S-phase arrested one-cell and of S-phase arrested zygotic nuclei of two-cell embryos, their nuclei were microinjected with plasmid DNA, and gene expression assayed as described above. Injected plasmid contained a heterologous yet sensitive reporter gene, luciferase, linked to the Herpes simplex virus (HSV) thymidine kinase (tk) promoter (ptkluc) uncoupled or coupled to the Py F101 enhancer (pF101tkluc) placed 600 base pairs upstream from the tk promoter. This well-characterized promoter contains four transcription factor binding sites: two for Sp1, one for CAAT-box binding transcription factor (CTF), and one for TATA-box binding protein (TBP). The F101 enhancer has been found to contain tandem duplication of DNA binding sites for transcription enhancer factor-1 (TEF-1) (Xiao et al. 1991; Melin et al. 1993). The tk promoter can be stimulated by a wide variety of *cis*-acting enhancers, like F101, as well as *trans*-activators like the HSV immediate early protein ICP4. The tk promoter and F101 enhancer were selected because they use cellular transcription factors exclusively, and function in a wide variety of cell types, including undifferentiated mouse embryonic stem cells and cleavage-stage mouse embryos (Martínez-Salas et al. 1989; Blatt and DePamphilis 1993; Melin et al. 1993). F101 is the strongest enhancer found so far for stimulating promoter activity in cleavage stage two-cell to eight-cell mouse embryos and in undifferentiated embryonic stem cells and embryonic carcinoma cells (Blatt and DePamphilis 1993; Melin et al. 1993). When embryos were injected with different amounts of ptkluc, the amount of luciferase activity observed was dependent on the amount of DNA injected (Fig. 19.2). At the optimum DNA concentration, about 90,000 relative light units (RLU) of luciferase activity per embryo were observed. This was the average activity observed for 50–100 individually assayed embryos. For comparison, a control construct containing the luciferase gene but without the tk promoter produced only 500 RLU under similar conditions (data not shown). This level of tk promoter activity was not affected by linking it to the F101 enhancer (pF101tkluc). In contrast, when mouse two-cell embryos are isolated and cultured under the same conditions, promoter activity was greatly reduced (as much as 1 %) relative to one-cell embryos, and the F101 enhancer was required to restore activity in two-cell embryos to the level observed in one-cell embryos.

The lack of enhancer function in the paternal pronuclei of one-cell embryos is found not due to a difference in transcriptional capacity, promoter requirements or the lack of enhancer-activation protein (Majumder et al. 1993). On the other hand, the lack of enhancer function in one-cell embryos is observed with diverse range of promoter and origin sequences that bear little homology and that interact with different initiation factors, indicating that it is a general phenomenon. Furthermore, repression of promoters injected into two-cell embryos does not result from changes in nuclear origin or their ploidy (Wiekowski et al. 1993), but instead appears to be mediated by changes in chromatin structure that result, in part, from factors present in two-cell cytoplasm and absent in one-cell cytoplasm. Promoters injected into paternal pronuclei in S-phase arrested one-cell embryos is not repressed, but transfer of the injected pronucleus to a two-cell embryo results in repression of the injected promoter that can be relieved by linking it to the F101 enhancer (Henery et al. 1995). In addition, DNA injected into two-cell mouse embryos is more readily assembled into nucleosomes than DNA injected into one-cell embryos (Martinez-Salas et al. 1989), and chromatin assembled in one-cell embryos appears to be organized differently than chromatin assembled in two-cell embryos. In other systems, nucleosome

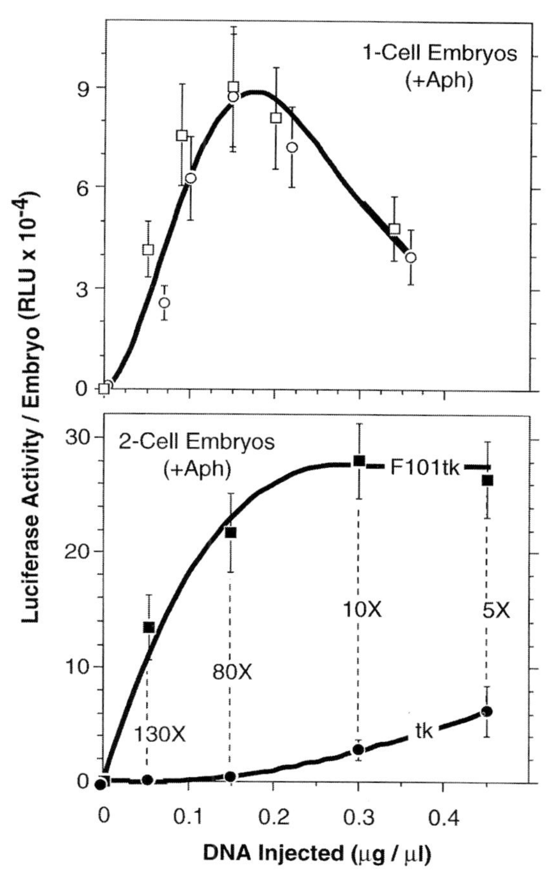

Fig. 19.2. Expression of luciferase following injection of plasmid DNA ptkluc (○, ●) or pF101tkluc (□, ■) into the paternal pronucleus of a one-cell embryo (*top panel*), and into one of the zygotic nuclei of a two-cell embryo (*bottom panel*) (Majumder et al. 1993). Firefly luciferase (luc) is a reporter gene that allows quantitative measurement of promoter/enhancer activity in a single embryo. The Herpes simplex virus thymidine kinase (tk) promoter is a well-characterized promoter that is utilized by a wide variety of cell types and responds strongly to enhancers. The polyomavirus F101 enhancer (F101) stimulates promoter activity in most mouse cells, particularly in embryonal carcinoma F9 cells, embryonic stem cells and mouse cleavage-stage embryos. Embryos were cultured in aphidicolin to arrest one-cell embryos at the beginning of S-phase, and two-cell embryos (which were isolated in their G2-phase) at the beginning of S-phase of the four-cell stage. Luciferase activity was determined 42 h postinjection (injected DNA is stable for several days) for 40 to 60 individual embryos per data point, and the mean value ± its standard error was expressed as relative light units (RLU). In the absence of a promoter, luciferase levels were 200 RLU (two-cell embryos) to 500 RLU (one-cell embryos). The extent of enhancer stimulation is indicated at each DNA concentration

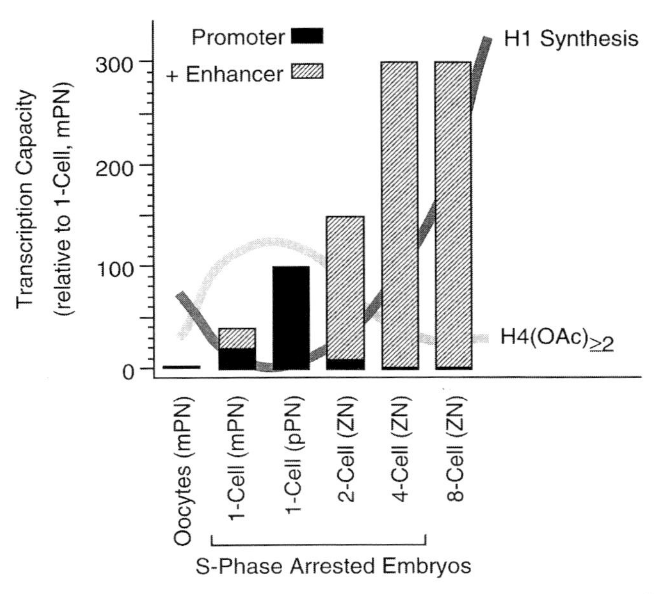

Fig. 19.3. Changes in promoter and enhancer utilization reflect changes in chromatin composition. Promoters and enhancers linked to the firefly luciferase reporter gene were injected into the nuclei (*m* maternal; *p* paternal; *z* zygotic) of mouse oocytes or early embryos, and the amount of gene expression in individual oocytes and embryos was quantified. The SEM is ~± 15 % for a pool of 40 to 60 successful injections (Majumder and DePamphilis 1995). Promoter (*filled bar*) and promoter + enhancer (*hatched bar*) activities are presented relative to [promoter + enhancer] activity in S-phase arrested one-cell embryos (= 100 %). F101 polyomavirus (PyV) enhancer activity is observed only from the two-cell stage onward. In the presence of saturating amounts of GAL4:VP16 protein, weak GAL4:VP16-driven enhancer activity is observed in oocytes and maternal pronuclei of S-phase arrested one-cell embryos but strong activity (equivalent to the F101 PyV enhancer) is observed only in two-cell embryos onward. The extent of enhancer stimulation depends on the strength of the promoter and the amount of DNA injected. Relative levels of histone H1 synthesis and hyperacetylated histone H4 (contains two or more acetyl groups) are also indicated

assembly has been found to repress both transcription and replication by interfering with the activity of transcription/replication proteins (Workman and Buchman 1993; Workman et al. 1991). Furthermore, treatment of two-cell embryos with butyrate, an agent that destabilizes chromatin structure by blocking histone deacetylase (Tazi and Bird 1990; Turner 1991; Majumder and DePamphilis 1995), strongly stimulates promoter activity, relieving repression and reducing the need for enhancers

(Majumder et al. 1993; Wiekowski et al. 1993). In contrast, butyrate does not stimulate promoter activity in paternal pronuclei of one-cell embryos. Thus butyrate appears to substitute for the function of enhancers in two-cell embryos.

Results described above are also obtained when two-cell embryos were cultured in the absence of aphidicolin, but the levels of activity were three- to four-fold less. Similar results were also obtained when various other promoters, both natural and synthetic, were tested in the paternal pronuclei of one-cell and two-cell embryos (Martinez-Salas et al. 1989; Wiekowski et al. 1991; Melin et al. 1993; Majumder et al. 1993). The extent of repression and accompanying stimulation by enhancers in two-cell embryos depends on promoter strength, the amount of DNA injected, and the extent of development (Majumder et al. 1993). Repressor and enhancer activities increase severalfold as two-cell mouse embryos develop into four-cell embryos (Henery et al. 1995). Similar results are observed with plasmid DNA replication, instead of transcription, driven by the polyomavirus replication origin (Martinez-Salas et al. 1988). Furthermore, in rabbit embryos, enhancer function first appears at the 8- to 16-cell stage (Delouise et al. 1992; Christians et al. 1994). Thus, stage-specific appearance of enhancer function may occur later during development of other mammals. Taken together, these results suggest that the activities of promoters and replication origins are repressed when injected into two-cell mouse embryos, but not when injected into one-cell embryos, and that the role of enhancers is to relieve this repression (Figs. 19.2, 19.3).

The appearance of repression in mouse two-cell embryos correlates with the appearance of histone H1 (Majumder and DePamphilis 1995; Fig. 19.3): promoter repression begins at two-cell embryos and becomes progressively stronger as the embryos develop to late four-cell to eight-cell stage. Histone H1 synthesis also begins in late one-cell embryos, but cannot be detected by anti-histone H1 antibodies until the late four-cell to eight-cell stage (Clarke et al. 1992). The repressive action of histone H1 is observed only when it is bound to the chromatin. Histone H1 binds poorly to hyperacetylated chromatin. The nascent histone H4 in one-cell embryos is hyperacetylated and the degree of hyperacetylation decreases in two-cell and four-cell embryos. Thus, the appearance of repression at the two-cell stage in mouse development may be due to the appearance of histone H1 or a novel histone H1 subtype (Smith et al. 1988; Ohsumi and Katagiri 1991).

The in vivo results described above are supported by in vitro results: enhancers are found to have a small effect, if any, on DNA replication or transcription when they are assayed in cell-free systems that cannot assemble chromatin (discussed in Majumder et al. 1993). Furthermore, enhancers can stimulate promoters in cell-free systems only when the DNA is packaged into chromatin containing histone H1 (Laybourn and Kadonaga, 1991, 1992). However, one publication suggests that, although the chromatin structure is required for enhancer function, the role of histone H1 may be less prominent than was thought before, when chromatin assembly is carried out with *Drosophila* nuclear extract S-190 as opposed to polyglutamic acid (Pazin et al. 1994). In addition, knockout experiments in *Tetrahymena thermophila* indicate that histone H1 regulates activated transcription, but not global transcription in vivo (Shen and Gorovsky 1996). Therefore, taken together, results of these experiments suggest that the repression of promoter activity observed in two-cell embryos is due to chromatin structure, and that the role of histone H1 in this process is still unclear. These results also suggest that enhancers, like butyrate, alleviate the repression by altering chromatin structure. Thus, the in vivo model described here can be utilized not only to decipher the mechanism of enhancer function, but also to resolve discrepancies posed by in vitro and in vivo experiments, as described above.

In summary, if the injection of preassembled chromatin templates injected into the paternal pronucleus of one-cell embryos can exhibit promoter repression, and subsequent relief of this repression by enhancers, this system would have a tremendous potential of examining important aspects of transcriptional regulations. This could be done by manipulating various components of the reconstituted chromatin. The chromatin template reconstituted in the absence of a particular component that is actually required in vivo for promoter repression and subsequent enhancer stimulation, would not exhibit these properties. Furthermore, such incompetent templates could possibly be rescued by the addition of the corresponding component (depending on the architectural properties of the chromatin, the timing of addition of that factor, and the presence of other components, like ATP) either in the reconstitution mixture, or by directly coinjecting the component in one-cell embryos. Thus, chromatin reconstituted with or without histone H1, for example, would indicate its structural as well as regulatory role in enhancer function in vivo. Acetylation of core histones has been hypothesized to be involved in transcriptional activation (Turner 1991). However, it is not clear if acetylation is a cause or effect. We can now reconstitute chromatin using core components like hyperacetylated core histones and underacetylated histones (Lee et al. 1993), and examine their importance in enhancer function. Similarly, the role of histone tails can now be determined by using trypsin-treated histones in the reconstitution experiment (Lee et al. 1993). It has been suggested that the stoichiometry of H2A-H2B relative to H3-H4 can affect transcription (Paranjape et al. 1994). Such questions also can now be answered in our system using H3-H4 tetramers in reconstituting the chromatin template. The biological role of other nonhistone proteins, like the high-mobility group proteins, although postulated to be important in transcriptional activation (Tremethick and Drew 1993), is not yet known. It will be very interesting to examine the role of these and other chromatin-remodelling factors on transcriptional activities of reconstituted chromatin templates inside a living cell.

- Expression of promoters and enhancers in oocytes and two-cell embryos

Unlike the paternal pronuclei of one-cell embryos, and similar to two-cell embryos, enhancer-responsive promoters are strongly repressed in oocytes and, to some extent, in maternal pronuclei of one-cell embryos. Furthermore, this repression appears to be chromatin-mediated, as the repression can be relieved by sodium butyrate. Therefore, one would expect the promoter activity to be stimulated by an enhancer, just like in two-cell embryos. However, this repression cannot be relieved by a Gal4-dependent enhancer, F101 enhancer, or the transactivator ICP4 (Majumder et al. 1997; Fig. 19.3). In order to determine if the lack of promoter stimulation was due to the absence of enhancer activating protein, we observed that even in the presence of functional Gal4-VP16 protein, the Gal4-dependent enhancer was inactive in oocytes, but active in two-cell embryos (Fig. 19.3). Furthermore, the lack of Gal4-dependent enhancer activity in oocytes can be partially rescued by expression of mRNA obtained from undifferentiated embryonic stem cells derived from the blastocyst stage embryos (Majumder et al. 1997). Furthermore, enhancer-competition experiments suggest that the coactivator activity is shared by different enhancers (Majumder et al. 1997). Therefore it appears that a coactivator activity is required for the optimal activity of an enhancer/transactivator (Fig. 19.4), and that the activity of this coactivator is missing in oocytes, and that it is first expressed in two-cell embryos. The presence of a coactivator, or a family of coactivators, that can activate several enhancers or transactivators can be of great biological importance, as it can serve as a master switch to regulate several genes at once during development. At this point, it is not known if the coactivator activity is regulated by a single regulator, or a family of regulators. However, since oocytes do not exhibit the coactivator(s) activity,

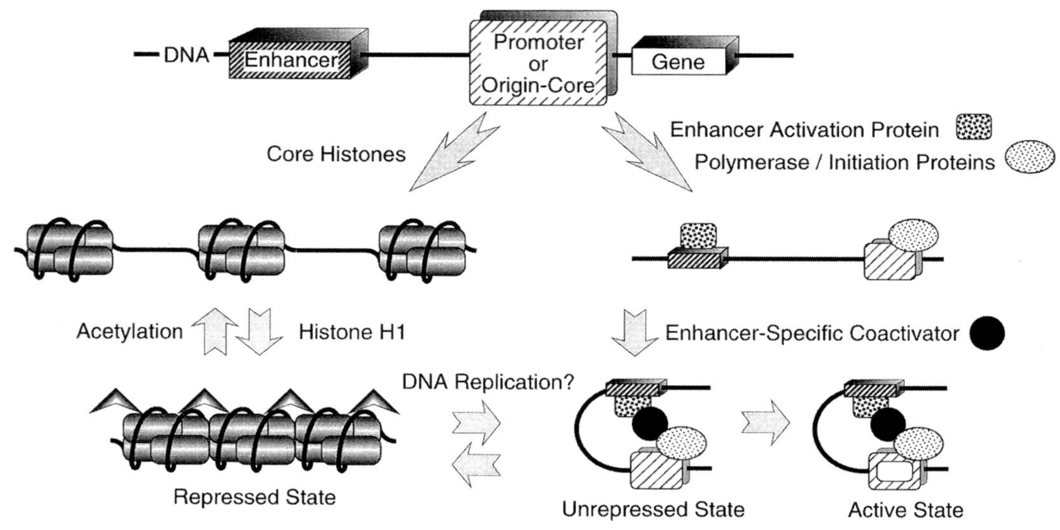

Fig. 19.4. A working model showing the repression of promoters and replication origins by chromatin structure and the role of enhancers. Core histones and transcription/replication proteins (including enhancer activation proteins) compete to bind to the plasmid DNA microinjected into mouse oocytes and early embryos. Depending on their relative affinities for DNA, there is a dynamic equilibrium between DNA bound to core histones and DNA bound to various transcription/replication factors. In the presence of histone H1, DNA bound to core histones can then interact with them, resulting in a condensed chromatin structure and a repressed state. The repressive action of histone H1 can be blocked by acetylation of core histones. Sodium butyrate is known to inhibit histone deacetylases, and thus increase the fraction of acetylated core histones, causing stimulated transcription. On the other hand, the equilibrium can be shifted to the other direction where, in the presence of enhancer specific coactivators, DNA bound to transcription/replication factors and enhancer activation proteins can interact with them and result in the prevention of repression and the formation of an active state. DNA replication at each cell division may provide the cell with a chance to reestablish the equilibrium between repressed and unrepressed states

they can be useful as an assay system for the missing function, and thus help identify the putative coactivator(s).

Acknowledgements: This work was supported in part by grants from the National Institutes of Health (GM53454), Pediatric Brain Tumor Foundation of the United States, and Association for Research on Childhood Cancer.

References

Bachvarova RF (1992) A maternal tail of poly(A): the long and the short of it. Cell 69:895–897

Blatt C, DePamphilis M (1993) Striking homology between mouse and human transcription enhancer factor-1 (TEF-1). Nucleic Acids Res 21:747–748

Carey M, Leatherwood J, Ptashne M (1990). A potent GAL4 derivative activates transcription at a distance in vitro. Science 247:710–712

Christians E, Rao VH, Renard JP (1994) Sequential aquisition of transcriptional control during early embryonic development in the rabbit. Dev Biol 164:160–172

Clarke HJ, Oblin C, Bustin M (1992) Developmental regulation of chromatin composition during mouse embryogenesis: somatic histone H1 is first detectable at the four-cell stage. Development 115:791–799

Conover JC, Gretchen LT. Zimmermann JW, Burke B. Schultz RM (1991) Stage-specific expression of a family of proteins that are major products of zygotic gene activation in the mouse embryo. Dev Biol 144:392–404

Delouis C, Bonnerot C, Vernet M, Nicolas J-F (1992) Expression of microinjected DNA and RNA in early rabbit embryos: chnages in permissiveness for expression and transcriptional selectivity. Exp Cell Res 201:284–291

DePamphilis ML, Herman SA, Martinez-Salas E, Chalifour LE, Wirak DO, Cupo DY, Miranda M (1988) Microinjecting DNA into mouse ova to study DNA replication and gene expression and to produce transgenic animals. BioTechniques 6:662–680

Felsenfeld G (1992) Chromatin as an essential part of the transcriptional mechanism. Nature 355:219–224

Henery CC, Miranda M, Wiekowski M, Wilmut I, DePamphilis M (1995) Repression of gene expression at the beginning of mouse development. Dev Biol 169:448–460

Hogan BL, Constantini F, Lacy E (1986) Manipulating the mouse embryo. Cold Spring Harbor Laboratory, Cold Spring Harbor

Kamakaka RT, Bulger M, Kadonaga JT (1993) Potentiation of RNA polymerase II transcription by Gal4-VP16 during but not after DNA replication and chromatin assembly. Genes Dev 7:1779

Latham KE, Solter D, Schultz RM (1992) Acquisition of a transcriptionally permissive state during the one-cell stage of mouse embryogenesis. Dev Biol 149:457–462

Laybourn PJ, Kadonaga JT (1991) Role of nucleosomal cores and histone H1 in regulation of transcription by RNA polymerase II. Science 254:238–245

Laybourn PJ, Kadonaga JT (1992) Threshold phenomena and long-distance activation of transcription by RNA polymerase II. Science 257:1682–1685

Lee DY, Hayes JJ, Pruss D, Wolffe A (1993) A positive role for histone acetylation in transcription factor access to nucleosomal DNA. Cell 72:73–84

Lira SA, Kinloch RA, Mortillo S, Wassarman P (1990) An upstream region of the mouse ZP3 gene directs expression of firefly luciferase specifically to growing oocytes in transgenic mice. Proc Natl Acad Sci 87:7215–7219

Majumder S, DePamphilis ML (1994a) Requirements for DNA transcription and replication at the beginning of mouse development. J Cell Biochem 55:59–68

Majumder S, DePamphilis ML (1994b) TATA-dependent enhancer stimulation of promoter activity in mice is developmentally acquired. Mol. Cell. Biol. 14:4258–4268

Majumder S, DePamphilis ML (1995) A unique role for enhancers is revealed during early mouse development. BioEssays 17:879–889

Majumder S, Miranda M, DePamphilis M. (1993) Analysis of gene expression in mouse preimplantation embryos demonstrates that the primary role of enhancers is to relieve repression of promoters. EMBO J 12:1131–1140

Majumder S, Zhao Z, Kaneko K, DePamphilis M (1997) Developmental aquisition of enhancer function requires a unique coactivator activity. EMBO J 16:1721–1731

Manley JL, Proudfoot NJ (1994) RNA 3' ends: formation and function-meeting review. Genes Dev 8:259–264

Martínez-Salas E, Cupo DY, DePamphilis ML (1988) The need for enhancers is acquired upon formation of a diploid nucleus during early mouse development. Genes Dev 2: 1115–1126

Martínez-Salas E, Linney E, Hassell J, DePamphilis ML (1989) The need for enhancers in gene expression first appears during mouse development with formation of a zygotic nucleus. Genes Dev 3:1493–1506

Mélin F, Miranda M, Montreau N, DePamphilis ML, Blangy D (1993) Transcription enhancer factor-1 (TEF-1) DNA binding sites can specifically enhance gene expression at the beginning of mouse development. EMBO J 12:4657–4666

Millar SE, Lader E, Liang L-F, Dean J (1991) Oocyte specific factors bind a conserved upstream sequence required for mouse zona pellucida promoter activity. Mol Cell Biol 11:6197–6204

Miranda M, DePamphilis M (1993) Preparation of injection pipettes. In: Wassermann PM, DePamphilis ML (eds) A guide to mouse development. Methods Enzymol 225:412–433

Miranda M, Majumder S, Wiekowski M, DePamphilis ML (1993) Application of firefly luciferase to preimplantation development. In: Wassarman PM, DePamphilis ML (eds) A guide to mouse development. Methods Enzymol 225:412–433

Ohsumi K, Katagiri C (1991) Occurrence of H1 subtypes specific to pronuclei and cleavage-stage cell nuclei of anuran amphibians. Dev Biol 147:110–120

Paranjape SM, Kamakaka RT, Kadonaga JT (1994) Role of chromatin structure in the regulation of transcription by RNA polymerase II. Annu Rev Biochem 63:265–297

Pazin MJ, Kamakaka RT, Kadonaga JT (1994) ATP-dependent nucleosome reconfiguration and transcriptional activation from preassembled chromatin templates. Sience 266:2007–2011

Richter JD (1991) Translational control during early development. BioEssays 13:179–183

Sambrook J, Fritsch EF, an Maniatis T (1989) Molecular cloning a laboratory manual. Cold Spring Harbor Laboratory Press, New York

Schickler M, Lira SA, Kinloch RA, Wassarman P (1992) A mouse oocyte-specific protein that binds to a region of mZP3 promoter responsible for oocyte-specific mZP3 gene expression. Mol Cell Biol 12:120–127

Schultz GA, Heyner S (1992) Gene expression in pre-implantation mammalian embryos. Mutat Res 296:17–31

Schultz RM (1993) Regulation of zygotic gene activation in the mouse. BioEssays 8:531–538

Seshagiri PB, McKenzie DI, Bavister BD, Williamson JL, Aiken JM (1992) Golden hamster embryonic genome activation occurs at the two-cell stage: correlation with major developmental changes. Mol Reprod Dev 32:229–235

Shen X, Gorovsky M (1996) Linker histone H1 regulates specific gene expression but not global transcription in vivo. Cell 86:475–483

Simerly C, Schatten G (1993) Techniques for localization of specific molecules in oocytes and embryos. Methods Enzymol 225:516–553

Smith RC, Dworkin-Rastl E, Dworkin MB (1988) Expression of a histone H1-like protein is restricted to early *Xenopus* development. Genes Devel 2:1284–1295

Struhl K (1996) Chromatin structure and RNA polymerase II connection: implications for transcription. Cell 84:179–182

Studitsky V, Clark D, Felsenfeld G (1995) Overcoming a nucleosomal barrier to transcription. Cell 83:19–27

Tazi J, Bird A (1990) Alternative chromatin structure at CpG islands. Cell 60:909–920

Telford NA, Watson AJ, Schultz GA (1990) Transition form maternal to embryonic control in early mammalian development. Mol Reprod Dev 26:90–100

Tremethick DJ, Drew H (1993) High mobility group proteins 14 and 17 can space nucleosomes in vitro. J Biol Chem 268:11389–11393

Turner BM (1991) Histone acetylation and control of gene expression. J Cell Sci 99:13–20

Walters M, Fiering S, Eidemiller J, Magis W, Groudine M, Martin D (1995) Enhancers increase the probability but not the level of gene expression. Proc Natl Acad Sci 92:7125–7129

Wassarman P, DePamphilis M (1993) Methods in Enzymology, vol 225. Guide to techniques in mouse development. Academic Press, New York

Wiekowski M, Miranda M, DePamphilis ML (1991) Regulation of gene expression in preimplantation mouse embryos: effects of zygotic gene expression and the first mitosis on promoter and enhancer activities. Dev Biol 147:403–414

Wiekowski M, Miranda M, DePamphilis ML (1993) Requirements for promoter activity in mouse oocytes and embryos distinguish paternal pronuclei from maternal and zygotic nuclei. Dev Biol 159:366–378

Wolffe AP, Pruss D (1996) Targeting chromatin disruption: transcription regulators that acetylate histones. Cell 84:817–819

Workman JL, Buchman, AR (1993) Multiple functions of nucleosomes and regulatory factors in transcription. Trends Biochem Sci 18:90–95

Workman JL, Taylor ICA, Kingston RE (1991) Activation domains of stably bound GAL4 derivatives alleviate repression of promoters by nucleosomes. Cell 64:533–544

Xiao JH, Davidson I, Matthes H, Garnier JM, Chambon P (1991) Cloning, expression, and transcriptional properties of the human enhancer factor TEF-1. Cell 65:551–568

Chapter 20

The Use of Mouse Preimplantation Embryos for the Identification of DNA-Binding Proteins Essential in Early Development

Valerie Botquin[1], Jörg R. Schlehofer[2], and Angel Cid-Arregui[3]*

Introduction

Mouse preimplantation embryos develop in vitro to the blastocyst stage, thereby providing a unique system which, conveniently manipulated by microinjection, may be very useful in preimplantation development and gene expression studies (for reviews see Rossant and Pederson 1986; Majumder and DePanphilis 1995; Majumder, Chap. 18, this Vol.), in addition to its application to transgenesis and knockout experiments.

Although several dozens of genetic abnormalities interfering with mouse embryogenesis have been described (Magnuson 1986; Lyon and Searle 1989), most of them affect postimplantation embryos and only a few mutations have been reported to cause developmental arrest and lethal defects in preimplantation embryos (see Cheng and Constantini 1993). None of the genes responsible for such effects has been identified to date. Indeed, very little is known about specific proteins indispensable for normal development of the early embryo. This is due to the complexity of the developmental mechanisms and the limited numbers of mouse embryos that can be obtained in a single experiment. However, parentally inherited DNA-binding proteins and those synthesized de novo by the embryo are likely to be among the essential components of preimplantation development. A possible way to overcome the difficulties in identifying these proteins is the combined use of preimplantation embryos and embryonic stem (ES) cells (Fig. 20.1), i.e., cell lines derived

* corresponding author: phone: +49-6221-548321; fax: +49-6221-548301; e-mail: cid@urz.uni-heidelberg.de

[1] European Molecular Biology Laboratory, Meyerhofstrasse 1, D-69117 Heidelberg, Germany

[2] Institute for Applied Tumor Virology, Deutsches Krebsforschungszentrum, Im Neuenheimer Feld 242, D-69120 Heidelberg, Germany

[3] Dept. of Neurobiology, University of Heidelberg, Im Neuenheimer Feld 364, D-69120 Heidelberg, Germany

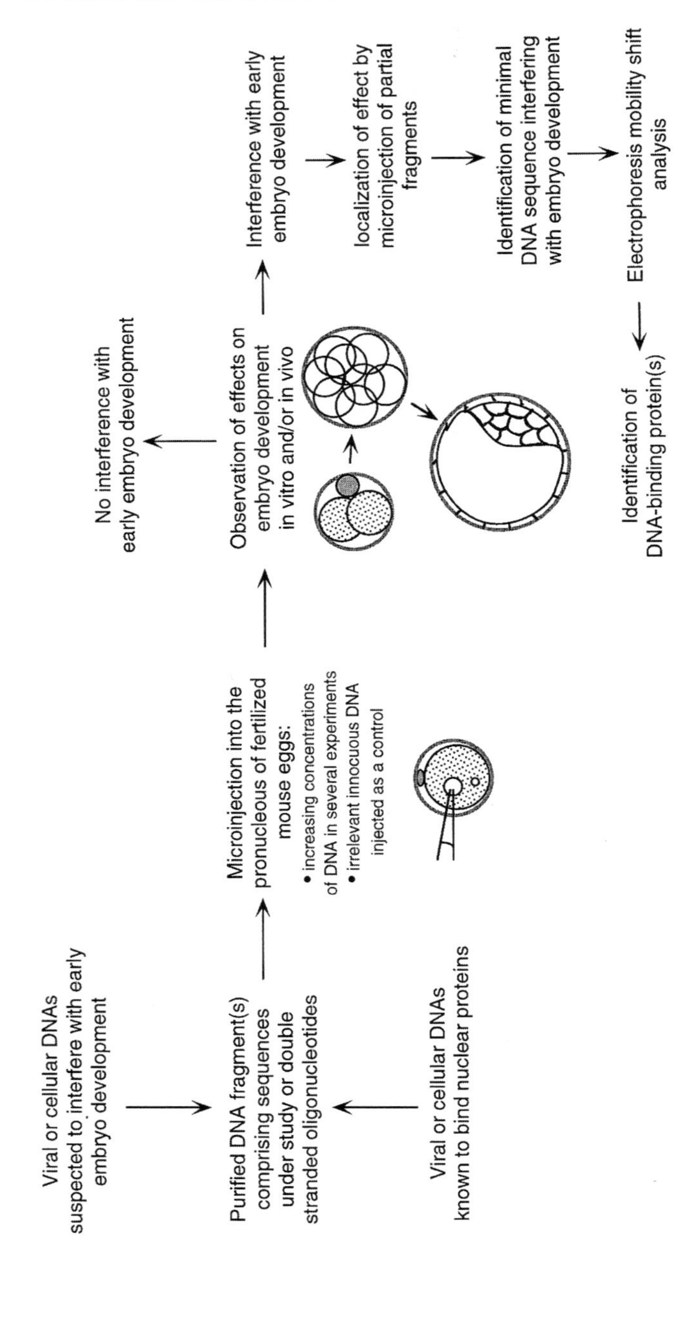

Fig. 20.1. Outline of a strategy for the identification of DNA binding proteins essential in preimplantation development

from the inner cell mass of blastocysts which conserve their totipotent capability. DNA fragments can be easily tested for lethal effects in embryos by microinjecting them at different concentrations into the egg pronuclei, and following their development in vitro. Further, once deleterious effects have been assessed, the cellular targets involved can be isolated and characterized in the ES cells.

A limiting step to this approach is the proper identification of DNA sequences interfering with embryo development. This has been accomplished focusing on DNA sequences that are known to bind nuclear factors expressed in early embryos. Thus, for instance, Rosner et al. (1991) showed that microinjection into mouse eggs of a small DNA fragment carrying the Oct-3 binding site caused arrest of embryo development; in different experiments, Blangy et al. (1991, 1995) found that pronuclear microinjection of double-stranded oligonucleotides containing a mouse telomeric region identical to the CDEI centromere sequence of yeast interferes with early embryo development.

However, early embryo lethality can be observed also fortuitously while attempting to produce transgenic mice with DNA fragments carrying single genes or more complex sequences, like the complete genome of a virus. In these cases, embryo lethality is often due to deleterious effects of the protein(s) encoded by the transgene, but it may be caused also by competitive interaction of its promoter-enhancer sequences with essential DNA-binding proteins of the embryonic cells. Exogenous DNA injected into mouse egg pronuclei interacts first with the maternal/paternal inherited nuclear proteins. As a result, a number of DNA molecules undergo degradation while the rest are processed by nuclear enzymes, thereby oligomerizing and integrating into the genome. Likewise, the integrated foreign DNA interacts with nuclear factors; if this occurs during development or later is specified by the DNA sequence itself, but also by epigenetic events such as methylation, integration locus, etc. Therefore, upon injection in high copy numbers, foreign DNA molecules may specifically sequester essential nuclear factors and thus interfere with embryo development.

We have shown previously that microinjection of the Adeno-associated virus type 2 (AAV-2) genome into the pronucleus of one-cell embryos blocks development at the morula stage, and that this effect maps to the viral terminal repeats and the P5 promoter region (Botquin et al. 1994). Here, we describe methods to identify and map DNA sequences interfering with early embryo development and to assess if they bind nuclear proteins of the early mouse embryo.

Materials

Equipment

- Animal facilities. We used mice supplied by the Zentral Institut für Versuchstierkunde (Hanover, Germany), which were maintained in a specific pathogen-free (SPF) unit
- Dissection materials: surgical forceps and scissors, fine forceps (Dumont no. 5)
- Tissue culture equipment: incubator (5 % CO_2/37 °C), flow hood, plastic dishes (3 and 6 cm∅)
- Microinjection setup: inverted microscope, microinjector, microforge, micromanipulators, glass capillaries to prepare microinjection pipettes (see Chap. 1, this Vol.), holding micropipettes and embryo transfer capillaries (see Hogan et al. 1994), stereomicroscope, gas and ethanol burners
- Electrophoresis equipment for agarose and acrylamide gels: power supply, electrophoresis chambers, glass plates, combs, spacers
- Vacuum gel dryer
- Exposure cassettes and film (Kodak X-Omat).

Reagents and solutions

For microinjection

- Embryo culture media (see Hogan et al. 1994): M2 (50 ml) and M16 media (10 ml), paraffin oil (Merck 7161.0500, spectroscopy grade) hyaluronidase (Sigma H-3884, stock solution 10 mg/ml in water)
- Gonadotropins

  Pregnant mare' serum (PMS): Sigma G-4527. Resuspend powder in 0.9 % NaCl or Dulbecco's PBS to a concentration of 500 IU/ml. Divide into 100-μl aliquots and store at -80 °C until needed, but no longer than 6 months. To treat ten females, thaw one aliquot, dilute with 0.9 ml 0.9 % NaCl or PBS and inject 0.1 ml (5 IU) intraperitoneally into each mouse

  Human chorionic gonadotropin (hCG): Sigma C-8554. Resuspend powder in deionized water to a concentration of 500 IU/ml. Deliver 100 μl aliquots into microfuge tubes, lyophilize and store at −80 °C until needed. For ten females, resuspend one aliquot in 1 ml of 0.9 % NaCl or PBS and inject 0.1 ml (5 IU) intraperitoneally into each mouse
- Microinjection buffer: 5 mM Tris.HCl pH 7.4, 0.1 mM EDTA
- DNA purification reagents (Geneclean kit, Dianova, cat. no. 1-800-424-6101; or Qiaex II gel extraction kit, Qiagen cat. no. 20021)
- Cell culture

  ES cells are cultured on gelatine-coated dishes: prepare 0.1 % gelatine in water, cover the dishes with this solution, and incubate at room

temperature for 1–2 h or overnight in the cold room. Remove the gelatine and let the dishes dry before use
ES culture medium: D-MEM supplemented with 15 % fetal calf serum (Pan Systems GmBH, cat. no. 30–0102), 1 % L-glutamin 200 mM (Gibco-BRL, cat. no. 25030–024), 1 % penicillin/streptomycin 10,000 U/ml (Gibco-BRL, cat. no. 15140–114), 1 % beta-mercaptoethanol (Sigma, M-7522), 1 % leukemia inhibitory factor (LIF) 1000 U/ml (Gibco-BRL, cat. no. 13275–029)

For the electrophoretic mobility shift assay (EMSA)
- Oligonucleotides
- Nucleotides (Pharmacia, dATP cat. no. 27–2050–01, dCTP cat. no. 27–2060–01, dGTP cat no. 27–2070–01, dTTP cat. no. 27–2080–01)
- Poly d[(I-C)] (Boehringer, cat. no. 108 812). Prepare a stock solution as follows:

1. Dissolve 50 OD Units (A_{260}) of poly d[(I-C)] in 1 ml of deionized water.
2. Denature at 95 °C for 15 min in a thermoblock.
3. Let anneal and let cool down slowly (e.g., by switching off the thermoblock and waiting untill it cools down to room temperature).
4. Dialyze overnight against TE buffer (10 mM Tris.HCl, 1 mM EDTA, pH 8).
5. Determine the concentration of dialyzed poly d[(I-C)] by preparing a dilution 1:100 in 1 ml of TE and reading its optical density at A_{260}. Calculate the concentration, assuming that 1 OD unit corresponds to 30 μg/ml.

- Protein extraction buffer: 150 mM NaCl, 0.2 mM EDTA, 0.5 mM DTT, 0.5 mM PMSF, 25 % glycerol, 20 mM Hepes pH 7.8
- Protein binding buffer (stock 5×): 2.5 mM EDTA, 2.5 mM DTT, 50 % glycerol, 250 mM NaCl, 125 mM Hepes pH 7.6
- TBE buffer (5×): 54 g Tris base, 27.5 g boric acid, 20 ml 0.5 M EDTA pH 8, adjust volume to 1 l with distilled water
- Nondenaturing acrylamide gel for bandshift experiments. A 5 % acrylamide gel is prepared by mixing the following solutions:
  25 ml 30 % acrylamid/bis-acrylamid
  30 ml 5× TBE buffer
  93.5 ml deionized water
  1.5 ml 10 % Ammonium persulfate
  150 μl TEMED

Pour into the space between two glass plates of 250 mm×300 mm separated by 2 mm with teflon spacers.

20.1 Recovery, Culture and Microinjection of Mouse Embryos

Microinjection experiments require large numbers of fertilized eggs. These are usually obtained from superovulated rather than naturally mated females. Efficient embryo recovery, however, depends on several variables which should be taken into account in order to optimize the number of healthy embryos per pregnant mouse. These variables are:

- Animal house. Mouse colonies maintained in pathogen free facilities (see Morrell, Chap. 11, this Vol.), or at least in a clean and antiseptic environment, are healthier, live longer, and reproduce better. Therefore, the numbers of pregnant females and hence of fertilized eggs that can be obtained are higher. The light-dark cycle should be as follows: 05:00 to 19:00 h light and 19:00 to 5:00 h dark.
- Strain of mouse. The number of eggs released from the ovaries after hormonal induction of ovulation varies among strains. Imbred strains like the C57BL/6J, BALB/CByJ, CBA/CaJ, and the F_1 hybrids (BALB/CByJ×C57BL/6J)F_1, (C57BL/6J×DBA/2J)F_1, (C57BL/6J×C3H/HeJ)F_1 are considered high ovulators (20–40 eggs per mouse), whereas other strains like DBA/2J and C3H/HeJ or 129/J have been classified as low ovulators (15 or less eggs per mouse) in response to superovulation treatment.
- Superovulation. A double hormonal treatment of the females permits inducion of synchronous maturation of a large number of follicles, their rupture, and subsequent release of the oocytes. This is achieved by intraperitoneal administration of follicle-stimulating and luteinizing hormones (PMS and hCG, respectively, see Materials, above) to females 3–6 weeks of age weighing more than 12 g. On a 05:00 h to 19:00 h light cycle, PMS is administered between 13:00 and 14:00 h, and hCG injected 46–48 h after PMS, between 12:00 and 13:00 h. Superovulation ensures that for each microinjection experiment an optimal number of fertilized eggs (about 150) is obtained with a smaller number of females (5–7) than would be required if naturally ovulating females were used (15–30). However, embryos from superovulated females develop somewhat more slowly to blastocysts, and

some of these are abnormal and stop developing. The reason for this phenomenon is not well understood, but it has to be taken into consideration if studies on developing preimplantation embryos are to be carried out. In our hands, upon superovulation, the hybrid strain we use (see below) yields relatively large numbers of healthy embryos that develop normally.

- Stud males. It is necessary to maintain a colony of at least 20 stud mice 8 weeks to 8–12 months of age separated in individual cages. For mating superovulated females, only one female is presented to each stud male. The mating performance of each animal should be noted, so that those that mate inefficiently can be identified and replaced. After mating, the number of sperm cells in the semen decreases transitorily. Therefore, mated males should be rested for at least 1–3 days to ensure the highest fertilization rates.

Procedure

For the experiments described here we used (C57BL/6J×C3H/HeJ)F_1 hybrids purchased from a supplier breeding pathogen-free stocks. The animals were maintained in strict barrier facilities.

For each microinjection experiment proceed as follows:

1. Order from a commercial supplier ten females 3 weeks of age. Let mice adapt for 1 week to the appropriate light/dark cycle and superovulate them as indicated above.
2. Mate the superovulated females with ten stud males of the same strain.
3. The following morning, sacrifice the plugged females by cervical dislocation. Open the abdominal cavity, dissect the oviducts and place them into M2 medium containing 300 μg/ml hyaluronidase. Break the wall of the ampulla with a pair of fine forceps to release the embryos. Separate embryos from each other and from cumulus cells by pipetting them up and down using a glass transfer capillar connected by tubing to the mouth of the experimentator. Do not incubate the embryos in the hyaluronidase-containing medium longer than necessary, as this may reduce viability.

5. Wash the embryos twice in M2 without hyaluronidase and transfer them to microdrop cultures of M16 medium (e.g., 5–6 drops of 30–50 μl on a 6-cm∅ dish, covered with 4 ml of paraffin oil).

6. Transfer groups of 15–20 embryos to the microscope stage for microinjection (see Hogan et al. 1994 for a detailed description of the procedure). The embryos are transferred to a drop of M2 on a slide, covered with paraffin oil to prevent evaporation, and placed onto the microscope stage. Microinject DNA into the male or female pronuclei.

7. Transfer each group of microinjected embryos to fresh M16 microdrops (not to those used for previous steps). Incubate at 37 °C / 5 % CO_2.

8. Use as controls: (a) several groups of noninjected embryos (control for suitability of culture media); and (b) several groups of embryos injected with an innocuous DNA (control for survival after microinjection).

Comments

- Some batches of reagents used to prepare the culture media contain impurities which may damage the embryos, so that they do not progress beyond the two-cell stage. Therefore, new batches of reagents and stock solutions should be tested for embryo toxicity. Sometimes the water contains impurities that account for such effects. Use deionized water of the best quality, and store it in a plastic container. Never change all reagents, solutions, and water at the same time, but do it stepwise to find out what the problem was.

- The day before obtaining embryos, prepare 50 ml of M2 and 10 ml of M16 in plastic tubes using plastic pipettes as described in Hogan et al. (1994). We recommend to omit phenol red, which may sometimes damage the embryos, from both media, and to filter them through 0.22-μm membranes. In addition, it is advisable to check the pH: (a) take an aliquot of M2 (2–3 ml) into a 15-ml plastic tube and measure its pH in the pH meter, discard the aliquot, if necessary add a small drop of 1 M NaOH and check the pH of a new aliquot, proceed as before until pH is 7.2–7.4; (b) pour the M16 medum into a small tissue culture flask, incubate at 37 °C/5 % CO_2 for 15–30 min, take an aliquot of 2–3 ml and proceed as above until pH is 7.4, then pour the M16 into a 15-ml plastic tube, gas with 5 % CO_2 / 95 % air for 1 min and close the tube tightly.

20.2
Identification of Short DNA Sequences Capable of Interfering with Early Embryo Development

When cultured under optimal conditions, mouse embryos develop in vitro to the blastocyst stage in 4–5 days (the day the vaginal plugs are found is considered as day 0 postcoitum). Thereafter, the fully expanded blastocysts break out their zona pellucida, escape from it, as they would do naturally for implantation in the uterus, and subsequently degenerate.

The development of the injected embryos should be observed during this period, preferably by photographing them in an inverted microscope and analyzing embryo morphology on the pictures (see Fig 20.2). This minimizes damage due to the microscope light and exposure to the room atmosphere and temperature conditions.

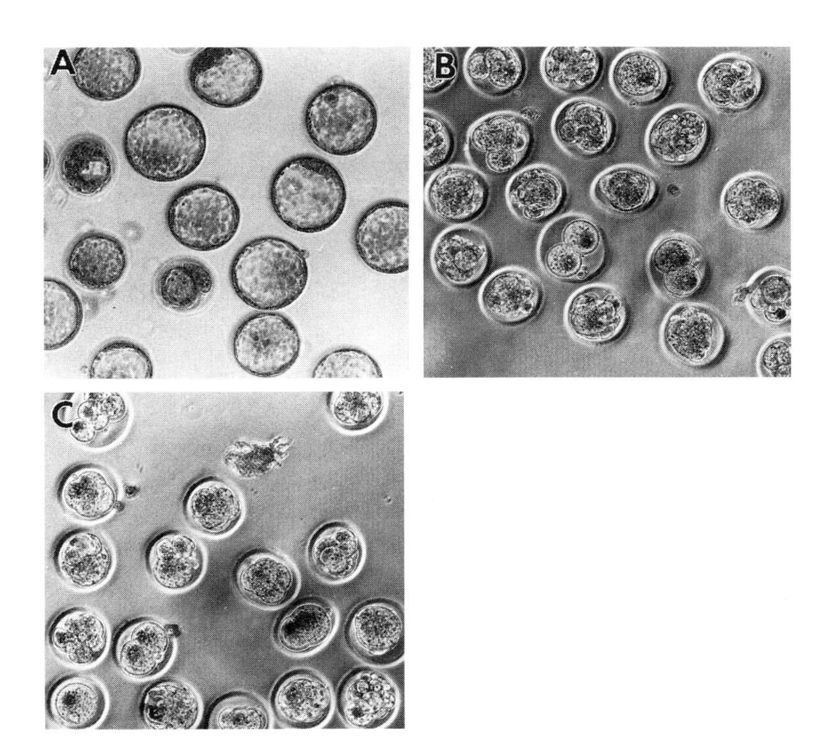

Fig. 20.2A–C. Developmental arrest at the morula stage of eggs microinjected with AAV subgenomic fragments. **A** Embryos injected with pBluescript DNA. **B** Embryos injected with AAV-2 complete genome. **C** Embryos injected with an AAV-2 fragment containing the p5 promoter sequence flanked by the TRs (see Results)

Procedure

1. At day 1 (20–24 h after injection), examine the embryos and discard those which did not progress to two-cell stage.

2. Examine the noninjected embryos briefly and check if they survive and develop normally. By day 2 they should be 6–8 cell morulas, by day 3 compacted morulas, and by days 4–5 early and expanded blastocysts. If they do not develop as expected, there is a problem with the media and the experiment should be repeated using new media.

3. Examine the embryos injected with control DNA. They should develop as the control noninjected embryos with no statistically significant difference. If they do not, the DNA used as a control is not innocuous and should be replaced in further experiments. We have used for our experiments a lacZ gene with a viral promoter which is inactive in mouse embryos, and also the plasmid Bluescript.

4. Examine the embryos injected with the DNA under study and look for statistically significant differences in the numbers of embryos developing normally as compared to the controls. If interference with development is noted, and the DNA fragment was large, seek a suitable restriction site to cut it into two smaller fragments and test their effect on development by separate injections into eggs. Repeat the procedure until mapping the interference effect to the smallest fragment possible. Likewise, test a series of overlapping oligonucleotides designed to cover this fragment. Finally, use the oligonucleotide(s) retaining the blocking effect for the isolation of the corresponding DNA-binding protein(s) using the ES cells (see below).

Viability To test the viability of the embryo cells proceed as follows:

1. Remove the zona pellucida by a short incubation with acid Tyrode's solution at 37 °C (see Hogan et al. 1994).

2. Wash embryos twice in M16 medium briefly.

3. Pipette embryos into a drop of 0.1 % Trypan blue in PBS.

4. Transfer the dish to the stage of the inverted microscope and photograph the embryos with phase contrast. The dye stains only dead cells, since it is excluded by living cells.

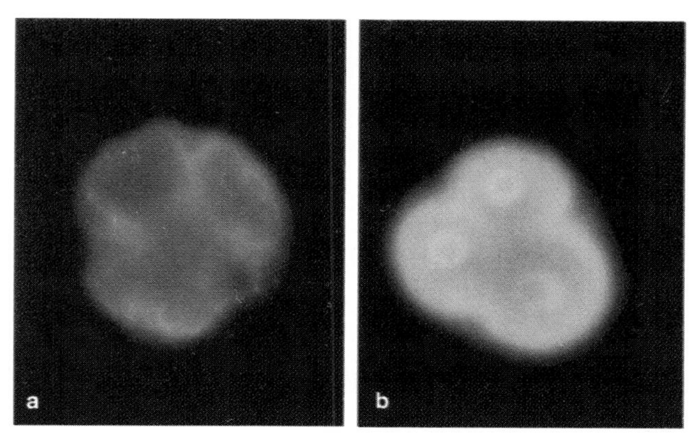

Fig. 20.3a, b. Immunofluorescence analysis of the expression of AAV-2 Rep proteins in 3-day-old embryos developing in vitro after pronuclear injection of the complete AAV-2 genome. **a** Embryos microinjected with pBluescript. **b** Embryos microinjected with the AAV-2 DNA

20.3 Use of ES Cells to Isolate DNA-Binding Proteins

Procedure

Preparation of Protein Extracts from ES Cells

1. Grow the ES cells to 80 % confluency on a 10 cm∅ gelatine-coated dish with 10 ml ES medium.
2. Remove the medium and wash the cells twice with ice-cold PBS.
3. Scrape cells in 1.4 ml cold PBS and transfer to a microfuge tube.
4. Pellet cells at 600 *g* at 4 °C for 5 min.
5. Remove the supernatant and resuspend the cells in 50 μl of protein extraction buffer.
6. Disrupt cells by three cycles of freezing-thawing (i.e., alternate incubations of 5 min into liquid nitrogen and a 37 °C water bath).
7. Centrifuge at 14,000 *g* for 15 min to remove cellular debris.
8. Transfer the supernatant to a new microfuge tube.

The concentration of protein in the extracts can be readily determined by the method described by Kalb and Bernlohr (1977) as follows:

9. Prepare different dilutions of the extract in 0.1 % SDS.
10. Read absorbance at 230 nm (A230) and 260 nm (A260).
11. Calculate concentration according to the following equation: concentration (μg/ml) = $(A_{230} \times 183) - (A_{260} \times 75.8)$.

Electrophoretic Mobility Shift Assay

The following protocol is based on that by Silverter and Schöler (1994).

Oligonucleotide annealing

1. Mix complementary oligonucleotides (50 μg of each) in a microfuge tube. Complete volume to 100 μl with deionized water.
2. Denature at 95 °C for 15 min in a thermoblock
3. For annealing, let the tube cool down slowly (e.g., by switching off the thermoblock and waiting until it reaches the room temperature).
4. Read optical density at A_{260} of a 2-μl aliquot diluted in 1 ml deionized water. Calculate the concentration of DNA taking 1 OD unit as the absorbance of a solution of 50 μg DNA per ml.

Radioactive labeling

Use 50 ng of a 50 bp double stranded oligonucleotide for radioactive labeling as follows:

5. Mix 50 ng of oligonucleotide with 4 μl of forward reaction buffer (Gibco, provided with T4 kinase), 5 μl of g[32P]dATP (5000 μCi/mmol, Amersham), 1 μl T4 kinase (Gibco, cat. no. 18004–026). Complete volume to 20 μl with deionized water.
6. Perform labeling at 37 °C for 30 min.
7. Stop reaction by addition of 5 μl of 0.5 M EDTA pH 8.

Purification of probes

Remove nonincorporated $\gamma[^{32}P]$dATP using a Sephadex G-50 quick-spin column (Boehringer, cat. no. 1273 973) as follows:

7. Compress Sephadex G-50 by centrifugation at 1100 *g* for 3 min.
8. Apply the radioactive-labeled DNA on the column and centrifuge at 1100 *g* for 3 min.

9. Save the eluate which contains the purified DNA probe.

10. Measure radioactivity: pipette 1 μl into a microfuge tube, introduce it into a vial, and count it in a scintillation β-counter.

Binding reaction

11. Prepare a reaction mixture in a 1.5-μl microfuge tube as follows:
    3.0 μl 5× binding buffer
    2.0 μl 30 mg/ml BSA (fraction V, Sigma, A2153)
    1.5 μl 1 % Triton X-100
    0.5 μl 1 mg/ml poly d[(I-C)]
    Add 30000 cpm of 32P-labeled oligonucleotide
    Complete volume to 15 μl with deionized water
    Incubate at 20–25 °C for 30 min.

Electrophoresis

12. Pre-run the nondenaturing gel (10 mA, 30 min)

13. Load the samples on the pre-run nondenaturing gel and run at 300 V (70–80 mA) for 2.5 h.

14. Take the gel from the glass plate by applying onto it a sheet of 3 mm thick Whatman paper.

15. Cover the gel with Saran wrap and dry it in a vacuum gel drier.

16. Expose the dried gel on a film overnight (see Fig. 20.4).

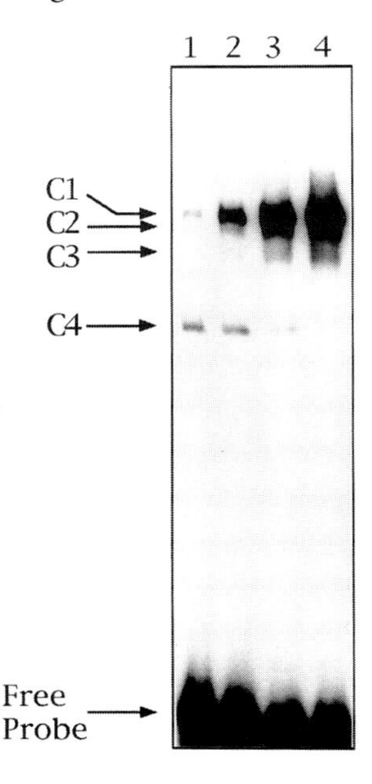

Fig. 20.4. Electrophoretic mobility shift assay designed for the detection of ES cell proteins complexed with AAV-2 DNA. The numbers *1*, *2*, *3* and *4*, correspond to the amounts in μg of protein extracts from ES cells, which were incubated with ^{32}P-labeled double-stranded oligonucleotides comprising the AAV-2 p5 promoter sequence (see Results)

Results

We have applied a strategy like that summarized in Fig. 20.1 to identify a small region of the AAV-2 genome which, when microinjected into the pronuclei of mouse eggs, causes developmental arrest at the morula stage.

AAV are small human viruses of the family of Parvoviridae which can propagate as an integrated provirus or by lytic infection (for review see Berns and Bohensky 1987; Siegl et al. 1985; Muzyczka 1992). Five AAV serotypes have been identified, but the best characterized is AAV-2. AAV-2 virions (icosahedral, 20 nm∅) consist of three capsid polipeptides (named VP 1, 2 and 3) and a ssDNA genome (either plus or minus strand) of 4681 bases which consists of a long unique region flanked by terminal inverted repeats (TRs) of 145 nucleotides.

The TRs are required for AAV DNA replication, for packaging of the viral genome into the virions, and for integration of proviral AAV into the host genome, apparently at preferential chromosomal loci (Kotin et al. 1990; Walz and Schlehofer 1992).

The large nonrepeated region contains two open reading frames (ORFs). The right ORF encodes the three VP proteins, while the left ORF encodes four nonstructural (Rep) proteins. For replication, AAVs depend strongly on host cell functions and usually on co-infection by helper viruses such as adenovirus, herpes virus, or vaccinia virus (see Berns and Bohensky 1987). In the absence of a helper virus, the infected cells carry the AAV DNA integrated into their genome, where it can remain latent for many cell generations. However, it retains the ability to replicate upon helper virus infection.

So far, AAVs have never been found associated to disease. Rather, AAVs have been reported to behave as tumor suppressors in several systems (see Rommelaere and Cornelis 1991; Schlehofer 1994), an effect that might be related to the ability of the virus to induce differentiation of certain cell types, including human keratinocytes and leukemia cells, and mouse ES cells (Klein-Bauernschmitt et al. 1992; Botquin 1994).

In order to study the effects of AAV-2 in the context of a living animal, we attempted to produce transgenic mice carrying the AAV-2 complete genome. In several rounds of injections, an abnormally low number of offspring was obtained. A total of 12 mice were born from about 400 embryos surviving injection which were transferred to the oviducts of 18 foster mothers. Southern blot hybridization analysis of DNA from the tails of these mice showed that none of them carried the AAV-2 DNA. These results suggest that the AAV-2 DNA might have a deletirous effect upon its injection (devoid of plasmid sequences) into egg pronuclei at concentrations routinely used for transgenic experiments (1–5 μg/ml).

To examine this possibility, we performed experiments in which the complete genome of AAV-2 was injected into the pronuclei of one-cell embryos and followed their development in vitro for 5 days. As shown in Fig. 20.2B, about 90 % of embryos microinjected with AAV-2 DNA were blocked at the morula stage, while 90 % of control embryos injected with pBluescript plasmid reached the blastocyst stage.

The expression of viral nonstructural proteins was analyzed in the injected embryos, because Rep proteins have been shown to induce cytotoxic effects in other systems (Heilbronn et al. 1994). Immunofluorescence analysis showed expression of Rep proteins into the nuclei of 3-day-old embryos (Fig. 20.3). To analyze if Rep proteins mediate the interference of AAV-2 DNA with embryo development, an AAV-2 DNA fragment lacking the left and right ORFs but containing the TRs plus a short sequence of the p5 promoter was injected into fertilized mouse oocytes. These embryos were also arrested at the morula stage (Fig. 20.2C), this suggesting that the expression of Rep proteins is not necessary to interfere with development

Such deletion mutants of AAV-2 retain the ability to integrate into the cellular genome (de la Maza and Carter 1981). A possible mechanism for disturbance of embryo development might be the integration of the AAV-2 sequences into specific loci leading to disruption of gene(s) which are critical for development. However, it is more likely that interaction of the AAV-2 DNA with nuclear factors leads to competition with the host genome for critical factors and thereby interference with embryo development. To check the latter hypothesis, we tested separately the TR and p5 promoter sequences for binding to cellular factors. To this end, we extracted proteins from ES cells and ran electrophoretic mobility shift assays (EMSA) by incubating the ES protein extracts in the presence of labeled oligonucleotides representing the terminal repeat sequence or the p5 promoter. No binding of cellular factors on the TR was observed (data not shown). In contrast, four different protein-DNA complexes were detected with the p5 oligonucleotide and protein binding increased in concentration-dependent manner (Fig. 20.4). Interestingly, microinjection of the p5 promoter sequence alone led to a dose-dependent lethal effect on embryonic development (see Botquin et al. 1994).

Taken together, these data suggest that the AAV DNA interferes with embryo development and that this effect may be due to inappropriate binding of nuclear factors present in critical amounts during early stages of development. Identification of such factors will help understand not only the mechanisms responsible for this effect by AAV-2 in mouse embryos, which might also occur in infected humans, but also important aspects of early embryo development.

Acknowledgment: We are grateful to Prof. H. zur Hausen for support and stimulating discussions.

References

Berns, KI, Bohensky RA (1987) Adeno-associated viruses: an update. Adv Virus Res 32:243–306

Blangy A, Leopold P, Vidal F, Rassoulzadegan M, Cuzin F (1991) Recognition of the CDEI motif GTCACATG by mouse nuclear proteins and interference with the early development of the mouse embryo. Nucl Acids Res 19:7243–7250

Blangy A, Vidal V, Cuzin F, Yang Y-H, Boulukos K, Rassoulzadegan M (1995) CDEBP, a site specific DNA-binding protein of the APP-like family, is required during the early development of the mouse. J Cell Sci 108:675–683

Botquin V (1994) Influence of the human adeno-asociated virus type 2 on development: a mouse model. PhD thesis Ruprecht-Karls-Universität. Heidelberg

Botquin V, Cid-Arregui A, Schlehofer J R (1994) Adeno-associated virus type 2 interferes with early development of mouse embryos. J Gen Virol 75:2655–2662

Cheng SS, Costantini F (1993) Morula decompaction (mdn), a preimplantation recessive lethal defect in a transgenic mouse line. Dev Biol 156:265–277

de la Maza LM, Carter BJ (1981) Inhibition of adenovirus oncogenicity in hamsters by adeno-associated virus DNA. J Natl Cancer Inst 67:1323–1326

Heilbronn R, Schlehofer J, zur Hausen H (1994) Selective killing of carcinogen-treated SV40-transformed Chinese hamster cells by a defective parvovirus. Virology 136:439–441

Hogan B, Beddington R, Costanini F, Lacy E (1994) Manipulating the mouse embryo, a laboratory manual, 2nd edition. Cold Spring Harbor Press, Cold Spring Harbor

Kalb JR, VF, Bernlohr RW (1977) A new spectrophotometric assay for protein in cell extracts. Anal Biochem 82:362

Klein-Bauernschmitt,P, zur Hausen H, Schlehofer JR (1992) Induction of differentiation-associated changes in established human cells by infection with adeno-associated virus type-2. J Virol 66:4191–4200

Kotin RM, Siniscalco M, Samulski RJ, Zhu XD, Hunter L, Laughlin CA, McLaughlin S, Muzyczka N, Rocchi M, Berns KI (1990) Site-specific integration by adeno-associated virus. Proc Natl Acad Sci USA 87:2211–2215

Lyon MF, Searle AG (1989) Genetic variants and strains of the laboratory mouse. Oxford University Press. Oxford, UK

Magnuson T (1986) Mutations and chromosomal abnormalities: how are they useful for studying genetic control of early mammalian development? In: Rossant J, Pederson RA (eds), Experimental approaches to mammalian embryonic development. Cambridge University Press, Cambridge, UK, pp 437–474

Majumder S, DePanphilis ML (1995) A unique role for enhancers is revealed during early mouse development. BioEssays 17:879–910

Muzyczka N (1992) Use of adeno-associated virus as a general transduction vector for mammalian cells. Curr Top Microbiol Immunol 158:97–129

Rommelaere J, Cornelis JJ (1991) Antineoplastic activity of parvoviruses. J Virol Meth 33:233–251

Rosner MH, De Santo RJ, Arnheiter H, Staudt LM (1991) Oct-3 is a maternal factor required for the first mouse embryonic division. Cell 64:1103–1110

Rossant J, Pederson RA (eds) (1986) Experimental approaches to mammalian embryonic development. Cambridge Univ. Press, Cambridge, UK

Schlehofer JR (1994) The tumor suppressive activities of adeno-associated viruses. Mut Res 305:303–313

Siegl G, Bates RC, Berns KI, Carter BJ, Kelly DC, Kurstak E, Tattersall P (1985) Characteristics and taxonomy of Parvoviridae. Intervirology 23:61–73

Sylverter I, Schöler, HR (1994) Regulation of the Oct-4 gene by nuclear receptors. Nucleic Acids Res. 22:901–911

Walz C, Schlehofer JR (1992) Modification of some biological properties of HeLa cells containing adeno-associated virus DNA integrated into chromocome 17. J Virol 66:2990–3002

Chapter 21

Application of Transgenes and Transgenic Mice to Study Gene Activity from the Oocyte to the Early Embryo

SYLVIE FORLANI[1], LUCILE MONTFORT[1], AND JEAN-FRANÇOIS NICOLAS[1]*

Introduction

Development is initiated during gametogenesis. The contribution of the oocyte has been precisely defined in a few species, such as in invertebrates, using genetic approaches (St Johnston and Nüsslein-Volhard 1992), and also in certain vertebrates, such as amphibians, using biochemical analyzes (Ruiz i Altaba and Melton 1990; Kessler and Melton 1994). However, the steps of embryogenesis that occur prior to gastrulation exhibit a surprisingly large diversity among the different zoologic groups. It is therefore impossible to extrapolate rules or strategies and, so far, no conserved developmental pathways have been found.

In mammals, the steps of embryogenesis prior to gastrulation are very poorly understood. Problems which will benefit from studies exploiting transgenes include the following:

- The ultimate contribution of the oocyte to extra embryonic and perhaps embryonic development
- The way in which the link between gametic and zygotic gene activity is obtained (Schultz 1993; Forlani and Nicolas 1996a)
- The molecular basis for the changes in transcriptional specificity during the transition from maternal to zygotic control of gene activity
- The importance of parental effects; that is, what of the parental origin persists and has biological significance in the zygote. This, among other phenomena, is at the basis of the failure of parthenogenetic and

* corresponding author: phone: +33-1-45688490; fax: +33-1-45688521; e-mail: jfnicola@pasteur.fr

[1] Unité de Biologie moléculaire du Développement, URA 1947 du Centre National de la Recherche Scientifique, Institut Pasteur, 25, rue du Docteur Roux, 75724 Paris Cedex 15, France

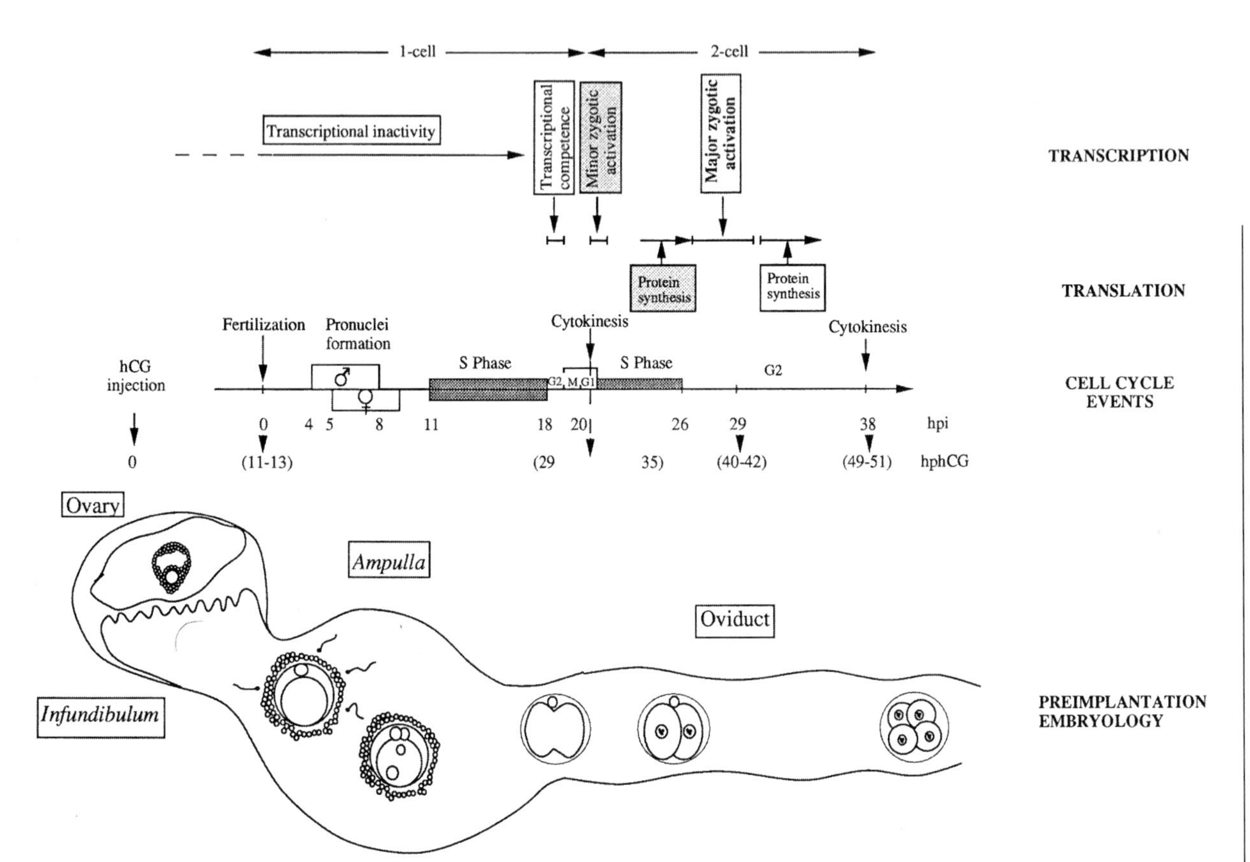

Fig. 21.1. Timing of some biochemical and biological events during the first three cell cycles following fertilization in the mouse. *Above* the chronology of some transcriptional and translational events during the first three cell cycles after fertilization in the mouse embryo are shematized. Successive phases of the cell cycle [G1, S phase (DNA replication), G2 and cytokinesis] are mentioned. *Below* are represented the corresponding morphological stages of development in the female genital tract. The time is in hours postinsemination (hpi) for in vitro fertilized eggs. Correspondance in hours postinjection of human chorionic gonadotrophin (hphCG), as used in the text, is given below in brackets for in vivo fertilized eggs. In vitro fertilization reduces the developmental asynchrony between eggs which is of several hours in in vivo fertilized eggs (see for example the delay from 29 to 35 hphCG for the first cytokinesis). The timing is from Vernet et al. (1993)

androgenetic development in mammals (McGrath and Solter 1984; Surani et al. 1986; Thomson and Solter 1988)

- The molecular nature of parental genetic imprints (on which parental effects are based) in the gametes and subsequently during development.

Until recently, these fundamental aspects of development have only been studied using biochemical methods which are strongly limited by the minute quantities of material available (Flach et al. 1982; Bensaude et al. 1983; Howlett and Bolton 1985; Conover et al. 1991; Latham et al. 1991). However, the general picture of early development is still mainly derived from these results (Fig. 21.1). Cell manipulations in particular, nuclei transfers and enucleation (McGrath and Solter 1984; Surani et al. 1986; Latham and Solter 1991; Reik et al. 1993) and genetic analyzis (Weng et al. 1995) have further indicated the existence of parental effects that regulate certain aspects of embryonic development (imprinting) and gene expression in the adult.

During the past few years, the use of transgenes has considerably enriched our understanding of the early steps of embryogenesis. This is mainly due to several advantageous features of transgenes which simplify experimental analysis:

- *cis* elements of transgenes are easily modified by mutagenesis or recombination

- In combination with reporters, it is feasible to visualize and measure gene activity at the single cell level, not only in the adult organism or fetus but also in the preimplantation embryo. The bacterial *LacZ* and the firefly luciferase genes (for sources, see Vernet et al. 1993a and Miranda et al. 1993, respectively) have been extensively used as reporter genes in eggs and preimplantation embryos (Bonnerot et al. 1991; Christians et al. 1995; Henery et al. 1995; Yeom et al. 1996) because, in addition to their developmental neutrality and cell autonomous expression, methods to detect these enzymes are highly sensitive and can therefore be used to detect the low quantities of material contained in these early embryos.

- Transgenes serve as probes even after integration into the host genome and therefore can be detected and reisolated

- Finally, transgenes can be analyzed in hemizygote animals, in which the parental origin of the transgene is known.

One of the first uses for transgenes was to study the general picture at the early steps of development that had previously emerged from biochemical analyzes. These experiments confirmed the existence of a period of genetic inactivity due to control of the transcriptional rather than translational apparatus (Vernet et al. 1992). More recently, it has also been confirmed that there is a resumption of gene activity in the late two-cell embryo, since the pattern of expression of transgenes, driven by promoters lacking tissue specificity, begins at this stage (L. Montfort, S. Forlani, unpubl. results). Finally, the minor zygotic activation observed at the early two-cell stage has been shown not to be an artifact due to manipulation of the egg, as had been previously suggested (Christians et al. 1995).

Moreover, transgenes have revealed novel aspects of genic activity in the embryo. The timing of the acquisition of transcriptional competence has been determined to occur at G2 of the late one-cell embryo (Latham et al. 1992; Vernet et al. 1992; Ram and Schultz 1993), which is well before the major zygotic activation and just prior to the minor zygotic activation. Differences in properties between the male and the female pronucleus have been revealed by microinjection experiments. The male pronucleus uses the minimal promotor (the TATA and CAT/Sp1 box-containing region) at an unusual rate, in comparison to oocytes (Dooley et al. 1989; Majumder et al. 1993; Wiekowski et al. 1993). Finally, microinjection experiments of a representative panel of promoters linked to reporter genes indicate that the transcriptional specificity of the oocyte, fertilized egg, and two-cell embryo prior to and after zygotic activation are similar: there is a materno-zygotic continuum (Bonnerot et al. 1991). However, this similarity does not correspond to complete identity. In fact, experiments from DePamphilis' laboratory may indicate that the susceptibility of a promotor to enhancer effect, acting at a distance, is not acquired before the late two-cell stage. They have thus postulated the establishment of a repressed state of the genome at the two-cell stage (Martinez-Salas et al. 1989; Wiekowski et al. 1991; Henery et al. 1995). However, it is also possible that stage-specific variations in transcriptional specificity exist in addition to the common element of the transcriptional specificity. For instance, certain transgene-dependent expression patterns start at the four-cell stage (S. Forlani, unpubl. results). This may be at the basis of the quantitative and qualitative changes observed in protein and RNA expression patterns (Bachvarova et al. 1989; Latham et al. 1991; Rothstein et al. 1992).

Altogether, these data lead to a more precise description of the sequence of events involved in the modulation of gene activity from the oocyte to the four-cell embryo. However, they also raise novel questions, whose elucidation will probably also be generated from studies involving transgenes. Among these questions are the following: what explains the delay between the acquisition of transcriptional competence and the expression of integrated genes only at the late two-cell stage? If transcriptional specificity evolved during the early stages of embryogenesis, is it because transcriptional factors become organized in a network? How is this network generated? What are the nuclear or chromatin changes which underlie the differences – if they receive confirmation – between pronuclei and zygotic nuclei in their ability to respond to enhancers? Have all events necessary to reach the definitive transcriptional state of the zygote already occurred in the two-cell embryo, or are there additional modifications later? A response to this latter problem will also require an understanding of the significance of demethylation and methylation of the genome at these stages (Monk et al. 1987; Reik et al. 1990; Kafri et al. 1992, 1993), as well as other potential modifications to the nuclear matrix and chromatin. The exploitation of YACs containing reporter genes should help define these levels of genetic organization, since it is more reasonable to compare expression of genes in the genome to expression of genes contained in YACs (which can be obtained following injection of no more than five copies, S. Forlani, unpubl. data) rather than to genes in transgenes. The answers to these questions will also help in defining which events relate to the transmission of developmental programs from the oocytic cytoplasm to the zygote, and which events are involved in the erasure of the differentiated state of the gametes (Forlani and Nicolas 1996a).

The second major contribution of transgenes concerns parental effects. For instance, the expression of a few transgenes is dependent on their parental origin and is reversed following their passage through the germline of the opposite sex, indicating true genomic imprinting. In addition, molecular analyzes have revealed a parental imprint based on reversible methylation patterns in the DNA of several transgenes (Sapienza et al. 1987, 1989; Swain et al. 1987; Engler et al. 1991; Ueda et al. 1992; Chaillet et al. 1995; Weng et al. 1995). More generally, numerous studies of transgenes have completely renewed our understanding of parental imprinting (Tables 21.1 and 21.2) and some conclusions are summarized below:

- Two kinds of imprints on transgenes have been observed. First, there are imprints that are dependent on genetic background and are only observed on certain outbred backgrounds. They are often transgene-

Table 21.1 Parental dependent methylation of transgenes and genetic background effects.

	Transgene	Parental dependent methylation: On outbred background	Parental dependent methylation: On inbred background	Parental dependent action of modifiers	Expression pattern
INDEPENDENT	HRD from ↑ from ↓	♀D2 ↓ ♂D2 ↑ ♀B6 ↑ ♂B6 ↑	B2: ♀↑ ♂↑ D2: ♀↓ ♂↓	No: ↓ Lw/B6 ↑; B6/Lw ↑; (B6 D2/Lw) 1 ↑ : ↓ 1	When DNA is un or hypomethylated expression is observed in the spleen and thymus
POSITION	RSVI-gmyc		FVB: ♀↑ ♂↓ B6: ♀↑ ♂↑	No: ↑ FVB/high ↓; high/FVB ↑	When the transgene is not methylated, expression is detected in adult heart
POSITION DEPENDENT	TKZ 751 [(1)] from ♂↓ from ♂↑* ♂↑ ^{tot}	♀D2 ↓ ♀BC ↑ ♀D2 ↓ ♀BC ↑ ♀D2 ↑ ♀BC ↑	Bc: ♀↑ ♂↑ D2: ♀↓ ♂↓	Yes: ↓ Lw/Bc ↓; Bc/Lw ↑; (D2Bc)/Lw 1 ↑ : ↓ 1; Lw/(D2Bc) ↓ D2 high: ↑* ↓	Position dependent pattern in post-implantation embryos. Relation between methylation and expression of the transgene is unknown

HRD and RSVIgmyc present a pattern of methylation independent of their integration site. For RSVIgmyc, the pattern of expression is also independent of the integration site. TKZ 751 has a pattern of expression dependent on the integration site. The pattern of methylation is indicated as follows: ↑ and high : hypermethylated transgene; ↓ and lw : hypomethylated transgene; ↓ * : partially methylated transgene and ↓ ^{tot} : completely methylated transgene. In the column, " parental dependent methylation", the genetic background is indicated above each methylation pattern of the offspring transgene. B6 : C57BL6; D2 : DBA/2; Bc : BalB/c; FVB : FVB/N. When the parental origin of the transgene is not indicated for example Lw/B6, the female origin is always noted first. The dependence on parental effects of modifiers is indicated by "no" or "yes" and beside this is noted the methylation pattern influenced by the modifier. Under each cross is indicated the methylation pattern of the offspring transgene. For example, the offsprings of the Lw/B6 crosses are hypermethylated and the offsprings of (B6D2)/Lw crosses are hypermethylated and hypomethylated in equal

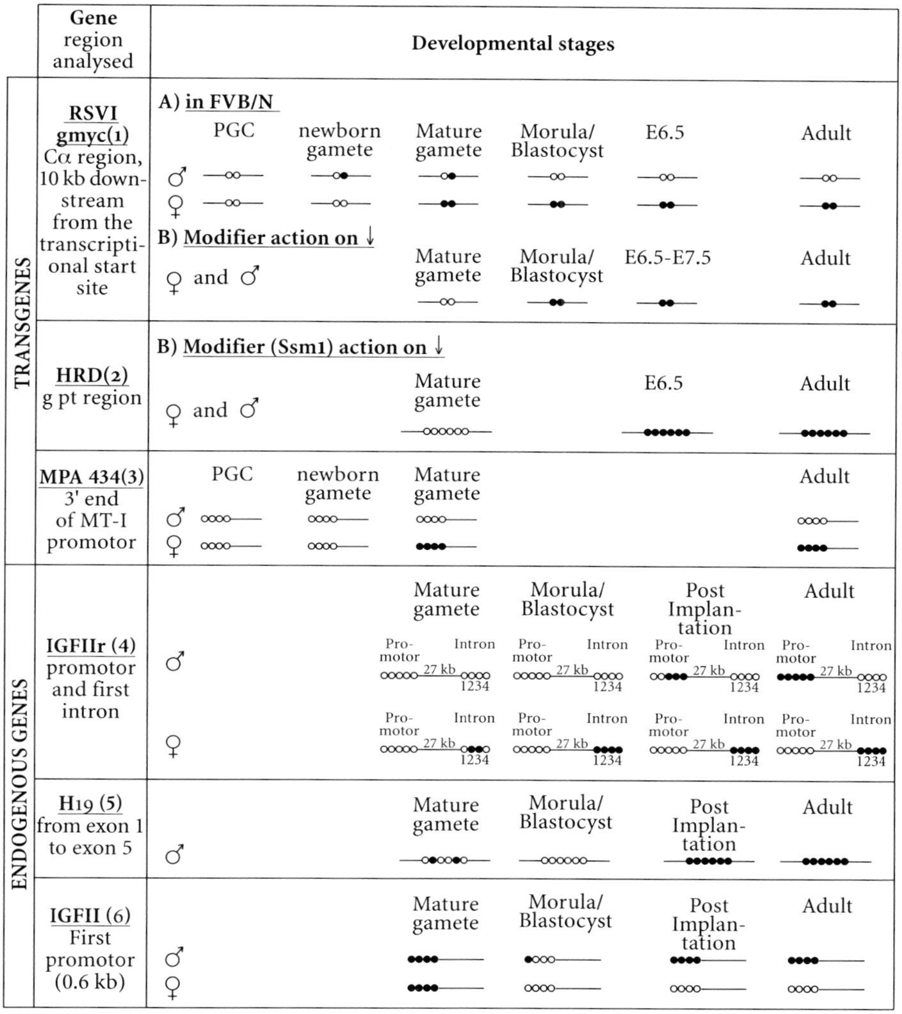

	Gene region analysed	Developmental stages					
TRANSGENES	**RSVI gmyc(1)** Cα region, 10 kb downstream from the transcriptional start site	A) in FVB/N	PGC	newborn gamete	Mature gamete	Morula/ Blastocyst	E6.5	Adult
		♂	—oo—	—o●—	—o●—	—oo—	—oo—	—oo—
		♀	—oo—	—oo—	—●●—	—●●—	—●●—	—●●—
		B) Modifier action on ↓			Mature gamete	Morula/ Blastocyst	E6.5-E7.5	Adult
		♀ and ♂			—oo—	—●●—	—●●—	—●●—
	HRD(2) g pt region	B) Modifier (Ssm1) action on ↓			Mature gamete		E6.5	Adult
		♀ and ♂			—oooooo—		—●●●●●●—	—●●●●●●—
	MPA 434(3) 3' end of MT-I promotor		PGC	newborn gamete	Mature gamete			Adult
		♂	oooo—	oooo—	oooo—			oooo—
		♀	oooo—	oooo—	●●●●—			●●●●—

	Gene region analysed		Mature gamete	Morula/ Blastocyst	Post Implantation	Adult
ENDOGENOUS GENES	**IGFIIr (4)** promotor and first intron	♂	Promotor ooooo —27 kb— oooo Intron 1234	Promotor ooooo —27 kb— oooo Intron 1234	Promotor oo●●● —27 kb— oooo Intron 1234	Promotor ●●●●● —27 kb— oooo Intron 1234
		♀	Promotor ooooo —27 kb— o●●o Intron 1234	Promotor ooooo —27 kb— ●●●● Intron 1234	Promotor ooooo —27 kb— ●●●● Intron 1234	Promotor ooooo —27 kb— ●●●● Intron 1234
	H19 (5) from exon 1 to exon 5	♂	—o●oo●o—	—oooooo—	—●●●●●●—	—●●●●●●—
	IGFII (6) First promotor (0.6 kb)	♂	●●●●—	●ooo—	●●●●—	●●●●—
		♀	●●●●—	oooo—	oooo—	oooo—

independent (transgenes HRD and TKZ 751, Table 21.1; Engler et al. 1991; Weng et al. 1995). Second, there are imprints that are observed on inbred backgrounds (RSVIgmyc, Table 21.1; Chaillet et al. 1995). They are transgene-dependent. These two kinds of imprints probably represent different genetic systems that control epigenetic gene expression. It is reasonable to assume that parental effects in a natural population of mammals generally use these two genetic systems.

- *cis* DNA sequences can confer imprinting to a gene. For instance, the RSVIgmyc transgene in FVB/N strain is imprinted if it contains the switch repeat elements of the IgA gene (Chaillet et al. 1995).
- Strains of mice belong to two categories, those that are hypomethylated or hypermethylated. One of the genes involved in this characteristic has been mapped to chromosome 4 (it is called Ssml, for strain specific modifier 1; Weng et al. 1995). Ssml is involved in the methylation of HRD transgene at E 6.5, which corresponds to the time of general methylation of the genome (Kafri et al. 1993).
- The action of modifiers can, themselves, depend on a parental effect (see the example of TKZ 751, Table 21.2) and may also be influenced by oocytic factors early during embryogenesis (this by itself constitutes an excellent model for early nucleocytoplasmic events during embryogenesis, Surani et al. 1990).
- Finally, it has been established that transgenes subjected to parental effects are hypo- or nonmethylated when transmitted from the father but are hypermethylated when transmitted from the mother (Table 21.2). The hypermethylation of the RSVIgmyc transgene which is imprinted by the mother persists during the embryonic stages corresponding to general demethylation of the genome (Chaillet et al. 1995).

Based on these observations, a molecular model for parental imprinting (which may also be relevant for endogenous imprinted genes, see Table 21.2) has been proposed (Chaillet 1994). A provocative aspect of this model is that it postulates that genetic imprints rely on embryonic events – that is, protection of maternal genes from demethylation during embryogenesis – rather than on gametic events.

This short summary shows how the exploitation of transgenes by reverse genetics and classic genetics has been invaluable for increasing our knowledge of the crucial period of preimplantation development in mammals, during which the major changes associated with cellular differentiation and development are prepared.

We describe here recent technical improvements about the use of transgenes in eggs and preimplantation embryos: first, a method of quantification of DNA constructs to be microinjected, and second, the use of YACs as expression vectors in one-cell and two-cell embryos. Third, we present the conditions of use for some inhibitors to explore regulations of expression. Fourth, a method to evaluate the effects of in vitro culture of early embryos on transgene expression is also described. Finally, we present an assay to quantify β-galactosidase activity using the MUG substrate in individual embryos which are microinjected with or transgenic for *LacZ*.

21.1 Preparation of Plasmids and Inserts for Microinjection

Purification

DNA constructs can be microinjected into one-cell or two-cell embryos as supercoiled or linearized plasmids or as inserts that have been deleted of bacterial sequences. For transient expression assays, as well as in transgenesis experiments, we advise the injection of inserts rather than entire plasmids, because the expression of plasmids lacking any eukaryotic promoter has been observed (our unpubl. results), demonstrating that bacterial sequences of the vector are not always neutral for expression. It should also be noted that for oocytes, DNA is most active in a circular form and therefore inserts must be religated prior to injection (Vernet et al. 1993a).

Plasmids

Extensive purification of DNA is necessary in order to obtain reproducible results. For plasmids, a large-scale preparation of material can be obtained by ultracentrifugation in CsCl-ethidium bromide gradients after lysozyme/alkaline or lysozyme/Triton X100-lysis (Maniatis et al. 1989). Some plasmids yield, for unknown reasons, very small amounts of material by this method. A single CsCl-ethidium gradient ultracentrifugation on a pool of 24 minipreps (lysis by boiling; Maniatis et al. 1989) is also a convenient way to obtain large amounts of plasmid for microinjection (our own results).

Inserts

Insert DNA is cut from plasmids by digestion with restriction enzymes, purified by electrophoresis in low melting point agarose gels in TAE buffer [40 mM Tris-acetate, 1 mM ethylenediaminetetraacetic acid (EDTA)] and isolated from agarose by glass bead purification (Vernet et al. 1993a). Elution is done in filtered 10 mM Tris HCl (pH 7.4), 0.1 mM EDTA (TE 10:0,1). The filtration of the buffer and the low concentration of EDTA are essential for a correct microinjection of DNA and the survival of the microinjected eggs.

Quantification of DNA by Slot Blot

Plasmid DNA concentrations are determined by measuring the OD_{260}. In the case of inserts, following glass bead purification, an aliquot of DNA is analyzed by gel electrophoresis to check for the purity of the fragment. This is also a means to quantify roughly its concentration by visual comparison with a range of known concentrations of a control fragment on the same gel (Maniatis et al. 1989). This semiquantitative method is precise enough to quantify DNA used for the generation of transgenic mice (approximately 2–10 ng/μl) or for a qualitative comparison of different promoters used in one- or two-cell embryos for transient expression assays (Bonnerot et al. 1991). However, this method lacks precision when constructs must be compared on the basis of quantitative expression (for instance, to test for the effects of enhancers or silencers).

Fluctuations of transgene activity from one embryo to another are inevitable because of the variations of the volume injected (a factor of 2) and above all, because of the absence of a control for the degradation of DNA which may occur after microinjection. Therefore, it is important to at least minimize the variability caused by an approximate estimate of DNA concentrations prior to microinjection. Because the concentration of inserts is generally too low to be measured by spectrophotometry, we quantify these by a slot blot technique.

Materials

- Nucleic acid nylon transfer membrane: Hybond-N+ (Amersham)
- Whatman 3 MM paper

Equipment

- Slot blot apparatus (Hybri-Slot Gibco/BRL no. 1052 MM)

Buffers

- Dilution buffer: 10 mM Tris HCl pH 7.4, 1 mM EDTA (TE 10:1)
- NaOH 0.4 M

- Wash buffer: SSC 2× (Stock solution ten-fold concentrated (20×): NaCl 175.3 g/l, Na citrate 88.2 g/l, pH 7) (Maniatis et al. 1989).

Procedure

1. Prepare a dilution range of the purified insert in TE 10:1 (we choose 1/5, 1/10, 1/25, 1/50, and 1/100).
2. Set the same dilution range for purified control plasmids of known concentration. For quantification of *LacZ* inserts using a *LacZ* probe, any *LacZ* containing plasmid is convenient as a positive control which will serve to establish a reference curve. A plasmid without the *LacZ* gene serves as a negative control for hybridization with the probe. Here, the pBluescript KS II plasmid is used (Stratagene). Plasmids are chosen as controls rather than inserts because their quantities permit quantification by spectrophotometry. They are used at concentrations of 35 ng/μl.
3. Add 1 μl of each dilution sample to 50 μl of NaOH 0.4 M and heat at 37–40 °C for 10 min.
4. Wet a precut piece of nylon membrane in distilled water. Place it above a Whatman 3 MM paper in the slot apparatus. Connect the apparatus to a vacuum pump.
5. Deposit samples in each well and filter by aspiration with the vacuum pump, whereby DNA is deposited on the membrane. Avoid forming bubbles when pipetting the samples.
6. Rinse each well with two drops of NaOH 0.4 M.
7. Continue applying the vacuum while lifting off the sample well block and recovering the membrane.
8. Rinse the nylon membrane in SSC (2×) 3×10 min.
9. Prehybridize and hybridize with a ^{32}P-labeled probe, as for a Southern blot (Maniatis et al. 1989).
10. Quantify the radioactivity in each sample with a PhosphorImager system and the supplied software ImageQuant (Molecular Dynamics). Signal intensity is estimated by the volume integration method.

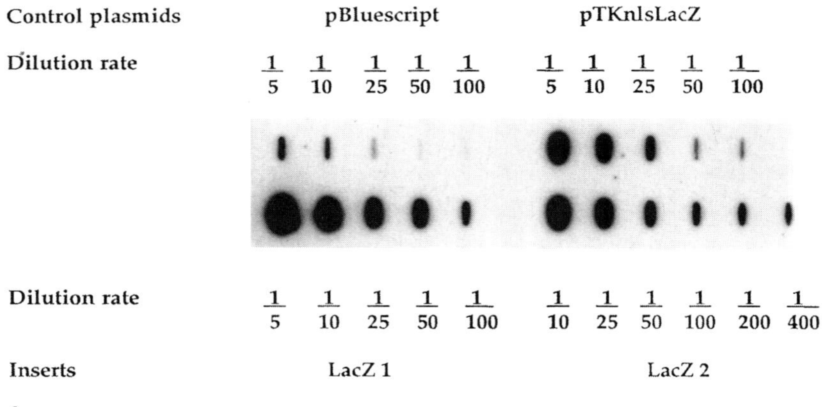

A

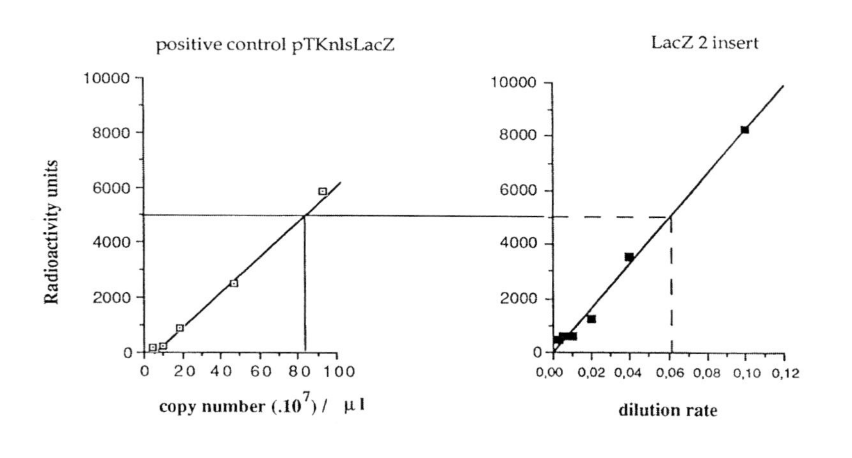

$$\text{Concentration of LacZ insert (ng/ul)} = \frac{\text{copy number X insert size (kb)}}{\text{dilution rate X 100}} = \frac{83 \text{ X } 7.6}{0.06 \text{ X } 100} = 105$$

B

◀ **Fig. 21.2A, B.** Slot blot method to determine the concentration of a *LacZ* insert. **A** Slot blot hybridized with a ^{32}P-labeled *LacZ* probe; 2.95-kb pBluescript and a 7.5-kb *nlsLacZ* plasmid (*pTknlsLacZ*) are, respectively, negative and positive control plasmids, both initially at 35 ng/μl. Values indicate the successive dilutions. LacZ 1 and LacZ 2 are the inserts to be quantified. **B** Quantification of the concentration of an insert. *Curves* represent the values of radioactivity generated by the positive control plasmid and the *LacZ* 2 insert, as quantified (in arbitrary units) by volume integration with the PhosphorImager, as a function of the dilution for the insert and of the copy number for the *nlsLacZ* control plasmid. Copy number is more relevant to compare insert to control plasmid as concentration is dependent on the size of DNA. The copy number is calculated from the relation that at 5 ng/μl, a 5-kb construct is present at 10^9 copies/μl, considering that each sample contains 1 μl of DNA. The plasmid *pTKnlsLacZ* is 7.5 kb in size, therefore a solution at 35 ng/μl contains 4.66×10^9 copies/μl. The copy number in 1 μl of *LacZ* 2 insert is determined from the reference curve taking the dilution into account. Concentration in ng/μl is calculated considering the size of the insert (here, 7.6 kb)

Results

Figure 21.2 shows an example of the quantification of a *LacZ* insert using a 7.5-kb *LacZ* plasmid at 35 ng/μl as a reference. The calculation is described in the legend (Fig. 21.1B). The principle is to compare the DNA of unknown concentration and the *LacZ* control plasmid on the basis of their copy number, since the concentration of a DNA fragment is dependent on its size. We demonstrate that this method can be used to reliably quantify a range of concentrations from 10 ng/μl to 1 μg/μl using 5- to 10-kb *LacZ* inserts. For less concentrated DNA solutions, it is preferable to set another reference curve with lower amounts of control plasmid. The sensitivity of the technique is limited to about 1 pg of DNA.

21.2 Preparation of YACs for Microinjection

Transgenic mice have been successfully generated after pronuclear microinjection of 250-kb YACs (carrying the human β-globin locus (Peterson et al. 1993) or the mouse tyrosinase gene (Schedl et al. 1993; see also Hiemisch et al., Chapt. 17 this Vol.) and recently from a 460-kb YAC containing the Xist region of the X chromosome and a *LacZ* gene linked to the cytomegalovirus (CMV) promoter (Heard et al. 1996). These results have demonstrated that YACs can be microinjected into nuclei of early embryos and remain functional after such treatment. Therefore, artificial chromosomes containing transgenes represent a new tool with additional advantages to plasmids (see Introduction) for use in future studies on transient expression assays and on the regulation of gene expression during preimplantation development. Protocols for preparation of YACs for microinjection can be found in the chapter of this book by Schütz and collaborators, focusing on the use of YACs in transgenesis and in Melanitou et al. (1996). Detailed information about methods in yeast genetics can be found in Rose et al. (1990). Here, we describe some specific features for the preparation of YACs for transient expression assays in one- and two-cell embryos.

Generation of Recombinant YACs

For transient expression assays, it is useful to introduce a reporter gene into the YAC. This gene can be linked to a heterologous promoter (Heard et al. 1996) or inserted distal to the promoter of a gene contained in the YAC. This latter situation is convenient to compare the activity of an endogenous promoter in a plasmidic environment or in its YAC context in the same type of assay.

Reporter constructs are inserted into YACs by homologous recombination in the yeast carrying the YAC of interest. The yeast are transformed with a linearized recombination plasmid that contains the reporter gene flanked on both sides by homologous regions of adequate length (minimum size: 400 bp). A selection marker can be used for the screening of yeast in which successful recombination has occurred, (for instance, complementation for one amino acid absent in the culture medium).

Yeast spheroplast transformation protocol

The highest transformation frequencies are obtained when yeasts are first converted to spheroplasts. The following protocol is slightly modified from Melanitou (Melanitou et al. 1996).

Materials

Yeast strains

- GMY16 or GMY19 convert to spheroplasts at a high efficiency

Media

- AHC: 20 g/l glucose, 6.7 g/l yeast nitrogen base without amino acids (Difco), 10 g/l casein hydrolysate-acid low salt (Sigma), 40 mg/l adenine hemisulphate. Add water, adjust pH to 5.8 with 4 M NaOH if necessary; for agar plates, add 17 g/l agar (Difco) and autoclave 20 min/1 l, 30 min/2 l, etc.
- SORB plates: synthetic medium (SD) plates (0.67 % yeast nitrogen base without amino acids, 2 % glucose, pH 5,8) containing 0.9 M sorbitol and supplemented with the appropriate amino acids (20 μg/ml each of adenine sulfate, L-arginine hydrochloride (HCl), L-histidine-HCl, L-methionine, L-tryptophane and uracil; 30 μg/ml each of L-isoleucine, L-lysine HCl and L-tyrosine; 50 μg/ml L-phenylalanine; 60 μg/ml L-leucine; 150 μg/ml L-valine). For example, a selection medium for the presence of the URA marker in YAC is an SD medium containing all of the above amino acids except uracil
- SCEM: 1 M Sorbitol, 0.1 M Na citrate (pH 5.8), 10 mM EDTA, 30 mM β-mercaptoethanol. STC: 1 M Sorbitol, 10 mM Tris- HCl (pH 7.5), 10 mM $CaCl_2$. SOS: 1 M Sorbitol, 6.5 mM $CaCl_2$, 0.25 % yeast extract, 0.5 % bactopeptone filter sterilized, supplemented with trytophan and uracil to 10 μg/ml. TOP: 1 M Sorbitol, 2.5 % agar in SD medium supplemented with the appropriate nutrients. PEG: 10 mM Tris-HCl (pH 7.5), 10 mM $CaCl_2$, 20 % (w/v) polyethyleneglycol (PEG) 8000. Lyticase (Sigma): stock solution of 10,000 U/ml in 50 mM KPO_4 (pH 7.5) stored in 20 % glycerol
- Carrier DNA: 10 mg/ml calf thymus or salmon sperm (stock).

Procedure

1. Inoculate 10 ml of liquid AHC medium with a loop full of cells from a single YAC-containing colony grown on AHC plate medium. Grow cells at 30 °C overnight with vigorous aeration.

2. Count a 1:10 dilution of the culture using a hematocytometer. They should have reached $1–2\times10^8$/ml. Dilute to 1×10^7/ml into 100 ml of AHC and grow at 30 °C with vigorous aeration for approximately 2 to 3 h (until cells reach 3×10^7/ml).

3. Transfer cells to sterile 50-ml tubes and centrifuge at 1500 rpm (in a Beckman centrifuge) for 10 min. Wash once with 20 ml of sterile water and once with 20 ml of 1 M sorbitol, pelleting cells for 5 min after each wash.

4. Resuspend cells in 20 ml of SCEM, add 400 U of lyticase and incubate at 30 °C, inverting the tube occasionally to keep the cells in suspension. Every 5 min read the OD_{800} for one drop of cell suspension. After 15–30 min, the OD should decrease to 10 % of the original (if this is not the case, an additional 500 U of lyticase is added and the incubation allowed to proceed for an additional 30 min). The spheroblasts are centrifuged at 800 rpm for 3–4 min. From this point on, the cells must be handled carefully.

5. Gently resuspend the cell pellet in 2 ml of 1 M sorbitol. Bring the volume up to 20 ml and centrifuge once more. Finally, resuspend the pellet in 1 ml of STC.

6. Set aside 200 µl aliquots of cells into several 12-ml tubes. In one tube, add 50 µg of carrier DNA alone (as a "no DNA" control). To the remaining cells, add carrier DNA plus 1 µg of linear recombination fragment (not whole linear plasmid) for recombination. The recombination fragment contains the appropriate selectable gene (e.g, LYS or HIS) for complementation of the absence of lysine or histidine in the culture medium. The total volume of DNA should not exceed 1/10 of the volume of the cells.

7. After a 10-min incubation at room temperature, gently resuspend the cells and carefully add 1 ml of PEG to the spheroplasts. Thoroughly and gently mix the suspension. After an additional 10 min at room temperature, pellet the spheroplasts at 800 rpm.

8. Aspirate the supernatant and gently resuspend the pellet in 150 µl of SOS.

9. Before plating the cells, make a ten-fold dilution by mixing 20 µl of spheroplasts with 180 µl of SOS. Add 4 ml of -HIS TOP agar (kept at 45–46 °C) to the tubes, invert quickly to mix and plate both the dilution and the remaining cells immediately on -HIS SORB plates that have been kept at 37 °C. Grow cells for 3–4 days.

10. Make replica plates of colonies, transferring them onto appropriate media to select for presence or absence of selection markers in the recombinant YAC. Maintain the selected clones on appropriate

medium (for example, His+ Trp+ Ura- clones are maintained on medium lacking Trp and His).

Results

We have observed for the insertion of the LacZ gene into a 400-kb YAC a frequency of 5 to 30 % for transformants with the appropriate phenotype on selection media (J. Mushler, unpub. results).

Before their purification for microinjection, YACs contained in selected clones are tested by PCR and must be further verified for appropriate recombination events by standard southern blot hybridization (after inclusion in plugs and pulse-field gel, see below) using all the available probes for the YAC. Comparison of the restriction maps of the recombinant YAC fragments with that of the original YAC can also be done (Pavan et al. 1990).

Purification of Recombinant YACs

The following protocols are modified from Melanitou (Melanitou et al. 1996) who adapted these techniques for cultured cells (Gnirke et al. 1993) to microinjection in fertilized eggs (see also Peterson et al. 1993).

Yeasts containing the recombinant YAC must be embedded in agarose plugs and the yeast cell wall digested in the agarose to liberate the DNA. Agarose blocks containing high molecular weight yeast DNA are then placed in wells of a pulse field gel.

Materials

- Filters: ultrafree-MC filter unit (nominal molecular weight limit, 30,000; Millipore)
- Agarose molds

Equipment

- Pulse-field gel apparatus (Pharmacia)
- Sea plaque agarose
- Low melting point agarose (LMP)

Buffers

- Solution I: 1 M sorbitol, 20 mM EDTA (pH 8), 14 mM 2-mercaptoethanol, 2 mg/ml zymolase-20 T (ICN)
- Solution II: 0.5 M EDTA, 1 % N-laurayl sarcosine (Sigma) 2 mg/ml proteinase K
- Tris- HCl 10 mM (pH 7.4), 1 mM EDTA (TE 10:1)
- Tris borate EDTA (TBE): stock solution 5× is 54 g/l Trizma Base, 27.5 g/l boric acid, and 20 ml 0.5 M EDTA (pH 8) in H_2O
- Agarase buffer: 10 mM Bis-Tris HCl (pH 6.5), 0.25 mM EDTA, 50 mM NaCl
- β-agarase (Biolabs)

Procedure

Low-density fast block preparation

1. Use one colony to inoculate 3–5 ml of an appropriate selective liquid medium and cells are grown overnight.
2. Pellet cells at 1500 rpm in a Beckman centrifuge.
3. Wash cells once with 0.5 ml 50 mM EDTA (pH 8), pellet at 1500 rpm and resuspend in 80 μl of solution I.
4. Add 100 μl of solution I lacking zymolase and containing 1.5 % Sea - plaque agarose (1 % agarose final).
5. Aliquot the solution into agarose molds (60 μl each) that have been sealed with clear tape, then chill the molds on ice.
6. Drop agarose plugs into 1 ml of solution I at 37 °C for 2 h.
7. Aspirate off the liquid and add 1 ml of solution II. Incubate overnight at 50 °C.
8. Wash plugs three times with 10 ml TE 10:1 at room temperature.
   Note: All steps of incubation of agarose plugs are done without agitation to reduce potential shearing of DNA molecules (Peterson et al. 1995).

Pulse-field gel

1. A 1 % agarose gel (low melting point: LMP) is prepared and electrophoresed in sterile 0.5× TBE. Yeast chromosomes markers are loaded on both sides of the gel. Run the gel at approximately 130 V (current should be between 80–100 mA). Parameters of the pulse should be adjusted to the size of the YAC to be isolated, the switching time increases with the size of the YAC. For example, we use a 45-s pulse for

a 400-kb YAC and a 25-s pulse for a 150-kb YAC. The same time has been used for a 590-kb YAC and 14-s used for a 220-kb YAC (Gnirke et al. 1993). An analytical gel is run for approximately 48 h. An LMP preparative gel is run 50 % longer.

2. Strips containing the yeast DNA markers are cut from either side of the gel and stained for 30 min in 0.5× TBE with 4 μl of 10 mg/ml ethidium bromide (EtBr) per 100 ml buffer.

3. Under UV, the bands of interest (relative to the size of YAC) are marked with a cutter. By replacing these marked gel portions beside the unstained central portion of the gel, the location of the desired YAC can be approximated and the corresponding gel slice (3 mm wide) cut out. The gel slice is further cut into 1–2 cm pieces and transferred to a 50-ml tube.

YAC purification from gel slice

1. Equilibrate a 2-cm gel fragment in 10 ml of 0.5× agarase buffer for 2 h.

2. Put the gel fragment in an 1.5-ml Eppendorf tube and weigh. Centrifuge the slice for 2 s to place it at the bottom of the tube. Melt the agarose at 68 °C for 10 min before adding 1 U β-agarase per 100 mg agarose. Gently mix with a tip without pipetting. Digest agarose at 40 °C for 2–3 h.
   Note: We do not centrifuge the solution after digestion, as this caused a significant loss of DNA, but the agarase digestion has to be optimal.

3. The YAC DNA can be injected directly after agarose digestion or concentrated by filtration as described (Gnirke et al. 1993). Solutions of YAC DNA are stored at 4 °C and can be conserved for 2 months.

Results

We obtained solutions of 100, 150 and 400 kb LacZ recombinant YACs at 6.5, 5.3, and 3 ng/μl, respectively, with a total yield of about 3 μg for each (from a 2-cm long-block of agarose gel) (J. Mushler, unpubl. results).
Note: Peterson et al. (1995) recommend agarose treatment of a 3- to 3.5-g gel slice overnight in a volume of 4 ml followed by concentration of the solution through a filter with a 100,000 nominal molecular weight limit cut off. This allows for the recovery of a greater quantity of full-length YAC molecules.

Quantification of YACs

Following purification, the quality (purity and integrity) of the YAC DNA must be checked by pulse-field gel electrophoresis.

The concentration of DNA can be estimated in a 0.8 % agarose minigel using λ phage DNA as a standard. For a precise determination of the concentration, an aliquot of the YAC solution is diluted four times in water (pipette DNA with a cutoff tip to avoid shearing) and the OD_{260} is read. To avoid sacrificing large quantities of YAC DNA, precise YAC concentration can be alternatively determined by a slot blot method.

Procedure

Slot blot quantification of YAC concentration

The procedure is the same as that described above for plasmids and inserts. However, two modifications should be made:

- It is important in this case to shear the YAC by heating each sample in 0.4 M NaOH at 65 °C for 10 min to assure its fixation on the nylon membrane, since DNA fragments which are too large (15 kb or more) are retained less by membrane filters. Shearing of DNA is allowed in this slot blot method as the parameter important for quantification is the copy number of *LacZ* gene.
- The *LacZ* control plasmid and the YAC to be quantified should be approximatively at the same molarity in order to assess copy number. For example, at the same concentration, the copy number of a 200-kb *LacZ* recombinant YAC will yield a value 30-fold less than that of a 7-kb *LacZ* plasmid. Therefore, for a relevant quantification of the concentration of such a YAC, the concentration of the control plasmid should be diluted 30-fold (Fig. 21.2.).

21.3 Preparation of Eggs and Embryos

Eggs and preimplantation embryos are isolated in vivo at the required stage (e.g. growing oocytes, full grown oocytes, fertilized one-cell embryo to blastocyst) for immediate use, or can be isolated at one stage and cultivated until they reach the required stage. In the mouse, it is possible to obtain preimplantation development in specific culture media with a high efficiency (i.e., 80% of cultured one-cell embryos reach the morula or blastocyst stage). This is an invaluable means to study this period of development and is necessary for the manipulation of embryos for purposes such as microinjection, nuclear transfer, or activation of oocytes to form parthenogenotes. Oocytes can also be fertilized with sperm cells in vitro, and this also reduces the asynchrony of development (as long as 8 h) which exists among embryos fertilized in vivo. However, despite the fact that there is a high efficiency in the development of eggs fertilized or cultured in vitro when transferred back into foster mothers, it is important to be aware that in vitro culture may not be neutral on embryos (see below).

Strains

Mice must be chosen based on their ability to give:

- a high number of eggs when superovulated with hormones,
- eggs displaying clearly visible pronuclei or nuclei,
- eggs that are able to develop in vitro.

These are necessary conditions for transient expression assays and transgenesis. Hybrid (F1) strains (C57BL/6J×DBA2), (C57BL/6J×SJL/J) and (C57BL/6J×CBA) are convenient for these purposes. Because the genetic background of embryos may influence the expression of the transgene, embryos from inbred FVB/N mice can also be used. However, it must be noted that the number of injectable embryos per female and the survival of embryos after microinjection are slightly lower for FVB/N than for outbred (C57BL/6J×SJL/J) embryos (56.3 and 58% vs. 63.5 and 66.7%, respectively; Paris et al. 1995). For fertilization, F_1 males of the same strain are used.

Superovulation

Hormonal injection increases the number of ovulated eggs. Injected hormones mimic the action of the FSH (follicule stimulating hormone) and of the LH (luteinizing hormone).

Materials

Buffers and hormones

- Phosphate-buffered saline (PBS): 138 mM NaCl, 2.7 mM KCl, 1.5 mM KH_2PO_4, 8.1 mM Na_2HPO_4, pH 7.3
- PMSG (pregnant mare serum gonadotrophin) or Folligon (Intervet S.A). Stock solution: 50 IU/ml in PBS
- hCG (human chorionic gonadotrophin) or Chorulon (Intervet S.A). Stock solution: 50 IU/ml in PBS.

Procedure

Hormonal treatments

A 4- to 10-week-old female is peritoneally injected with 5 IU (in 100 μl) of PMSG and 42–48 h later with 5 IU (100 μl) of hCG.

The timing of injection is relatively flexible, but the time of day should be considered, since it is necessary to inject hCG before the release of endogenous LH, the timing of which is regulated by the dark-light cycle of the animal facility. For example, during a 06–20 h dark-light cycle, PMSG can be injected between 14.00 and 19.00 h and hCG between 10.00 and 17.00 h 2 days later. For in vivo fertilization, females are bred immediately after the injection of hCG. This injection serves as a reference for the timing of development of preimplantation embryos (Fig. 21.2).

Results

The number of ovulated eggs is around 30 for a 4- to 10-week-old superovulated female compared to 5–10 eggs from untreated females.

Recovery and Culture of Eggs and Embryos

Until the eight-cell stage, the embryo needs lactate and pyruvate as exogenous sources of energy since it does not use glucose in an efficient way. Embryos do not need added amino acids, vitamins or serum. Many studies have been directed at optimizing the media for in vitro culture (Chatot et al. 1989; Lawitts and Biggers 1993). Two types of media are employed: one is used whenever embryos have to be manipulated at ambient temperature (M2 or PBI) and the other for culture at 37 °C (M16 or Whitten). We routinely use PBI and Whitten (described in Hoppe 1985) (for composition of M2 and M16; see Vernet et al. 1993).

Note: Protocol for in vitro fertilization and required media can be found in Vernet et al. (1993a).

Materials

Equipment

- An eroded needle (Microlance 0.3×13 mm, Becton Dickinson)
- Hard glass capillary tubing (1.5 mm O.D., BDH Chemicals) flame drawn to a diameter of approximately 200 μl
- Sterile four-well plate (Nunc)
- Stereomicroscope with understage illumination, e.g., Wild MZ8
- Incubator (37 °C, 5 % CO_2 in air)

Media

- PBI
  Stock A_{PBI} solution (10× concentrated): NaCl 1.030 M (5.97 g/100 ml), KCl 27 mM (0.203 g/100 ml), KH_2PO_4 14 mM (0.194 g/100 ml), $MgSO_4$. 7 H_2O 11.8 mM (0.290 g/100 ml), glucose 55 mM (1 g/100 ml), penicillin (1×10^5 IU).
  PBI is a stock of A_{PBI} solution diluted ten-fold, plus Na_2HPO_4 8 mM (0.114 g/100 ml), sodium pyruvate 0.39 mM (0.0043 g/100 ml), $CaCl_2$. $2H_2O$ 0.92 mM (0.010 g/100 ml), phenol red (0.1 g/100 ml) and bovine serum albumin (BSA) (4 mg/ml) (BSA (Fraction V) Sigma no. A-9647).
- Whitten
  Stock Aw solution (10× concentrated): NaCl 1.100 M (0.640 g/100 ml), KCl 48 mM (0.360 g/100 ml), KH_2PO_4 11 mM (0.160 g/100 ml), $MgSO_4$. 7 H_2O 11.8 mM (0.290 g/100 ml), glucose 55 mM (1 g/100 ml), penicillin (1×10^5 IU) and streptomycin (0.050 g/100 ml).
  Whitten is a stock of Aw solution diluted ten-fold plus $NaHCO_3$ 22.6 mM (0.190 g/100 ml), sodium pyruvate 0.22 mM (0.0025 g/100 ml), Ca lactate. $2H_2O$ 4.1 mM (0.045 g/100 ml), phenol red (0.1 g/100 ml), and BSA (4 mg/ml).
  PBI and Whitten solutions are sterilized by filtration through a 0.22-μM Millipore membrane before use. Stock solutions can be stored at −20 °C for several months. PBI and Whitten media are kept at 4 °C for no longer than 2 weeks.
- Phosphate-buffered saline (PBS): 138 mM NaCl, 2.7 mM KCl, 1.5 mM KH_2PO_4, 8.1 mM Na_2HPO_4, pH 7.3
- Pancreatic Trypsin (type II S from porcine pancreas, Sigma no. T-7409). Stock solution is at 3000 U/ml in PBS and stored at 2−8 °C.
- N^6, 2’0-dibutyryladénosine 3’, 5’-cyclic monophosphate (dbc AMP, Sigma)

- Hyaluronidase (Type IV-S from bovine testes, Sigma no. H3884). Stock solution is at 10 mg/ml in H_2O, filter sterilized, aliquoted and stored at −20 °C.

Procedure

Embryos at any given stage are recovered from the mother after death by cervical dislocation, or asphyxia in a container containing dry ice. Death by asphyxia has no effect on the subsequent development of embryos in culture. Eggs and embryos are transferred by mouth pipetting and cultured in 0.5 ml Whitten (37 °C, 5 % CO_2 in air) in sterile plastic four-well-plates. We recommend this low volume and these plates for culture because the movement of the embryos is minimized when the plate is taken in and out of the incubator. Note that for an immediate analyzis of transgene expression in transgenic eggs or embryos, recovery can be done in PBS.

Recovery of Oocytes

1. For the recovery of growing or fully grown oocytes, nonsuperovulated 12–14 day or 8–10-week old F_1 females, respectively, are sacrified.
2. Ovaries are dissected and placed in PBI supplemented with 300 U/µl pancreatic trypsin and kept at 37 °C for 25 min.
3. Ovaries are then transferred into PBI alone. Under the stereomicroscope, the envelope is discarded and oocytes are released from follicles by puncturing them with forceps or tungsten needles.
4. Oocytes are rinsed three times by successive transfer into PBI to eliminate follicular cells before culture. This operation is necessary for the oocytes to be microinjected correctly.

Note: Fully grown oocytes undergo spontaneous germinal vesicle breakdown 1 to 2 h after recovery and therefore must be injected immediately after isolation, unless the postdissection media is supplemented with 100 µg/ml of dbc AMP. Growing oocytes do not mature spontaneously in the presence of dbc AMP, and DNA can be microinjected 1–4 h after their isolation. Injection of DNA has been reported to disturb the maturation process, but this was not observed after injection of RNA into the cytoplasm (Vernet et al. 1992).

Isolation of One- to Four-Cell Embryos After Fertilization In Vivo

Preimplantation embryos are obtained from superovulated pregnant females. The latter are sacrificed at the required stage of embryonic development, the timing of which is established using the injection of hCG as a reference (Fig. 21.1).

One-cell embryos

1. Sacrifice females at 18–20 h post-hCG (hphCG).
2. Isolate the oviducts (located between ovary and uterus) in PBI containing 300 μg/ml hyaluronidase and dissect the swollen part named the *ampulla* (see Fig. 21.1) with forceps. Hyaluronidase releases eggs from the cumulus cells within 2–3 min.
3. Rinse eggs by pipetting in PBI (without hyaluronidase) before culture in Whitten (37 °C, 5 % CO_2 in air).

Note: The best time to isolate one-cell embryos for microinjection is at about 20 hphCG because the two parental pronuclei are clearly visible and this verifies that fertilization has occurred. Microinjection can be done from 18 to 28 hphCG, before the time of amphimixis (Fig. 21.1). Note that one-cell embryos that are too old are injected with difficulty, since the cytoplasm becomes more granular and pronuclear envelops are less visible.

Two-cell and four-cell embryos

1. Sacrifice females at 42–43 hphCG or at 52 hphCG for recovery of two-cell or four-cell embryos, respectively (Fig. 21.1).
2. Place oviducts in PBI (without hyaluronidase) and release the embryos by flushing the oviducts using a syringe filled with PBI attached to a 30-gauge, blunt-ended needle plunged into the infundibulum (opening of the oviduct).
3. Rinse eggs by pipetting in PBI before culture in Whitten (37 °C, 5 % CO_2 in air).

Note: The period during which these stage embryos can be microinjected is from 38 to 47 hphCG for two-cell embryos and from 51 to 57 hphCG for four-cell embryos. At these stages, in general only one blastomere is injected.

21.4
Microinjection of DNA

Injection Volume

The tolerated volume of injection is 1–4 pl in the germinal vesicles of oocytes, 1–2 pl in the pronuclei of one-cell embryos, and 0.4–0.8 pl in the nuclei of two-cell and four-cell embryos, in accordance with the estimated increase in nuclear diameter (Vernet et al. 1993). Cytoplasmic volumes of injection are 6 to 10 pl for one- and two-cell embryos and 2 to 3 pl in growing oocytes, as measured by injection of a solution of [^{3}H] thymidine (Vernet et al. 1993a).

Quantity of DNA

Plasmids and inserts

Concerning the quantity of DNA to inject, two options have been reported:

- injecting the highest amount of DNA tolerated by the embryos which produces the highest activity of the reporter enzyme (Martinez-Salas et al. 1989; Majumder et al. 1993; Ram and Schultz 1993). This corresponds to a concentration of 100–200 ng/μl or 50,000–100,000 injected copies for one- and two-cell embryos, for a 5-kb construct at 5 ng/μl, containing 1000 copies/pl;
- injecting low quantities of DNA at concentrations of 2–10 ng/μl, which corresponds to approximately 1000–5000 injected copies depending on the length of the DNA (Bonnerot et al. 1991; Vernet et al. 1992; Christians et al. 1995; Yeom et al. 1996) and produces sufficient reporter gene enzymatic activity. These quantities of DNA are slightly higher than the concentration used to generate transgenic mice.

We recommend the second strategy, since too high a concentration (0.1 to 1 μg/μl) of DNA can have toxic effects in embryos and may titer out some of the putative regulatory or structural molecules required for the function of the chosen promoter, leading to misinterpretations of results. DNA is diluted in TE (10 mM Tris-HCl, 0.1 mM EDTA pH 7.4). Solutions of inserts are stored at −20 °C.

YACs

Low concentrations are absolutely required for injection of YACs, as their size makes them fragile and solutions are viscous at much lower concentrations than those of plasmids or inserts. For transgenesis, a 2 ng/μl solution is usually used (Schedl et al. 1993; Heard et al. 1996), corresponding to 10 copies/pl for a 200-kb construct. β-galactosidase activity has been detected in transient expression assays after injection of 150- and 200-kb *LacZ* recombinant YACs at 5 to 30 copies/pl (our unpubl. results). DNA is diluted in 50 mM NaCl, 0.25 mM EDTA, 10 mM Bis-Tris HCl, pH 6.5.

Note: Protocols and materials for microinjection of plasmids and inserts are described in Hogan et al. (1986). For injection of YACs, caution must be taken to preserve DNA from shearing. Be specially careful to avoid using too thick a microinjection pipette and sudden changes in the pressure of injection (inject in continuous flow).

Survival and Development of Microinjected Embryos

Non manipulated control one-cell embryos develop to two-cell embryos in culture at a frequency of 90 to 98 %. The frequency is slightly lower (70 to 82 %) for injected one-cell embryos that survive the injection. Similarly, 60 to 80 % of injected two-cell embryo blastomeres go on to cleave normally.

It must be noted that the timing of cleavage of injected one-cell embryos or two-cell blastomeres is retarded by a few hours relative to that of non injected controls (Vernet et al. 1992). This must be taken into account as biochemical events may be unrelated to morphological events, such as cleavage. To bypass this temporal delay, one-cell embryos can be synchronized before microinjection relative to the formation of pronuclei and timed after microinjection relative to the development of uninjected embryos from the same group (Ram and Schultz 1993).

21.5 Study of Expression in Cultured Embryos

The ability to obtain preimplantation development in vitro allows for the study of this developmental period. However, it is important to realize that embryos cultured in vitro may not represent their in vivo counterparts. Embryos in culture are subject to numerous perturbations. For instance, the culture media can vary for parameters such as pH, temperature or concentration of free oxygen radicals, etc., to which embryos are sensitive (Johnson and Nasr 1994). This sensitivity of embryos to culture and manipulation can be measured by assessing their developmental potential beyond the two-cell stage or after culture when transferred back to foster mothers. Stress due to culture is reduced by using a culture medium which has been previously warmed and equilibrated in CO_2 in an incubator. We usually observe good developmental potential in embryos cultured in Whitten. Recently, Lawitts and Biggers (Lawitts and Biggers 1993) have developed a new medium, called SOM, which reduces the osmotic stress which may result, in part, from free oxygen radicals (Johnson and Nasr 1994).

Effect of in Vitro Culture of Preimplantation Embryos on Transgene Expression

The effect of culture on gene expression is also an important question, as most studies regarding zygotic genome activation and its regulation have been conducted on cultured embryos. Transgenic mice containing an exogenous reporter gene have provided a means to compare gene expression in embryos developed both in vivo and in vitro; this is in contrast to analyzis of endogenous protein synthesis, which necessarily involves a period of culture to incorporate radiolabeled amino acids (Flach et al. 1982; Bolton et al. 1984). The influence of culture on gene expression depends on the promoter assayed. Culture induces the activity of a β-gal transgene driven by the long-term repeat of human immunodeficiency virus 1 (HIV1-LTR) in two- and four-cell embryos as a result of oxidative stress (Vernet et al. 1993b). This promoter is also activated by UV radiations (Cavard et al. 1990). Modulations in gene activity by in vitro culture of embryos have also been reported for *hsp70-1* (Christians et al. 1995). The influence of in vitro culture may also depend on the genetic background of the embryo (Kothary et al. 1992). We propose below a method to evaluate these in vitro effects on the expression of a transgene in embryos issued from a single female. Surgery of the animal follows the protocol for oviduct transfer of embryos, described in (Hogan et al. 1986).

Materials

- Animal balance
- Two fine blunt forceps
- Dissection scissors

- Serafine (1.5 in or smaller)
- Wound clips and clip applicator
- Tissues (for soaking up any blood).

Equipment

- Stereomicroscope (SZ, Olympus).

Anesthetic

- Avertin 2.5 %. A 100 % stock solution is prepared by mixing 10 g of tribromoethyl alcohol with 10 ml of tertiary amyl alcohol (Hogan et al. 1986). Dilution to 2.5 % is done in 37 °C PBS.
- 70 % and 100 % ethanol.

Procedure

1. Superovulate one 8–10-week female and mate her (one animal or both carrying a transgene).
2. At 20–22 hphCG, weigh the mouse and anesthesize her with a peritoneal injection of Avertin 2.5 % (0.015–0.017 ml/g of body weight).
3. Sterilize all instruments with 100 % ethanol.
4. Wipe the back of the mouse with 70 % ethanol and make a small transverse incision with the dissecting scissors at the level of the spinal column, slightly above the virtual line which joins the top of the two legs. Remove any loose hairs by wiping with a tissue soaked in 70 % ethanol.
5. Slide the skin to the right until the incision comes over the ovary which appears orange through the body wall. Make a small incision just over the ovary. Pick up the fat pad of the ovary with blunt forceps and pull out the ovary, oviduct, and uterus, which are held outside of the body by clipping the serafine onto the fat pad and laying it down over the back.
6. Under the stereomicroscope, gently raise the ovary with blunt forceps and cut the oviduct out from the ovary and uterus. Gently soak up blood with tissue soaked in 70 % ethanol.
7. Put the cut oviduct in PBI containing 300 μg/ml hyaluronidase. Push the uterus and ovary back inside the body wall. Sew up the body wall (optional) and close the skin with wound clips. Leave the mouse in its cage in a warm place to recuperate.

8. Recover and place the one-cell embryos in culture as previously described. Verify the fertilization of the embryos by checking for the presence of pronuclei.
9. At 46 hphCG, sacrifice the same mouse and recover the two-cell embryos from the left oviduct as previously described.
10. Treat these two-cell embryos in parallel with those developed in vitro for qualitative or quantitative analyzis of reporter enzymatic activity.

Results

Avertin apparently causes neither pertubations in the development of embryos in the left oviduct nor aberrant transgene expression. In an operated female, embryos show a correct development at least until 14.5 days of gestation (our own results). This method is also appropriate to compare the activity of a transgene at two different preimplantation stages in the same female.

Use of Inhibitors in Preimplantation Embryos

Transient expression of microinjected transgenes has shown that certain promoters are active in the embryo at an earlier stage (one- to two-cell stage) than for the corresponding endogenous promoters (Bonnerot et al. 1991). Therefore, the newly fertilized embryo seems to contain the transcriptional machinery required for the expression of various genes, but the genome may exert additional regulation which delays the endogenous expression. To study this regulation, one strategy is to investigate the influence of certain epigenetic events on expression such as those related to the cell cycle (DNA replication, cell cleavage) or to chromatin structure. It is possible to alter these events by treatment of transgenic embryos in vitro with appropriate inhibitors.

Inhibition of DNA replication

DNA replication can be blocked by aphidicolin, which inhibits the DNA polymerase α (Ikegami et al. 1978), or by 5-fluorodeoxyuridine (FUdR), a thymidine analog (Cozzarelli 1977). Embryos cultured in the presence of aphidicolin or FUdR before or during the S phase of each cell cycle do not cleave as a consequence of the inhibition of DNA replication. The effect of the inhibitors is reversible if the embryos are treated for less than 2 h. It has been reported that the inhibition of S phase during the first cell cycle blocks the major phase of zygotic genome activation (ZGA) at the two-cell stage (Bolton et al. 1984), as interpreted from the profile of protein synthesis in arrested embryos.

Inhibition of the cell cycle

Methyl [5-(2-thienylcarbonyl)] -1,4-benzimidazole 2-carbamate (nocodazole) (Hoebeke et al. 1976) inhibits spindle formation by disrupting microtubules and causes arrest of embryos in metaphase. Cytochalasin D inhibits actin filaments and therefore causes failure of cytokinesis without affecting karyokinesis (DNA replication and division of nuclei). The effects are reversible if drugs are removed from the culture medium. Both can be added to the culture medium at any time before mitosis. Regarding the effects of these inhibitors on expression, it has been reported that nocodazole-treated one-cell embryos failed to undergo ZGA, probably as a consequence of the maintenance of the genome as metaphasic chromosomes. Cytochalasin D has no obvious effect on ZGA (Bolton et al. 1984).

Inhibition of histone deacetylases

The effects of chromatin structure on the regulation of gene expression in preimplantation embryos can be studied by modifying the activity of the histone deacetylases. It has been proposed that lysine acetylation of histones destabilizes chromatin structure (for review see Wolffe and Pruss 1996). Drugs which inhibit histone deacetylase, the most widely used being sodium n-butyrate (NaB) and trichostatine A (TSA) (Yoshida et al. 1995), are thought to act on the chromatin structure by keeping histones in an acetylated state. Both drugs block the cell cycle in G1 or G2 beyond the two-cell stage. Treatment of microinjected one-cell and two-cell embryos with NaB results in expression of DNA plasmids which are normally repressed in untreated embryos (in the female pronucleus (Wiekowski et al. 1993) and in the zygotic nucleus of two-cell embryos (Majumder et al. 1993; Wiekowski et al. 1993). We have observed the same effects for an integrated transgene of maternal origin in one-cell embryos treated either with NaB or TSA (our unpubl. results).

Inhibition of Transcription

α-Amanitin is an irreversible inhibitor of RNA polymerase II (Levey and Brinster 1978). It has been extensively used to time the ZGA (Flach et al. 1982; Telford et al. 1990). This inhibitor can also be used to obtain information about the transcription of integrated transgenes whose expres-

Table 21.3. Conditions of use of some inhibitors in preimplantation embryos

Product	Inhibition of	Concentration in culture medium	Stock solution
α-amanitin	Transcription (RNA polymerase II)	11 μg/ml	1,1 mg/ml in H_2O (stored at −20 °C)
Aphidicolin	Replication (DNA polymerase α)	2 μg/ml	2 mg/ml in DMSO (stored at −20 °C)
FUdR (5 fluorodeoxy-uridin)	Replication analog of thymidine)	200 μM	10 mM in H_2O (stored at −20 °C)
Nocodazol (Methyl[5-(2-thienylcarbonyl)]-1,4-benzimidazol 2-carbamate)	cleavage (disruption of microtubules of the spindle)	10 μM	10 mM in DMSO (stored at 4 °C)
Cytochalasin D	cytokinesis (actin filaments)	0,5–1 μg/l	0,5–1 mg/ml in H_2O (stored at −20 °C)
NaB (Sodium *n*-butyrate)	Histone deacetylase (pleiotropic effects)	2,5 mM	2,5 M in H_2O (stored at −20°)
TSA (Trichostatin A)	Histone deacetylase	66 nM	4 mM (1 mg/ml) in DMSO and 66 μM in H_2O (stored at −20 °C)

see text for details and references

sion is usually analyzed in an indirect way at the protein level by the enzyme activity encoded by reporter genes.

Materials

The conditions of use of these inhibitors and the composition of stock solutions are summarized in Table 21.3.

Procedure

1. Sacrifice superovulated and pregnant females at the appropriate time for the cellular event to be inhibited (Fig. 21.1).

Note: For the inhibition of the S phase in two-cell embryos, it must be considered that the G2 of the first cell cycle and the G1 of the second cell cycle are very short (Fig. 21.1) and that in vivo fertilized eggs show developmental asynchrony. To block the second S phase entirely, embryos can be recovered at the very late one-cell stage (30 hphCG) as the first S phase is completed. Alternatively, embryos can be synchronized at the first cleavage: for this, embryos are cultured from the one-cell stage, examined at half-hour intervals, and any two-cell embryos that have formed within the previous half hour are harvested and cultured separately (Bolton et al. 1984).

2. Recover the embryos as previously described.
3. Separate embryos into two pools. One group is cultured in Whitten (37 °C, 5 % CO_2 in air) supplemented with the inhibitor. The other is cultured in Whitten alone and serves as the control for development and gene expression. It is recommended to set up a control for each experiment even if the same experiment is done several times.

21.6 Qualitative and Quantitative Analysis of β-Galactosidase Activity in Preimplantation Embryos

Here, we describe methods developed to analyze the activity of β-galactosidase, the enzyme encoded by the *LacZ* gene. These methods allow for the detection of β-galactosidase activity at the cellular level in individual embryos, and they can be applied to microinjected as well as to transgenic embryos. Methods for the detection of luciferase activity can be found in Miranda et al. (1993) and Thompson et al. (1995).

Delay Between Microinjection and Analysis of Expression

The expression of DNA that has been microinjected into the male pronucleus has been observed at the very late one-cell stage (30 hphCG), at the time when transcriptional competence has just been reestablished (Vernet et al. 1992; Ram and Schultz 1993).

In two-cell embryos, expression has been observed as soon as 2–3 h after nuclear injection of a one single blastomere (Vernet et al. 1992). In four-cell embryos, expression can be detected from 4 h onward following microinjection of one blastomere (our own results).

Qualitative Analyzis: X-Gal Staining

The lower limit of sensitivity of β-galactosidase detection by this method has been estimated to be approximately 10^3–10^4 molecules per egg based on cytoplasmic microinjection of known quantities of pure β-galactosidase in fertilized eggs (Vernet et al. 1993).

Materials

Buffers

- PBS
- Fixation medium: 1 % (v/v) formaldehyde and 0.2 % (v/v) glutaraldehyde in PBS, stored at room temperature and made fresh every 2 months.
- X-Gal reaction mixture: 4 mM $K_4Fe(CN)_6\ 3H_2O$ (potassium ferrocyanide), 4 mM $K_3Fe(CN)_6$ (potassium ferricyanide), 2 mM $MgCl_2$ and 1 mg/ml 4-chloro-5-bromo-3-indolyl β-galactoside (X-Gal) in PBS (pH 7.3). The X-Gal reaction mixture is kept at 4 °C and made fresh every 2 weeks. Stock solutions are 0.2 M potassium ferricyanide and 0.2 M potassium ferrocyanide in H_2O (both stored at room temperature and protected from light) and 40 mg/ml X-Gal in DMSO (stored at −20 °C).

Procedure

1. Rinse cultured or freshly recovered oocytes, one-cell, two-cell or four-cell embryos in PBS and transfer for 5 min in a drop of fixation medium in a petri dish.
2. Rinse cells three times (1–2 min each) by successive transfer in drops of PBS.
3. Transfer cells to one drop of the X-Gal reaction mixture on a petri dish.
4. Incubate cells at 37 °C in a humidified chamber (a plastic box containing a water-soaked Kleenex) for 20 h (the most positively staining eggs may be stained after 1–3 h of incubation).

Troubleshooting

Problems of embryos sticking to glass pipettes and to petri dishes may occur during transfer. To avoid this, siliconized pipettes and petri dishes designed for tissue culture can be used. In addition, all media for the staining procedure can be supplemented with 1 % (v/v) serum, since the presence of proteins reduces adhesion.

For correct staining of preimplantation embryos, it is necessary to use a fixation medium containing both formaldehyde and glutaraldehyde and to respect the 5-min fixation time. If the fixation is too short, eggs rise to the surface of the PBS drop and are difficult to submerge.

Quantitative Analysis of β-Galactosidase Activity: the MUG Assay

Although the histochemical detection of β-galactosidase activity is highly sensitive, visual observation of X-Gal-stained embryos does not allow for a quantification of the activity, and qualitative differences in staining are not distinguishable above a certain level.

Two quantitative assays for β-galactosidase activity, using the fluorogenic substrate 4-methylumbelliferyl β-D galactoside (MUG), have been recently applied to embryos (Vernet et al. 1993a; Bevilacqua et al. 1995). These assays measure the accumulation of a fluorescent reaction product after a set period of time and the fluorescence is enhanced by adding basic glycine and/or Na_2CO_3 to samples. We have introduced modifications to the assay developed by Vernet et al. (1993a) which increase both its sensitivity and reproducibility and make it ideal for the quantification of minute levels of β-galactosidase in transgenic or microinjected individual preimplantation embryos (Forlani and Nicolas 1996b). The most important modifications to the original protocol are

- we determine the kinetics of the enzymatic reaction rather than measuring the accumulation of the reaction product after a fixed period of time and
- we do not add glycine to samples to stop the reaction, since we observed that, for an unknown reason, glycine increases fluorescence in a non linear way at higher fluorescence values.

Materials

- Flat-bottomed 96-well plate.

Equipment

- Fluoroskan fluorimeter (Fluoroskan II, computer controlled by the software Δ Soft II program, Lifescience).

Enzymes and buffers

- *E. coli* β-galactosidase (Sigma no. G5635 1000 U/mg solid). Stock solution is 1000 U/ml in water, stored at −20 °C in 10-μl aliquots. The latter are thawed only once. A range of 10^{-2} to 10^{-7} β-Gal units/μl is prepared prior to the MUG assay by a ten-fold dilution in ice cold dilution buffer (10 mM HEPES, pH 7.5, 1 mM EDTA, 2 mg/ml BSA). The number of units is determined by spectrophotometric assay (one unit of β-galactosidase generates one μmol of o-nitrophenol from o-nitrophenyl-β-D-galactoside (ONPG) in 1 min)
- Lysis buffer is: 60 mM Na_2HPO_4, 40 mM NaH_2PO_4, 10 mM KCl, 1 mM $MgSO_4$, pH 7.2 (Z buffer), 0.5 % Triton X-100, 100 μg/ml BSA and 0.1 % Na azide (stored at −20 °C) (BSA is added to the lysis buffer to stabilize proteins and sodium azide is added to neutralize possible microbial contamination. These have no influence on β-galactosidase activity). Stock solutions of Z buffer, BSA, and Na azide are 10× concentrated and stored at −20 °C (for Z buffer and BSA) or at 4 °C (for Na azide)
- 4 methylumbelliferyl β-D galactoside (MUG) (Sigma no. M1633) stock solution is 5 mM in Z buffer (stored at −20 °C). The powder is dissolved at 100 °C.

Procedure

1. Individual microinjected or transgenic embryos are transferred into 10 μl of lysis buffer in a 96-well microplate. A range of 10^{-2} to 10^{-7} units of pure β-galactosidase is used as a standard. One uninjected or nontransgenic embryo and a microtiter well with no embryo serve as controls.

2. The solubilization of the precipitated MUG in its 5 mM stock solution is achieved by heating for 10–15 min at 100 °C (invert the tube occasionally to mix). A 0.44 mM MUG solution is then made by dilution in lysis buffer; 90 μl of this solution (40 nmol of MUG) is added to each well to start the reaction, followed by two drops of paraffin oil to prevent evaporation. Plates are incubated at 37 °C.

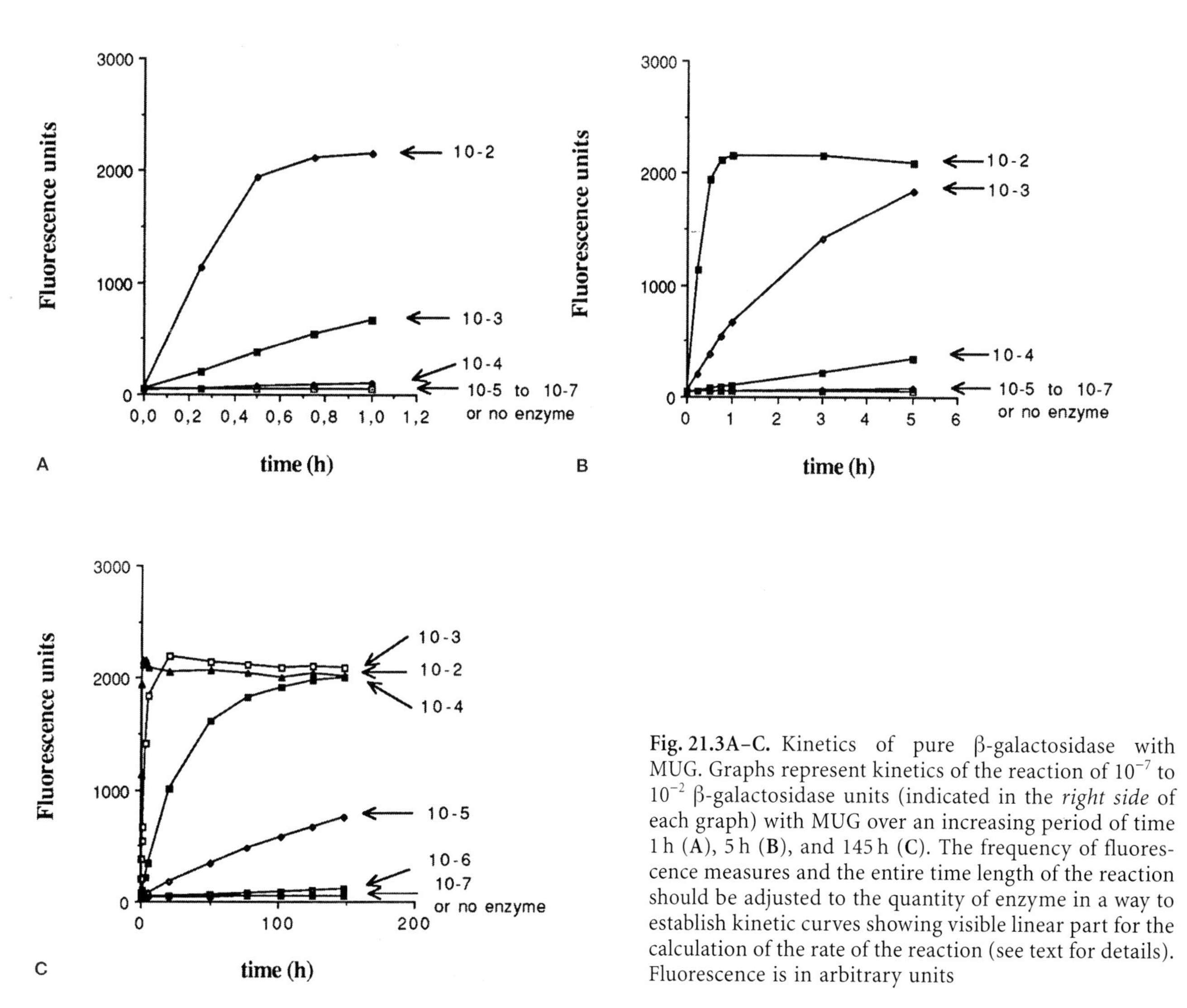

Fig. 21.3A–C. Kinetics of pure β-galactosidase with MUG. Graphs represent kinetics of the reaction of 10^{-7} to 10^{-2} β-galactosidase units (indicated in the *right side* of each graph) with MUG over an increasing period of time 1 h (**A**), 5 h (**B**), and 145 h (**C**). The frequency of fluorescence measures and the entire time length of the reaction should be adjusted to the quantity of enzyme in a way to establish kinetic curves showing visible linear part for the calculation of the rate of the reaction (see text for details). Fluorescence is in arbitrary units

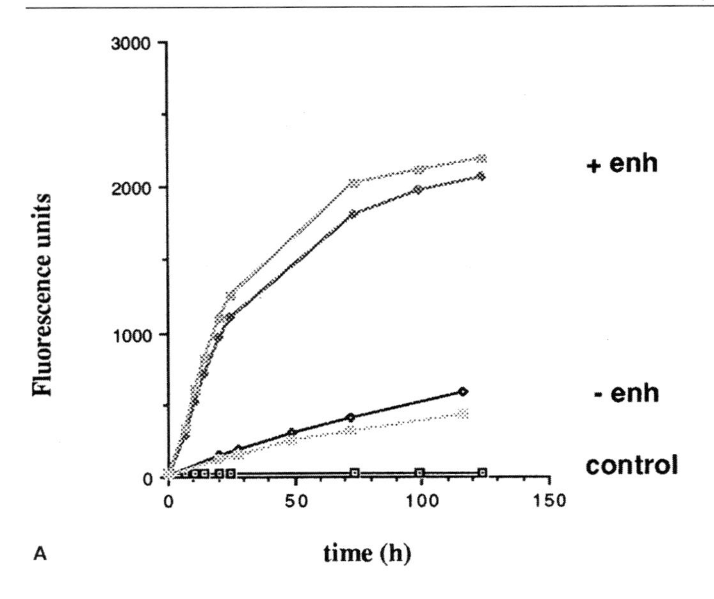
3000
2000
1000
0
Fluorescence units
+ enh
- enh
control
0
50
100
150
time (h)
A

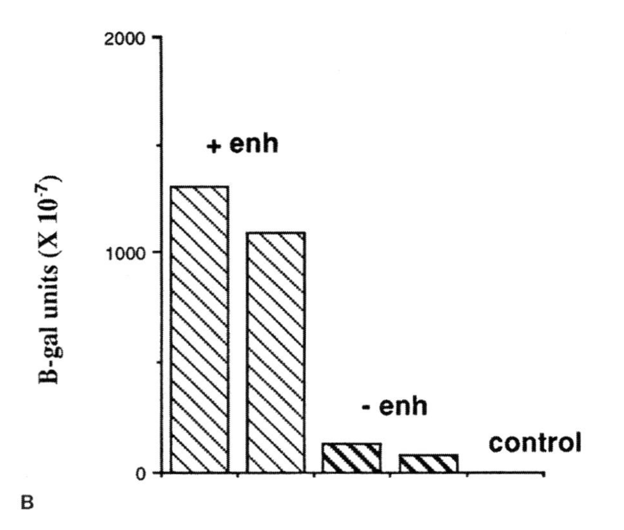
2000
1000
0
B-gal units (X 10^{-7})
+ enh
- enh
control
B

◀ **Fig. 21.4A, B.** MUG assay in single microinjected two-cell embryos. Two-cell embryos are microinjected into the nucleus of one blastomere with 1000–2000 copies of *LacZ* inserts either containing (+ enh) or lacking (– enh), a regulatory sequence tested for enhancer effect. Quantification of β-Gal activity is done with MUG 24 h following injection. **A** Kinetics of β-galactosidase reaction with MUG in individual embryos. Graph represents two representative kinetics for both inserts over a period of 120 h. The control embryo is not injected. Fluorescence is in arbitrary units. **B** β-Gal activity in single microinjected embryos β-Gal units produced by each embryo are calculated from the linear portion of the respective kinetic curve shown in **A** (see text for details of the calculation). Comparison of the mean of β-Gal units produced by several embryos microinjected with each insert will allow to quantify the enhancer effect of the tested sequence on expression

3. The hydrolysis of MUG by β-galactosidase produces 4-methylumbelliferone, whose fluorescence is measured by its emission at 480 mm (excitation at 355 mm) with a fluorimeter. During incubation at 37 °C, the kinetics of the reaction is determined by successively measuring methylumbelliferone fluorescence. For low values, readings are done once every 24 h for 120 h. For higher values, fluorescence is read frequently during the first few hours of the reaction and once every 24 h for the next 4 days (see below and Fig. 21.3).

Note: The MUG assay can also be applied to 1 μl of protein or cell extract in 100 μl of lysis buffer 0.4 mM MUG.

Results

The MUG assay was first used with a range of 10^{-2} to 10^{-7} units of pure β-galactosidase in order to describe the characteristics of the reaction. Results showed that the time period of reaction and the frequency of readings must be adjusted with respect to the quantity of enzyme assayed (Fig. 21.3). Results further revealed that β-galactosidase is stable for at least 120 h under the assay conditions used, and this permits reliable quantification and comparison over a large range of β-galactosidase activity (at least four orders of magnitude). The detection limit of this MUG assay is between 10^{-7} and 10^{-6} units of β-galactosidase (Fig. 21.3C) which corresponds to 5.10^5–5.10^6 molecules of enzyme. Under our assay conditions, one unit of pure enzyme hydrolyzes 0.136 μmol of MUG in 1 min, which corresponds to 7×10^3 fluorescence units.

Applications

An application for the MUG assay is illustrated in Fig. 21.4. We tested the effect of an enhancer regulatory sequence on the activity of a promoter linked to the *LacZ* gene in a transient expression assay, following microinjection of 1000–2000 copies of *LacZ* inserts in two-cell embryos. Figure 21.4A shows the kinetics of the MUG reaction in individual embryos, 24 h after microinjection. The different embryo samples are compared on the basis of the level of β-galactosidase activity (in β-Gal units) they produce (Fig. 21.4B). The results demonstrate the ability of the MUG assay to discriminate between relative levels of *LacZ* transcription in single embryos. Similar conclusions have been drawn from quantification of β-Gal activity with MUG in *LacZ* transgenic embryos (Forlani and Nicolas 1996b).

Note: A chemiluminescent assay also exists to quantify β-galactosidase activity in which a phenylgalactose-substituted dioxethane substrate is used (Lumi-Gal 530, Lumigen or Clontech Laboratories) (Beale et al. 1992). The sensitivity of this assay is similar to that of the MUG assay. Note that the quantification of β-galactosidase activity with a chemiluminescent assay has not yet been reported in single *LacZ* preimplantation embryos.

Acknowledgments. We thank Joan Shellard for careful reading of the manuscript, John Muschler for his permission to mention unpublished data, Françoise Kamel for help in typing the manuscript. JFN is from the Institut National de la Recherche Medicale.

References

Bachvarova R, Cohen EM, De Leon V, Tokunaga K, Sakiyama S, Paynton BV (1989) Amounts and modulation of actin mRNAs in mouse oocytes and embryos. Development 106:561–565

Beale EG, Deeb EA, Handley, RS, Akhavan-Tafti H, Schaap AP (1992) A rapid and simple chemiluminescent assay for *Escherichia coli* β-galactosidase. BioTechniques 12:320–323

Bensaude O, Babinet C, Morange M, Jacob F (1983) Heat shock proteins, first major products of zygotic gene activity in mouse embryo. Nature 305:331–333

Bevilacqua A, Kinnunen, LH, Bevilacqua S, Mangia F (1995) Stage-specific regulation of murine Hsp68 gene promoter in preimplantation mouse embryos. Dev Biol 170:467–478

Bolton VN, Oades PJ, Johnson MH (1984) The relationship between cleavage, DNA replication, and gene expression in the mouse two-cell embryo. J Embryol Exp Morphol 79:139–163

Bonnerot C, Vernet M, Grimber G, Briand P, Nicolas JF (1991) Transcriptional selectivity in early mouse embryos – a qualitative study. Nucleic Acids Res 19:7251–7257

Cavard C, Zider A, Vernet M, Bennoun M, Saragosti S, Grimber G, Briand P (1990) In vivo activation by ultraviolet rays of the human immunodeficiency virus type 1 long terminal repeat. J Clin Invest, Inc 86:1369–1374

Chaillet JR (1994) Genomic imprinting: lessons from mouse transgenes. Mutat Res 307:441–449

Chaillet JR, Bader DS, Leder P (1995) Regulation of genomic imprinting by gametic and embryonic processes. Genes Dev 9:1177–1187

Chatot CL, Ziomek CA, Bavister BD, Lewis JL, Torres I (1989) An improved culture medium supports development of random-bred one-cell mouse embryos in vitro. J Reprod Fertil 86:679–688

Christians E, Campion E, Thompson EM, Renard JP (1995) Expression of the *HSP 70.1* gene, a landmark of early zygotic activity in the mouse embryo, is restricted to the first burst of transcription. Development 121:113–122

Conover JC, Temeles GL, Zimmermann JW, Burke B, Schultz RM (1991) Stage-specific expression of a family of proteins that are major products of zygotic gene activation in the mouse embryo. Dev Biol 144:392–404

Cozzarelli NR (1977) The mechanisms of action of inhibitors of DNA synthesis. Ann Rev Biochem 46:461–468

Dooley TP, Miranda M, Jones NC, DePamphilis ML (1989) Transactivation of the adenovirus Ella promoter in the absence of adenovirus E1A protein is restricted to mouse oocytes and preimplantation embryos. Development 107:945–956

Engler P, Haasch D, Pinkert CA, Doglio L, Glymour M, Brinster R, Storb U (1991) A strain-specific modifier on mouse chromosome 4 controls the methylation of independent transgene loci. Cell 65:939–947

Flach G, Johnson MH, Braude PR, Taylor RAS, Bolton VN (1982) The transition form maternal to embryonic control in the two-cell mouse embryo. EMBO J 1:681–686

Forlani S, Nicolas J (1996a) Gene activity in the preimplantation mouse embryo. In: Transgenic animals: generation and use. Harward academic publishers gmbh. Editor Houdebine JM. Amsterdam pp 345–359

Forlani S, Nicolas J (1996b) Quantification of min levels of B-galactosidase. The example of individual 2- to 16-cell stage mouse embryos. Trends Genet 12:498–500

Gnirke A, Huxley C, Peterson K, Olson MV (1993) Microinjection of intact 200- to 500-kb fragments of YAC DNA into mammalian cells. Genomics 15:659–667

Heard E, Kress C, Mongelard F, Courtier B, Rougeulle C, Ashworth A, Vourc'h C, Babinet C, Avner P (1996) Transgenic mice carrying an *Xist*-containing YAC. Hum Mol Genet 5:441–450

Henery CC, Miranda M, Wiekowski M, Wilmut I, DePamphilis, ML (1995) Repression of gene expression at the beginning of mouse development. Dev Biol 169:448–460

Hoebeke J, Van Nigen G, De Brabander M (1976) Interaction of oncodazole (R17934), a new anti-tumoral drug, with rat brain tubulin. Biochem Biophys Res Commun 69:319–342

Hogan B, Costantini F, Lacy E (1986) Manipulating the mouse embryo. Cold Spring Harbor Laboratory Press, Cold Spring Habor, New York

Hoppe PC (1985) Technique of fertilization in vitro. In: Dixon RL (ed) Reproductive toxicology. Raven Press, New York, pp 191–199

Howlett SK, Bolton VN (1985) Sequence and regulation of morphological and molecular events during the first cell cycle of mouse embryogenesis. J Embryol Exp Morphol 87:175–206

Ikegami S, Taguchi T, Ohashi M, Oguro M, Nagano H, Mano Y (1978) Aphidicolin prevents mitotic cell division by interfering with the activity of DNA polymerase. Nature 275:458–460

Johnson MH, Nasr EM (1994) Radical solutions and cultural problems: could free oxygen radicals be responsible for the impaired development of preimplantation mammalian embryos in vitro? Bioessays 16:31–38

Kafri T, Ariel M, Brandeis M, Shemer R, Urven L, McCarrey J (1992) Developmental pattern of gene-specific DNA methylation in the mouse embryo and germ line. Genes Dev 6:705–714

Kafri T, Gao X, Razin A (1993) Mechanistic aspects of genome-wide demethylation in the preimplantation mouse embryo. Proc Natl Acad Sci USA 90:10558–10562

Kessler DS, Melton DA (1994) Vertebrate embryonic induction: mesodermal and neural patterning. Science 266:596–604

Kothary RK, Allen ND, Barton SC, Norris ML, Surani MAH (1992) Factors affecting cellular mosaicism in the expression of a lacZ transgene in two-cell stage mouse embryos. Biochem Cell Biol 70:1097–1105

Latham KE, Solter D (1991) Effect of egg composition on the developmental capacity of androgenetic mouse embryos [published erratum appears in Development 1992 Apr; 114(4):preceding table of contents]. Development 113:561–568

Latham KE, Garrels JI, Chang C, Solter D (1991) Quantitative analyzis of protein synthesis in mouse embryos. I. extensive reprogramming at the one- and two-cell stages. Development 112:921–932

Latham KE, Solter D, Schultz RM (1992) Acquisition of a transcriptionally permissive state during the one-cell stage of mouse embryogenesis. Dev Biol 149:457–462

Lawitts JA, Biggers JD (1993) Culture of preimplantation embryos. In: Methods in Enzymology: guide to techniques in mouse development, Wassarman PM, DePamphilis ML (eds) Academic Press, vol 225 pp 153–164

Levey IL, Brinster RL (1978) Effects of alpha-amanitin on RNA synthesis by mouse embryos in culture. J Exp Zool 203:351–360

Majumder S, Miranda M, DePamphilis ML (1993) Analysis of gene expression in mouse preimplantation embryos demonstrates that the primary role of enhancers is to relieve repression of promoters. EMBO J 12:1131–1140

Maniatis T, Fritsch E, Sambrook J (1989) Molecular cloning: a laboratory manual. Cold Spring Harbor Laboratory, Cold Spring Harbor, New York

Martinez-Salas E, Linney E, Hassell J, DePamphilis ML (1989) The need for enhancers in gene expression first appears during mouse development with formation of the zygotic nucleus. Genes Dev 3:1493–1506

McGrath J, Solter D (1984) Completion of mouse embryogenesis requires both the maternal and paternal genomes. Cell 37:179–183

Melanitou E, Simmler M C, Heard E, Rougeulle C, Avner P (1996) Selected methods related to the mouse as a model system. In: Kenneth WA (ed) Methods in molecular genetics: human molecular genetics. pp 439–469

Miranda M, Majumder S, Wiekowski M, DePamphilis ML (1993) Application of Firefly luciferase to preimplantation development. In: Methods in enzymology: guide to techniques in mouse development, Wassarman PM, DePamphilis ML (eds) Academic Press, vol 225 pp 412–433

Monk M, Boubelik M, Lehnert S (1987) Temporal and regional changes in DNA methylation in the embryonic, extraembryonic and germ cell lineages during mouse embryo development. Development 99:371–382

Paris D, Toyama K, Sinet PM, Kamoun P, London J (1995) A comparison between an inbred strain and hybrid lines to generate transgenic mice. Mouse Genome 93:1038–1040

Pavan WJ, Hieter P, Reeves RH (1990) Generation of deletion derivatives by targeted transformation of human-derived yeast artificial chromosomes. Proc Natl Acad Sci USA 87:1300–1304

Peterson KR, Clegg CH, Huxley C, Josephson BM, Haugen HS, Furukawa T, Stamatoyannopoulos G (1993) Transgenic mice containing a 248-kb yeast artificial chromosome carrying the human beta-globin locus display proper developmental control of human globin genes. Proc Natl Acad Sci USA 90:7593–8597

Peterson KR, Li QL, Clegg CH, Furukawa T, Navas PA, Norton EJ, Kimbrough TG, Stamatoyannopoulos G (1995) Use of yeast artificial chromosomes (YACs) in studies of mammalian development: production of beta-globin locus YAC mice carrying human globin developmental mutants. Proc Natl Acad Sci USA 92:5655–5659

Ram PT, Schultz RM (1993) Reporter gene expression in G2 of the one-cell mouse embryo. Dev Biol 156:552–556

Reik W, Howlett S, Surani M (1990) Imprinting by DNA methylation: from transgenes to endogenous sequences. Development Suppl 99–106

Reik W, Romer I, Barton SC, Surani MA, Howlett SK, Klose J (1993) Adult phenotype in the mouse can be affected by epigenetic events in the early embryo. Development 119:933–942

Rose M, Winston F, Hieter P (1990) Methods in yeast genetics: a course manual. Cold Spring Harbor Laboratory Press, Cold Spring Harbor, New York.

Rothstein J L, Johnson D, DeLoia J A, Skowronski J, Solter D, Knowles B (1992) Gene expression during preimplantation mouse development. Genes Dev 6:1190–1201

Ruiz i Altaba A, Melton D (1990) Axial patterning and the establishment of polarity in the frog embryo. Trends Genet 6:57–64

Sapienza C, Peterson AC, Rossant J, Balling R (1987) Degree of methylation of transgenes is dependent on gamete of origin. Nature 328:251–254

Sapienza C, Paquette J, Hang Tran T, Peterson A (1989) Epigenetic and genetic factors affect transgene methylation imprinting. Development 107:165–168

Schedl A, Montoliu L, Kelsey G, Schütz G (1993) A yeast artificial chromosome covering the tyrosinase gene confers copy number-dependent expression in transgenic mice. Nature 362:258–261

Schultz RM (1993) Regulation of zygotic gene activation in the mouse. BioEssays 15:531–538

St Johnston D, Nüsslein-Volhard C (1992) The origin of pattern and polarity in the *Drosophila* embryo. Cell 68:201–219

Surani MA, Barton SC, Norris ML (1986) Nuclear transplantation in the mouse: heritable differences between parental genomes after activation of the embryonic genome. Cell 45:127–136

Surani MA, Kothary R, Allen ND, Singh PB (1990) Genome imprinting and development in the mouse. Development Suppl 89–98

Swain JL, Stewart TA, Leder P (1987) Parental legacy determines methylation and expression of an autosomal transgene: a molecular mechanism for parental imprinting. Cell 50:719–727

Telford NA, Watson AJ, Schultz GA (1990) Transition from maternal to embryonic control in early mammalian development: a comparison of several species. Mol Reprod Dev 26:90–100

Thompson EM, Adenot P, Tsuji FI, Renard JP (1995) Real time imaging of transcriptional activity in live mouse preimplantation embryos using a secreted luciferase. Proc Natl Acad Sci USA 92:1317–1321

Thomson JA, Solter D (1988) The developmental fate of androgenetic, parthenogenetic, and gynogenetic cells in chimeric gastrulating mouse embryos. Genes Dev 2:1344–1351

Ueda T, Yamazaki K, Suzuki R, Fujimoto H, Sasaki H, Sakaki Y, Higashinakagawa T (1992) Parental methylation patterns of a transgenic locus in adult somatic tissues are imprinted during gametogenesis. Development 116:831–839

Vernet M, Bonnerot C, Briand P, Nicolas JF (1992) Changes in permissiveness for the expression of microinjected DNA during the first cleavages of mouse embryos. Mech Dev 36:129–139

Vernet M, Bonnerot C, Briand P, Nicolas J-F (1993a) Application of LacZ gene fusions to preimplantation development. Methods Enzyol 225:434–451

Vernet M, Cavard C, Zider A, Fergelot P, Grimber G, Briand P (1993b) In vitro manipulation of early mouse embryos induces HIV1-LTRlacZ transgene expression. Development 119:1293–1300

Weng A, Magnuson T, Storb U (1995) Strain-specific transgene methylation occurs early in mouse development and can be recapitulated in embryonic stem cells. Development 121:2853–2859

Wiekowski M, Miranda M, DePamphilis, ML (1991) Regulation of gene expression in preimplantation mouse embryos: effects of the zygotic clock and the first mitosis on promoter and enhancer activities. Dev Biol 147:403–414

Wiekowski M, Miranda M, DePamphilis ML (1993) Requirements for promoter activity in mouse oocytes and embryos distinguish paternal pronuclei from maternal and zygotic nuclei. Dev Biol 159:366–378

Wolffe A, Pruss D (1996) Targeting chromatin disruption: transcription regulators that acetylate histones. Cell 84:817–819

Yeom YI, Fuhrmann G, Ovitt CE, Brehm A, Ohbo K, Gross M, Hubner K, Scholer HR (1996) Germline regulatory element of Oct-4 specific for the totipotent cycle of embryonal cells. Development 122:881–884

Yoshida M, Horinouchi S, Beppu T (1995) Trichostatin A and trapoxin: novel chemical probes for the role of histone acetylation in chromatin structure and function. Bioessays 17:423–430

Chapter 22

Transitory Transgenic Analysis as an In Vivo System to Study Promoter Regulatory Elements

Diana Escalante-Alcalde[1] and Luis Covarrubias[1]*

Introduction

During development, gene expression is tightly regulated in time and space. The specific expression pattern of a particular gene results from the combination of several parameters, which include:

- The presence in its promoter of sequences binding specific *trans*-acting proteins
- The expression of these *trans*-acting proteins which, when bound to their target sequence, determine the activation or repression of transcription
- An adequate chromatin conformation that, in its open conformation, allows binding of transcription factors to their target sequences.

The RNA polymerase II complex binds to promoter sequences usually found immediately upstream of the gene coding region. However, initiation of transcription by this complex is strongly influenced by a concerted action of *cis*– and *trans*-regulatory factors. Although *cis*-regulatory elements are most frequently located 5' to the gene coding region, they can also be found in introns and even in the 3' sequences of the gene.

Transgenic mouse technology offers a powerful tool to study, in vivo, different molecular aspects of gene expression. One of them is the identification of regulatory sequences that determine temporal and tissue-specific gene expression. In transgenic mice, the recombinant gene is carried by all cells, and during the entire process of development. There-

* corresponding author: phone: 525-622-7636/7631; fax: 52-73-172388; e-mail: covs@ibt.unam.mx
[1] Departamento de Genética y Fisiología Molecular, Instituto de Biotecnología, Universidad Nacional Autónoma de México. Apdo, Postal 510-3, Cuernavaca, Morelos 62271, México

fore, evaluation of the influence of *cis*-acting DNA sequences in transgenic mice may be a definitive strategy to identify the elements determining the specific pattern of transcription in the developing animal.

To date, the most widely used method to generate transgenic mice has been the microinjection of recombinant DNA into the pronucleus of fertilized eggs. In order to identify the relevant *cis*-acting sequences in a given gene, its promoter and possible regulatory sequences to analyze are conveniently fused to a reporter gene (i.e., a gene encoding an innocuous and easily detectable protein) which contains the necessary transcription processing signals (Fig. 22.1). The temporal and tissue-specific pattern of reporter gene expression in the transgenic animal should be expected to reflect that specified by the elements directing transgene expression. However, an intrinsic limitation of this approach is the quasi-random insertion of transgenes into the host genome. Thus, it is frequently observed that endogenous *cis*-regulatory sequences influence transgene expression in an unpredictable way, which may obscure the elements under analysis. Therefore, reproducible expression patterns have to be obtained, by characterizing animals from three or more transgenic lines, in order to generate reliable data (see below).

The establishment of stable transgenic lines of mice and the analysis of transgene expression in their embryos requires several months (Fig. 22.1). Furthermore, it should be considered that, for a particular reporter gene construct, each transgenic line represents a single integration event and consequently a unique expression pattern. This time-consuming approach can be simplified by the production and direct analysis of transgenic embryos (i.e., transitory transgenic approach) (Fig. 22.1). In this case, the analysis is made on transgenic embryos developed in utero after pronuclear egg injection and transfer into the oviducts of pseudopregnant mice. This approach overcomes the necessity to generate and maintain different lines of transgenic mice and, since each transgenic embryo represents a distinct integration event, it provides a representative pattern of expression. Moreover, the expression of a particular promoter-reporter gene construct can be tested in several independent transgenic embryos in no longer than 20 days. A prerequisite of this approach is to optimize the technique in order to provide a high efficiency of transgenesis.

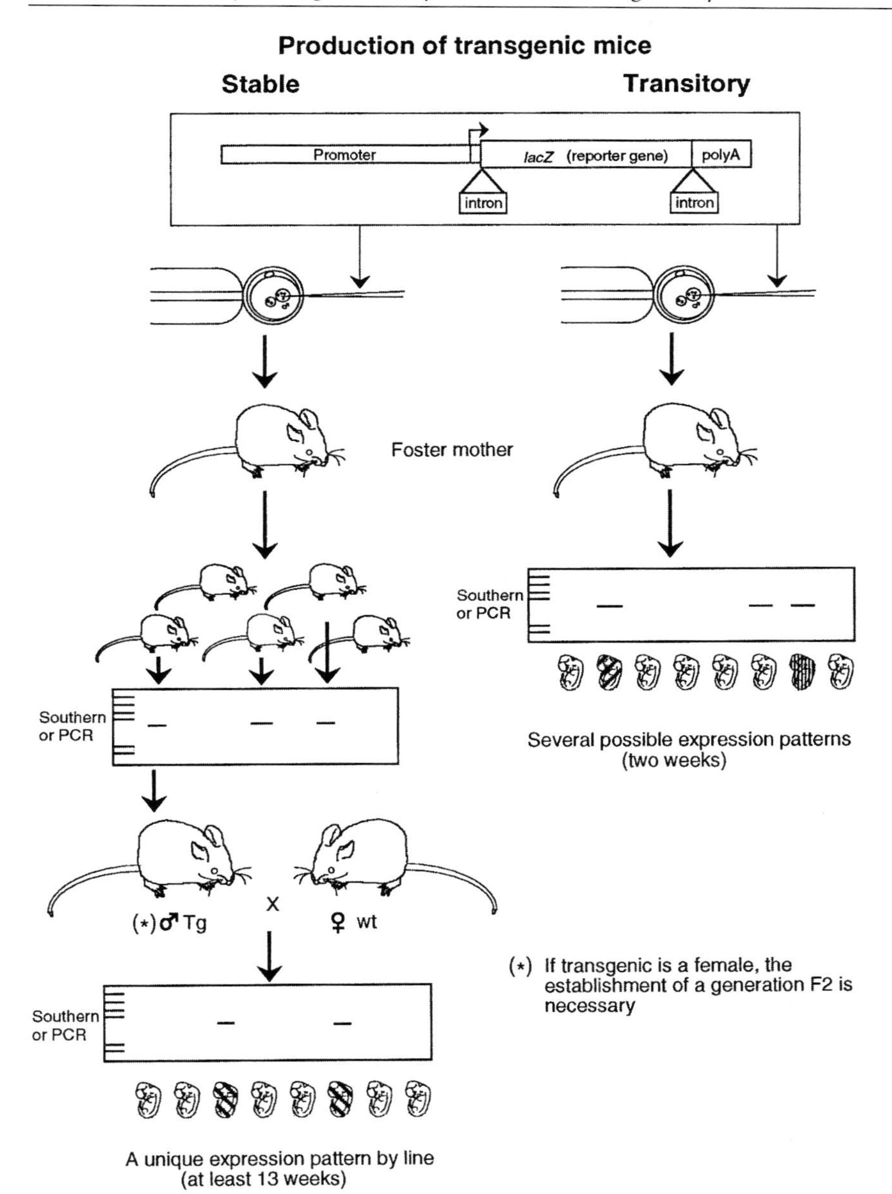

Fig. 22.1. Two alternative routes to analyze promoter sequences in transgenic mouse embryos. Transgenics could be derived from stable lines (*left*) or directly from injected eggs (*right*). In both strategies, the eggs are microinjected with promoter-reporter fusion constructs (*box*). Promoter fragments must contain the transcription start site (*bent arrow*), polyadenilation signals in the 3' end and generally splice donor/splice aceptor sequences (intron) before or after the reporter gene

Materials

Equipment

- Dissecting microscope and optic fiber illumination (Nikon, Japan)
- Inverted microscope with interference optic (Diaphot, Nikon, Japan)
- Antivibration table (i.e., marble) for manipulation
- Micropipette puller (P-97, Sutter Instrument, Co, USA)
- Microforge (Nikon, Japan)
- Microburner
- Two hydraulic micromanipulators (Narishige, Japan)
- Two robotic joysticks (Narishige, Japan)
- One syringe micrometer for micromanipulation suction system
- One 50-ml syringe for micromanipulation injection system
- Microcentrifuge.

Reagents and supplies

- Microinjection buffer (5 mM Tris-HCl , 0.1 mM EDTA, pH 7.5)
- M2 and M16 media (Hogan et al. 1994)
- Mineral oil (M8410, Sigma, USA)
- Avertin (Hogan et al. 1994; 24,048–6 and T4,840–2, Aldrich, USA)
- Hyaluronidase (H4272, Sigma, USA)
- Bovine serum albumin (A3311, Sigma, USA)
- Water for embryo transfer (W1503, Sigma, USA)
- High purity agarose (GIBCO/BRL, USA)
- Qiaex DNA purification kit (Qiagen, Inc., USA)
- Rubber tubing with mouthpiece (A7177, Sigma, USA)
- Watchmakers' forceps (no. 5, Fine Scientific Tools, USA)
- 5–0 silk thread
- Microcapillary tubing for holding pipettes (1B100–6, World Precision Instruments, Inc., USA)
- Microcapillary with filament for injection needles (TW100F-6, World Precision Instruments, Inc., USA)
- Pasteur pipettes.

Procedure

The first step in the analysis of a promoter region is to determine the pattern of expression of the gene under study. This can be done by in situ hybridization, immunofluorescence, or inmunohistochemistry and, in a few cases, by histochemistry (the example shown below). After preparing adequate fusion constructs containing the promoter sequences, a series of pronuclear egg microinjections are performed.

Purification of DNA Fragments

Higher gene expression and integration frequencies are obtained using lineal constructs, devoid of vector sequences, than with circular plasmids (see for review Hogan et al. 1994). The purity of DNA is perhaps the most important consideration to obtain high frequency of transgenesis. Purification of the DNA fragment to be used for microinjection can be achieved by different methods (e.g., zonal sucrose gradient centrifugation, electroelution, low melting agarose). In our experience, the use of chaotropic agents and activated silica gel particles (Qiaex, Qiagen Inc.) is the fastest procedure to obtain high integration frequencies (25–50 %). The protocol described below allows the recovery, with high yield, of DNA fragments of large size (i.e., 20–50 kb) without significant shearing.

1. 20–30 μg of plasmid construct are digested with the appropriate restriction enzymes to release the fragment for injection.
2. The fragment is separated by agarose gel electrophoresis. A 0.7 % agarose gel is recommended for fragments between 5 and 20 kb.
3. Agarose slice (up to 100 mg), containing the DNA fragment of interest, is loaded in a microcentrifuge tube and the Qiaex gel extraction protocol described by the manufacturer is followed.
4. The DNA fragment is eluted from the silica gel beads by incubating in the microinjection buffer for 2 min at 50 °C.

Production of Postimplantation Transgenic Embryos

The production of transgenic mice by pronuclear microinjection of eggs has been described extensively elsewhere (Gordon 1993; Mann and McMahon 1993; Hogan et al. 1994). Therefore, in this section we describe a brief stepwise protocol to obtain transgenic embryos.

1. Obtaining fertilized eggs. Fertilized eggs, surrounded by cumulus cells, are recovered from the ampulla of day 0.5 pregnant females. For the outbred CD-1 strain superovulation is not required.
2. Hyaluronidase treatment. To release eggs from the cumulus cells, they are incubated for 1 min at RT in M2 medium containing 300 μg/ml hyaluronidase. Somatic cells are removed by several washes in M2 medium. The eggs are placed in the microinjection chamber.

3. Pronuclear microinjection. In general, the male pronucleous is selected for injection because it is larger than the female one. We use DNAs which are diluted to a concentration of 2.5 μg/ml in microinjection buffer.

4. Transference. The eggs surviving microinjection (generally over 80 %) are transferred immediately to the oviduct of pseudopregnant mice (CD-1). Alternatively, the embryos may be transferred after overnight incubation in M16 medium, when they are at the two-cell stage, without any difference in the outcome. For best results, the number of eggs transferred to each oviduct should be around 15. About half of these are expected to develop normally (at least up to the E12.5 stage), and 25 to 50 % of them should be transgenics (i.e., 25–50 % transgenesis).

Analysis of Reporter Gene Expression

Among the different reporter genes which are available, those allowing in situ analysis are recommended. The bacterial β-galactosidase (*lacZ*) is the most widely used, because in situ detection of enzyme activity is very simple and the endogenous β-galactosidase activity is very low. Whole mount analysis using X-gal as substrate gives good results for embryos up to 15.5 days postcoitum (dpc). For later stages, organs or embryo fragments of interest should be dissected, as the diffusion of substrate is limited. Alternatively, staining can be performed in serial sections of frozen embryos.

1. After transfer to the foster mother, the embryos are left to develop until the stage of choice, and then the mother sacrificed and the uterus dissected.

2. Embryos are fixed separately with 4 % paraformaldehyde (in PBS), and permeabilized with a 0.02 % NP-40 solution (in PBS).

3. Staining is performed with 1 mg/ml of X-Gal and 5 mM potassium ferricyanide/ferrocyanide in PBS. We recommend incubating the embryos immersed in staining solution overnight at 30 °C to avoid significant background. In some cases, in which reporter expression is strong, 4 h of incubation may be enough to see the blue color of the X-Gal staining.

4. Embryos carrying the transgene are identified by extracting DNA from a fragment of the placenta or from the yolk sac of each embryo,

and analyzing it by Southern or dot blot hybridization and/or by PCR using labeled specific probes or the appropriate oligonucleotides, respectively. These techniques are described in detail in molecular biology manuals (see, for instance Sambrook et al. 1989).

Results

Cis-acting promoter/enhancer elements are in general very complex. Therefore, the interpretation of results obtained from the analysis of transgenic animals will be more or less difficult, depending on the complexity of the promoter being studied. In general, if at least two transgenic embryos are obtained that exhibit the expected pattern of expression, it can be considered as functional in terms of promoter specificity, but insufficient for a definitive conclusion. Only more than two lines with the expected pattern can be taken as conclusive.

As an example of the use of transitory transgenic mice, in this section we show the results that we obtained with promoter fragments of the tissue non-specific alkaline phosphatase gene (*TNAP*; Terao et al. 1990; Studer et al. 1991; Escalante-Alcalde et al. 1996). At 12.5 dpc, *TNAP* is expressed in the primordial germ cells, the neural tube, the apical ectodermal ridge of limbs, the stomach and lung epithelia, and muzzle. Although this gene has two promoters, during the embryonic life, transcription takes place only from the exon 1A (E1A) promoter (Escalante-Alcalde et al. 1996). Table 22.1 and Fig. 22.2 show the expression patterns obtained with several reporter constructs containing the E1A promoter of *TNAP*.

With a construct containing the foremost 1.9 kb of E1A promoter, the influence of sequences flanking the integration site is very evident. Thus, the expression patterns observed for the reporter were diverse and not coincident with the *TNAP* expression pattern (Table 22.1; Fig. 22.2A–C). Regions of tissue-specific expression (apical ectodermal ridge and stomach and lung epithelium) start to be seen with constructs containing 4.3 and 5.5 kb of sequences immediately upstream E1A transcription start site (Table 22.1; Fig. 22.2D,E). However, it should be noted that the expected pattern of expression appeared infrequently (i.e., although reproducible, only a few embryos showed such expression pattern; Table 22.1), and overlapped to an unspecific expression pattern, probably due to regulatory elements around the integration site. Interestingly, the combination of proximal (up to −1.9 kb) and distal elements (between −5.5 and −4.5 kb) resulted in reproducibility of the specific expression in

Table 22.1 Reporter gene expression in 12.5 dpc transgenic embryos

	pAP/ 1A1.9	pAP/ 1A4.3	pAP/ 1A5.5	pAP/ 1A8.5	pAP 1A5.5Δ 2.6	pAP/ 1A8.5Δ 1.2
Primordial germ cells[a]	–/3	–/4	–/2	–/7	–/9	–/6
Stomach epithelium[a]	–/3	1/4	1/2	1/7	**4/9**	–/6
Lung epithelia[a]	–/3	1/4	1/2	1/7	**4/9**	–/6
Intestine[a]	–/3	–/4	–/2	–/7	2/9	–/6
Neural tube[a]	–/3	–/4	–/2	–/7	1/9	1/6
Apical ectodermal ridge[a]	–/3	1/4	–/2	–/7	–/9	1/6
Mesonephric ducts[b]	–/3	–/4	1/2	–/7	–/9	–/6
Developing skeletal elements[b]	–/3	1/4	1/2	–/7	2/9	–/6
None	1/3	–/4	1/2	6/7	3/9	3/6

[a] Tissues that express *TNAP*. [b] Tissues that do not express *TNAP*. The number in each construct corresponds to the size of the promoter in kb. Internal deletions are indicated by a Δ followed by the size in kb of the fragment deleted.

lung and stomach epithelia, but yet included within an unspecific expression pattern (Table 22.1, numbers in bold; Fig. 22.2G–I). Note that either the 1.9 kb or the 4.3 kb fragments proximal to the transcription start alone were not able to efficiently direct specific expression. Element(s) directing the specific expression to primordial germ cells seem not to be present in the promoter fragment analyzed (until –8.5 kb).

Although less frequently reported, transgenic embryo analyses can also be applied to uncover the presence of elements regulating negatively gene expression. This can clearly be seen in the results shown for the construct containing 8.5 kb of *TNAP* E1A promoter. Here only one out of seven individual embryos expressed the reporter gene (Table 22.1, underlined numbers). This observation led us to determine that promoter E1A contains a *cis*-methylation regulatory elements which act negatively on *TNAP* expression (Escalante-Alcalde et al. 1996). In this case the negative effect was observed in all developing tissues, but tissue-specific negative elements such as silencers have also being identified in transgenic mice (Wuenschell et al. 1990).

Troubleshooting

As mentioned above, when analyzing promoter regulatory sequences in transgenic embryos, it is a prerequisite to obtain high efficiency of transgenesis (in the range of 25–50 %). Low integration frequencies are obtained if microinjection is not performed properly. Microinjection of very small volumes of DNA solution is the cause of low efficiency in transgenesis. However, care should be taken, as microinjection of an excessive volume would result in pronuclear rupture. Thus, an optimal pronuclear swelling is important for both egg survival and DNA integration frequency. To determine egg damage due to microinjection, we recommend evaluating egg survival after an overnight culture, as this relates directly to the number of recovered embryos. Additional relevant parameters are DNA purity and its concentration, composition of injection buffer, and mouse strain.

Weak expression driven by tissue-specific elements could be limiting for their identification. To overcome this problem, one possibility is to use alternative reporter genes. However, although several reporter genes have been used to date (e.g., chloramphenicol acetyltransferase and luciferase genes), direct in situ detection of their expression is not possible. Recently, the gene encoding the green fluorescent protein (GFP) has been introduced as a reporter. The GFP gene should allow direct confocal analysis of its expression in the intact embryo (up to 9.5 dpc) or in dissected tissues.

If specific expression of the reporter gene is not observed, it could be due to the fact that important regulatory elements are missing in the sequence analyzed. Therefore, it should be considered that regulatory elements are distributed along the entire sequence of the gene, and in some cases covering long distances. Then, analysis of regulatory elements using transgenic embryos should be first tried using the largest possible fragment, comprising 5' and 3' flanking sequences as extended as possible. However, there is an inverse correlation between the length of the injected fragment and integration frequency. This effect starts to be obvious with fragments larger than 20 kb. To solve this limitation, it is possible to coinject overlapping fragments and select for those where homologous recombination has occurred, as has been reported previously (Zambrowicz et al. 1994). Alternatively, it is now possible to introduce very large fragments into embryos using YAC constructs (Schedl et al. 1993; see also Hiemisch et al., Chapt. 17, this Vol.).

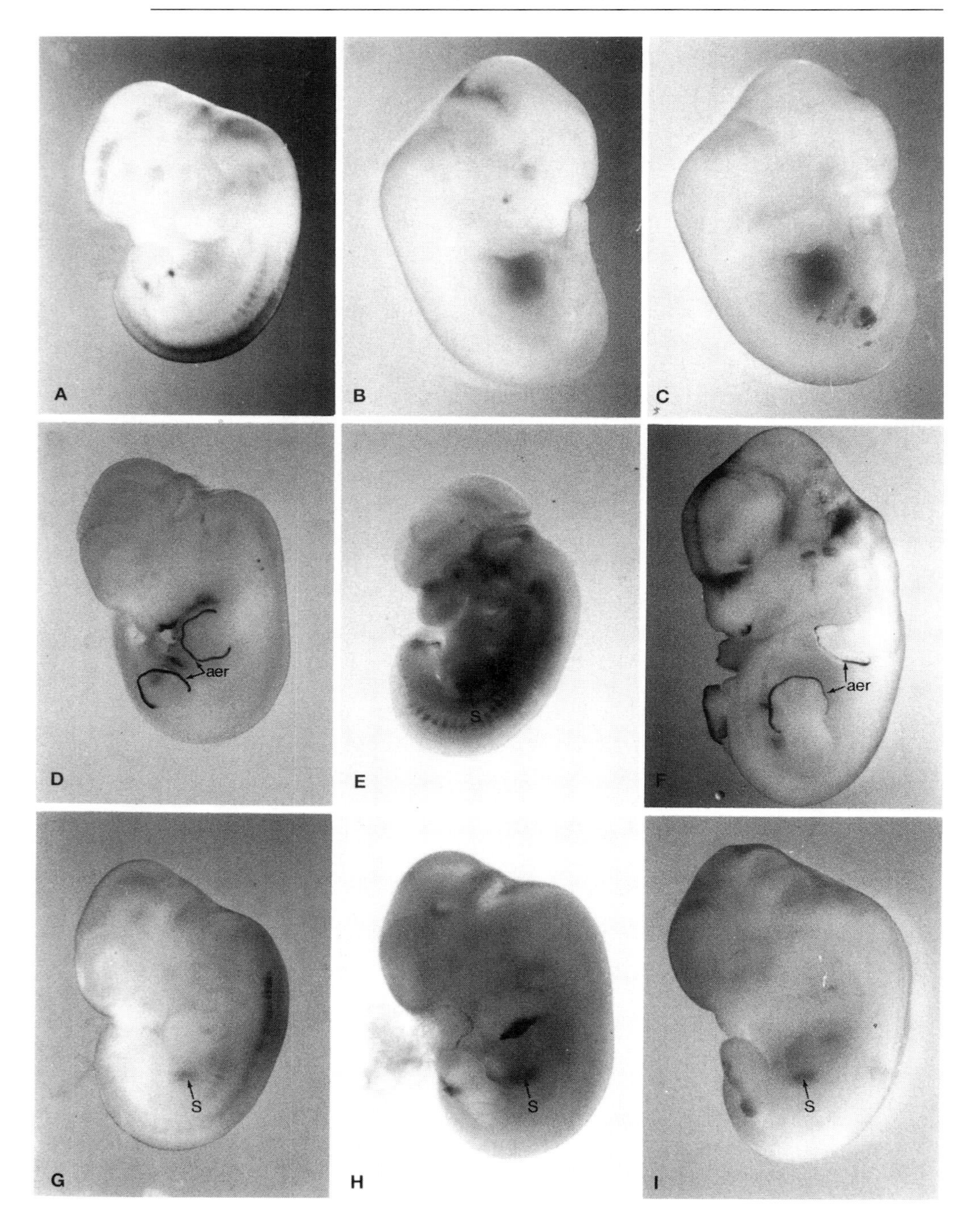
A
B
C
aer
D
S
E
aer
aer
F
S
G
S
H
S
I

◀ **Fig. 22.2A–I.** Patterns of β-galactosidase expression observed in 12.5 dpc transgenic embryos carrying *TNAP* E1A promoter-*lacZ* constructs. (**A–C**) Transgenic embryos bearing the construct pAP/1A1.9. Independent lines of transgenic mice with this construct show different expression patterns: regions of the cefalic and dorsal areas, and in the base of limbs (**A**); a spot in the lateral region of the face and a difuse pattern in the posterior region of the head (**B**); no expression in any tissue (**C**). The dark appearance of the liver in **B** and **C** is due to its high blood content. **D–I** Transgenic embryos showing reporter gene expression in some specific tissues. Transgenic mouse embryos bearing constructs pAP/1A4.3 (**D**), and pAP/1A8.5Δ1.2 (**F**), show expression in the apical ectodermal ridge (*aer*), whereas pAP/1A5.5 (**E**) and pAP/1A5.5Δ2.6 (**G-I**) show strong X-Gal staining in the stomach (**S**) and lung epithelia (not shown). Expression in the distal tip of muzzle is observed in a trangenic embryo bearing the construct pAP/1A4.3 (**D**). Note, however, that unspecific expression is also observed in the transgenic embryos such as: in several skeletal elements (**E**), regions of the head (**F**), some paraxial elements (**G**), the posterior region of limbs (**H**) and some spots in the base of limbs (**I**)

Comments

One of the major problems when the activity of a promoter is being studied in transgenic mice is the noise caused by ectopic transgene expression. This problem is mainly due to the regulatory elements around the integration site; therefore, isolation of transgenes from the influence of endogenous sequences would be desirable. This can be achieved by flanking the transgene with elements that prevent the transmission of chromatin structural features associated with repressed or active chromosomal domains. These elements, known as insulators, define the limits of transcriptionally active chromatin, protecting it against the repressive influence of neighboring heterochromatin (Chung et al. 1993; Wolffe 1994). Then, insulators might have important practical implications for the efficiency and accuracy of transgene expression. However, this approach would not be suitable to study sequences which promote changes in chromatin structure such as some elements within long control regions (LCRs) that regulate certain loci (Festenstein et al. 1996).

Ideally, *cis*-regulatory elements should be studied in their natural chromosomal location. This is now possible by targeting mutations to specific elements by homologous recombination in embryonic stem cells. These cells can be used to produce chimeric and, finally, the mutant mice (for details on the gene targeting technology, see for example Ramírez-Solis et al. 1993). The gene targeting approach has the advantage that independent regulatory elements can be studied in a context of integrity, where all remaining regulatory elements and the appropriate chromosomal environment are preserved (Fiering et al. 1993, 1995).

Acknowledgments. This work was supported by The Program of United Nations for Development (PNUD/MEX/93/019), CONACyT (1663M9209), and PAPIIT/UNAM (IN-201991 and IN-206194).

References

Chung JH, Whiteley M, Felsenfeld G (1993) A 5' element of the chicken β-globin domain serves as an insulator in human erythroid cells and protects against position effect in *Drosophila*. Cell 74:505–514

Escalante-Alcalde D, Recillas-Targa F. Hernández-García D, Castro-Obregón S, Terao M, Garattini E, Covarrubias L (1996) Retinoic acid and *cis*-methylation regulatory elements control the mouse tissue non-specific alkaline phosphatase gene expression. Mech Dev 57:21–32

Festenstein R, Tolain I N, Corbella P, Mamalaki C, Parrington J, Fox M, Miliou A, Jones M, Kioussis D (1996) Locus control region function and heterochromatin-induced position effect variegation. Science 271: 1123–1125

Fiering S, Kim CG, Epner E, Groudine M (1993) An "in-out" strategy using gene targeting and FLP recombinase for the functional dissection of complex DNA regulatory elements: Analysis of the β-globin locus control region. Proc Natl Acad Sci USA 90:8469–8473

Fiering S, Epner E, Robinson K, Zhuang Y, Telling A, Hu M, Martin DJK, Enver T, Levy TJ, Groudine M (1995) Targeted deletion of 5' HS2 of the murine β-globin LCR reveals that it is not essential for proper regulation of the β-globin locus. Genes Dev 9:2203–2213

Gordon J (1993) Production of transgenic mice. Methods Enzymol 225:747–771

Hogan B, Beddington R, Costantini F, Lacy E (1994) Manipulating the mouse embryo: a laboratory manual, 2nd edn. Cold Spring Harbor Laboratory Press, Cold Spring Harbor

Mann JR, McMahon AP (1993) Factors influencing frequency production of transgenic mice. Methods Enzymol 225:771–781

Ramírez-Solis R, Davis AC, Bradley A (1993) Gene targeting in embryonic stem cells. Methods Enzymol 225:855–878

Sambrook J, Fritsch EF, Maniatis T (1989) Molecular cloning: a laboratory manual, 2nd edn. Cold Spring Harbor Laboratory Press, Cold Spring Harbor

Schedl A, Larin Z, Montoliu L, Thies E, Kelsey G, Lehrach H, Schutz G (1993) A method for the generation of YAC transgenic mice by pronuclear microinjection. Nucleic Acids Res 21:4783–4787

Studer M, Terao M, Gianní M, Garttini E (1991) Characterization of a second promoter for the mouse liver/bone/kidney-type alkaline phosphatase gene: cell and tissue specific expression. Biochem Biophys Res Commun 179:1352–1360

Terao M, Studer M, Gianní M, Garattini E (1990) Isolation and characterization of the mouse liver/bone/kidney-type alkaline phosphatase gene. Biochem J 268:641–648

Wolfe AP (1994) Insulating chromatin. Curr Biol 4:85–87

Wuenschell CW, Mori M, Anderson DJ (1990) Analysis of SCG10 gene expression in transgenic mice reveals that neural specificity is achieved through selective derepression. Neuron 4:595–602

Zambrowicz BP, Zimmermann JW, Harendza CJ, Simpson EM, Page DC, Brinster RL, Palmiter RD (1994) Expression of a mouse *Zfy-1/lacZ* transgene in the somatic cells of embryonic gonad and germ cells of the adult testis. Development 120: 1549–1559

Chapter 23

Transgenic Analysis of Embryonic Gene Expression Using LacZ as a Reporter

Catherine E. Ovitt[1*], Young Il Yeom[2], and Hans R. Schöler[1]

Introduction

Determining the expression pattern of a gene is often the first step in defining its function. The analysis of expression during mammalian embryonic development can present a challenge if the gene of interest is expressed for only a short interval of time, or in a small number of cells. In situ hybridization and immunological staining are the established methods by which expression analysis is usually accomplished, but both are laborious, require a purified probe for optimal detection, and often require sectioning of the embryo. An alternative method for determination of the expression pattern involves the use of a reporter to mark the gene of interest. Detection of the reporter protein itself, or of its enzymatic activity following histochemical staining, provides a way to trace expression of the marked gene. The bacterial lacZ gene, which encodes the enzyme β-galactosidase (Wallenfels and Malhotra 1960), has proved to be very useful as a reporter gene. β-galactosidase cleaves its substrate 5-bromo-4-chloro-3-indolyl-β-D-galactoside (X-Gal) to yield a blue reaction product which precipitates in the cell. β-galactosidase is well suited as a reporter for a number of reasons including:

- Its rapid and simple enzymatic assay
- Its ability to be assayed and detected in situ
- Its cell autonomy, allowing detection in single cells
- Its sensitivity as a marker of gene activity.

* corresponding author: phone: +49-6221-387557; fax: +49-6221-387306; e-mail: ovitt@embl-heidelberg.de
[1] European Molecular Biology Laboratory, Meyerhofstrasse 1, 69117 Heidelberg, Germany
[2] Genetic Engineering Research Institute, Korea Institute of Bioscience and Biotechnology (KIST), PO Box 17, Taeduk Science Town, Taejon 305-606, Korea

The ease with which this marker can be used is exemplified by the number of different strategies in which it is employed to study gene expression during mouse embryonic development. By integrating the lacZ gene into an endogenous gene locus through homologous recombination, the expression pattern of the targeted gene can be followed utilizing intact regulatory sequences in their native location (LeMouellic et al. 1990). Functional analysis and identification of *cis*-acting regulatory elements is often done by linking them to lacZ and analyzing the lacZ transgene expression following microinjection into oocytes. LacZ has also proved to be an invaluable tool in fusion transgenes which have been designed to act as gene- or enhancer traps when randomly integrated into the genome (reviewed by Joyner 1991). Finally, a method for clonal analysis in embryos has been devised in which single cells are marked by the introduction of a defective lacZ-containing retrovirus (reviewed by Sanes 1994).

Basic Protocol and Notes

The use of lacZ as a reporter in a transgenic approach permits the analysis of whole embryos from preimplantation stages through E13.5, and of dissected embryos after this stage, thus providing a simple method for the dynamic visualization of a gene expression pattern through development. A caveat, however, when using lacZ as a reporter gene is the stability of its protein product. Because the half-life of the lacZ protein may be considerably longer than that of the gene product under study, the kinetics of observed changes in gene expression may be slower.

The activity of a lacZ fusion gene can be determined in either transient or stable transgenic lines. Often, confirmation of the ability of the transgene to express lacZ is most rapidly achieved through a transient approach, i.e., direct analysis of F_0 mice. The use of transient transgenics is also convenient when analyzing expression in preimplantation embryos, due to the ease with which the embryos can be maintained in vitro, and their ability to undergo normal differentiation in culture. These features permit rapid screening of large numbers of early embryos for expression of the transgene.

In this chapter we describe the standard protocols used for the analysis of lacZ expression in both pre- and postimplantation mouse embryos. Basic protocols are given for the isolation, fixation, and staining of embryos with X-Gal, as well as for the culture of preimplantation embryos. In order to observe staining of internal organs in the whole embryo, we have adapted a method for clearing the embryos, and have introduced a postfixation step to prevent fading of the stain during the clearing process.

LacZ Reporter Transgene Constructs

LacZ constructs are usually generated from parent plasmids such as pCH110 and pMC1871 (both available from Pharmacia). A modified form of the lacZ gene containing a nuclear localization signal is also available and gives distinct staining of the nucleus, which may be advantageous in some cell types (Bonnerot et al. 1987). To optimize translation initiation, the prokaryotic intiation codon is often replaced by an oligonucleotide containing a consensus Kozak translation intiation sequence. Alternatively, an in-frame fusion links the lacZ coding sequence to the amino terminal portion of the gene of interest. In our experience, it is best to keep the length of an amino terminal fusion at a minimum to ensure full activity of the β-Galactosidase protein (Y. Yeom and H. Schöler, unpubl. observations). A second alternative involves the introduction of an internal ribosome entry site (IRES) at the 5' end of the lacZ gene, eliminating problems associated with fusion proteins (reviewed by Mountford and Smith 1995). Transcriptional termination of the transgene is dependent on the endogenous 3' sequences of the gene, or more frequently, on the SV40 polyadenylation signal included in the pCH110 plasmid. It should be noted that truncation of the lacZ gene at the carboxyl terminal end will result in an inactive gene product.

Materials

CZB culture media for preimplantation embryos

Final concentration	Stock solution	Storage (°C)
82 mM NaCl	1.0 M	4
4.86 mM KCl	0.1 M	4
1.17 mM KH_2PO_4	0.1 M	4
1.18 mM $MgSO_4$	0.1 M	4
30.1 mM sodium lactate	0.1 M	4
0.26 mM sodium pyruvate	0.1 M	4
5.0 mg/ml BSA	100 mg/ml	4
25 mM $NaHCO_3$	1.0 M	4
1.71 mM $CaCl_2$	0.171 M	4
1.0 mM glutamine	0.1 M	–20
0.1 mM EDTA	0.1 M	4
100 units/ml penicillin	10,000 units/ml	–20
50 μg/ml streptomycin sulfate	5.0 mg/ml	–20

Adjust pH to 7.4 with 0.2 N NaOH.

The culture media are made with sterile water (Sigma). All other components are cell culture grade. The solutions are stored at 4 °C and used within 1–2 weeks.

Solutions

- Fixing solution

Volume to make 5 ml	Stock solution	Final concentration
0.135 ml	37 % formaldehyde (Merck)	1 % formaldehyde
0.04 ml	25 % glutaraldehyde (Sigma)	0.2 % glutaraldehyde
0.01 ml	10 % Nonidet P-40	0.02 % NP-40
4.815 ml	1× PBS	

When fixing preimplantation embryos, add 1 % (v/v) serum to the fixing solution.

Store fixative solution at 4 °C and use within 48 h.

Stock solutions for X-Gal staining

- X-Gal: 40 mg/ml 5-bromo-4-chloro-3-indolyl-β-galactoside (Sigma) in DMSO; store at −20 °C in aliquots until use.
- $K_3Fe(CN)_6$: 500 mM potassium ferricyanide in water; store at room temperature in the dark.
- $K_4Fe(CN)_6.3H_2O$: 500 mM potassium ferrocyanide in water; store at room temperature in the dark.
- X-Gal staining solution

Volume to make 10 ml	Stock solution	Final concentration
9.35 ml	1× PBS	
0.25 ml	X-Gal (40 mg/ml)	1 mg/ml
0.1 ml	500 mM $K_3Fe(CN)_6$	5 mM
0.1 ml	500 mM $K_4Fe(CN)_6$	5 mM
0.2 ml	0.1 M $MgCl_2$	2 mM

It is important to mix the components in the order given, and to keep the solution at room temperature to avoid the formation of a precipitate. X-Gal staining solution should be made fresh just before use.

- 4 % Paraformaldehyde

  Add 4 g paraformaldehyde to 90 ml water. Add 100 μl of 1 N NaOH
  Incubate at 37 °C with shaking for 1–2 h

Adjust pH to 7 with 1 M HCl
Add 5 ml of 20× PBS
Adjust final volume to 100 ml with water
Store at 4 °C and use within 2 days
Aliquots may also be stored at −20 °C.

- BABB clearing solution

Mix 2 vol benzyl benzoate (Merck) to 1 vol benzyl alcohol (Merck) just before use.

Note: This solution dissolves plastic, and should be mixed and used in glassware only.

Procedure

Detection of the Integrated Transgene

Screeening for the presence of the transgene in embryos is done on DNA isolated from extra-embryonic membranes. The method of DNA preparation is described in Murphy and Ruether Chapter 15, this Volume. Screening is performed by either PCR or Southern blot analysis. Southern blot analysis is carried out as described in Sambrook et al. (1989). We routinely use an internal 617 bp BstXI fragment (nucleotides 2232 to 2849) of the lacZ coding sequence as a probe. The methods employed for PCR screening of DNA, as well as the primers used to detect lacZ, are described by Hanley and Merlie (1991). Screening of transient transgenic preimplantation embryos is only feasible by X-Gal staining.

Analysis of LacZ Expression in Transient Transgenic Preimplantation Embryos

Following microinjection of the lacZ transgene, the embryos are maintained in culture, where they continue to develop up to the blastocyst stage. Several different media have been demonstrated to permit in vitro development of mouse embryos (reviewed by Lawitts and Biggers 1993). We routinely use CZB media described by Chatot et al. (1990).

Embryo culture

1. Isolate fertilized oocytes at 0.5 dpc and microinject the transgene DNA using standard procedures described elsewhere in this manual.
2. Prepare the culture dishes by placing drops (20–40 µl) of CZB media on a 35-mm tissue culture dish, overlay with light paraffin oil (Mallinckrodt) and preequilbrate at 37 °C in a 5 % CO_2 incubator.

3. Transfer embryos by mouth pipetting to the drops, placing only five to ten embryos per drop to prevent aggregation. Incubate at 37 °C in 5 % CO_2 for up to 3 days.
   Note: Embryos cultured in this manner normally generate a mixed population randomly arrested at different stages of preimplantation development.

Fixing embryos

4. Transfer the embryos from the culture medium by mouth pipetting using a pulled micropipette prefilled with 1× PBS containing 1 % serum (to reduce sticking of the embryos to the pipette and each other). Place five to ten embryos into single wells of a 24-well tissue culture plate containing 1× PBS with 1 % serum.
5. Transfer the embryos in the same manner to a second well containing 1 ml of fixing solution and 1 % serum. Incubate at 4 °C for no longer than 5 min.
6. Wash the embryos twice by transferring 2 times into wells containing 1× PBS, 1 % serum at room temperature for 5 min each.

Staining embryos

7. Place freshly prepared staining solution into wells of the 24-well plate and transfer the embryos as above.
8. Cover the plate and incubate overnight (up to 2 days) at 30 °C.
9. Wash the embryos by transferring to a well containing 1× PBS, 1 % serum at room temperature.
10. Photography of preimplantation embryos is usually done using an inverted microscope. The embryos are easily visualized if transferred in 1× PBS to a 35-mm tissue culture dish. We use Kodak tungsten film 64T for bright- and dark-field images.
11. The embryos may be stored in 1× PBS at 4 °C.

Analysis of LacZ Expression in Transgenic Postimplantation Embryos

The analysis of postimplantation embryos is usually done using timed pregnancies, with day of the vaginal plug designated as the day 0.5 dpc. Detailed procedures for the isolation of early postimplantation embryos are given in Hogan et al. (1994).

Collecting post-implantation embryos

1. Sacrifice the pregnant female on the appropriate day of gestation by cervical dislocation.
2. Isolate and place the uterus in a 10-cm petri dish containing cold 1× PBS on ice.
3. The individual conceptuses should be easily distinguishable. Cut away a section containing an individual embryo (still enclosed in the decidua) and place in a second dish of 1× PBS for isolation.
4. Remove the uterine wall lining and discard. Separate the placenta and extraembryonic membranes from the embryo, place in a marked 1.5 ml Eppendorf tube and set aside for DNA extraction.
5. Dissect and transfer the embryo to a well of a 24-well tissue culture plate containing 1× PBS. Embryos up to E8 can be transferred using a micropipette. Embryos older than E8 are best transferred using the wide end of a Pasteur pipette with suction from a rubber bulb. For embryos E12 and older, very gentle handling with forceps is easiest.
6. When analyzing embryos older than E12, a sagittal cut through the midline using a scalpel blade will allow more efficient fixation and staining of internal organs.

Fixing embryos

7. When all the embryos have been collected, replace the 1× PBS with 1–2 ml of cold fixative solution. Incubate the plate at 4 °C for the length of time indicated: E3.5–4.5, 5 min; E5.5–7.5, 15 min; E8.0–E10, 30 min; E10.5–E11, 45 min; E11.5–E12.5, 60 min; E13 and older, 120 min.
8. Remove the fixative solution and wash the embryos twice with 1× PBS for 20 min each at room temperature. Washes of E13 and older embryos should be extended to 30 min each.

Staining embryos

9. Prepare the staining solution and keep the mixture at room temperature to avoid precipitation. When analyzing embryos E12.5 and older, add 0.3 % Triton X-100 to the staining solution to improve penetration of the stain.
10. Remove the 1× PBS and add 2 ml of staining solution to each well containing an embryo. Incubate the plate in the dark for 1–2 days at 30 °C.
11. After staining, remove the staining solution and wash the embryos twice for 20–30 min with 1× PBS. The embryos may be stored at 4 °C in 1× PBS after staining.

Postfixation of embryos

12. Replace the 1× PBS with 1–2 ml of 4 % paraformaldehyde. Incubate embryos up to E11.5 at 4 °C for 30 min; 1 h for E12 and older embryos.
13. Rinse the embryos twice in 1× PBS for 30 min each at room temperature.

Clearing embryos

14. Dehydrate the embryos serially by incubating sequentially in 25, 50, 70, 80 and, finally, 100 % ethanol. The length of incubation for each step is the same as the fixation times given in step **7** above.
15. **Caution:** The components of the clearing solution dissolve plastic! Transfer the embryos to a glass petri dish in 100 % ethanol. Replace the ethanol with a 1:1 solution of 100 % ethanol:BABB. Allow the embryos to equilibrate for 10 min at room temperature.
16. Replace with the BABB solution. **Caution:** This solution corrodes metal forceps and dissolves plastic. Transfer and arrange embryos using glass pipettes.
17. Photograph embryos as soon as possible after clearing. We use either 64T tungsten Kodak film or Provia 100 Fujichrome for both bright and dark field exposures.
18. The embryos can be stored in glass vials in BABB solution, but fading of the β-galactosidase staining will occur after several weeks.

Results

Using lacZ as a reporter in a transgenic analysis, we have obtained a detailed picture of the embryonic expression pattern and regulation of the transcription factor Oct-4 (Yeom et al. 1996). Initially, a genomic fragment of the Oct-4 gene was used to drive the expression of lacZ and was found to reproduce the endogenous expression pattern of the gene. Once the construct was shown to recapitulate the correct expression pattern, it was truncated and dissected in order to identify and analyze regulatory elements responsible for the germ-cell-specific expression of Oct-4. Using the techniques described in this chapter, we identified a unique enhancer element which drives expression of the lacZ transgene exclusively in the pluripotent cells of the preimplantation embryo, and later in the migrating primordial germ cells (see Fig. 23.1A,B).

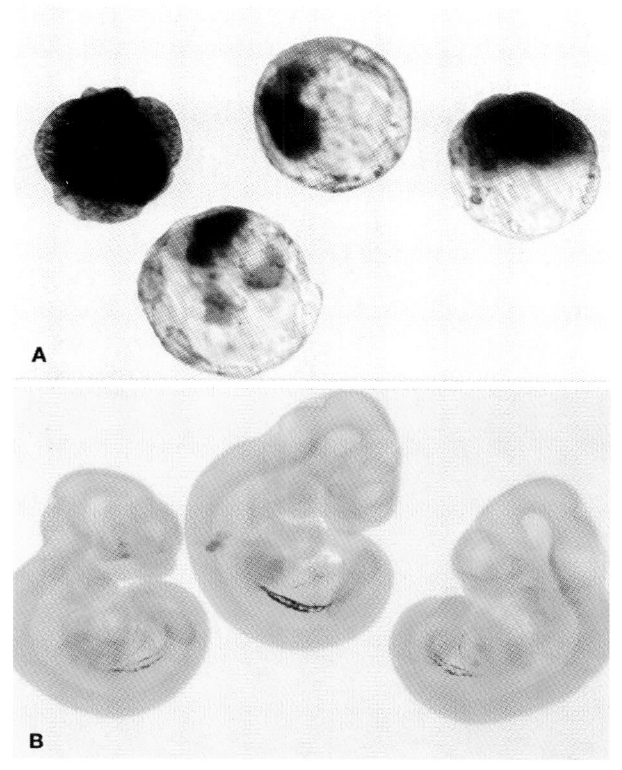

Fig. 23.1A, B. The expression of an Oct-4/lacZ fusion transgene in pre- and postimplantation embryos. **A** Analysis of the transgene activity in transient transgenic preimplantation embryos. Oocytes were injected with the transgene and cultured in vitro to the morula or blastocyst stage. Regulatory sequences of the Oct-4 gene limit expression of lacZ to only pluripotent cells of the morula and the inner cell mass of the blastocysts. **B** Expression of the Oct-4/lacZ fusion transgene is confined to the germ cell lineage of postimplantation embryos. In cleared E11.5 embryos, lacZ staining is visible in germ cells located within the primordial genital ridges

Troubleshooting

- The temperature and pH of the staining step are critical factors to control in order to avoid endogenous β-galactosidase activity. Nontransgenic animals of the same stage should always be included in the analysis as a negative control for endogenous β-galactosidase activity. Many protocols call for staining at 37 °C rather than 30 °C, but we find that at this temperature the level of endogenous staining is often high.

- When mixing the staining solution, it is critical to add the items in the order listed to prevent precipitation of the X-Gal and K_3/K_4. It has also been our experience that an occasional batch of X-Gal will always form precipitates. If this is observed, it is best to obtain a new batch.
- When staining older embryos, it is advisable to completely remove the outer membranes, as the yolk sac in particular can display endogenous β-galactosidase activity.
- Significant differences in pattern and intensity of staining are often observed among the embryos derived from in vitro cultures (Yeom et al. 1996; Kothary et al. 1989). These are most likely due to differences in the site of transgene integration. It is assumed that up to the eight-cell stage most lacZ expression is derived from episomal copies of injected DNA (Kothary et al. 1989). However, by the blastocyst stage, any lacZ activity is presumed to reflect expression of the integrated transgene.
- Clearing postimplantation embryos with BABB must be preceded by postfixation with 4% paraformaldehyde to avoid loss of the staining signal. The embryos may be stored after clearing, but obtaining a photographic record as soon as possible is recommended.

References

Bonnerot C, Rocancourt D, Briand P, Grimber G, Nicolas J-F (1987) A β-galactosidase hybrid protein targeted to nuclei as a marker for developmental studies. Proc Natl Acad Sci USA 84(19):6795–6799

Chatot CL, Lewis JL, Torres I, Ziomek CA (1990) Development of one-cell embryos from different strains of mice in CZB medium. Biol Reprod 42(3):432–440

Hanley T, Merlie JP (1991) Transgene detection in unpurified mouse tail DNA by polymerase chain reaction. Biotechniques 10(1):56

Hogan B, Beddington R, Constantini F, Lacey E (1994) Manipulating the mouse embryo: a laboratory manual. 2nd edn. Cold Spring Harbor Laboratory Press, Cold Spring Harbor

Joyner AL (1991) Gene targeting and gene trap screens using embryonic stem cells: new approaches to mammalian development. Bioessays 13(12):649–656

Kothary RK, Allen ND, Surani MAH (1989) Transgenes as molecular probes of mammalian developmental genetics. In: Maclean M (ed) Oxford Surveys on Eukaryotic Genes, Vol 6. Oxford University Press, pp 145–178

Lawitts JA, Biggers JD (1993) Culture of preimplantation embryos. In: Wassarman PM, DePamphilis ML (eds) Methods in Enzymology, vol 225: Guide to techniques in mouse development, pp 153–164

Le Mouellic H, Lallemand Y, Brulet P (1990) Targeted replacement of the homeobox gene Hox-3.1 by the *Escherichia coli* lacZ in mouse chimeric embryos. Proc Natl Acad Sci USA 87(12): 4712–4716

Mountford PS, Smith AG (1995) Internal ribosome entry sites and dicistronic RNAs in mammalian transgenesis. Trends Genet 11(5):179–184

Sambrook J, Fritsch EF, Maniatis T (1989) Molecular cloning. A laboratory manual, 2nd edn. Cold Spring Harbor Laboratory Press, Cold Spring Harbor

Sanes JR (1994) Lineage tracing. The laatest in lineaage. Curr Biol 4(12):1162–1164

Wallenfels K, Malhotra OP (1960) In The enzymes, vol 4. Academic Press, New York, 409 pp

Yeom YI, Fuhrmann G, Ovitt CE, Brehm A, Ohbo K, Gross M, Hübner K, Schöler HR (1996) Germline regulatory element of Oct-4 specific for the totipotent cycle of embryonal cells. Development 123:881–894

Chapter 24

Autonomous Cell Labeling Using a *laacZ* Reporter Transgene to Produce Genetic Mosaics During Development

Luc Mathis[1] and Jean-François Nicolas[1]*

Introduction

Analyses performed in invertebrates have revealed the central role of cell behavior for the progressive positioning and determination of cells within developing organisms (Garcia-Bellido et al. 1973; Sulston et al. 1983; Shankland 1991; Sternberg 1991). However, the study of cell behavior in organisms for which the individual sequence of embryonic cell divisions is not stereotyped remains a very complex issue and this is particularly true for vertebrates (Lawson et al. 1991; Quinlan et al. 1995). Analyses are hampered by the difficulty of labeling cells in the embryo and the lack of stable cell autonomous markers (Beddington and Lawson 1990), but also by the characteristics inherent in the prospective methods generally used.

Classical methods used to describe the behavior of groups of cells include cell transplants and tissue or cell marking. For example, the transplantation of quail grafts into chicken (LeDouarin 1982), as well as the orthotopic and heterochronic transplantation of ^{3}H thymidine-labeled cells (Copp 1990) or grafts ubiquitously expressing *lacZ* in mouse (Beddington 1994; Quinlan et al. 1995) have been efficient tools to describe global morphogenetic movements at a multicellular level and to fate map territories during development. In situ labeling within a single cell by iontophoretic reagents, such as horseradish peroxidase or lysine-rhodamine-dextran, has been extensively used to analyze the cell behavior in more detail during early development and gastrulation in chick (Selleck and Stern 1991) and mouse embryos (Lawson et al. 1991). These studies have shown that the cellular basis of development is probabilis-

* corresponding author: phone: 01 45 68 84 90; fax: 01 45 68 85 21;
e-mail: jfnicola@pasteur.fr

[1] Unité de Biologie moléculaire du Développement, URA 1947 du Centre National de la Recherche Scientifique, Institut Pasteur, 25, rue du Docteur Roux, 75724 Paris Cédex 15, France

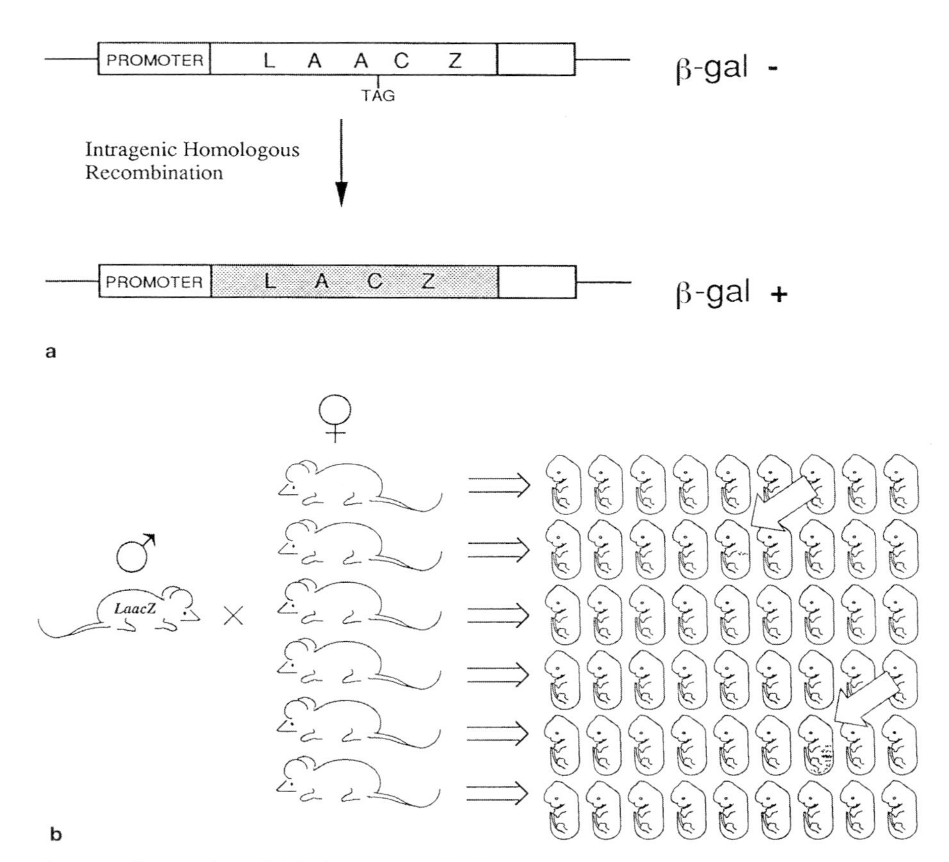

Fig. 24.1a,b. Single-cell labeling using the *laacZ* method. **a** The *lacZ* gene was inactivated by a duplication that introduces several stop codons. The expression of the resulting *laacZ* gene is controlled by the chosen promoter in a given cell type, cell lineage, or topographical domain. A homologous intragenic recombination generates a functional *lacZ* gene in a cell hemizygous for the transgene, whose descendants in the targeted compartement are visualized histochemically. **b** Mosaic *lacZ* animals are found in the progeny of homozygous transgenic males crossed with nontransgenic females

tically determined from the onset of gastrulation, and fate maps of the mouse embryo have been drawn as early as at the prestreak stage. However, the rapid dilution of the chosen marker along cell divisions (after six to seven divisions), and the necessity to use only embryos grown in vitro, restrict analyses to short time periods and limit their precision.

Another method for lineage analysis is a retrospective approach that uses genetic marking by replication-defective retroviral vectors, expressing a reporter gene such as *lacZ* (Sanes et al. 1986; Bonnerot et al. 1987; Bonnerot and Nicolas 1993a) or alkaline phosphatase (PAL) (Ryder and Cepko 1994). Cells in the embryo are infected in utero or in vitro by the recombinant retrovirus in which the viral sequences (gag, pol, and env) have been replaced by the *lacZ* (Nicolas and Rubenstein 1987) or PAL gene. The analysis can be performed at any subsequent stage of development, including postnatal animals. This method, allowing an autonomous cell labeling, is a powerful way to determine the potentialities of progenitor cells during development (Galileo et al. 1990; Franck and Sanes 1991). However, this method is not completely adapted to describe morphogenetic organization, for numerous reasons such as the rapid down-regulation of the expression of the reporter gene in vivo, and the difficulty in determining whether a cluster of marked cells are descendents of one or several infected progenitors (Cepko et al. 1993). Other complications arise from the fact that not all cells and periods of development are amenable to retroviral transfers, because of either problems of accessibility or lack of infectability. Furthermore, injection experiments are independent of other and are therefore subject to a number of variabilities.

The method of cell labeling we describe here is a retrospective approach based on genetic marking with a *laacZ* reporter gene in unmanipulated embryos (Bonnerot and Nicolas 1993b; Nicolas et al. 1996). The *laacZ* gene is generated by introducing a direct duplication of 289 bp in the *lacZ* gene, such that it encodes a truncated protein without β-galactosidase activity (Fig. 24.1a). First, mouse transgenic lines are created with the appropriate *laacZ* transgene. Next, mouse embryos hemizygous for *laacZ* are produced by breeding. If a homologous intragenic recombination has occurred within the *laacZ* gene, generating a functional *lacZ* gene during the development of these embryo, a genetic mosaic is created. The descendants of the labeled cell can then be visualized for β-galactosidase activity histochemically (Fig. 24.1b) at any stage of development. The choice of the promoter driving *laacZ* expression allows for restriction of the analysis to a preselected cellular domain. The promoter can be tissue-specific, with the transgene being expressed in only one given cell type, or lineage-specific, whereby the transgene is expressed in several cell types of a given tissue or system (e.g., all neurons of the nervous system, all cells of the hematopoietic system, etc.). In our laboratory, mouse lines have been produced that express the *laacZ* gene specifically in the myotomal part of the somite (using the αAChR promoter, a tissue-specific promoter; Klarsfeld et al. 1991; Bonnerot and Nicolas 1993b) and other mouse lines have been established that express the *laacZ* gene ubiquitously in neurons of the central nervous system (our unpubl. results; using the neuron-specific enolase promoter, a lineage-specific promoter; Forss-Petter et al. 1990).

In comparison with other methods of cell labeling, a main advantage of the present method is the ease with which it can be used, since mosaic embryos or animals are produced by simple breeding and harvesting of embryos. More importantly, the *laacZ* method does not require any manipulation or injury to the embryo, such as that resulting from in vitro culture or transplantation of territories or microinjection, whose impact on the development is not assessed. Another important feature of this method is that the marker is stably inherited during cell divisions, which allows for long-term experiments in the developing embryo or even at postnatal or adult stages. The low frequency of intragenic homologous recombination events (10^{-5} to 10^{-6} per cell and per generation) results in the labeling of single cells; thus, there is no difficulty in establishing the clonality of labeled cells within one embryo. This genetic recombination is a spontaneous mutational event and therefore occurs at random in terms of cell targeting and timing during development. Consequently, the probability in obtaining a marked cell is, at any given time during development, mainly dependent on the size of the pool of cells at the origin of the structure analyzed. In particular, there is no spatial or temporal limitation imposed by the marking event. Therefore this method is an objective way of defining levels of cellular organization. For instance, it allows for a definition of the time at which progenitors of a structure are first organized into a pool of cells with similar properties (Nicolas et al. 1996).

In the prospective methods, the experimentator decides on the territory and time of cell labeling. Because all cells within a territory may not possess an equivalent developmental contribution, the choice of a cell to label may introduce a bias by over- or underestimating the importance of certain types of cell behavior. This bias does not exist in the *laacZ* method because of the random aspect of the marking event and because only the clones effectively participating in a given structure are analyzed. The relative developmental importance of distinct cell behaviors giving rise to clones with particular characteristics can be directly estimated by the frequency at which they occur. In the *laacZ* method, the random aspect of the labeling event, which results in labeled cells at all stages of development, and its stability during cell divisions, permits an analysis of the genealogical relationships between precursor cells that are spatially and temporally distant in the embryo. When the promoter chosen is lineage-specific, the resulting recombinant *lacZ* clones allow for an analysis of the genealogical link between different cell types whose production might be distant in both time and space.

In conclusion, the application of this method of random labeling and retrospective analysis may help to define important aspects of early cell behavior (that is, cell growth and cell repartition) in the cellular systems of organisms whose complete lineage cannot be otherwise described due to high cell numbers and the nonstereotyped sequences of cell divisions. In addition, an extension of this method to mutant cells will help in the study of the effects of a genetic perturbation, since the altered cells can be followed in a normal embryo.

Production of the *laacZ* Founder Transgenic Animal

Several methods can be used to produce *laacZ* transgenic animals (Wassarman and DePamphilis 1993). One method consists of microinjecting DNA constructs into the pronuclei of one-cell embryos. A second method is based on the introduction of DNA into embryonic stem cells by electroporation. The transduced cells are then introduced into blastocysts or fused to the inner cell mass. In both methods the number of integrated copies of the transgene is not controlled and might vary largely (between one to hundreds). One way to control the number of copies is to use recombinant retroviruses (Nicolas and Rubenstein 1984). However, the expression of transgenes in retroviruses is subject to in vivo downregulation, since viral DNA contains several negative regulatory sequences located in and between the LTRs. Recently, a new family of recombinant retroviruses that produce proviral structures deleted for these inhibitory sequences (the so-called solitary LTR recombinant retrovirus, sLTR) has been described (Choulika et al. 1996), which may be of help in the future for introducing a single copy of the transgene into ES cells or embryos, and thus alleviate these problems of expression. Another method under consideration is one in which a gene whose deletion produces no phenotype is replaced by homologous recombination with the *laacZ* gene. Such a method would circumvent the potential problems related to random integration into the genome and would permit better targeting of transgene expression (our unpublished studies).

In addition to the *laacZ* transgenic lines, control mice expressing the *lacZ* gene under control of the chosen promoter should also be produced. These control mice help in defining the very precise spatiotemporal pattern of expression, and in selecting the appropriate stages of development to analyze the *lacZ* clones. These mice also provide for an analysis of specific issues, that cannot be learned from in situ hybridization experiments, such as cell numbers in the targeted compartment.

This information is necessary for the calculation of the size of the polyclone at the origin of a given structure, as well as other parameters. The control mice also provide information about the dynamics of the compartment under analysis which can be subsequently correlated with the *lacZ* clonal distributions.

24.1 Analysis of *laacZ* Expression Pattern

In almost all the methods used to generate the transgenic founder animal, the integration site of *laacZ* in the genome is random. Thus the expression of the transgene can be influenced, at least in part, by host genetic elements at the integration site, which could lead to improper expression within the targeted lineage. It is therefore necessary to control for the expression pattern of each transgenic line. Another reason for such a control is that there might be variegation of expression related to the genetic background of animals. Because the elements controlling the variegation are not understood, we have chosen to select individual homozygous transgenic males on the basis of the intensity of their expression pattern before generating clones. This can be best achieved by in situ hybridization using a *lacZ* RNA probe (Fig. 24.2a). We present here only results obtained from in situ hybridization on whole mount E13 embryos (Henrique et al. 1995). For E>13 embryos, in situ hybridization is performed on cryostat slides (Groves et al. 1995)

Materials

- Synthesis buffer : 0.6 mM CTP,ATP, and GTP, 0.4 mM UTP, 0.2 mM digoxygenin-UTP, 10 mM dithiothreitol (DTT), 40 u RNasin and 30 u RNA polymerase in 1× transcription buffer
- 5× transcription buffer: 200 mM TrisHCl pH 7.5, 30 mM $MgCl_2$, 10 mM Spermidine, 50 mM NaCl
- Hydrolysis solution : 12.5 mM DTT, 100 mM $NaHCO_3$, 150 mM Na_2CO_3
- Neutralizing solution : 12.5 mM DTT, 250 mM NaAc, 1.25 % v/v acetic acid
- TE: 10 mM Tris HCl pH 7.5, 1 mM EDTA
- PBS : 138 mM NaCl, 2.7 mM KCl, 1.5 mM KH_2PO_4, 8.1 mM Na_2HPO_4, pH 7.3

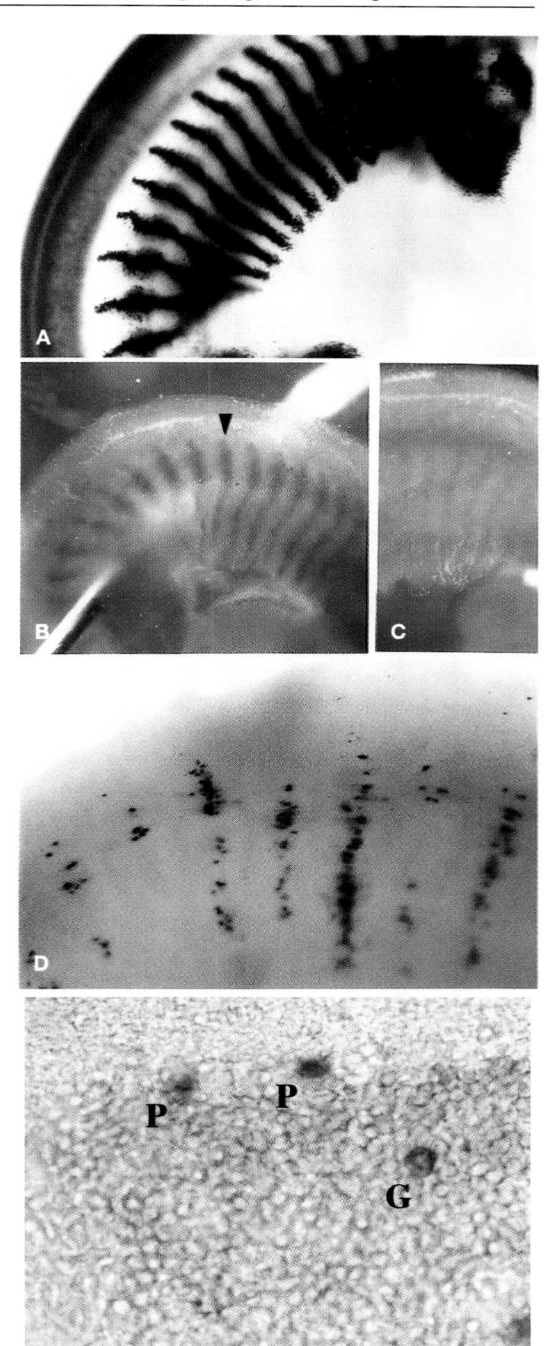

Fig. 24.2A–E. Analysis of the expression pattern of the reporter *laacZ* transgenes. **A** Control *lacZ* expression pattern in a E11.5 embryo, showing the targeted and specific expression specifically in the myotomal part of the somite by the αAChR promoter. **B** Whole mount in situ hybridization using a digoxygenin labeled antisense *lacZ* RNA probe in a E11.5 embryo of the α-2 αAChR-*laacZ* transgenic line. The expression pattern of the *laacZ* transgene is the same as for the control *lacZ* embryo. **C** Control, in situ hybridization using a nontransgenic embryo. **D** *lacZ* recombinant clone derived from the α-2 transgenic line. **E** Analysis of common precursors for different cells types. Frontal cryostat section of a P15 NSE-*laacZ* cerebellum shows *lacZ* expression in clonally related Purkinje (*P*) and Golgi (*G*) neurons. The cell types are identified by their position and size

- PTW : PBS, 0.1 % Tween 20
- Hybridization mix (one embryo requires a volume of ~11 ml) : 50 % formamide, 1.3× SSC (stock: 20× SSC, pH 5, with citric acid), 5 mM EDTA (stock: 0.5 M EDTA pH 8), 50 μg/ml total yeast RNA, 0.2 % Tween 20, 0.5 % CHAPS, 100 μg/ml Heparin. Can be stored at −20 °C
- TBST : 0.15 M NaCl, 3 mM KCl, 25 mM Tris-HCl pH 7.5, 1 % Tween 20. Stock: 10× salt, add Tween 20 on the day of use
- Total yeast RNA : treat with phenol-chloroform a solution of total yeast RNA, and store 10 mg/ml aliquots at −20 °C
- MABT : 100 mM maleic acid, 150 mM NaCl, pH 7.5, 0.1 % Tween 20. Stock: 10× salt, add Tween 20 on the day of use
- NTMT: 100 mMNaCl, 100 mM TrisHCl pH 9.5, 50 mM $MgCl_2$, 1 % Tween 20. Make from stocks on day of use
- AP substrate : BM Boehringer no. 1442074 with 1 % Tween 20
- Sheep serum (Sigma) is first heat-treated at 55−60 °C, 30 min, then stored in aliquots at −20 °C. Thawed aliquots can be stored at 4 °C with addition of Na azide to 0.1 %
- Make 10 % blocking reagent (Boehringer Blocking Reagent no. 1096 176) stocks in MAB (MABT without Tween 20) by heating to dissolve, then autoclave, aliquot, and freeze.

Procedure

1. The RNA antisense probe is synthesized from a linearized plasmid carrying the *lacZ* gene under the control of the T3 or T7 promoter. The synthesis is performed on 100 ng of linearized plasmid in 20 μl of synthesis buffer for 1.5 h at 37 °C. Check the RNA/DNA ratio on a 1 % agarose minigel (the ratio should be 50:100). Add 7.5 μ of RNase-free DNase RQ1 and digest 15 mn at 37 °C. Add 30 mg of total yeast RNA (previously extracted with phenol-chloroform and stored in 10 mg/ml aliquots at −20 °C).

2. The probe size must be 200 bp to allow for penetration diffusion into tissues. The probe is first partially digested : add 50 μl of hydrolysis solution, digest at 60 °C. The time of digestion depends on the length of the probe: 200−400 bp; 20 mn; 400−1000 bp; 45 mn; >1200 bp; 1 h. Add 50 μl of neutralizing solution, then 2 vol ethanol, 1/10 vol 3 M NaAc and leave for 30 min at −80 °C to precipitate the probe. Spin 30 mn at 13,000 rpm at 4 °C, wash pellet gently with 70 % ethanol, dry rapidly, and resuspend in 100 μl TE. The concentration of the probe is typically 100 μg/ml. Store at −20 °C.

Note: RNA is highly sensitive to degradation, so use sterile tubes, water, and salt solutions; wear gloves, at least until the hybridization step.

3. Dissect embryos into PBS; remove the extraembryonic membranes. For stages post-E10.5 the embryo must be bilaterally sectioned using a surgical chisel and the structures then dissected away in order to avoid background activity. Fix the embryo in 10 ml of 4 % formaldehyde (HCHO) in PBS +2 mM EGTA, 1–2 h at room temp; or at 4 °C, 2 h overnight. Wash twice in PTW. Wash once with 50 % MeOH/PTW, then twice with 100 % MeOH; embryos can be stored at this point at −20 °C (no longer than 1 month). Rehydrate embryos through sequential washes of 75, 50, 25 % MeOH/PTW (allowing embryos to settle), followed by two washes with PTW.

4. Treat embryos with 10 μg/ml proteinase K in PTW. The time of incubation must be adapted to the stage of the embryo (between 10 mn for E6.5 embryo and 30 mn for E12.5 embryo). Remove proteinase, by rinsing briefly (take care, as embryos are fragile at this stage) with PTW, then postfix for 20 min in 4 % HCHO +0.1 % glutaraldehyde, in PTW. Rinse and wash once with PTW. Transfer embryos to a 2-ml, sterile, round-bottomed microtube. Rinse once with 1:1 PTW/hybridization mix. Let embryos settle.

5. Rinse embryos with 1 ml of hybridization mix. Let embryos settle. Replace the solution with 1 ml of hybridization mix and incubate horizontally ≥1 h at 70 °C.

6. Add 1 ml of prewarmed hybridization mix containing 1 μg/ml DIG-labeled RNA probe. Incubate horizontally at 70 °C overnight. Mix after the first 20–30 min.

Note: Steps **3–4** are carried out using a roller, and 3–5 ml of solution in 6 ml round bottomed Falcon tubes. Thereafter, use 1.5–2 ml of solution in a 2-ml microtube, and rocking at room temperature unless otherwise stated. Rinses are immediate, and washes are for 5 min; 4 % formaldehyde/PBS should be made on the day of use by diluting 1 vol 40 % HCHO with 9 vol PBS. Add NaOH 1 M to pH 7.5. Hybridization mix can be stored at −20 °C. A stock of 25 % glutaraldehyde is stored in aliquots at −20 °C.

7. Rinse twice with prewarmed (70 °C) hybridization mix, wash 2×30 min/70 °C with prewarmed hyb mix, wash 20 min/70 °C with prewarmed 1:1 hyb mix/TBST.

Rinse 2× with TBST, wash 2×30 min with TBST. Rinse 2× with MABT.

8. Preincubate >1 h in MABT +2 % blocking reagent +20 % heat-treated serum. Incubate overnight at 4 °C (or 4 h at RT) in (fresh) MABT +2 % blocking reagent +20 % serum +1/2000 dilution of anti-DIG antibody.
9. Rinse three times with MABT; transfer to a 12 ml round-bottomed Falcon tube (it is possible to reuse the antibody solution; store with 0.1 % Na azide). Wash 3×1 h with 10 ml MABT using a roller. Wash 2×10 min with NTMT.
10. Incubate with 2 ml of AP substate and agitate the solution by rocking. Keep in the dark. When the color has developed to the desired extent (30 min to 20 h), wash 3x with PTW. Store in 70 % glycerol in PBS including 0.1 % Na azide.

Note: After each 70 °C wash, let embryos settle by incubating the tube vertically at 70 °C, then change the solutions individually so samples do not cool. Keep wash solutions at 70 °C.

24.2 Production of *laacZ/lacZ* Mosaic Embryos

Materials

Sterile stock solutions of Foligon and Chorulon (Intervet Laboratories Ltd) are made as 50 IU/ml in PBS. Aliquot and store at −20 °C for no longer than 2 months.

Procedure

1. Animals are obtained by crossing homozygous males with nontransgenic females such as F_1 (B6:D2), whose progeny are subsequently analyzed. Because of the low frequency of recombination, it is necessary to produce a large number (one to several thousand) of transgenic animals at a given developmental or adult stage to obtain β-Gal + recombinant clones. For embryos prior to stage E14.5, the number of embryos per female can be increased by superovulating the females

prior to breeding. Superovulation is performed as follows: inject a female (4 to 12 weeks old) intraperitoneally with 5 UI of Foligon, a gonadotropin that mimics the action of FSH. 40–45 h later, inject the female intraperitoneally with 5 UI of Chorulon, a gonadotropin that mimics LH. Place the female with a male for a single night. The morning of the copulation is taken as E0.3.

2. Dissect embryos into PBS and remove the extraembryonic membranes to increase the efficiency of the coloration. For E>13 embryos, perform a bilateral section under the microscope using a razor blade, or dissect out organs to increase accessibility to the structure to be analyzed.
3. Transfer the embryo or anatomic pieces to a fresh solution of 4 % (w/v) paraformaldehyde pH 7.4 in PBS. Incubate at 4 °C for 20 mn on a roller. Wash 3x in PBS, 5 mn each.

Note: Rocking increases the quality of the fixation; 8 % paraformaldehyde stock can be stored at 4 °C. Intracardiac perfusion of the fixative is not necessary if the structures are isolated by dissection.

24.3
In Situ β-Galactosidase Histochemistry

Materials

Histochemical mixture: 1 mg/ml 4-chloro-5-bromo-3-indolyl-b-galactoside (X-Gal), 4 mM $K_4Fe(CN)_6.3H_2O$ (potassium ferrocyanide), 4 mM $K_3Fe(CN)_6$ (potassium ferricyanide), 2 mM $MgCl_2$ in PBS. Stock solutions are 40 mg/ml in dimethyl sulfoxide (DMSO) for X-Gal (storage at −20 °C) and 0.2 M in water for potassium ferrocyanide and potassium ferricyanide; both stocks must be protected from light and made fresh every 2 weeks.

Procedure

1. Whole embryos or anatomic pieces are then transferred into histochemical reaction mixture.
2. Samples are incubated at 30 °C in a humidified incubator. The histochemical reaction proceeds slowly, and can be prolonged for 2 days to

obtain adequate labeling of individual cells. If the samples were correctly fixed, no background activity should develop.

3. Postfix in 4 % paraformaldehyde at 4 °C for 1 h on a roller. Wash 2× in PBS, 5 mn each.

Note: To avoid background activity, the pH (7.3) and temperature (30 °C or below) of the reaction are important. Do not touch the samples with metallic forceps while they are in the histochemical mixture. Metal provoques a blue precipitate at the surface of the embryo that can be mistaken for a β-Gal+ recombinant clone. Leave the sample for 1 h in PBS before manipulating with forceps.

24.4 Detection of LacZ Clones

For a description of the cellular behavior, it is essential to obtain all the β-Gal+ clones, even the smallest ones (single or few cells clones) contributing to the formation of a structure at a given stage of development. To facilitate the examination of the samples, they can be cleared first by changing the optical properties of the cell membranes. Several products (benzyl-benzoate mixtures or methyl salycilate) which are toxic should be avoided for the screening step; we routinely use 70 % glycerol.

Procedure

1. Transfer the samples to 70 % glycerol in PBS +0.1 % Na azide for 24 h at 4 °C. The volume of glycerol should be at least ten times more than the volume of the embryos to avoid dilution. Glycerol might crumple the embryo; in this case transfer the embryo through a progressive (25–50–70 %) in PBS series of glycerol concentrations incubating 2 h each on a roller at 4 °C.
2. The embryos can be stored at 4 °C for at least 6 months; simply add Na azide every 3 months to avoid microbial contamination.
3. If the structures to be analyzed are not apparent (e.g., central nervous system, viscera, heart, etc.), the structure should be dissected from each embryo to have a direct visual access to the structure. Bilateral section is generally sufficient, and should be performed under the

microscope at ×20 with a razor blade (if the structure is at the midline, such as the CNS, use a surgical chisel to improve the accuracy of the section).

4. Carefully observe the embryos at ×60 using a stereomicroscope.
5. Store embryos containing β-Gal+ clone in 70 % glycerol in PBS with Na azide. Do not discard the β-Gal-embryos, as it is generally necessary to reexamine embryos to find small β-Gal+ clones.

Modification

A more complete clearing of the embryos can be achieved as follows:

1. Dehydrate through ethanol series (70, 80, 90, 95, 100 %). The steps should be as short as possible, e.g., 15 min for E9 embryo and up to 30 min. for E12 embryo, so as not to dilute the X-Gal blue precipitate in ethanol.
2. Remove 100 % ethanol and add sufficient benzyl alcohol/benzyl benzoate (1:2) mixture or methyl salicilate so that the embryo is just floating. Refresh mixture after embryo sinks if incompletely transparent.
3. Return the embryos for storage to aqueous medium via the ethanol series.

Note: Benzyl-benzoate mixtures and methyl salycilate are highly toxic. It is essential to work cleanly, to take necessary precautions – gloves, fume hood – and to dispose of the waste responsibly. The X-Gal blue precipitate will eventually dissolve in the clearing mixture and ethanol, but no loss of staining has been seen over a period of 2 h.

Result

Description of the Clones

The rules governing the formation of a structure analyzed at one stage of development can be safely derived only when the number of clones produced corresponds to a saturation, that is, when new clones produced at that stage have characteristics similar to those already obtained. It is worth noting that saturation is not an absolute feature and concerns individual parameters only, because the lineage relationships between cells are, in general, not fully stereotyped in mammals. Thus, saturation for an individual parameter can be reached before all possible labeled clones have been generated.

The principle of the retrospective analysis rests first on a geometrical description of each clone and its record. This can be achieved by camera lucida drawing of the clones (Fig. 24.3a,b). This realistic representation can then be shifted into a more practical representation involving statistical analysis of the data to study given relationships (Fig. 24.3c,d,e). Computer analysis of the data permits rapid grouping and sorting of clones according to individual or combined parameters (Fig. 24.3d,e). The classification of clones is useful, since clones within a given class may derive from pools of progenitor cells with similar characteristics. The random aspect of the labeling implies that the pattern of clones gives a picture of the structure that relates to the genealogical history of its precursor cells. This characteristic allows for testing of different models for the production of a structure, since each model makes specific predictions about the expected distribution and frequency of clones (Fig. 24.4; Nicolas et al. 1996). For instance, an important parameter of a structure is its clonal complexity, which is defined by the number of times a given sector is populated by the pattern of the clones (Fig. 24.3e). Oriented or temporal modes of formation for a structure imply a correlative increase in the clonal complexity (Fig. 24.4).

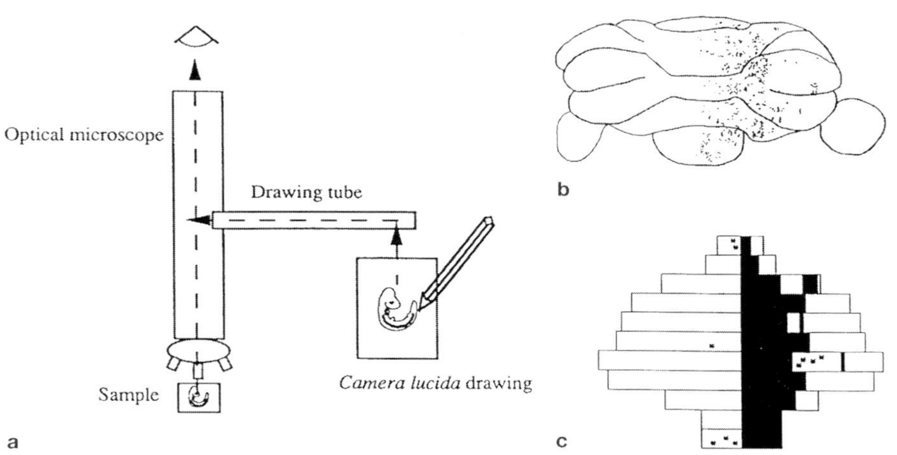

Fig. 24.3a–e. Representation and computer analysis of the clones. **a** Principle of the camera lucida. A drawing tube adapted to a microscope allow for visual superimposition of the sample and the representation. **b** Example of a camera lucida drawing, showing the distribution of neurons derived from thecerebellar plates in the cerebellum of a P12 NSE-*laacZ* transgenic animal. **c** Schematic representation of the same clone, in which the distribution of labeled cells is adapted into a flattened cerebellum (**d** and **e** see pages 453 and 454)

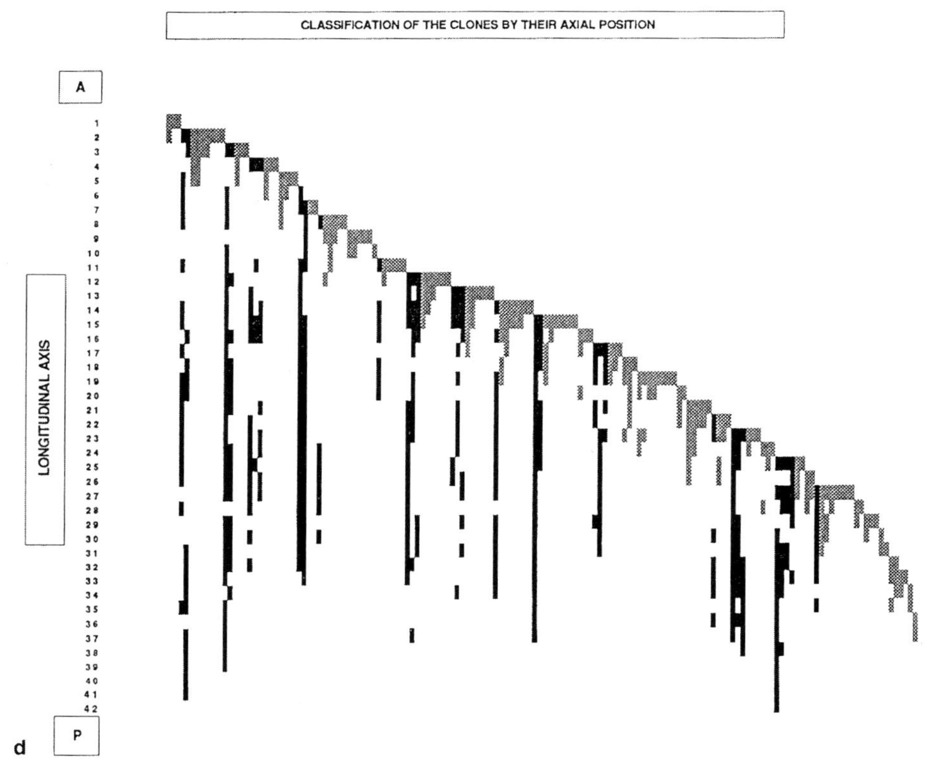

Fig. 24.3d. Computer analysis of 153 myotomal clones derived from the α-2 transgenic line. Large clones are represented by *black boxes* and small clones by *gray boxes*. Clones are classified according to their most anterior border. Note that the number of large clones (clonal complexity) increases from anterior to posterior

A comparison between the size of the precursor cell pools for the different classes of clones is achieved by comparing the frequency of clones in different classes. For example, because the largest clones are necessarily composed of smaller subclones, for which the intragenic recombination can occur at a predictable frequency, this comparison makes it possible to determine which class of small clones derived from the large clones.

Cell counting is another important parameter for comparing clones. It might be misleading to take cell number as a direct indication of the birth date of the clone. For instance, in the myotome, the anterior border and bilaterality of a clone are the two parameters that give the best

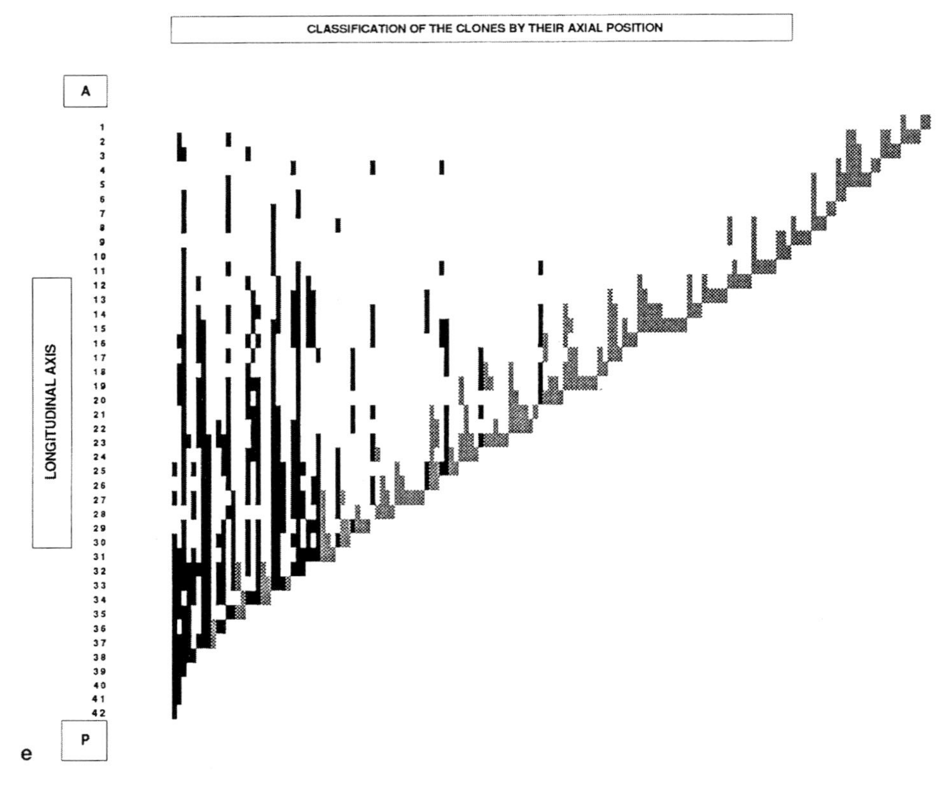

Fig. 24.3e. Same clones classified by their posterior border.

estimation of its birth date (Nicolas et al. 1996), regardless of the cell number comprising the clone. Cell counting allows for an estimation of the size of the polyclone that gives rise to the structure, after calculating the fraction of the total compartment (derived from *lacZ* control mice) populated by the clones. Cells of glycerol-cleared embryos are counted under a stereomicroscope at ×60.

24.5 Determination of the Cell Types

When *laacZ* expression is driven by a lineage-specific promoter in a variety of cell types, an analysis of the cell types composing of the clones is used to define these cell types deriving from common progenitors, and to establish genealogical relationships.

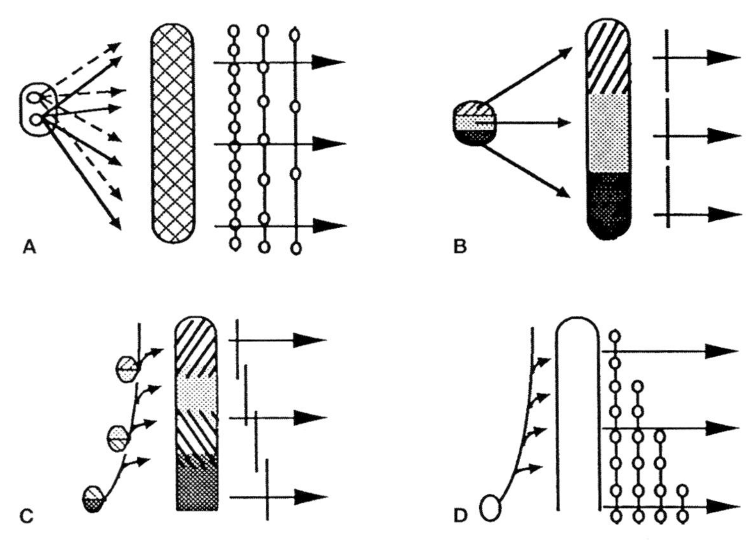

Fig. 24.4A–D. The models of formation for a structure can be tested by the analysis of *lacZ* recombinant clones. **A** Predictions derived from different models. Four possible models of formation of a longitudinal structure represented in relation to its pool of precursor cells. **A** Extension and intercalation. **B** Early regionalization. **C** Temporal production of the precursors from a changing pool of cells. **D** Temporal production from a unique, self-renewing pool of stem cells. On the *left* is a representation of the pool of precursor cells at an early stage and, in the *middle*, a representation of the longitudinal axis after its formation. Shown on the *right* side of each figure is the longest clone expected from the labeling of a cell in the pool of precursors. The *horizontal arrows* represent clonal complexity in relation to axial level. In the formation of the myotome, the clonal continuity observed and the increase in clonal complexity from anterior to posterior (Fig. 24.3d) favored a model involving a pool of permanent stem cells progressing from anterior to posterior. (Nicolas et al. 1996)

Determination of the cell types present in each clone is achieved histologically on cryostat sections of the samples.

Materials

Neutral red: 1 % in 50 mM sodium acetate, pH 3.3

Procedure

1. Prepare, in advance, gelatinized glass slides for microscopy. Dissolve 5 g of gelatin (Merck) and 1 g of chromium potassium sulfate dodecahydrate (chrome alum) in 1 l distilled water heat at 65 °C with water

with constant stirring until the reagents are completely dissolved. Filter the hot solution through Whatman (Clifton, NJ) 3-MM paper. Immediately immerse clean microscopy slides in the solution for 1 min at 45 °C. Dry the slides at 37 °C overnight. Store the glass slides at 4 °C.

2. Transfer the anatomic piece from 70 % glycerol to 30 % sucrose (30 g/ 100 ml) in PBS, pH 7.3 and incubate overnight at 4 °C. Sucrose is a cryoprotector. Embed the anatomic piece in Tissue-tek (OCT medium, Miles Scientific) and immerse slowly in cold isopentane. Cold isopentane is obtained by placing it in a beaker placed on dry ice for 10 min.

3. Place the cold anatomic piece on the cryostat holder and let it equilibrate at −20 °C in the cryostat for 1 h. Sections are cut to a thickness of 5 to 100 μm, depending on the tissue type and subsequent manipulations.

4. Stain sections briefly (<30 s) with neutral red. Dehydrate rapidly and successively in ethanol : 70, 90, and 99.9 % (v/v). Wash twice in pure toluene and mount slides in Eukitt. These last operations must be performed in a chemical fume hood (toluene is extremely dangerous).

Acknowledgments. We thank Shahragim Tajbakhsh and Denis Houzelstein for advice in in situ hybridization, Joan Shellard for careful reading of the manuscript, Françoise Kamel for help in typing the manuscript. JFN is from the Institut National de la Recherche Medicale.

References

Beddington R, Lawson KA (1990) Clonal analysis of cell lineages. In: Copp AJ, Cockroft DL (eds) Postimplantation mammalian embryos – a practical approach. Oxford University Press, Oxford, pp 267–291

Beddington RS (1994) Induction of a second neural axis by the mouse node. Development 120:613–620

Bonnerot C, Nicolas J.-F (1993a) Application of LacZ gene fusions to post-implantation development. Methods Enzymol 225:451–469

Bonnerot C, Nicolas JF (1993b) Clonal analysis in the intact mouse embryo by intragenic homologous recombination. C R Acad Sci 316:1207–1217

Bonnerot C, Rocancourt D, Briand P, Grimber G, Nicolas JF (1987) A β-galactosidase hybrid protein targeted to nuclei as a marker for developmental studies. Proc Natl Acad Sci 84:6795–6799

Cepko CL, Ryder EF, Austin CP, Walsh C, Fekete DM (1993) Lineage analysis using retrovirus vectors. In: Wassarman PM, DePamphilis ML (eds) Methods in enzymology: guide to techniques in mouse development. pp 933–960

Choulika A, Guyot V, Nicolas JF (1996) Transfer of single gene-containing long terminal repeats into the genome of mammalian cells by a retroviral vector carrying the *cre* gene and the *loxP* site. J Virol 70:1792–1798
Copp AJ (1990) Studying developmental mechanisms in intact embryos. In:. Cockroft AJC a.D.L. (ed) Postimplantation mammalian embryos. A practical approach Oxford University Press, Oxford, pp 293–315
Forss-Petter S, Danielson PE, Catsicas S, Battenberg E, Price J, Nerenberg M, Stucliffe JG (1990) Transgenic mice expressiong β-galactosidase in mature neurons under neuron-specific enolase promoter control. Neuron 5:187–197
Franck E, Sanes JR (1991) Lineage of neurons and glia in chick dorsal root ganglia: analysis in vivo with a recombinant retrovirus. Development 111: 895–908
Galileo DS, Gray GE, Owens GC, Majors J, Sanes JR (1990) Neurons and glia arise from a common progenitor in chicken optic tectum: demonstration with two retroviruses and cell type-specific antibodies. Proc Natl Acad Sci USA 87:458–462
Garcia-Bellido A, Ripoll P, Morata G. (1973) Developmental compartmentalization of the wing disc of *Drosophila*. Nat New Biol 245:251–253
Groves AK, George KM, Tissier SJ, Engel JD, Brunet JF, Anderson DJ (1995) Differential regulation of transcription factor gene expression and phenotypic markers in developing sympathetic neurons. Development 121:887–901
Henrique D, Adam J, Myat A, Chitnis A, Lewis J, Ish-Horowicz D (1995) Expression of a Delta homologue in prospective neurons in the chick. Nature 375:787–790
Klarsfeld A, Bessereau J-L, Salmon A-M, Triller A, Babinet C, Changeux J-P (1991) An acetylcholine receptor alpha-subunit promoter conferring preferential synaptic expression in muscle of transgenic mice. EMBO J 10:625–632
Lawson KA, Meneses JJ, Pedersen RA (1991) Clonal analysis of epiblast fate during germ layer formation in the mouse embryo. Development 113:891–911
LeDouarin H (1982) The neural crest. Developmental and cell biology. Cambridge University Press, Cambridge
Nicolas JF, Rubenstein J (1987) Retroviral vectors. Butterworths, London
Nicolas JF, Mathis L, Bonnerot C (1996) Evidence in the mouse for self-renewing stem cells in the formation of a segmented longitudinal structure, the myotome. Development 122:1–14
Quinlan GA., Williams EA, Tan SS, Tam, PL (1995) Neuroectodermal fate of epiblast cells in the distal region of the mouse egg cylinder : implication for body plan organization during early embryogenesis. Development 121:87–98
Rubinstein JRL, Nicolas JF, Jacob F (1984) Construction of a retrovirus capable of transducing and expressing genes in multipotential embryonic cells. Proc Natl Acad Sci USA 81:7137–7140
Ryder EF, Cepko CL (1994) Migration patterns of clonally related granule cells and their progenitors in the developing chick cerebellum. Neuron 12:1011–1028
Sanes J, Rubenstein J, Nicolas JF (1986) Use of a recombinant retrovirus to study post-implantation cell lineage in mouse embryos. EMBO J 5:3133–3142
Selleck MAJ, Stern CD (1991) Fate mapping and cell lineage analysis of Hensen's node in the chick embryo. Development 112:615–626
Shankland M (1991) Leech segmentation : cell lineage and the formation of complex body patterns. Dev Biol 144:221–231
Sternberg PW (1991) Control of cell lineage and cell fate during nematode development. In: Bode HR (ed) Current topics in developmental biology. Academic Press, San Diego, pp 177–225

Sulston JE, Schierenberg E, White JG, Thomson JN, Von Ehrenstein G (1983) The embryonic cell lineage of the nematode *Caenorhabditis elegans*. Dev Biol 100:64–119

Wassarman PM, DePamphilis ML (1993) Methods in enzymology: guide to techniques in mouse development. Academic Press, New York

Multiple Developmental Functions of the Zinc Finger Gene *Krox-20* Are Revealed by Targeted Insertion of the *E. coli lacZ* Coding Sequence

Sylvie Schneider-Maunoury[1*], Piotr Topilko[1], Tania Seitanidou[1], Giovanni Levi[1,2], Paula Murphy[1,3], and Patrick Charnay[1]

Background

The inactivation of the *Krox-20* gene in the mouse has allowed us to study the involvement of this gene in the regulation of several processes as diverse as hindbrain segmentation, peripheral nervous system (PNS) myelination and endochondral bone formation (Schneider-Maunoury et al. 1993; Topilko et al. 1994; Levi et al. 1996). The aim of the present chapter is to describe the procedure we selected to inactivate the *Krox-20* gene and to show how this strategy, which involved the use of a *lacZ* tracer gene, has helped us in understanding the phenotypes associated with the mutation and has enabled us to discover novel *Krox-20* functions.

Krox-20 was originally identified as a serum response immediate-early gene (Chavrier et al. 1988). It encodes a transcription factor with a DNA-binding domain consisting of three C_2H_2-type zinc fingers (Chavrier et al. 1988, 1990; Vesque and Charnay 1992). The interest in *Krox-20* as a regulatory gene in development originated from the finding that *Krox-20* was segmentally activated in the developing mouse hindbrain. *Krox-20* is expressed in two stripes, which prefigure and then coincide with rhombomeres 3 and 5 (Wilkinson et al. 1989). Rhombomeres are periodic swellings observed transiently along the anteroposterior axis in the hindbrain region. In the past few years, rhombomeres have been shown to play an essential role in establishing the developmental pattern governing cranial nerve organisation and cranio-facial morphogenesis (reviewed in Wilkinson 1993). Neurons of the brain stem, as well as neu-

* corresponding author: phone: +33-1-44323714; fax: +33-1-44323988; e-mail: maunoury@wotan.ens.fr

[1] Biologie Moléculaire du Développement, Unité INSERM 368, Ecole Normale Supérieure, 46 rue d'Ulm, 75005 Paris, France

[2] Unita di Morfogenesi Molecolare, Advanced Biotechnology Center, I.S.T., Viale Benedetto XV no.10, 16132 Genova, Italy

[3] Biotechnology Center of Oslo, PO box 1125, N0316 Oslo, Norway

ral crest cell precursors which migrate to generate neurons and glial cells of the PNS and bones of the skull, are produced in the hindbrain neuroepithelium in a repeated pattern which follows the segmental organization of rhombomeres. The observation that *Krox-20* expression precedes morphologic segmentation suggested that this gene might have an important role in the segmentation process (Wilkinson et al. 1989). This hypothesis was reinforced by the discovery that Krox-20 regulates the transcription in rhombomeres 3 and 5 of at least one member of the *Hox* family (Sham et al. 1993), which appears to play a fundamental role in pattern formation in vertebrates (reviewed in Krumlauf 1994). We therefore decided to inactivate the *Krox-20* gene by homologous recombination in embryonic stem (ES) cells in order to analyze its role during hindbrain segmentation.

Strategy

An important aspect of the strategy was to create an in-frame insertion of the *E. coli lacZ* gene sequence into the *Krox-20* coding sequence in order to follow the expression of the targeted allele in the embryo. The insertion of the *lacZ* reporter had already proven to be useful in several cases (Le Mouellic et al. 1990; Mansour et al. 1990). Moreover, insertion of nonhomologous sequences does not seem to reduce the efficiency of homologous recombination significantly (Mansour et al. 1990). It was expected that the resulting Krox-20/β-galactosidase fusion protein would have β-galactosidase activity but no Krox-20 activity. Homologous recombination events would place the fusion gene under the control of *Krox-20* regulatory elements, and its expression during embryogenesis was expected to reproduce the *Krox-20* expression profile.

To favor a high frequency of homologous recombination events, we engineered long homology regions on both sides of the targeting vector. The extent of homology between the targeting vector and the target locus has indeed been shown to have a crucial impact on the frequency of homologous recombination (Hasty et al. 1991; Deng and Capecchi 1992). We intended to perform a negative selection of random integration events. The promoter trap strategy described by Joyner and collaborators (1989) was excluded for *Krox-20* since the gene is not expressed in ES cells. The positive-negative selection based on the introduction of a Herpes Simplex Virus type 1 thymidine kinase expression cassette at the extremity of the vector has been shown to improve the proportion of homologous recombinant clones among G418-resistant clones by a fac-

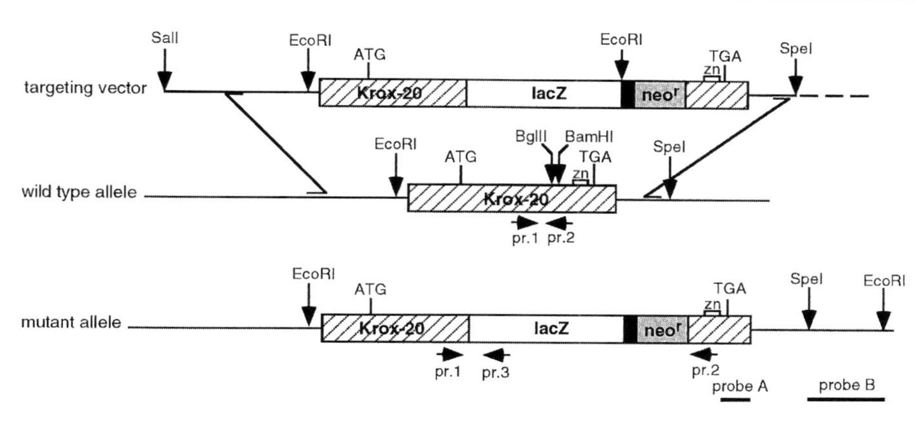

Fig. 25.1. Targeted inactivation of the *Krox-20* gene. The targeting vector contains an in-frame insertion of *lacZ* and a cassette containing the *neo^r* gene under the control of the phosphoglycerate kinase promoter. The *dotted line* corresponds to plasmid sequences. The diagrams of the wild-type and mutant *Krox-20* alleles show the sites used for the cloning and for the Southern blotting experiment. Also given are the positions of the PCR primers *pr1, 2,* and *3,* used for determining the genotype of the embryos, and of the probes used in Southern blotting (*probe A*) and in situ hybridization experiments (*probe B*). *zn* Region coding for the zinc-finger DNA binding domain of the Krox-20 protein. The expected positions of the cross-over between the replacement vector and the wild-type allele are indicated

tor of up to 2000 (Mansour et al. 1988). In our hands, however, the use of this strategy for *Krox-20* did not lead to a significant increase in the proportion of homologous recombinant clones. This is presumably because in most cases of random integrations the extremities of the construct were deleted, including part or all of the TK gene. Therefore, in our final targeting vector (Fig. 25.1) we did not include the TK gene.

25.1 X-Gal Staining on Whole Mount Embryos and Mice

There are some variations in the staining protocols depending on the age of the embryo or animal, but the basic X-Gal staining solution is the same for all stages.

Materials

Solutions

X-Gal staining solution

- PBS solution containing:
  Potassium hexacyanoferrate (II) (ferrocyanide) 4 mM (stock 200 mM)
  Potassium hexacyanoferrate (III) (ferricyanide) 4 mM (stock 200 mM)
  $MgCl_2$ 4 mM
  X-Gal (5-bromo-4-chloro-3-indolyl-β-D-galactopyranoside) 400 μg/ml (stock 40 mg/ml in dimethylsulfoxide, kept frozen in aliquots).

Procedure

Dissect out the embryos from the decidua. Keep yolk sacs for genotyping by PCR amplification.

Embryos up to 9.5 dpc

1. Rinse the embryos in PBS.
2. Fix for 5 to 15 min, depending on the embryonic stage, in 4 % paraformaldehyde (PFA) in PBS.
3. Rinse three times in PBS.
4. Transfer into X-Gal staining solution. Incubate for at least 1 h (up to overnight) at 30 °C. Agitation is not necessary at these early stages.

Embryos from 10.5 to 14.5 dpc

1. Rinse the embryos in PBS.
2. Fix in 4 % PFA, 0.2 % glutaraldehyde in PBS for 15 to 30 min.
3. Rinse three times in PBS.
4. Incubate at 30 °C in X-Gal staining solution supplemented with 0.02 % Nonidet P40 (NP40) and 0.01 % sodium deoxycholate, under agitation.

Embryos after 15.5 dpc and postnatal animals

Embryos older than 15.5 dpc are stained better by perfusion through the umbilical vein. Newborn to adult animals are fixed and stained by transcardial perfusion. This technique is also used to fix the embryos and animals for immunocytochemistry. The basic system is the same for transcardial and umbilical vein perfusions, although the source of pressure is different. For transcardial perfusion, gravitation pressure can be used. For perfusion through the umbilical vein, it is better to use a perfusion pump (Bioblock, cat. no. B88801) to obtain an accurate adjustment of the pressure. Before each perfusion, remove any air bubbles form the perfusion apparatus by passing PBS through all the tubing.

Anesthesize the embryos on ice and the animals with CO_2. Pin down the animal. For transcardial perfusion, open and pin out the rib cage. Open the left ventricle with fine scissors and introduce the needle (blunt-ended). Open the right auricle with forceps in order to let the blood flow out of the body. Perfuse with the following solutions in succession:

1. PBS containing 20 u/ml heparin for 5 min.
2. Fixative (4 % PFA, 5 mM EGTA, 2 mM $MgCl_2$ in PBS), for 8–10 min.
3. PBS for 2 min.
4. X-Gal staining solution for 5 min.

After perfusion, dissect out the organs and soak them overnight in staining solution at 30 °C under agitation.

For perfusion through the umbilical vein, insert the needle into the vein and proceed as for transcardial perfusion. The perfusion times given above are for adult animals and can be reduced for newborns or embryos.

25.2 Clearing Embryos with Benzyl Benzoate/Benzyl Alcohol

Whole embryos can be cleared in order to see the staining of internal tissues. After staining, dehydrate the embryo in a graded series of ethanol (75, 95, 100 %), then transfer into the clearing solution.

Materials

Clearing solution
- 2/3 benzyl benzoate
- 1/3 benzylalcohol

Procedure

The clearing reaction takes from 30 min to 3 h, depending on the size of the sample. Control the reaction by checking under a dissecting microscope. When the specimen is cleared, take pictures immediately. Pro-

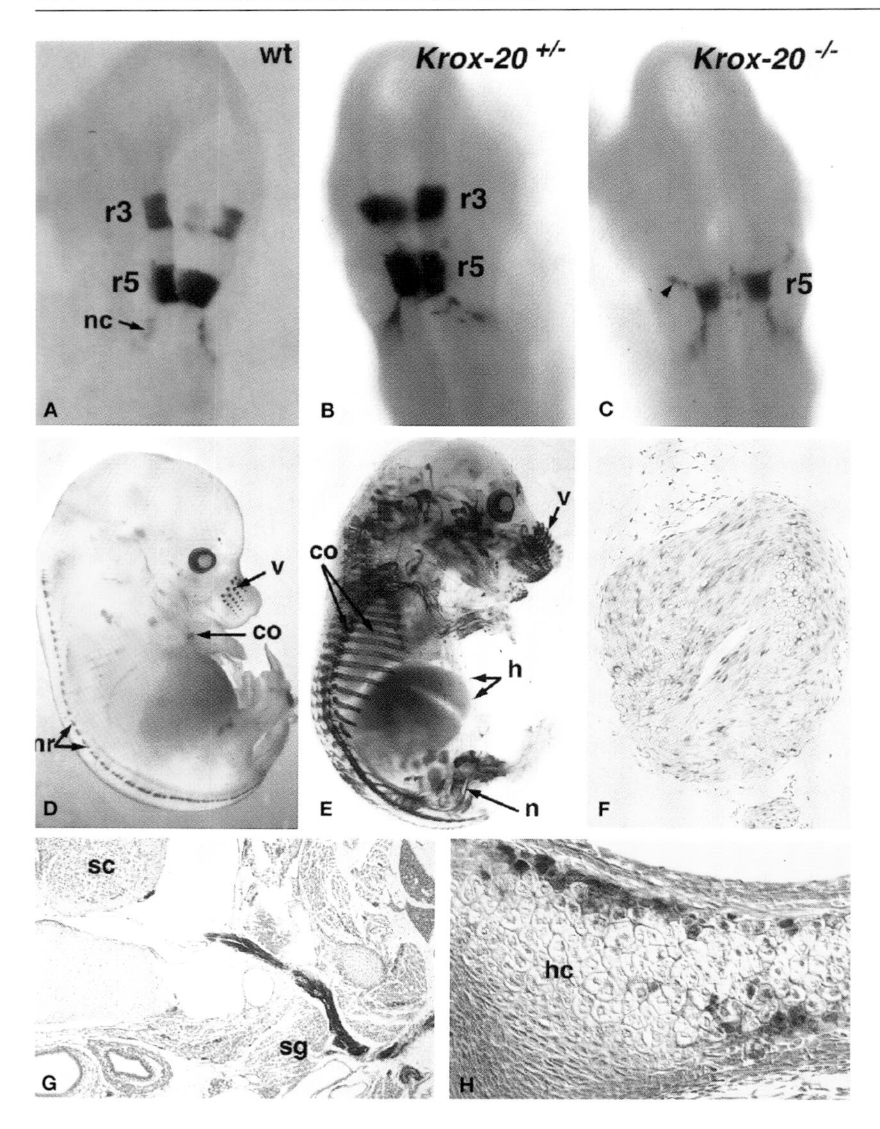
wt
Krox-20 +/-
Krox-20 -/-
r3
r5
nc
r3
r5
r5
A
B
C
v
co
nr
D
co
v
h
n
E
F
sc
sg
G
hc
H

◀ **Fig. 25.2A–H.** Expression of the Krox-20/β-galactosidase fusion protein in heterozygous and homozygous mutant embryos at different stages of development. **A** Whole-mount in situ hybridization analysis of a 9-dpc embryo with the *Krox-20* probe. Labeling is observed in r3 and r5 and in a stream of cells, presumably migrating neural crest cells (*nc*), caudal to r5. **B** Whole-mount X-Gal staining of a 9-dpc heterozygous mutant embryo. Note that the staining is very similar to the *Krox-20* mRNA pattern in **A.** **C** Whole-mount X-Gal staining of a 9-dpc homozygous mutant embryo. The r3 stripe has already disappeared at this stage. A second stream of putative neural crest cells is detected rostral to r5 (*arrowhead*). **D** Whole-mount X-Gal staining of a 14.5 dpc *Krox-20*$^{+/-}$ embryo. The staining is restricted to the follicles of the vibrissae (*v*), to the spinal and cranial nerve roots (*nr*) and to centers of ossification in the diaphysis of long bones and ribs (*co*). **E** Whole mount X-Gal staining of a 16.5-dpc *Krox-20*$^{-/-}$ embryo. This embryo was cleared after X-Gal staining, as described in the protocol section. Peripheral nerves (*n*), follicles of hair (*h*), vibrissae (*v*), and centers of endochondral ossification (*co*) are stained. **F** Transverse section through the sciatic nerve of a 3-month old *Krox-20*$^{+/-}$ mouse, stained with X-Gal and hematoxylin. Krox-20/β-galactosidase activity is localized within the cytoplasm of myelinating Schwann cells. **G** Transverse section through a 17.5-dpc *Krox-20*$^{-/-}$ embryo. At this stage the spinal nerves are stained along their entire length. *sc* Spinal cord; *sg* sympathetic ganglion. **H** Section through the hypertrophic cartilage of a 17.5-dpc *Krox-20*$^{-/-}$ vertebral process. Krox-20/β-galactosidase is detected in hypertrophic (*hc*) and calcifying chondrocytes

longing the clearing process leads to decoloration of the sample. The reaction can be stopped by returning the embryo to ethanol.

Results

Targeted Insertion of the *E. coli lacZ* Coding Sequence into the *Krox-20* Gene in ES Cells and Mice

- Construction of the vector for homologous recombination
  The replacement vector pK20LacZneo is shown in Fig. 25.1. It contains a 10-kb SalI-SpeI genomic DNA fragment spanning the *Krox-20* open reading frame (ORF). The *E. coli lacZ* coding sequence is inserted in frame with the *Krox-20* ORF, at the BglII site, and a cassette containing the G418 resistance gene (*neo*^{r}), under the control of the phosphoglycerate kinase promoter (PGKNeo; Adra et al. 1987), is present just downstream of *lacZ* at the BamHI site (Fig. 25.1). The hybrid *Krox-20-lacZ* gene gives rise to a chimeric protein consisting of the 128 N-terminal amino acids of Krox-20 fused to almost the entire β-galactosidase protein. This protein does not include the Krox-20 DNA

binding domain, which consists of three zinc fingers. Therefore, the targeted allele is expected to be null in terms of Krox-20 activity. The absence of *Krox-20* transcripts corresponding to the region 3' to the neo^r insertion site was confirmed by in situ hybridization on homozygous embryos between 8.5 and 9 days postcoitum (dpc) (Fig. 25.1, probe B).
Since *Krox-20* is not expressed at detectable levels in mouse ES cells (our unpubl. data), the β-galactosidase activity of the fusion protein could not be checked directly in homologous recombinant ES clones. The activity of the hybrid protein was therefore tested by transfection of pK20lacZneo into fibroblasts and subsequent serum stimulation of the cells. A strong X-Gal staining was observed in transfected cells after serum induction and not in uninduced controls, demonstrating that the fusion protein was active and that its expression was reflecting that of *Krox-20* in cultured fibroblasts (Chavrier et al. 1989).

- Transfection and selection of homologous recombinant clones
The pK20LacZneo construct was transfected into mouse D3 ES cells by electroporation and the recombinant clones were selected with G418. Recombination events at the *Krox-20* locus were identified by Southern blotting analysis of ES cell genomic DNA digested with EcoRI and revealed with probe A (Fig. 25.1). Among 137 G418-resistant clones analyzed, 22 presented a pattern indicating a modification of one *Krox-20* allele, giving a frequency of 16 % of homologous recombination. These clones fell into two classes: seven clones had the predicted modification for a replacement event, while the other clones had a smaller fragment, suggesting a rearrangement during the recombination event, and were not further analyzed. However, since this rearrangement was obtained with a high frequency, the stability of the transgene following transmission in mice was later checked by Southern blotting analysis.

- Production of chimeric, heterozygous, and homozygous mice
One ES cell clone corresponding to a replacement event led to germ line transmission after injection into C57Bl6 blastocysts. The chimeras obtained were used to establish a pool of heterozygous carriers by crossing with C57Bl6 × DBA/2 F1 females. Presence of wild-type and *Krox-20*$^-$ alleles was determined by polymerase chain reaction (PCR) amplification using three primers, located in the *Krox-20* (pr1 and pr2) and *lacZ* (pr3) sequences (Fig. 25.1). In the course of the analysis of the *Krox-20/lacZ* expression pattern in heterozygous animals, we noticed a strong expression in hair follicles throughout the life of the

animals. Thus X-Gal staining was also used to determine the genotype of F_1 offsprings in outcrosses, which is both faster and cheaper than PCR amplification.

Mice heterozygous for the *Krox-20* mutation were healthy and fertile, and did not show any phenotype. In contrast, mice homozygous for the *Krox-20* mutation died shortly after birth. About two thirds of the homozygotes died during the first 48 h after birth. The others gained weight more slowly than their wild-type and heterozygote littermates and usually died around the end of the second postnatal week.

β-Galactosidase Activity in Early Embryos: a Role for *Krox-20* in Hindbrain Segmentation

- Since *Krox-20* had been shown to be expressed in the developing hindbrain between 8 and 10.5 dpc and was suspected to play a role in the segmentation process, we first focused our analysis on this period of embryogenesis (Schneider-Maunoury et al. 1993). The β-galactosidase pattern in heterozygous embryos reproduced faithfully the *Krox-20* expression pattern, both temporally and spatially, except that the fusion protein persisted slightly longer than the RNA (Fig. 25.2A,B). X-Gal staining first appeared in prospective r3 around 8 dpc, and a second, more caudal, stripe corresponding to prospective r5 was detected around 8.5 dpc. By 9 dpc both stripes were of similar intensity (Fig. 25.2B). At 9.5 dpc the intensity of the anterior stripe started to decrease, and by 10.5 dpc only the posterior stripe was barely visible.

In homozygous mutant embryos, both β-galactosidase positive stripes appeared normally in the hindbrain, at the right time and position, suggesting that r3 and r5 did form in the absence of the Krox-20 protein. However, X-Gal staining decreased much earlier than in heterozygous embryos. At 9 dpc, the r3 stripe had almost disappeared (Fig. 25.2C), and at 9.5 dpc the r5 stripe had also decreased dramatically. Two possible, nonexclusive explanations could be proposed for the early decrease in expression of the fusion protein in homozygous embryos: first, *Krox-20* may require its own product for the maintenance of its transcription in r3 and r5; alternatively, *Krox-20* disruption may lead to the disappearance of r3 and r5.

To discriminate between these two possibilities, we performed a detailed anatomical and molecular analysis of the hindbrain and of the cranial nerves in mutant embryos. Our results favored the disappearance of r3 and r5: the total length of the rhombencephalon was reduced by about 30 %, the Vth and VIIth nerves roots and ganglia, which exit the brain at the level of r2 and r4, respectively, were brought together. The pattern of expression of rhombomere-specific genes was also consistent with an absence of r3 and r5. More recently, a careful observation of motor neuronal populations within the hindbrain and of rhombomere boundaries confirmed the total disappearance of rhombomeres 3 and 5 in homozygous mutant embryos (Schneider-Maunoury et al. 1997). Together, these data suggest that *Krox-20* is required for the maintenance of r3 and r5 but not for their initial formation. The possible mechanisms for the disappearance of r3 and r5 in *Krox-20*$^{-/-}$ embryos will be discussed below.

β-Galactosidase Activity in Older Embryos: Involvement of *Krox-20* in Additional Developmental Processes

- Expression pattern of *Krox-20/lacZ*

  To identify late *Krox-20* expression sites, the activity of the Krox-20/β-galactosidase protein was analyzed in whole mount heterozygous embryos from 10.5 to 17.5 dpc. The fusion protein was detected in three main sites: peripheral nerves, centers of endochondral ossification, and developing whisker and hair follicles. No difference was observed between heterozygous and homozygous mutants, except for the intensity of the staining which was consistently much higher in homozygous embryos. The expression pattern of a 14.5 dpc heterozygous embryo and of a 16.5 dpc homozygous embryo are presented in Fig. 25.2D,E).

  Expression in the peripheral nervous system was observed from 15.5 dpc onwards along the whole length of peripheral nerves, excluding the sensory ganglia and the sympathetic system (Fig. 25.2E,G). In cranial and spinal nerve roots, an expression was detected as early as 10.5 dpc (Fig. 25.2D). In developing bones, β-galactosidase activity was first detected in the central region of the cartilage model of long bones at around 14.5 dpc (Fig. 25.2D), in a population of chondrocytes located at the periphery of the diaphyseal part of the cartilage model. At later stages of development, β-galactosidase activity appeared progressively in all bones undergoing endochondral ossification in a temporal sequence similar to that described for the initiation of ossification. The staining was localized in hypertrophic and calcifying chondrocytes (Fig. 25.2G). In whisker follicles, *Krox-20/lacZ* expres-

sion was detected from 13.5 dpc and followed the spatial and temporal gradient of maturation of the whisker pad (Fig. 25.2D). Expression in hair follicles appeared later, at 15.5 dpc (Fig. 25.2E). The fusion protein was also detected in tendons, some muscle cells, and in developing teeth (data not shown).
Since most of these expression sites had not been reported previously, we performed in situ hybridizations with *Krox-20* probe A (Fig. 25.1) on embryos at different stages, to establish that the expression of the fusion gene in heterozygous embryos did indeed reflect the expression of *Krox-20* in wild-type embryos (Topilko et al. 1994 and unpubl. data).

- β-Galactosidase expression in Schwann cells and the observation of a myelination defect in *Krox-20* homozygous mutants
  Krox-20/lacZ expression in the peripheral nerves was of particular interest since we had noticed that *Krox-20*$^{-/-}$ animals that survived up to 15 days trembled, suggesting a possible effect of the mutation on the peripheral nervous system. We therefore analyzed the expression in peripheral nerves in further detail (Topilko et al. 1994). Observation of sections of sciatic nerves of 15 days postnatal (dpn) heterozygous offsprings demonstrated that the expression was localized to the cytoplasm of myelinating Schwann cells (Fig. 25.2F). Expression in these cells persisted throughout the life of the animal.
  The PNS myelin is an extension of the Schwann cell plasma membrane that wraps around axons tightly and allows rapid conduction of nerve impulses. The first step in myelin formation is the engulfment of the axon by cytoplasmic processes of the Schwann cell. Then the inner cytoplasmic lip of the Schwann cell progresses inward by displacing the outer lip from the axon and thus makes successive turns around it. A later step, called compaction, leads to the elimination of Schwann cell cytoplasm from the myelin sheath. An analysis performed by electron microscopy of sciatic nerves of 15 dpn *Krox-20* $^{-/-}$ animals showed a total absence of myelin, while control offsprings at the same age presented a fully compacted myelin. In the homozygous nerves, the axon was surrounded by a Schwann cell that had made one and a half turns around it and had thus initiated myelination. Therefore the mutation seemed to block myelination after the initial recognition and engulfment of the axon by the Schwann cell, but before spiralization and compaction. This interpretation was confirmed using antibodies directed against different myelin and Schwann cell proteins. Immunocytochemistry experiments showed a large reduction in the late com-

ponents of the peripheral myelin, protein P0, and myelin basic protein (MBP), while the level of an early myelin marker, myelin-associated glycoprotein (MAG), and of the general Schwann cell marker S100 were not affected (Topilko et al. 1994).
These data suggest that Krox-20 is involved in the control of the myelination process and directly or indirectly activates several late myelin genes. However, it is not yet known whether the downregulation of late myelin proteins in homozygous mutants is a cause or a consequence of the myelination defect.

- β-Galactosidase expression in developing bones points to a role for *Krox-20* in endochondral bone formation
  The demonstration of the expression of *Krox-20* in regions of bone formation led us to examine skeletal development in *Krox-20*$^{-/-}$ mice in greater detail (Levi et al. 1996). Long bones of *Krox-20*$^{-/-}$ mutant mice were shorter and thinner than those of their wild-type littermates, more porous and more transparent to X-rays, suggesting a calcium deficiency. Histological analysis revealed a lesion characterized by a compression of the hypertrophic growth plate, a very strong reduction in the number and length of calcified trabeculae and abnormalities in the endosteal part of the diaphyseal bone. Abnormal development of the growth plate is already observed at birth and persists during postnatal development. It is difficult to know from these data which cell type(s) is affected in homozygous mutants. One possibility is that disruption of *Krox-20* prevents the last step of chondrocyte differentiation into "osteoblast-like" cells. Alternatively, the mutation might affect the differentiation of endochondral osteoblasts, which express the gene during their terminal differentiation into osteocytes.

Use of the *Krox-20/lacZ* Reporter to Analyze the Regulation of *Krox-20* Expression in Schwann Cells

- It is known that Schwann cells cocultured with axons can initiate the myelination program, including synthesis of myelin proteins like P0 and MBP, and even formation of compacted myelin sheets. Therefore, we used cultures of Schwann cells from heterozygous animals as a tool to analyze the regulation of *Krox-20* expression in these cells (Murphy et al. 1996). When cultured in standard conditions, Schwann cells from sciatic nerves of heterozygous newborn animals rapidly lost *Krox-20/lacZ* expression. However, when these cells were cocultured with primary neurons from dorsal root ganglia explants of wild-type 17.5 dpc embryos, *Krox-20/lacZ* expression was reactivated within 36 h. Effective induction of β-galactosidase expression required axonal contact. Using this system, we were able to analyze the characteristics of *Krox-20* induction in Schwann cells and their precursors. In vivo, *Krox-20* expression is first detected from 10.5 to 14.5 dpc in nerve roots (Fig. 25.2D). A major transition occurs at 15.5 dpc, when *Krox-20* is activated along the whole length of the peripheral nerves (Fig. 25.2E, G). No *Krox-20* expression is detected in the sensory ganglia or in the sympathetic system. We have shown that the induction of *Krox-20* expression in peripheral nerves at 15.5 dpc is due to a change in the responsiveness of Schwann cells, which corresponds to the transition of Schwann cell precursors to mature Schwann cells. We also showed that induction of *Krox-20* expression in nerve roots as early as 10.5 dpc is due to a diffusible factor synthesized by the neural tube and whose effect can be mimicked by Neu differentiation factor β (NDFβ). Finally, we have shown that in sensory ganglia, the microenvironment is capable of negatively regulating *Krox-20*, an inhibition which is removed when the ganglia are placed in culture. Therefore the analysis of β-Galactosidase expression in cultured *Krox-20*$^{+/-}$ Schwann cells revealed three levels of regulation of *Krox-20* expression during glial cell development (Murphy et al. 1996).

Comments

We have obtained mice in which the *Krox-20* gene has been replaced by a mutant form containing an in frame insertion of the E. coli *lacZ* gene. Analysis of these mutant mice has shown that *Krox-20* is involved in developmental events as diverse as central nervous system segmentation, peripheral nervous system myelination, and endochondral bone formation. This study has revealed a dual role for the Krox-20 transcription factor in mouse development: during myelination and endochondral ossification, the gene seems to be necessary for turning on a differentiation program, whereas in the hindbrain it is involved in the maintenance of its expression territories. There are many such examples of one regulatory gene being used in several developmental pathways.

The insertion of *lacZ* as a reporter in the *Krox-20* targeting vector proved to be useful in many different ways: (1) the observation of the early disappearance of the *Krox-20/lacZ*-expressing cells in the rhombencephalon of *Krox-20*$^{-/-}$ embryos suggested that the inactivation of *Krox-*

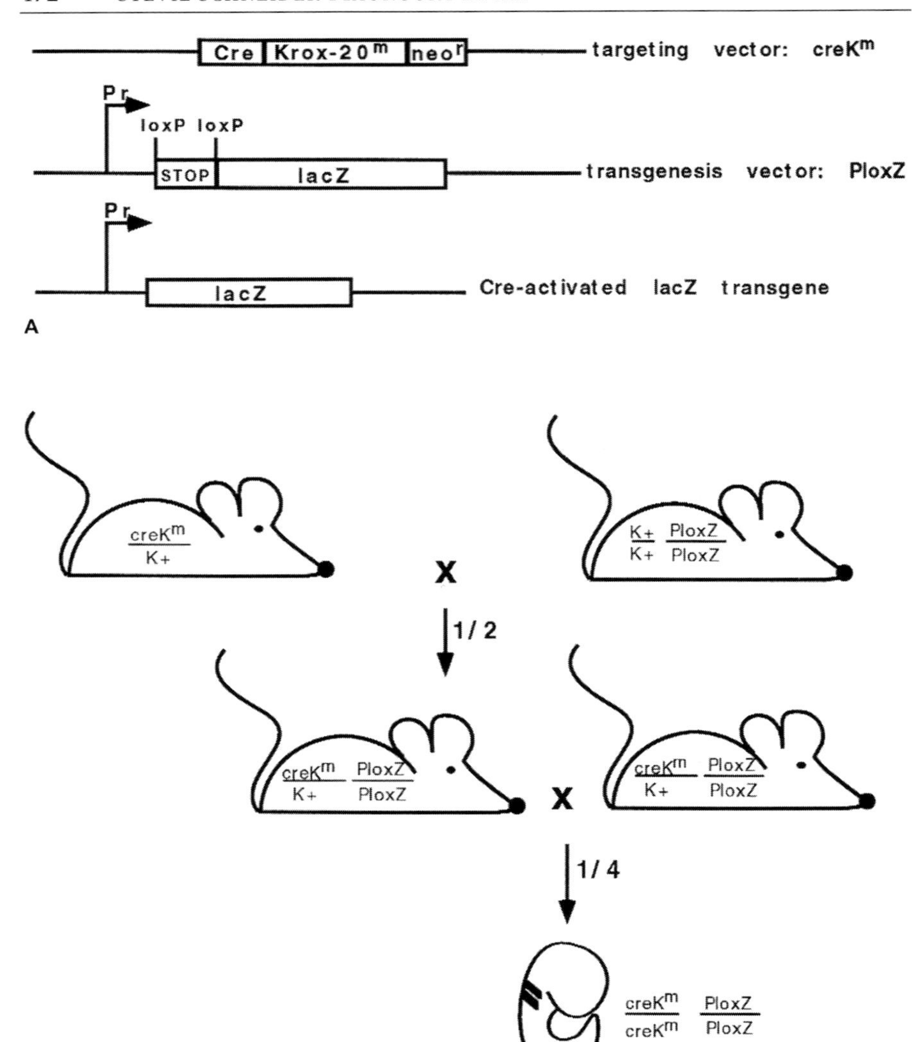
Cre
Krox-20m
neor
targeting vector: creKm
Pr
loxP loxP
STOP
lacZ
transgenesis vector: PloxZ
Pr
lacZ
Cre-activated lacZ transgene
A
creKm
K+
K+ PloxZ
K+ PloxZ
X
1/2
creKm PloxZ
K+ PloxZ
X
creKm PloxZ
K+ PloxZ
1/4
creKm PloxZ
creKm PloxZ
B

◀ **Fig. 25.3A,B.** Use of the cre/lox system combined to detection of β-galactosidase activity to study cell fate in *Krox-20* mutant embryos. **A** The targeting vector *creK*^{m} contains the *Krox-20* gene in which a mutation has been introduced (*Krox-20*^{m}). The Cre gene is inserted upstream of the *Krox-20* open reading frame, under the control of the regulatory sequences of the *Krox-20* gene. A *neo*^{r} expression cassette is inserted dowstream of *Krox-20*. The transgenesis vector *PloxZ* is composed of the *E. coli lacZ* coding sequence placed under the control of an ubiquitous promoter (*Pr*). A translational stop sequence is placed upstream of *lacZ* and is flanked by two loxP sites. The Cre-activated *lacZ* transgene corresponds to the stable modification of the transgenesis vector in cells in which the Cre recombinase has been expressed: the stop sequence between the two loxP sites has been removed and the *lacZ* gene is now expressed in these cells and all their descendants. **B** Strategy to obtain *Krox-20* homozygous mutant embryos expressing *lacZ* in r3 and r5 cells and all their descendants

20 led to the disappearance of rhombomeres 3 and 5, which could be confirmed by a careful anatomical and molecular analysis of the hindbrain and cranial nerves. (2) X-Gal stainings in heterozygous embryos allowed us to detect several potential *Krox-20* expression sites which had not been identified by previous in situ hybridization analyses. In two of these sites, the absence of a functional Krox-20 protein was shown to induce severe defects: absence of mature myelin in the peripheral nervous system and perturbation of endochondral bone formation. (3) Primary cultures of Schwann cells from heterozygous offsprings provided a tool to study the regulation of *Krox-20* expression in the peripheral nervous system (Murphy et al. 1996). Other knockout experiments have benefited from the use of the *lacZ* reporter in a similar manner (see, for instance, Tajbakhsh and Buckingham 1994; Tajbakhsh et al. 1994; Acampora et al. 1995).

Many questions concerning the function of the *Krox-20* gene remain unanswered, partly because the analysis of the mutant is not complete, and partly because of the limitations of the strategy we used to inactivate the gene. For instance, although several target genes of the Krox-20 transcription factor have been identified in the hindbrain (Sham et al. 1993; Nonchev et al. 1996; Seitanidou et al., 1997), it is not yet known whether the different functions of Krox-20 reflect differences in its direct targets, or whether there is a common pathway in response to Krox-20 activation in different cell types.

Another unanswered question is the fate of the r3 and r5 cells in the absence of Krox-20. Analysis of *Krox-20* mutant embryos showed that the r3 and r5 territories disappear, but the mechanisms responsible for this disappearance are not known. At least three hypotheses can be envisaged: cell death, cessation of cell proliferation, and change in cell identity. We could not detect any increased cell death in the hindbrain of *Krox-20* $^{-/-}$ embryos between 8.5 and 10.5 dpc (O. Voiculescu, unpubl. data). To address the question more generally, we are introducing another targeted mutation within the *Krox-20* locus. This experiment is described in Fig. 25.3, and is based on the use of the *lacZ* reporter and of the Cre/lox recombination system (Sauer and Henderson 1988). A targeting vector was constructed in which the *Cre* gene was introduced into the *Krox-20* locus. Mice carrying this mutation are currently being generated and will be crossed with transgenic mice carrying the *lacZ* gene under the control of an ubiquitous promoter. This reporter contains a translational stop flanked by two loxP sites, and expression of *lacZ* is released upon excision of this fragment by the Cre recombinase (Araki et al. 1995). The *Krox-20* homozygous embryos resulting from these crosses should therefore express *lacZ* permanently in r3/r5 cells and in all their descendants, thus allowing the determination of the fate of these cells. This strategy should enable further understanding of *Krox-20* function during hindbrain segmentation.

Acknowledgments. We thank Dr. Julie Adam for critical reading of the manuscript.

References

Acampora D, Mazan S, Lallemand Y, Avantaggiato V, Maury M, Simeone A, Brûlet P (1995) Forebrain and midbrain regions are deleted in *Otx-2* $^{-/-}$ mutants due to a defective anterior neuroectoderm specification during gastrulation. Development 121:3279–3290

Adra CN, Boer PH, McBurney MW (1987) Cloning and expression of the mouse *pgk-1* gene and the nucleotide sequence of its promoter. Gene 60:65–74

Araki K, Araki M, Miyazaki J, Vassalli P (1995) Site-specific recombination in fertilized eggs by transient expression of the Cre recombinase. Proc Natl Acad Sci USA 92:160–164

Chavrier P, Zerial M, Lemaire P, Almendral J, Bravo R, Charnay P (1988) A gene encoding a protein with zinc fingers is activated during G0/G1 transition in cultured cells. EMBO J 7:29–35

Chavrier P, Janssen-Timmen U, Mattei MG, Zerial M, Bravo R, Charnay P (1989) Structure, chromosome location, and expression of the mouse zinc finger gene *Krox-20*: multiple gene products and coregulation with the proto-oncogene *c-fos*. Mol Cell Biol 9:787–797

Chavrier P, Vesque C, Galliot B, Vigneron M, Dolle P, Duboule D, Charnay P (1990) The segment-specific gene Krox-20 encodes a transcription factor with binding sites in the promoter region of the Hox-1.4 gene. EMBO J 9:1209–1218

Deng C, Capecchi MR (1992) Reexamination of gene targeting frequency as a function of the extent of homology between the targeting vector and the target locus. Mol Cell Biol 12:3365–3371

Hasty P, Rivera Perez J, Bradley A (1991) The length of homology required for gene targeting in embryonic stem cells. Mol Cell Biol 11:5586–5591

Joyner AL, Skarnes WC, Rossant J (1989) Production of a mutation in mouse En-2 gene by homologous recombination in embryonic stem cells. Nature 338:153–156

Krumlauf R (1994) Hox genes in vertebrate development. Cell 78:191–201

Le Mouellic H, Lallemand Y, Brulet P (1990) Targeted replacement of the homeobox gene Hox-3.1 by the *Escherichia coli* lacZ in mouse chimeric embryos. Proc Natl Acad Sci USA 87:4712–4716

Levi G, Topilko P, Schneider-Maunoury S, Lasagna M, Mantero S, Cancedda R, Charnay P (1996) Defective bone formation in *Krox-20* mutant mice. Development 122:113–120

Mansour SL, Thomas KR, Capecchi MR (1988) Disruption of the proto-oncogene int-2 in mouse embryo-derived stem cells: a general strategy for targeting mutations to non-selectable genes. Nature 336:348–352

Mansour SL, Thomas KR, Deng CX, Capecchi MR (1990) Introduction of a lacZ reporter gene into the mouse int-2 locus by homologous recombination. Proc Natl Acad Sci USA 87:7688–7692

Murphy P, Topilko P, Schneider-Maunoury S, Seitanidou T, Baron-Van Evercooren A, Charnay P (1996) The regulation of *Krox-20* expression reveals important steps in the control of peripheral glial cell development. Development 122:2847–2857

Nonchev S, Vesque C, Maconochie M, Seitanidou T, Ariza-McNaughton L, Frain M, Marshall H, Sham MH, Krumlauf R, Charnay P (1996) Segmental expression of *Hoxa-2* in the hindbrain is directly regulated by *Krox-20*. Development 122:543–554

Sauer B, Henderson N (1988) Site-specific DNA recombination in mammalian cells by the Cre recombinase of bacteriophage P1. Proc Natl Acad Sci USA 85:5166–5170

Schneider-Maunoury S, Topilko P, Seitanidou T, Levi G, Cohen-Tannoudji M, Pournin S, Babinet C, Charnay P (1993) Disruption of Krox-20 results in alteration of rhombomeres 3 and 5 in the developing hindbrain. Cell 75:1199–1214

Schneider-Maunoury S, Seitanidou T, Charnay P, Lumsden A (1997) Segmental and neuronal architecture of the hindbrain of *Krox-20* mouse mutants. Development 124:1215–1226.

Seitanidou T, Schneider-Maunoury S, Desmarquet C, Wilkinson DG, Charnay P (1997) Krox-20 is a key regulator of rhombomere-specific gene expression in the developing hindbrain. Mech Dev 65:31–42

Sham MH, Vesque C, Nonchev S, Marshall H, Frain M, Gupta RD, Whiting J, Wilkinson D, Charnay P, Krumlauf R (1993) The zinc finger gene *Krox20* regulates HoxB2 (Hox2.8) during hindbrain segmentation. Cell 72:183–196

Tajbakhsh S, Buckingham ME (1994) Mouse limb muscle is determined in the absence of the earliest myogenic factor myf-5. Proc Natl Acad Sci USA 91:747–751

Tajbakhsh S, Vivarelli E, Cusella De Angelis G, Rocancourt D, Buckingham M, Cossu G (1994) A population of myogenic cells derived from the mouse neural tube. Neuron 13:813–821

Topilko P, Schneider-Maunoury S, Levi G, Baron-Van Evercooren A, Chennoufi AB, Seitanidou T, Babinet C, Charnay P (1994) Krox-20 controls myelination in the peripheral nervous system. Nature 371:796–799

Vesque C, Charnay P (1992) Mapping functional regions of the segment-specific transcription factor Krox-20. Nucleic Acids Res 20:2485–2492
Wilkinson DG (1993) Molecular mechanisms of segmental patterning in the vertebrate hindbrain. Perspect Dev Neurobiol 1:117–125
Wilkinson DG, Bhatt S, Chavrier P, Bravo R, Charnay P (1989) Segment-specific expression of a zinc-finger gene in the developing nervous system of the mouse. Nature 337:461–464

Chapter 26

Transgenic Mouse Strategies in Virus Research

Angel Cid-Arregui[1*], Nestor Morales-Peza[2], Prasert Auewarakul[3], Maria Jose Garcia-Iglesias[4], Maria Victoria Juarez[5], Gina Diaz[2], and Alejandro Garcia-Carranca[2]

Introduction

The use of viruses and viral genes for infection and microinjection experiments with mouse embryos has greatly contributed to the development of transgenic technologies. Likewise, these techniques have proved very useful in providing insights into basic aspects of virus research, specially into those related to viral pathogenesis and virus-host interactions. Indeed, an impressive number of studies involving transgenic mice carrying viral sequences have been reported during the past 10 years (nearly 900 records can be retrieved from the Medline databases under the entries transgenic mouse and virus). The following is a summary of the main topics on virus research using transgenic mice, including some illustrative references:

- Studies on the oncogenic activity of viral proteins (see for review Adams and Cory 1991; Merlino 1994). Oncogenes of the so-called tumor viruses (zur Hausen 1991) have been expressed in mouse tissues from either homologous or heterologous viral control regions, or with the help of eukaryotic promoters. In the latter case, promoters of

* corresponding author: phone: +49-6221-548321; fax: +49-6221-548301; e-mail: cid@urz.uni-heidelberg.de
[1] Dept. of Neurobiology, University of Heidelberg, Im Neuenheimer Feld 364, D-69120 Heidelberg, Germany
[2] Dept. of Molecular Biology, Institute for Biomedical Research, 04510 Mexico City, Mexico
[3] Dept. of Microbiology, Faculty of Medicine, Mahidol University, 10700 Bangkok, Thailand
[4] Dept. of Animal Pathology: Animal Medicine. Faculty of Veterinary Science, University of Leon, 24071-Leon, Spain
[5] European Molecular Biology Laboratory, Meyerhofstrasse 1, D-69117 Heidelberg, Germany

genes expressed selectively in specific tissues have been preferred. Thus, for instance, the α-A crystallin promoter and the keratin 10 promoters have been used to target transgene expression to lens and skin, respectively (e.g., Griep et al. 1993; Auewarakul et al. 1994). In some cases, two or more viral oncogenes were combined in transgenic mice for studies on oncogene cooperation (e.g., van Lohuizen et al. 1991; see also Berns 1991), as was done previously in studies in vitro. In addition, transgenes containing tissue-specific promoters fused to viral oncogenes (most frequently the SV40 large T-antigen), have been used to assess many aspects of tumorigenesis, apoptosis, and cell cycle control. Recently, male germ cells have been reported to undergo transmeiotic differentiation in vitro by expression of the SV40 large T-antigen (see Rassoulzadegan et al., Chap. 12, this Vol.).

- Immunology research using viral genes. A great number of studies have used expression of viral proteins in transgenic mice to assess different aspects of immunology, such as antigen presentation, auto-antibody production, peptide-induced tolerance, ablation of tolerance, T-cell selection, T-cell trafficking into the central nervous system, etc. (see, e.g., Ohashi et al. 1991; Oldstone and Southern 1993; Ashton-Rickardt et al. 1994; see also Moreno et al. Chap. 27, this Vol.). As an example, mice expressing viral proteins in the β-islets of pancreas have proved useful to study autoimmune diabetes since they have allowed to test induction of T-cell tolerance by peptides (Aichele et al. 1994).

- Specific cell ablation using viral products. Induction of lineage ablation was first accomplished by means of the cell-type-specific expression of a herpes simplex thymidine kinase gene in transgenic mice followed by treatment of animals with acyclovir (Borrelli et al. 1988). In other studies, tumor ablation has been achieved in transgenic mice using antisense inhibition of factors required for viral gene expression (e.g., Kitajima et al. 1992). Further, recombinant retroviruses can be used for cell-lineage studies in embryos (Sanes et al. 1986).

- Mouse models for human viral diseases. Transgenic mice are providing invaluable information on the mechanisms of viral pathogenesis. Thus, for instance, dozens of different transgenic mice carrying hepatitis B virus sequences have been generated which have provided new insights into the disease (see Chisari 1995 for a review). Mice transgenic for human immunodeficiency virus (HIV-1) have been useful in understanding the pathogenesis in the central nervous system which,

in the form of dementia, occurs in some patients suffering from acquired immunodeficiency syndrome (AIDS), while the virus itself is hardly detectable in the brain (reviewed in Klotman et al. 1995).
Transgenesis is a unique tool for in vivo studies on viruses that, challenging the Koch postulates, cannot be propagated in vitro or in laboratory animals. For example, the multistep carcinogenesis induced by specific types of human papillomaviruses in the uterine cervix is being addressed in transgenic mice expressing the viral oncogenes (see below). Transgenic mice have also provided important clues into the pathogenesis of the still poorly understood prion diseases (Prusiner 1993; Collinge et al. 1996).
Further the identification of cellular receptors for viruses is opening new avenues in virus research. Thus, for instance, the expression of the cellular receptor for poliovirus in the nervous system of transgenic mice has made it possible to generate an animal model for this disease (Koike et al. 1991).

- Vaccine and gene therapy studies. Although no transgenic mouse that will allow vaccine development has been generated to date, transgenic mice may be useful to test viral vaccine vectors (e.g., Andino et al. 1994) as well as new strategies for antiviral therapy, like the use of antisense oligodeoxynucleotides as inhibitors of viral gene expression (see Güimil Garcia and Eritja, Chap. 5, this Vol.).

- Studies on the regulation of viral gene expression. Transgenic mice can be used to dissect in vivo the mechanisms involved in tissue specificity, hormonal regulation, enhancement of expression by physical or chemical agents, and interaction with nuclear proteins essential during early development (e.g., Cid et al. 1993; see also Chap. 20, this Vol.).

In summary, a great variety of strategies involving transgenic mice have been devised during the past few years, and new ones are emerging, which will help improve our understanding of the mechanisms of viral pathogenesis and virus-host interactions in aspects that cannot be addressed in vitro. Here, we describe the strategies we have used to develop animal models for the oncogenesis by human papillomaviruses (HPVs), together with detailed descriptions of the protocols.

HPV-Transgenic Mouse Models

HPVs are small nonenveloped DNA viruses (72 capsomers, 55 nm diameter), of the papovaviridae family. Their genome consists of a circular double-stranded DNA molecule of about 8 kb (MW 5×10^6). To date, more than 70 HPV types have been isolated from benign and malignant epithelial neoplasias of the genital mucosae and the skin.

Concerning cervical cancer, which is an important cause of mortality in women worldwide, a large number of epidemiological and clinical studies implicate HPV 16 and 18 in its etiology (see zur Hausen and de Villiers 1994 for a review). HPV 16 and 18 are present in more than 90 % of primary cervical carcinomas, and also in their metastases and in cell lines derived from them, like HeLa cells, which carry multiple copies of the HPV18 DNA stably integrated into their genome. Interestingly, integration of the HPV DNA into the host genome leads to enhanced expression of the early viral genes E6 and E7 from their common upstream regulatory region (URR).

Many different in vitro studies have demonstrated that E6 and E7 are oncogenes (zur Hausen 1991). Although the mechanisms by which the E6 and E7 oncoproteins induce cellular transformation are not fully understood, this appears to be the consequence of E6 binding the tumor suppressor p53, thereby promoting its degradation via a ubiquitin-dependent pathway (Werness et al. 1990, Scheffner et al. 1990), and E7 interacting with pRB (the retinoblastoma susceptibility protein) which loses its capacity to functionally inactivate the cellular transcription factor E2F (Dyson et al. 1989; Chellappan 1992). It is, however, intriguing why the precursor lesions infected by HPVs need long latency periods of years or decades to develop cervical cancers, despite the fact that the E6 and E7 genes are expressed in these lesions. In fact, in vitro experiments have shown that E6 and E7 are able to immortalize human cells, but malignant transformation requires additional expression of other oncogenes, such as v-ras.

The study of these viruses has been complicated by the fact that they do not fulfill the Koch postulates, since they cannot be propagated in vitro. However, transgenic mouse technology has proved useful in providing a system in which these aspects of HPV pathogenesis can be addressed. Recently, we and others have demonstrated the oncogenic potential of the E6 and E7 genes and the expression capability of the HPV URR in transgenic mouse tissues (see Cid et al. 1993; Auewarakul et al. 1994 and references therein). To this end, we pursued a double strategy:

- Test expression from the HPV 16 and 18 promoters in mice harboring fusions of the respective URRs to heterologous reporter genes, namely the bacterial β-galactosidase (lacZ) and the chloramphenicol acetyl transferase (CAT) genes, both enzymes known to be innocuous for the animals and readily detectable in expressing tissues.
- Target expression of the HPV-16 E6/E7 genes to the epidermis. The E6/E7 open reading frame, which is expressed as a policystronic mRNA in infected cells, was fused to the cytokeratin 10 (K10) promoter to target its expression to the suprabasal (terminally differentiated) cells of the epidermis, which resemble the natural hosts of these viruses.

Materials

Equipment

- DNA and RNA work
  Materials for plasmid propagation and purification (there is no need for CsCl gradient purification, commercial kits like Qiagen or Geneclean can be used instead to purify DNA for microinjection experiments), electrophoresis chambers and glass plates for agarose and acrylamide gels, power supply, UV transilluminators (250 and 355 nm) coupled to an adequate photography system (e.g., Polaroid), microwave oven, thermoblock, UV cross-linker, slot blot apparatus, nitrocellulose or nylon membranes
- Tissue and embryo culture, biological safety cabinet, CO_2 incubator, inverted microscope, plastic dishes (3 and 6 cm∅)
- Production and analysis of transgenic mice
  Alcohol burner or small bunsen burner
  Surgical instruments: fine dissection forceps and scissors, blunt fine forceps, two pairs of watchmaker's no. 5 forceps, serrefine, surgical silk suture (size 5×0), curved surgical needle, wound clips, metal ear tags
  Microinjection setup (see Fig. 1.1A, Chap. 1, this Vol.): inverted microscope (Nikon TMS and Diaphot 300/2; Zeiss Axiovert 135), left and right-side Leitz micromanipulators, Eppendorf microinjector (Mod. 5242), micropipette puller, glass capillaries (see Materials in Chap. 1), microforge, holding pipettes
  Leitz stereomicroscope
  Fiber optic illumination

Embryo transfer pipettes, prepared with the burner by pulling Pasteur pipettes at their narrow side at some point 3–4 cm distant from the opening. It is convenient to polish the tips of the transfer pipettes by flaming them briefly.
Small container (1–2 l) for liquid nitrogen, cork sheet, tissue homogenizer (Polytron or similar), cryostat (Reichert-Jung)
Racks for slides, glass beakers, slide jars. Water bath with shaker tray
Glass slides. It is necessary to coat slides with silane or poly-L-lysine to ensure that tissue sections adhere firmly to their surface and do not become detached while performing X-Gal staining or in situ hybridizations. TESPA coating is advisable because it is easy to perform and the treated slides can be stored for several months (see below).

Reagents and solutions

- DNA and RNA work
  Injection buffer: 10 mM Tris-HCl (pH 7.6), 0.25 M EDTA
  Reagents and solutions for DNA and RNA extraction, purification, restriction enzyme digestions, Southern and Northern blots. Plasmids and reagents for preparation of antisense riboprobes
- Production of transgenic mice (see Hogan et al. 1994)
  Hormones for superovulation; hyaluronidase; media for embryo handling and culture (M2 and M16); paraffin liquid (Merck 7161.0500) anesthetics (Ketavet 100 mg/ml, Parke-Davis; Rompun 2 %, Bayer; inject i.p. 0.1 mg of each per g of body weight)
- Tissue handling
  Dulbecco's phosphate-buffered saline (PBS): 8 g NaCl, 0.2 g KCl, 1.44 g Na_2HPO_4, 0.24 g KH_2PO_4. Adjust pH with HCl to 7.2 (if it is for tissue culture and biopsies) or 7.3–7.4 (for histochemistry and in situ hybridization). Make up to 1 l with deionized water
  70 % ethanol, liquid nitrogen, isopentane
  Embedding medium: tissue freezing medium, Leica Instruments GmbH, cat. no. 0201 08926; or Tissue Tek, OCT compound, Miles Inc., cat. no. 4583.
  TESPA: 3-aminopropyl triethoxy-silane (Sigma A-3648). Store at 4 °C for up to 1 year. **Note:** Discard when it becomes yellow! Working solution: 2 % TESPA (v/v) in 100 % acetone. Prepare immediately before use!
- Slides

Coat glass slides as follows:

1. Cleaning: wash slides by consecutive immersion for 1–5 min into each of the following solutions:
   a) 10 % HCl, 70 % ethanol in deionized H_2O

b) deionized H_2O
c) 95 % ethanol.

Drain and incubate slides at 80 °C for 15 min. Then let them cool down to room temperature in a dust-free container. Alternatively, slides for in situ hybridization can be treated with DEPC as described below.

2. Coating: immerse slides for 10 s into each:
a) 2 % TESPA solution (see Reagents and Solutions, below)
b) 2× 100 % acetone
c) 1× deionized H_2O.

Dry at 37–42 °C (maximum) for 4 h or overnight at room temperature. Keep subbed slides in a clean plastic box until needed. If they remain stored for longer than 3 months, repeat coating before use

- β-Galactosidase histochemistry:
  PBS pH 7.3
  Fixative: 2 % paraformaldehyde (PFA), 0.2 % glutaraldehyde, 2 mM $MgCl_2$, in PBS pH 7.3.
  Always use freshly prepared fixative. Dissolve 1 g of PFA in 50 ml PBS in a fume hood, with stirring and heating at 60 °C. Add 15 μl of 10 N NaOH per min until the solution clears. Chill on ice. Add 0.1 ml of $MgCl_2$ and 0.4 ml of 25 % glutaraldehyde (Merck)
  Stock solutions: 50 mM potassium ferrocyanide ($K_4[Fe(CN)_6]$) and 50 mM potassium ferricyanide ($K_3[Fe(CN)_6]$), both in PBS pH 7.3. Store at 4 °C protected from the light. They are stable for several months
  Wash solution: 0.01 % sodium deoxycholate, 0.02 % NP-40, 2 mM $MgCl_2$, in PBS pH 7.3
  X-Gal (5-bromo-4-chloro-3-indolyl-β-D-galactopyranoside; Sigma B-4252). Use glass pipettes. Dissolve to 20 mg/ml in N,N-dimethylformamide and store in a glass container (never polycarbonate or polystyrene) at −20 °C in the dark
  X-Gal staining solution: 1 mg/ml X-Gal, 2 mM $MgCl_2$, 0.01 % sodium deoxycholate, 0.02 % NP-40, 5 mM potassium ferrocyanide, 5 mM potassium ferricyanide, in PBS pH 7.3 (for long incubations, add Tris pH 7.3 to 50 mM). Can be reused for months if stored at 4 °C protected from light (filter through 0.22 μm membranes after use)
- Bromodeoxyuridine (BrdU) labeling. Stock solution of BrdU 5 mg/ml in PBS. Anti-BrdU-fluorescein (Boehringer, Mannheim, cat. no. 1202 693)
- In situ hybridization

Diethyl pyrocarbonate (DEPC; Sigma D-5758). DEPC is a strong denaturing agent. It should be handled with care as it is suspected of being a carcinogen. DEPC is inactivated by autoclaving solutions or baking glassware. Handle it in a fume hood and wear gloves. Treat solutions (except those containing Tris, because DEPC reacts with Tris) with 0.1 % DEPC, incubate for at least 1 h at room temperature and then autoclave. Treat glassware with 0.1 % DEPC in water for 1 h, pour water off into a glass bottle for autoclaving, and bake glassware at 220 °C for 2 h
Fixative (prepare fresh): 4 % paraformaldehyde in PBS pH 7.4
20× SSC (3 M NaCl, 0.3 M sodium cytrate)
Fixogum glue or similar
Proteinase K (Merck 7393.0010, 20 mg/ml)
1 M glycine
Deionized formamide (stored at −20 °C)
Triethanolamine hydrochloride: 0.1 M solution, pH 8
Acetic anhydride
- Hybridization solution:

Stock Solution	**Final conc.**	**For 1 ml**
Deionized formamide	50 %	0.5 ml
20× SSC	2×	0.1 ml
Dextransulfate (MW = 50,000)	10 % (w/v)	0.1 g
1 M Tris.HCl (pH 7.5)	10 mM	10 µl
50× Denhardt's	1×	20 µl
1 % t-RNA	0.5 mg/ml	50 µl
5 mg/ml salmon sperm DNA	100 µg/ml	20 µl
10 % SDS	0.1 %	10 µl
Deionized water	to 1 ml	

Filter through a 0.22-µm membrane.

- FMS solution: 50 % formamide, 20 mM β-mercaptoethanol (β-ME), 2× SSC
- NTE buffer: 0.5 M NaCl, 10 mM Tris, 5 mM EDTA, pH 8
- Kodak NTB-2 emulsion, D19 developer
- 30 % sodium thiosulfate or Kodak Agefix
- 0.02 % toluidine blue (Sigma T-3260)
- Giemsa stock solution: 0.85 g Giemsa, 50 % glycerol, 50 % methanol (store at room temperature). Staining solution (prepare just before

use): 4 ml Giemsa stock solution, 2 ml 0.2 M sodium phosphate pH 6, to 100 ml with dionized water
- Xylene or Histoclear (National Diagnostics)
- DPX (BDH) or Permount (Fisher Scientific) mounting medium.

Procedures

[**Note:** for detailed descriptions of the molecular biology and transgenic mouse techniques and protocols, the reader should consult Sambrook et al. (1989), Wassarman and DePamphilis (1993), and Hogan et al. (1994)].

26.1 Preparation and Analysis of DNA and RNA for Transgenesis Experiments

The correct design and preparation of the DNA constructs to be micro-injected into embryos is crucial for the success of transgenesis experiments. Whenever possible, it is quite convenient to test such constructs, by transfection into suitable cultured cells, before using them to produce transgenic mice.

Further, appropriate RNA studies should allow quantitative determination of the levels of transgene expression in different tissues and, more importantly, identification of the mRNA species that are transcribed from the transgene. This is of particular importance when transgenes are composed of viral sequences which express polycistronic mRNAs, as is the case of HPVs.

Design of DNA constructs

In our studies we have used the following constructs:

- Plasmid pURR18LacZ (Cid et al. 1993), obtained by insertion of the HPV-18 URR (1050 bp) into the SmaI site of a promoterless lacZ expression vector. A similar construct containing the HPV-18 URR in front of the bacterial CAT gene was also used
- Plasmid pK10E6E7 (Auewarakul et al. 1994). This plasmid was obtained by inserting a XhoI-EcoRI fragment of the HPV16 DNA (nucleotides 58 to 867), carrying the E6 and E7 oncogenes, into the plasmid pBKVI (Blessing et al. 1989), which contains the bovine homolog of the human keratin 10 (K10) promoter. An SV40 poly(A)

sequence follows the 3' end of the E6/E7 fragment. The K10 sequence contains a TATA box but no cap site, which is provided by the E6/E7 sequence.

These constructs were partially sequenced to uncover possible mutations that might have been introduced accidentally during the subcloning procedure. We tested the pURR18lacZ plasmid for expression in vitro. However, it was not possible to test the pK10E6E7 plasmid, because the K10 promoter drives expression exclusively in terminally differentiated keratinocytes of the epidermis, which cannot be cultured.

Preparation of DNA samples for microinjection

DNA fragments for transgenic mouse experiments should contain the promoter and gene of interest devoid of bacterial sequences of the vector, because they might interfere with transgene expression (see Hogan et al. 1994). In our experiments, the URR18LacZ and K10E6E7 fragments were excised from the respective plasmids (see Fig. 26.1), separated electrophoretically and purified from agarose gels using glass powder. Finally, they were eluted with injection buffer and diluted to a concentration of 1–2 μg/ml for injection.

DNA isolation

At the age of 3–4 weeks, the mice are labeled with numbered ear tags and small biopsies of their tail tips (0.5–1 cm) taken and introduced into microfuge tubes labeled with the same numbers as the ear tags. This tissue is used for extraction of genomic DNA using a simplified but reliable protocol for genomic DNA extraction from mouse tissues described in detail by Murphy et al. (Chap. 15, Sect. 15.4, this Vol.). The extracted DNA is used to detect the presence of foreign injected sequences by Southern blot analysis. In general, 10–25 % of the offspring carry the transgene (see Hogan et al. 1994). Digestion of the genomic DNA with a restriction enzyme which does not cut within the transgene sequence allows estimate in the blots the number of copies integrated into the genome and assess if it is integrated at more than one site. The use of an enzyme which cuts into the transgene discloses rearrangements of the trangene which may occur during integration into the genome.

RNA isolation

Among the various methods for total RNA isolation from tissues we recommend those using chaotropic agents, because both purity and quantity of the recovered RNA are high. In particular, we recommend the purification of RNA from tissues extracted with guanidinium thiocyanate by centrifugation onto a highly dense solution of CsCl (see Sambrook et al. 1989). Since this method requires ultracentrifugation of

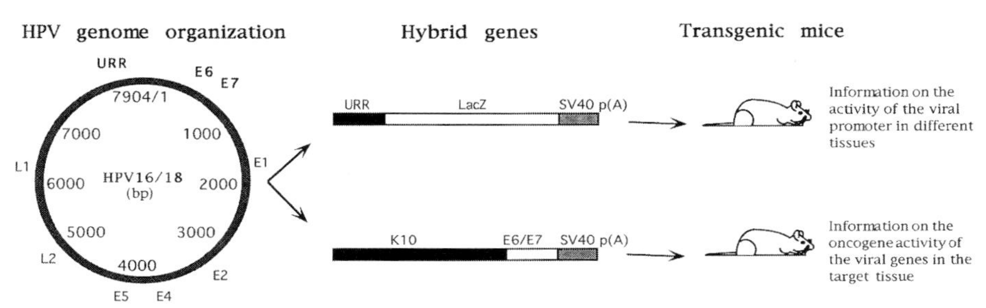

Fig. 26.1. Scheme of the genomic organization of HPV genomes and a double strategy to study HPV expression and tumorigenesis in transgenic mice. *Left* Genomic organization of HPVs. The viral genome consists of a circular double-stranded DNA of approximately 8 kb. It consists of a late region coding for capsid proteins (*L1, L2*), and an early region (E1 to E7 genes) coding for oncogenic (E6, E7) and regulatory proteins (*E1, E2*). A region of about 1 kb, known as URR, controls early gene expression. It harbors several DNA motifs recognized by cellular and viral factors. *Right* Representation of two different strategies used to generate transgenic mouse models carrying either the HPV URR fused to reporter genes (LacZ or CAT), or the E6 and E7 oncogenes under the control of the K10 promoter

samples for several hours, it may not be advisable for large numbers of samples. For this purpose, more simple methods should be tried (e.g., see Chomczinski and Sacchi 1987).

26.2 Production of Transgenic Mice

We have used the protocols described by Hogan et al. (1994) for the production of transgenic mice by means of DNA injection into the pronuclei of fertilized eggs. For our experiments, mouse fertilized eggs were obtained from superovulated (C57BL/6×C3H)F_1 females that were mated to (C57BL/6×C3H)F_1 males. Typically, 60–70 % of embryos survived injection and were transferred into the oviducts of pseudopregnant NMRI or CD1 recipient females on day 1 of gravidity. About 10–20 % of the transferred embryos developed normally to terminus, 15–25 % of these carrying one or more copies of the transgene. Permanent lines were established and maintained by crossing the founder animals and their progeny to F_1 nontransgenic hybrids. In about 50 founder mice obtained for 12 different constructs, 1 carried the transgene into the X chromosome and 2 had transgene concatemers integrated in two different sites. In one case, transgene integration occurred in two different chromosomes, resulting in two different transgenic lines from the same founder.

26.3 Processing of Tissue Samples

Expression or histopathological studies on transgenic animals is performed using whole organs or small samples of different tissues which, in most cases, are taken from mice sacrificed for the purpose. Samples may also be taken from organs of anesthesized mice if it is absolutely necessary to keep them alive for breeding or further examination. The latter, however, requires the assistance of a veterinary surgeon or experienced person.

Tissue sample preparation

Organs and tissues for RNA extractions and biochemical studies should be cut with the help of fine forceps and a scalpel into small pieces of about 0.5 cm^3, which are introduced separately into cryotubes labeled with the number of the ear tag, the name of the tissue, and the date. The tube is then closed and dropped carefully into liquid nitrogen. Tissue samples for histopathology and immunohistochemistry analysis should be immersed into formalin or PFA-fixative in isopentane, saturated in 30 % sucrose, embedded and frozen liquid nitrogen. Alternatively, the latter step can be performed using unfixed specimens.

1. Sacrifice mouse by cervical dislocation.
2. Soak it in 70 % ethanol.
3. Shave a small area of the skin and take 0.5–1 cm^2 of it for analysis. Skin samples may also be taken from other areas, like the tail or the paw.
4. Open the abdominal cavity: make a small transversal cut in the skin at the midpoint of the abdomen, pull the skin at both sides of the incision firmly in opposite directions (towards head and tail, respectively). Using fine scissors, cut the abdominal wall. The peritoneal cavity with its organs is then exposed.
5. Dissect organs carefully: remove the liver and take a piece of it, take the kidneys, the suprarrenal glands, the spleen, the pancreas, the stomach (cut longitudinally along the narrow curvature and wash it in a beaker filled with PBS), pieces of small and large intestine (cut longi-

tudinally and wash in a beaker filled with PBS). In females: take both ovaries, the uterus (one uterine horn cut longitudinally) and the vagina (cut longitudinally).

6. Open the thoracic cavity and dissect the heart (cut in two pieces), the lungs, the thymus and the mediastinum (contains trachea, esophagus and vessels).
7. Dissect the brain. If necessary, cut it to separate away the different parts (cerebellum, optic nerves, tracts, etc.).
8. Dissect the tongue, submaxilary glands, the eyes, a piece of muscle, and a bone (e.g., from the leg).

Fixation and sucrose embedding

Immerse tissue samples to be fixed into a 50-ml plastic tube containing 50 ml of fixative (4 % paraformaldehyde in PBS, prepare fresh as described above, without $MgCl_2$ and glutaraldehyde). Let fixation proceed at 4 °C for 1–2 h. Then drain the specimens on paper towels and transfer them to a 50-ml tube containing 50 ml of 30 % sucrose in PBS. Incubate on ice for 3–4 h or until the specimens sink deeply into the sucrose.

Deep freezing

For each sample proceed as follows: place the specimen on a piece of cork sheet (1.5–2 cm^2) that has a square glass coverslip inserted perpendicularly with aid of an scalpel to its upper surface (write on its back side the number of the sample), add embedding medium avoiding bubble formation, and place the specimen conveniently oriented using fine forceps. Then, drop the specimen into isopentane/liquid nitrogen. When all specimens have been frozen, take them from the liquid nitrogen with forceps, remove the glass coverslips, and keep them in a 50-ml plastic tube (labeled with the numbers of the samples) precooled by brief immersion into liquid nitrogen. Store at −80 °C.

Cryosections

The day before cutting, the specimens should be transferred to the −20 °C freezer. Transfer specimens to the cryostat chamber (usually set at −20 °C). Put a drop of water onto the sample holder and place a specimen on it (the embedded tissue facing up) before the water freezes. Wait 1 min until the specimen is stuck firmly. Cut sections of 5 μm and place them onto coated slides. Dry at room temperature for 1 h or longer. The sections can be stored at −80 °C in a box with dessicant for several months.

26.4 β-Galactosidase Histochemistry

The bacterial *lacZ* gene encodes a glycoside hydrolase, β-D-galactosidase (β-Gal) that is innocuous, very stable, and readily detectable in eukaryotic cells. This makes *lacZ* an ideal reporter gene which has been widely used for expression studies in transgenic mice as well as in cultured cells. A simple histochemical assay using the chromogenic substrate X-Gal, which is cleaved by β-Gal to yield blue insoluble indigo, enables quick and sensitive detection of β-Gal in tissue sections and whole organs. The following protocols are based on those described in Sanes et al. (1986), MacGregor et al. (1991), and Hogan et al. (1994). The reagents and solutions needed are described above. Note that PBS should have pH 7.3, which is the pH optimum of the bacterial β-Gal. In contrast, the two main groups of mammalian galactosidases, those located in the brush border of the small intestinal mucosa and those localized in lysosomes, have pH optima in the range of pH 6 and pH 3–4, respectively (see Wallenfels and Weil 1972).

X-Gal staining of cryosections

1. Immediately after dissection, fix tissue samples (pieces of approximately 0.5 cm^3, see above) by immersion into 50-ml fixative in a 50-ml plastic tube, at room temperature for 1–2 h or at 4 °C overnight. Keep fixative at 4 °C for step 7.
2. Wash with PBS/2 mM $MgCl_2$, at 4 °C, 2×5 min.
3. Incubate in 30 % sucrose in PBS/2 mM $MgCl_2$, at 4 °C for 1–3 h (avoid overnight incubations). The tissues, which initially float in this solution, migrate progressively towards the bottom of the tube as they become saturated with sucrose.
4. Embed specimens and freeze in isopentane/liquid nitrogen. Store blocks at −80 °C.
5. Section frozen organs on a cryotome onto coated slides.
6. Rinse slides in 50–100 ml of wash solution in a slide jar, at 4 °C for 5 min.
7. Refix sections in the same fixative solution used for step **1**, at 4 °C for 5 min.
8. Rinse in wash solution at 4 °C for 10 min.

9. Incubate in X-Gal stain solution at 37 °C for 2–24 h in the dark. For incubations longer than 4 h add Tris pH 7.3 to a final concentration of 50 mM.
10. Wash with PBS / 2 mM Mg_2Cl, 2×5 min.
11. Rinse in deionized water.
12. Counterstain and mount as described in Section 26.7. Reduce incubation in alcohols and solvents to a minimum as they may wash out the indigo precipitate.

Whole-mount X-Gal staining of tissues, organs and embryos

1. Dissection. For best results, large organs should be sliced into smaller pieces according to the region and plane of interest for the experiment to facilitate penetration of fixative and staining solutions. Embryos should be released from extraembryonic membranes. In addition, embryos from 13 days postcoitum to birth should be sliced sagitally in two halves using a razor blade.
2. Fix in 50 ml fixative (in a 50-ml plastic tube) supplemented with 0.01 % sodium deoxycholate and 0.02 % NP-40 to enhance permeability of tissues. Incubate at 4 °C for 1–2 h.
3. Remove fixative by aspiration and wash 3× [PBS/2 mM $MgCl_2$/0.01 % sodium deoxycholate/0.02 % NP-40] at room temperature for 5 min each.
4. Transfer specimens using fine forceps to a small glass container filled with X-Gal supplemented with 50 mM Tris pH 7.3. Incubate at 30 °C for 4–24 h in the dark.
5. Remove the staining solution. Wash 2×5 min with PBS/2 mM $MgCl_2$. Rinse in 3 % DMSO in PBS. Store at 4 °C in PBS.
6. To determine expression at the cellular level, proceed as follows:
   a) After step 4, rinse briefly in cold PBS and saturate tissues in 30 % sucrose in PBS/2 mM $MgCl_2$/0.01 % sodium deoxycholate/0.02 % NP-40, at 4 °C for 1–3 h.
   b) Embed samples and freeze in isopentane/liquid nitrogen. Section onto coated slides and air-dry for 30–60 min.
   c) Counterstain and mount (see Sect. 26.7). Minimize incubation in solvents to prevent losses of the blue indigo.

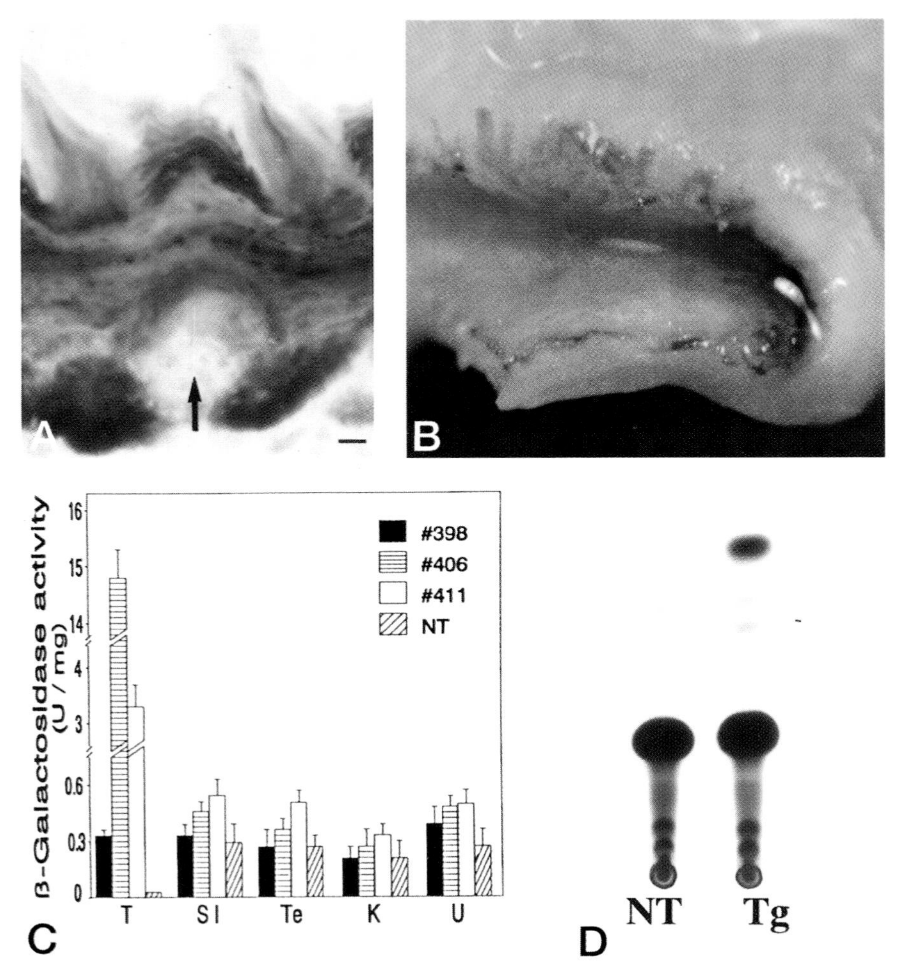

Fig. 26.2A–D. Reporter activities revealed in different tissues of HPV18-URR-[LacZ or CAT] transgenic mice. **A** X-Gal staining of a sagital section of the tongue of a male transgenic mouse. The activity is observed throughout the epithelium except for the basal layer. The section was counterstained with hematoxilin-eosin. **B** Whole mount X-Gal staining of the vagina and uterus of a transgenic mouse. The β-Gal activity is more intense in the region of the uterine cervix. **C** Quantitation of β-Gal activity in different tissues of transgenic and control mice using the colorimetric method. **D** Detection of CAT activity in crude extracts from the tongues of a transgenic mouse (*Tg*) and a non-transgenic control (*NT*). β-Gal and CAT assays were performed as described in Sections 26.5 and 26.8 using equal amounts of protein

26.5 Quantitation of β-Gal Activity in Tissue Extracts

Reporter β-Gal activity can be quantitated in protein extracts by three different methods (MacGregor et al. 1990; Jain and Magrath 1991):

- Colorimetric assay using o-nitrophenyl-β-D-galactopyranoside (ONPG), a galactoside analog. ONPG is cleaved by β-Gal to yield galactose and the chromophore o-nitrophenol whose concentration can be determined by measuring absorbance at 420 nm with a spectrophotometer. This method has a detection limit of approximately 1 ng of β-Gal (1×10^9 molecules).
- Fluorometric assay using 4-methyl-umbelliferyl-β-D-galactopyranoside (MUG), a non-fluorescent galactoside analog, which is cleaved into galactose and fluorescent methylumbelliferone that can be measured using a fluorometer by excitation at 350 nm and detection of fluorescence emission at 450 nm. This assay is highly sensitive and can detect 1 pg (1×10^6 molecules).
- Chemiluminiscent assay using methoxyspiro-dioxetane-tricyclodecanyl-phenyl-β-D-galactopyroside (AMPGD). Cleavage of AMPGD by β-Gal yields dioxetane anion that, at $pH > 9$, undergoes further decomposition to metaoxybenzoate anion which is in an excited state and emits light. The assay can detect up to 2 fg (4×10^3 molecules) of β-Gal.

We mainly have used the ONPG assay because it has proved to be sensitive enough for quantitation of β-Gal activity in tissues of our transgenic mice (Cid et al. 1993).

Preparation of tissue homogenates

This procedure can be performed with either fresh or frozen organs. A small piece of each organ or tissue to be analyzed is usually enough to obtain a measurable enzymatic activity.

1. Homogenize tissues on ice for 1 or 2 min with Polytron, in 1 ml of PM2 buffer (33 mM NaH_2PO_4, 66 mM Na_2HPO_4, 0.1 mM $MnCl_2$, 2 mM $MgSO_4$, pH 7.3). Just before use, add 2-mercaptoehanol to 40 mM. It is recommendable to use plastic tubes with rounded bottom.
2. Transfer to microfuge tubes and centrifuge at 14 000 *g* for 10 min. Carefully, remove the upper layer (enriched in lipids) using a Pasteur pipette, or a yellow tip, connected to a vacuum system. Recover the

supernatant and transfer to a new tube avoiding taking the pellet. Repeat the procedure 2× or 3× to completely eliminate every cellular debris.

3. Use 2 to 4 μl of protein extract, to determine protein concentration.

ONPG assay

4. Carry out the reactions in spectrophotometer plastic cuvettes, in a final volume of 1 ml, using 100 to 400 μg of protein and 800 μg of ONPG substrate in PM2 buffer. Immediately after adding ONPG, measure the absorbance (A_{420}) of each sample. Then incubate at 37 °C for 60 min (or longer, depending on the activity), so that values range from 0.2 to 1.5 O.D. Control reactions, including crude extracts from non-transgenic mice and blanks, should be run in parallel.
5. β-Gal activity is calculated for each sample after subtracting the initial value. Use the equation: Units = (380× A_{420})/time (in min); 380 is a constant such that 1 unit is equivalent to the conversion of 1 nanomole of ONPG per min at 37 °C.

26.6 In Vivo Labeling of Proliferating Cells with BrdU

1. Inject a mouse intraperitoneally with 100 μg BrdU/g of body weight (see Materials, above).
2. Sacrifice the mouse 1 h after injection. Dissect and freeze tissue samples quickly in isopentane/liquid nitrogen.
3. Cut 5-μm sections onto slides coated with albumin or gelatine. Let dry at room temperature for 30–60 min. Slides can be stored at −80 °C, preferably in a sealed box with dessicant.
4. Fix sections by immersing the slides in methanol at 4 °C for 5 min. Air-dry.
5. Rehydrate in PBS for 1 min and incubate in 2 M HCl for 1 h at 37 °C to denature DNA. Neutralize in 0.1 M borate buffer pH 8.5 for 10 min, shaking gently.
6. Wash 3× PBS, 5 min each.
7. Incubate with anti-BrdU-fluorescein antibody diluted in 0.1 % (w/v) BSA in PBS to a final concentration of 20 μg/ml, for 1 h at room temperature in the dark.

8. Wash 3× PBS, 5 min each.

9. Mount with Mowiol. Store at 4 °C in the dark.

26.7
In Situ RNA-RNA Hybridization

In situ hybridization with radiolabeled antisense RNA probes is a highly sensitive method for detection of specific transcripts in tissue sections (see Fig. 26.3C). Moreover, it allows a precise identification of the expressing cell types within tissues. However, it should be noted that detection of specific RNA does not ensure that the corresponding protein is being expressed properly. Therefore, whenever possible, antibodies specific to such protein should be used preferentially for expression analysis. If antibodies are not available, however, a combination of in situ hybridization and RNase protection or reverse transcriptase-polymerase chain reaction (RT-PCR) assays will help determine the pattern of transgene expression and the precise mRNA species transcribed.

We have used the protocol described below, prepared on the basis of Gibson and Pollak (1990), Wilkinson and Nieto (1993) and Hogan et al. (1994), to perform expression analyses in our transgenic mice. To avoid degradation of cellular RNA due to RNase contamination, all solutions equipment and materials should be handled with special care and reserved exclusively for this technique. Autoclave PBS (use PBS pH 7.4), SSC and triethanolamine solutions. Bake glassware (220 °C, 4–6 h) and keep separately those jars used for RNase treatment of specimens. If, in spite of these precautions, RNA degradation is suspected, treat solutions (except Tris) and glassware with DEPC (see Materials).

1. Perform cryosections (5–7 μm thick) and collect them onto coated slides (see Materials).
   a) Let dry at room temperature for at least 1–2 h protected from dust.
   b) Fix tissue in freshly prepared 4 % paraformaldehyde, 2× SSC, for 15 min at room temperature.

2. Dehydration.
   a) Immerse slides into ice-cold 70 % ethanol for 15 min.
   b) Transfer to 100 % ethanol at room temperature for 2 min, air-dry.
   c) Fence each section with a ring of Fixogum (or similar) glue. Dry in a fume hood for 30–60 min.

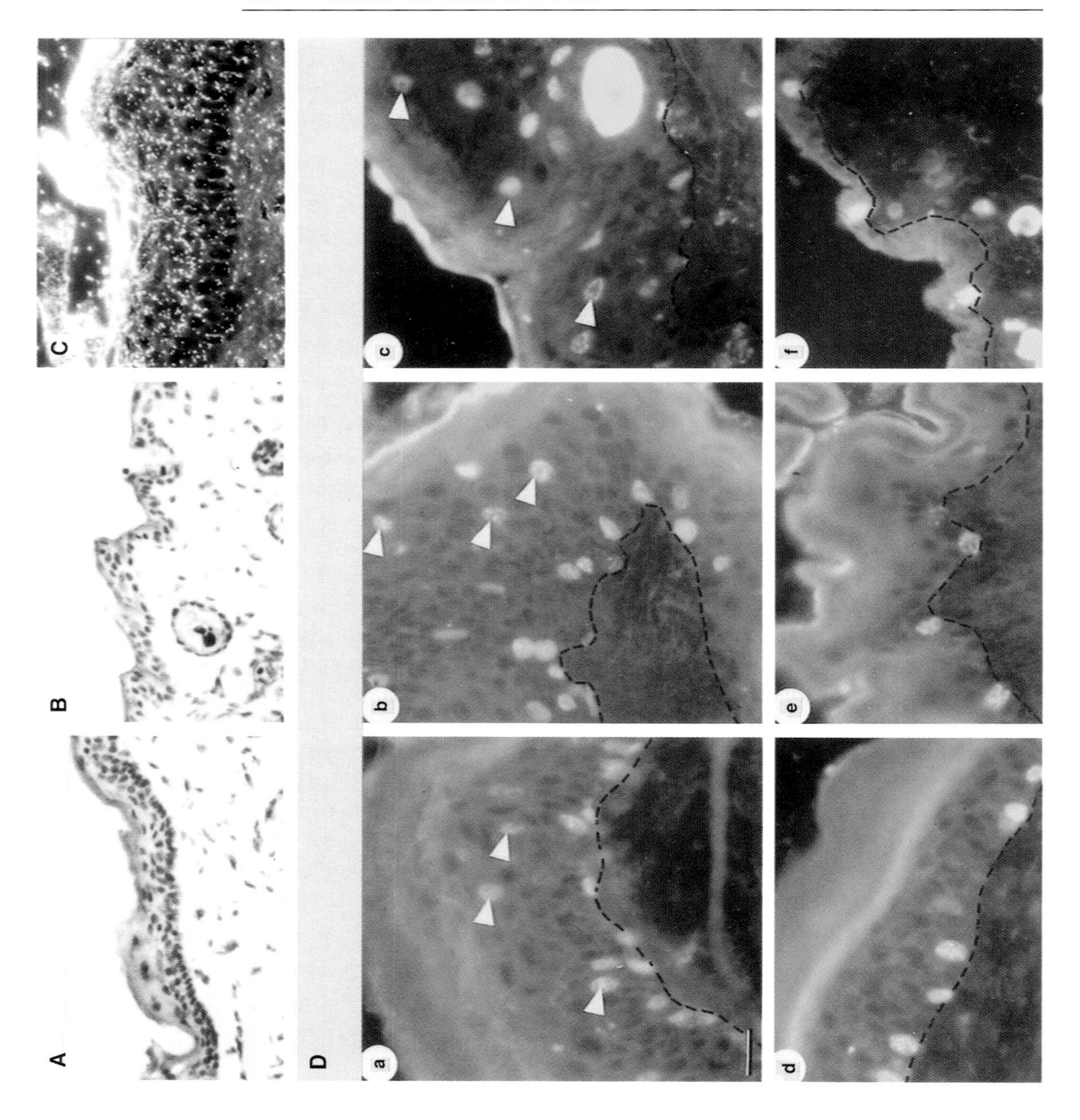
A
B
C
D
a
b
c
d
e
f

◀ **Fig. 26.3A–D.** Phenotype of the K10-E6/E7 transgenic mice. **A** and **B** Histological appearance of the skin of a nontransgenic control (**B**), and a K10-E6/E7 transgenic littermate (**A**) showing hyperthickening of the epidermis and an increased number of basal cells. **C** In situ hybridization of the epidermis of a K10-E6/E7 transgenic mouse with an antisense mRNA probe against the entire E6/E7 region; the E76/E7 mRNA expression is restricted to the suprabasal layers of the epidermis. No signal was seen, using the same probe, in a non-transgenic control (not shown). **D** Representative BrdU staining of proliferating cells in the epidermis of K10-E6/E7 transgenic mice (a,b,c) and nontransgenic controls (d,e,f) from plantar skin (a,d), forestomach (b,e) and ventral skin (c,f)

3. Proteinase K treatment.
   a) Prepare 20 ml of proteinase K solution: 2 ml of 20× SSC, 0.2 ml 10 % SDS, 1 µl of 10 mg/ml proteinase K, 18.8 ml H_2O.
   b) Pipette 1–2 drops of this solution onto each section and incubate for 10 min at 37 °C.
   c) Inactivate proteinase K: wash sections in 0.1 M glycine, 2× SSC (for 200 ml: 20 ml 1 M glycine, 20 ml 20× SSC, 160 ml H_2O).

4. Repeat fixation.
   a) Immerse slides in 4 % paraformaldehyde, 2× SSC, 5 mM $MgCl_2$, for 5–10 min at room temperature.
   b) Transfer to 50 % formamide, 2× SSC for 5–15 min.

5. Acetylation. This step is intended to block amino groups, thus preventing unspecific binding of probe to tissue.
   a) Prepare 0.1 M triethanolamine (TEA), 2x SSC (e.g., 2.25 g TEA in 150 ml of 2× SSC).
   b) Immerse slides into a glass beaker containing 150 ml of 0.1 M TEA/ 2× SSC supplemented with 375 µl of acetic anhydride, incubate at room temperature for 5 min. (**Note:** Add anhydride just before immersing the slides).
   c) Add fresh acetic anhydride (375 µl), mix by shaking gently, and repeat incubation.

6. Wash in 50 % formamide, 2× SSC at room temperature for 5–15 min. Drain slides by vigorous shaking.

7. Prehybridization.
   Add 20 µl of prewarmed (42 °C) hybridization solution (see Materials) to each section, incubate for 1–3 h at 42 °C in a humid chamber.

8. Hybridization.

a) Add $1-2\times10^6$ cpm of radiolabeled antisense riboprobe in a volume of 5 μl (final volume of radioactive hybridization solution per section = 25 μl). Mix by pipetting up and down a few times.

b) Add a comparable amount of radioactive sense probe to another specimen which serves as control for nonspecific hybridization.

c) Incubate overnight at 42 °C in the humid chamber.

Notes: Prepare a specific antisense riboprobe, ideally of about 0.1–0.5 kb, from a cDNA cloned into a suitable plasmid such as Bluescript or Gemini using an in vitro transcription kit commercially available. These vectors contain two promoters flanking the cDNA sequence, which allow synthesis of the antisense and sense probes from the same clone. Prior to using a riboprobe for in situ hybridization, it is advisable to test its specificity in Northern blots because, in some cases, the antisense probe hybridizes unspecifically to unknown RNA species, very often to the abundant ribosomal RNAs.

9. Washes.

a) Place slides into a container filled with 100 ml of 5× SSC, 20 mM β-ME prewarmed at 60–65 °C. Incubate at 60–65 °C for 15 min.

b) Transfer slides to a new (smaller) container with 50 ml of FSM solution (see Materials) prewarmed at 60–65 °C. Incubate in a water bath at 60–65 °C for 30 min, shaking slowly.

c) Transfer container to a 37 °C water bath and shake for 2–4 h.

d) Wash slides 3×100 ml NTE buffer, at 37 °C, 5 min each.

10. RNase treatment.

a) Incubate slides in NTE with 20 μg/ml RNase A at 37 °C for 15 min.

b) Wash with 100 ml NTE, at 37 °C for 15 min.

c) Wash with FSM, at 65 °C for 15 min.

d) Wash with 2× SSC, at room temperature for 15 min.

e) Wash with 0.1× SSC, at room temperature for 15 min.

11. Dehydration.

a) Immerse slides for 2 min into each solution of the following ethanol series: 30, 50, 70 % ethanol/0.25 M ammonium acetate/PBS, 80, 90, 95, 100 % ethanol.

b) Air-dry.

12. Autoradiography. Perform the following steps in a dark room, avoiding any leaks and control lights (water bath, etc.). Set water bath at 42–45 °C.

a) Melt Kodak NTB-2 emulsion in water bath (30–45 min). Dilute emulsion 1:1 in prewarmed 2 % glycerol in deionized water. Aliquot 10 ml in slide dippers, let solidify, and keep in a black box or plastic bag at 4 °C.
b) Melt an aliquot in the water bath and dip a test (clean) slide. Check in the light that covering is smooth and that there are no bubbles.
c) Dip slides containing the specimens, one by one, into the emulsion, hold vertically to allow excess film to drop back, place them vertically on bench paper and let dry for 2 h.
d) Transfer to a plastic slide box, introduce a small bag with desiccant, tape the seal, and keep box in a black plastic bag at 4 °C for 1–3 weeks.
e) Just before developing, set water bath in the dark room to 15 °C and prepare the solutions required (see below). All solutions should be at this temperature. Leave slides warm at room temperature for 5–10 min. Develop slides in racks by incubation in staining jars as follows:
- Kodak D19 developer (20 g/500 ml), 2–4 min
- 1 % acetic acid, 1 min
- 30 % sodium thiosulfate or Kodak Agefix fixative, 5 min
- Tap water, at room temperature, 3× 5 min

13. Counterstaining. Any of the following dyes can be used to stain cells in the hybridized tissues:
- Hematoxylin/eosin. Although rather more complicated to perform, this is the stain that gives best results. Consult manufacturer's instructions or a histology department.
- Toluidin blue. Stain in 0.02 % toluidin blue at room temperature for 1 min. Dehydrate by passing slides quickly through 30, 70, 90, and 100 % ethanol. Immerse in xylene or Histoclear 2×5 min. Mount.
- Giemsa. Transfer slides to Giemsa staining solution in a glass container for 20 min at room temperature. Wash out the staining solution with tap water. (**Note:** Do not remove slides from the staining solution during staining and do not use deionized water for washing). Drain slides and let dry vertically. Mount.
- Orange G. Incubate slides for 30 s in Orange G staining solution. Wash 3×5 min in deionized water. Dehydate through 50, 70, and 100 % methanol (1–2 min each). Immerse in xylene or Histoclear 2×5 min. Mount.

14. Mounting. Use DPX or Permount mounting medium.
15. View slides in dark field (silver grains appear intense white) and bright field (silver grains appear black on the stained tissue).

26.8 Detection of CAT Activity in Tissue Extracts

Transgenic mice carrying the CAT reporter gene can be used to determine the transcriptional activity of a promoter in different organs.

1. Prepare crude homogenates from fresh or frozen tissues as described in Section 26.5.
2. Incubate the homogenates at 65 °C for 10 min. This step is introduced to inactivate endogenous deacetylases, abundant in some tissue extracts, which may interfere with the CAT assay (see Sleigh 1986; Crabb and Dixon 1987).
3. Centrifuge at 14,000 g in a microfuge, at 4 °C for 5 min, to eliminate debris and aggregated proteins.
4. Measure protein concentration in the extracts.
5. Use 400 μg of protein to perform a typical CAT assay as described in Chapter 1, this Volume (Sect. 1.2 , steps **6** to **15**).

Results

Following the strategy described in Introduction we have established two types of HPV transgenic mouse models:

- Transgenic Mice Expressing Reporter Genes from the HPV-16 and HPV-18 URRs

As revealed by X-Gal staining, in these mice the HPV URR is transcriptionally active in epithelial cells of different tissues, most importantly the stratified epithelia of tongue and vagina (Fig. 26.2A,B). In contrast, no expression was observed in the epidermis. In addition, some reporter activity could be detected also in certain epithelial cells of the small intestine, uterus, ovary and testis. Such pattern of expression was observed with either HPV-16 or HPV-18 URRs. However, quantitation of β-Gal activity in protein extracts from these tissues, using ONPG as substrate, shows that the HPV-18 promoter drives about five-fold more expression than that of HPV-16, a difference that is also observed at the RNA level (P. Auewarakul, L. Gissmann, A. Cid-Arregui, unpubl. results). In addition, β-Gal quantitation shows that the stratified epithelium of the tongue supports the highest level of expression of the URR promoter (Fig. 26.2C). Similar results are obtained if the CAT gene is used as reporter (Fig. 26.2D).

Interestingly, the promoter activity of the HPV URR in the genital tract of transgenic females exhibits a clear dependence on the hormonal variations accompanying the estrus cycle, with a strong activity in estrus (Fig. 26.2B), when circulating levels of progesteron reach a maximum, and nondetectable reporter expression at diestrus. Indeed, experiments treating transgenic animals with antiprogestron drugs suggest that this hormone governs the activity of the viral promoter (N. Morales-Peza, A. Garcia-Carranca, A. Cid-Arregui, unpubl. results). These observations suggest that antiprogesteron drugs may be helpful in the treatment of cervical cancers.

- Transgenic Mice Expressing the E6 and E7 Oncoproteins in the Epidermis

Expression from a keratin 10 (K10) promoter of the E6 and E7 oncoproteins of HPV-16 in the suprabasal cells of the mouse epidermis induces hyperplasia, hyperkeratosis and parakeratosis in this tissue, affecting the different types of skin (from body, tail, ears, plantar) and also the forestomach, which are all known to be sites of K10 expression (Fig. 26.3A,B). Hyperthickening of the epidermis is seen in newborn as well as in adult animals. The expression of E6 and E7 in these tissues was verified by Northern blots and in situ hybridizations (Fig. 26.3C). In addition, the presence of proper transcripts was confirmed by PCR analysis.

In vivo labeling of proliferating cells with BrdU shows that the epidermal hyperplasia in the K10-E6/E7 transgenic mice is due to two distinct effects (Fig. 26.3D):

- The presence of an increased number of proliferating cells in the basal layer (two to three fold increase relative to nontransgenic controls), and
- The presence of BrdU-labeled cells in the suprabasal layers (about 10–15 % of total proliferating cells) which express the K10 keratin, indicating that these are indeed differentiated cells with proliferating capacity.

In contrast, none of the proliferating cells in the basal layer contains K10. Suprabasal proliferation is an abnormal phenomenon which is never observed in naturally occurring epidermal hyperplasia, nor it is induced by topical TPA treatment during experimental tumor promotion. Therefore, the presence of suprabasal differentiating cells retaining the proliferating capacity normally restricted to the basal cells seems to be a specific effect of the E6/E7 transgene.

An interesting finding was that TGF-α is overexpressed in the suprabasal cells of the epidermis of E6/E7 transgenic mice, thus suggesting that autocrine mechanisms may be involved in the maintenance of the epidermal hyperplasia. In addition, c-myc expression in the transgenic skin is strongly increased in the hyperproliferative tissues. It is not yet clear, however, if c-myc mediates the effects of the viral proteins in inducing proliferation in the suprabasal cells.

Finally, old animals develop a limited number of benign skin papillomas in areas of mechanical abrasion, like the ear tag attachment site, suggesting that the E6/E7 transgene provides a genetic predisposition to neoplasia. However, as in human lesions caused by these viruses, additional secondary events are necessary for progression from hyperplasia to papilloma formation. Therefore, these mice constitute a useful system for the study of the mechanisms and factors involved in the neoplastic transformation induced by the E6 and E7 oncoproteins.

Acknowledgments. We are grateful to Prof. H. zur Hausen for constant support and advise and Prof. L. Gissmann for helpful discussions and support. This work was supported by grants of the Mexican/German Joint Research Program, CONACYT, and Miguel Aleman Foundation.

References

Adams JM, Cory, S (1991) Transgenic models of tumor development. Science 254:1161–1167

Aichele P, Kyburz D, Ohashi PS, Odermatt B, Zinkernagel RM, Hengartner H, Pircher H (1994) Peptide-induced T-cell tolerance to prevent autoimmune diabetes in a transgenic mouse model. Proc Natl Acad Sci USA 91:444–448

Andino R, Silvera-D Suggett SD, Achacoso PL, Miller CJ, Baltimore D, Feinberg MB (1994) Engineering poliovirus as a vaccine vector for the expression of diverse antigens. Science 265:1448–1451

Ashton-Rickardt PG, Bandeira A, Delaney JR, Van-Kaer L, Pircher HP, Zinkernagel RM, Tonegawa S (1994) Evidence for a differential avidity model of T-cell selection in the thymus. Cell 76:651–663

Auewarakul P, Gissmann L, Cid-Arregui A (1994) Targeted expression of the E6 and E7 oncogenes of human papillomavirus type 16 in the epidermis of transgenic mice elicits generalized epidermal hyperplasia involving autocrine factors. Mol Cell Biol 14:8250–8258

Berns A (1991) Tumorigenesis in transgenic mice: identification and characterization of synergizing oncogenes. J Cell Biochem 47:130–135

Blessing M, Jorcano JL, Franke WW (1989) Enhancer elements directing cell-type-specific expression of cytokeratin genes and changes of the epithelial cytoskeleton by transfection of hybrid cytokeratin genes. Embo J 8:117–126

Borrelli E, Heyman R, Hsi M, Evans RM (1988) Targeting of an inducible toxic phenotype in animal cells. Proc Natl Acad Sci USA 85:7572–7576

Chellappan S, Kraus VB, Kroger B, Münger K, Howley PM, Phelps WC, Nevins JR (1992) Adenovirus E1A, simian virus 40 tumor antigen, and human papillomavirus E7 protein share the capacity to disrupt the interaction between transcription factor E2F and the retinoblastoma gene product. Proc Natl Acad Sci 89:4549–4553

Chisari FV (1995) Hepatitis B virus transgenic mice: insights into the virus and the disease. Hepatology 22:1316–1325

Chomczinski P, Sacchi N (1987) Single-step method of RNA isolation by acid guanidinium thiocyanate-phenol-chloroform extraction. Anal Biochem 162:156–159

Cid A, Auewarakul P, García-Carrancá A, Ovseiovich R, Gaissert H, Gissmann L (1993) Cell-type-specific activity of the human papillomavirus type 18 upstream regulatory region in transgenic mice and its modulation by tetradecanoyl phorbol acetate and glucocorticoids. J Virol 67:6742–6752

Collinge J, Sidle KCL, Meads J, Ironside J, Hill AF (1996) Molecular analysis of prion strain variation and the aetiology of 'new variant' CJD. Nature 383:685–690

Crabb DW, Dixon JE (1987) A method for increasing the sensitivity of chloramphenicol acetyltransferase assays in extracts of transfected cultured cells. Anal Biochem 163:88–92

Dyson N, Howley PM, Münger K, Harlow E (1989) The human papillomavirus-16 E7 oncoprotein is able to bind the retinoblastoma gene product. Science 243:811–814

Gibson SJ, Pollak JM (1990) Principle and applications of complementary RNA probes, p84–91. In: Pollak JM, McGee JOD (ed) In situ hybridization, principles and practice. Oxford University Press, Oxford

Griep AE, Herber R, Jeon S, Lohse JK, Dubielzig RR, Lambert PF (1993) Tumorigenicity by human papillomavirus type 16 E6 and E7 in transgenic mice correlates with alterations in epithelial cell growth and differentiation. J Virol 67:1373–1384

Hogan B, Beddington R, Costantini F, Lacy E (1994) Manipulating the mouse embryo, a laboratory manual, 2nd edn. Cold Spring Harbor Laboratory Press, Cold Spring Harbor, New York

Jain VK, Magrath IT (1991) A chemiluminiscent assay for quantitation of β-galactosidase in the femtogram range: application to quantitation of β-galactosidase in lacZ transfected cells. Anal Biochem 199:119–124

Kitajima I, Shinohara T, Bilakovics J, Brown DA, Xu X, Nerenberg M (1992) Ablation of transplanted HTLV-I Tax-transformed tumors in mice by antisense inhibition of NF-kappa B. Science 258:1792–1795

Klotman PE, Rappaport J, Ray P, Kopp JB, Franks R, Bruggeman LA, Notkins AL (1995) Transgenic models of HIV-1. AIDS 9:313–324

Koike S, Taya C, Kurata T, Abe S, Ise I, Yonekawa H, Nomoto A (1991) Transgenic mice susceptible to poliovirus. Proc Natl Acad Sci USA 88:951–955

MacGregor GR, Nolan GP, Fiering S, Roederer M, Herzenberg LA (1991) Use of *E coli lacZ* (β-galactosidase) as a reporter gene. In: Murray EJ (ed) Methods in molecular biology, vol. 7, p 217–135. Humana Press, Inc., Clifton, New Jersey
Merlino G (1994) Transgenic mice as models for tumorigenesis. Cancer Invest 12:203–213
Ohashi PS, Oehen S; Buerki K, Pircher H, Ohashi CT, Odermatt B, Malissen B, Zinkernagel RM, Hengartner H (1991) Ablation of "tolerance" and induction of diabetes by virus infection in viral antigen transgenic mice. Cell 65:305–317
Oldstone MB, Southern PJ (1993) Trafficking of activated cytotoxic T lymphocytes into the central nervous system: use of a transgenic model. J Neuroimmunol 46:25–31.
Prusiner SB (1993) Transgenic investigations of prion diseases of humans and animals. In: Flavell RB, Heap RB (eds) Transgenic modification of gemline and somatic cells. Chapman and Hall, London, pp 101–116
Sambrook J, Fritsch EF, Maniatis T (1989) Molecular cloning: a laboratory manual, 2nd edn. Cold Spring Harbor Laboratory Press, Cold Spring Harbor, New York
Sanes JR, Rubenstein JLR, Nicolas JF (1986) Use of a recombinant retrovirus to study post-implantation cell lineage in mouse embryos. EMBO J 5:3133–3142
Scheffner M, Werness BA, Huibregtse JM Levine AJ, Howley P (1990) The E6 oncoprotein encoded by human papillomavirus types 16 and 18 promotes the degradation of p53. Cell 63:1129–1136
Sleigh MJ (1986) A nonchromatographic assay for expression of the chloramphenicol acetyltransferase gene in eucaryotic cells. Anal Biochem 156:251–256
van Lohuizen M, Verbeek S, Scheijen B, Wientjens E, van der Gulden H, Berns A (1991) Identification of cooperating oncogenes in E μu-myc transgenic mice by provirus tagging Cell 65:737–752
Wallenfels K, Weil R (1972) β-Galactosidase. In Boyer P D (ed), The enzymes, vol 7, p 617–663. Academic Press, New York
Wassarman PM, DePamphilis ML (1993) Guide to techniques in mouse development. Methods in Enzymology, vol. 225. Academic Press, New York
Werness BA, Levine AJ, Howley PM (1990) Association of human papillomavirus types 16 and 18 E6 proteins with p53. Science 248:76–79
Wilkinson DG, Nieto MA (1993) Detection of messenger RNA by in situ hybridization to tissue sections and whole mounts. Methods Enzymol 225: 361–373
zur Hausen H (1991) Viruses in human cancers. Science 254:1167–1173
zur Hausen H, de Villiers EM (1994) Human papillomaviruses. Annu Rev Microbiol 48:427–447

Chapter 27

Isolation of Mutated Genes from Transgene Insertion Sites

MIRIAM H. MEISLER[1*], JAMES GALT[1], JOHN S. WEBER[1], JULIE M. JONES[1], DANIEL L. BURGESS[1], AND DAVID C. KOHRMAN[1]

27.1 How to Determine Whether You Have an Insertional Mutation in Your Transgenic Mice, and How to Proceed if You Do

After microinjection of fertilized eggs, the choice of the chromosomal site of transgene insertion appears to be random. Most insertions sites are located in the regions between genes, and do not influence gene expression. However, roughly 15 % of transgenic insertion sites are located *within* an endogenous mouse gene and result in loss of gene expression and phenotypic abnormalities. These fortuitous insertions can be extremely useful for isolation of the mutated gene. The transgene is used as a molecular marker for the disrupted gene, providing a direct route from mutant phenotypic to the underlying genetic defect. By comparison with positional cloning approaches, gene isolation from a transgene insertion site eliminates the need for high resolution genetic crosses, facilitates evaluation of positional candidate genes, and restricts gene identification efforts to a region containing only 1 or 2 genes rather than 10 to 20. Several interesting genes have already been isolated by this method, including the mutant genes *fused* (Perry et al. 1995), *pygmy* (Zhou et al. 1995), *microopthalmia* (Hodgkinson et al. 1993; Hughes et al. 1993), *limb deformity* (Woychik et al. 1990), *reeler* (D'Arcangelo et al. 1995), *dystonia* (Brown et al. 1995), and *motor endplate disease* (Burgess et al. 1995).

Visible mutations have been detected in 3 % of transgenic lines examined, and the frequency of prenatal lethal mutations is approximately 10 % (Meisler 1992 and unpubl. observations). Since hundreds of trans-

* corresponding author: phone: +1-313-7635546; fax: +1-313-7639691; e-mail: meislerm@hg-basic1mail.hg.med.umich.edu

[1] Department of Human Genetics, University of Michigan, 4708 Medica Science II, Ann Arbor, MI 48109-0618

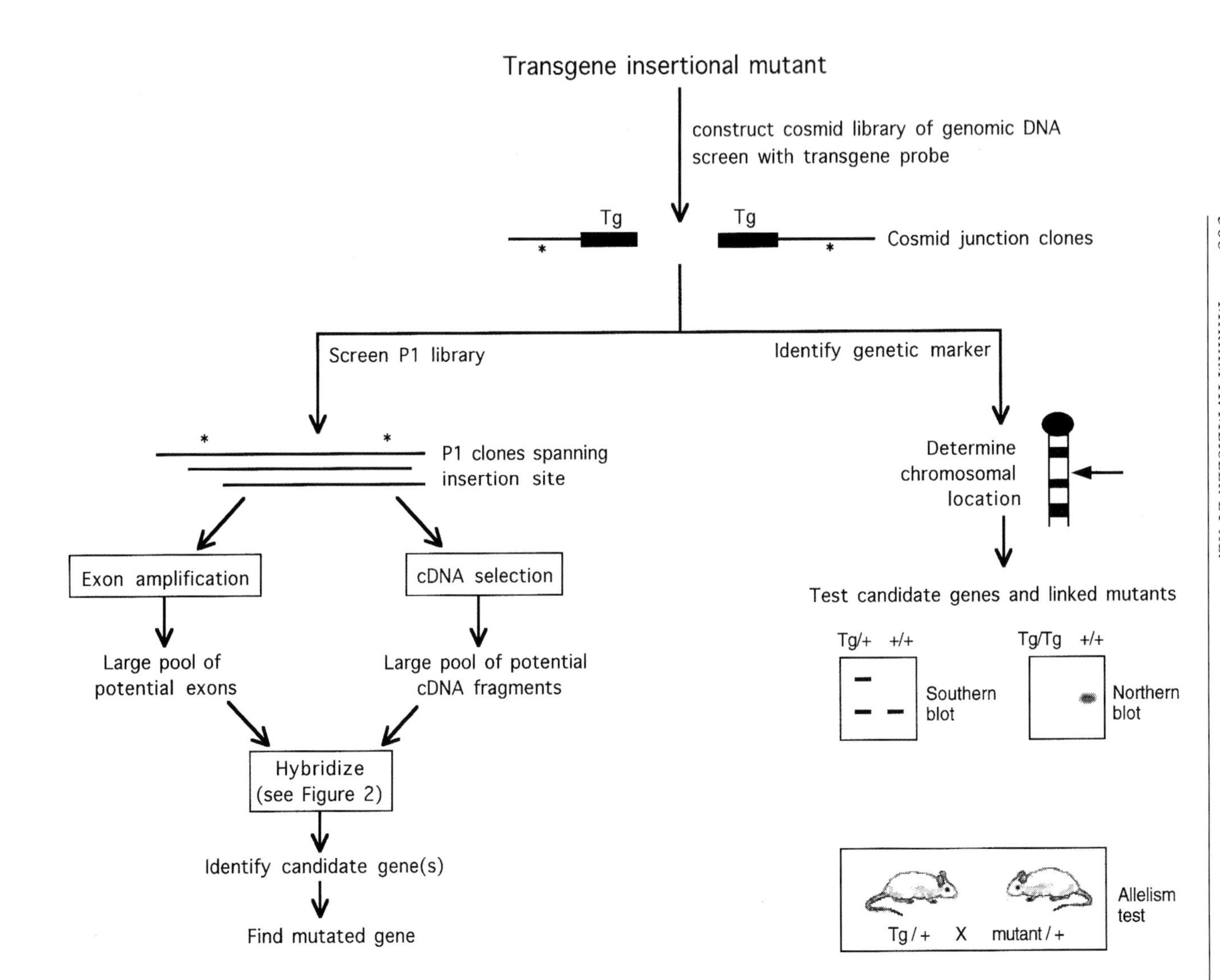

Fig. 27.1. Strategy for isolation and chromosomal mapping of insertional mutants

genic lines are being generated every year, there will be many opportunities to use this approach for isolating novel genes associated with interesting phenotypes. In this chapter, we describe our methods for recognition of transgene induced mutations, chromosomal mapping, evaluation of cloning feasibility, and isolation of mutated genes. The overall strategy is outlined in Fig. 27.1.

27.2 Mutant Characterization

Recognition of Transgene-Induced Mutations

Each independent transgenic line is descendent from a single founder animal with a unique transgene insertion site. (In the rare case of multiple transgene insertions in the same founder, independent lines carrying each insertion site can be generated by breeding.) Dominant transgene-induced mutations are easily recognized because all the transgenic offspring of one founder exhibit the mutant phenotype. Detection of autosomal recessive mutations requires matings between heterozygous transgenic male and female mice from the same transgenic line. Twenty five percent of the offspring of such matings will be homozygous for the transgene and will display a recessive mutant phenotype. When screening for recessive mutations, we examine at least 15 offspring (3 or 4 litters) from each transgenic line. Mutations may be detected by visual inspection or with tests for balance, swimming ability, hearing, and other phenotypes of specific interest. If a mutant phenotype is detected, it is necessary to demonstrate consistent cosegregation of transgene and phenotype in the offspring of crosses with nontransgenic mice. Individual animals are tested for the transgene by Southern blotting or PCR of genomic DNA prepared from tail samples (Gendron-Macguire and Gridley 1993). It may be difficult to distinguish reliably between homozygous and heterozygous transgenic mice at this stage of the investigation, using only the two-fold difference in concentration of the transgene in genomic DNA. After the insertion site is cloned, specific assays for the wild-type and transgenic chromosomes can be developed.

Before initiating a cloning project it is important to investigate the possibility that the mutant phenotype results from *expression* of the transgene, rather than insertional mutation. Phenotypes caused by expression will also cosegregate with the transgene. Comparison with additional transgenic lines that carry the same transgene *and have the same pattern of expression* is the most commonly used control. If only one of a group of similar transgenic lines has a mutant phenotype, then the phenotype is likely to be unrelated to transgene expression. Occasionally, an enhancer element located close to an insertion site may result in an unusual expression pattern in one transgenic line only. For this reason, it is important to demonstrate experimentally that the pattern of transgene expression in the mutant line does not differ from other lines. In some cases, transgene expression can be assayed directly in the affected organ. In general, dominant mutations are more likely to be caused by transgene expression than are recessive mutations, since there is only a twofold difference in the concentration of transgene product between homozygous and heterozygous transgenic mice.

Chromosomal Mapping of the Transgene Insertion Site

Mapping the transgene

The two main purposes for chromosomal mapping of the insertion site are

- to identify appropriate candidate genes located in the same chromosome region or in the human conserved linkage group, and
- to determine whether the transgene-induced mutation is a new allele of a previously mapped mutant.

Fluorescent in situ hybridization (FISH) with a transgene probe can rapidly localize the insertion site to one chromosome band (Ting et al. 1994). Genetic mapping provides higher resolution than FISH and is more useful for comparison with the map positions of candidate genes and other mutants. Genetic mapping can be most efficiently carried out after single copy DNA adjacent to the transgene has been isolated (Sect. 27.3).

FISH

In principle this is a straightforward approach to chromosomal localization of a transgene, but in practise the preparation of chromosome spreads from cells isolated from the transgenic line of interest can be technically challenging. Overnight cultures of spleen cells have been used for this purpose, and a protocol for the preparation of chromosomes from mouse peripheral blood has been developed (Shi et al. 1994). The sensitivity of signal will depend on the length of the transgene probe and the number of transgene copies present at the insertion site. Only a few laboratories currently have the expertise required for karyotyping of mouse chromosomes. However, the commercial development of multicolor spectral karyotying probes for mouse chromosomes as well as kits containing chromosome-specific probes will soon eliminate the need for traditional karyotyping. Mouse FISH is likely to become widely available in the near future.

Genetic mapping

Nonrepetitive fragments from DNA adjacent to the transgene (Sect. 27.3) can be used as probes to detect restriction fragment length polymorphisms between mouse strains. PCR primers can be used to detect polymorphic microsatellites or single-stranded conformation polymorphisms (SSCPs). The genetic location of a transgene insertion site can be determined by using these assays to type DNA samples from one of several cumulative mapping panels that have already been typed for large numbers of markers throughout the genome. To use these panels, DNA samples are provided to the investigator, and the genotype of each animal in the panel is determined and reported back to the mapping resource. Comparison with data for hundreds of other markers then identifies co-segregating and closely linked markers. Map positions are calculated and returned to the investigator. Information on use of the 200 animal Jackson Laboratory Interspecific Backcross (Rowe et al. 1994) is available through the internet at (http://www.jax.org/resource/docutments/cmdata/). Information on use of the 1000-animal European Collaborative Interspecific Backcross (Anonymous 1994) can be obtained at (http://www.hgmp.mrc.ac.uk/MBx/MBxHomepage.html). These resources provide reliable mapping information rapidly and efficiently, and eliminate the need to generate a separate mapping cross for each new mutant.

Identification and Evaluation of Positional Candidate Genes near the Insertion Site

After the insertion site has been localized on the mouse chromosome map, information about nearby genes can be obtained from the Mouse Chromosome Committee reports which are published annually in the journal *Mammalian Genome* and are also available electronically through the *Mouse Genome Database* (MGD) (http://www.informatics.-jax.org/mgd.html). The *Mouse Locus Catalog* contains descriptions of more than 500 characterized mouse mutants with references to the primary literature (Doolittle et al. 1996) and is also available on line at MGD. Information about homologous regions of the human genome, which often contain appropriate candidate genes for the mouse mutants, can be obtained from the *Genome Database* (GDB) (http://gdbwww.gdb.org/), the *Online Mendelian Inheritance of Man* (OMIM) (http://www3.ncbi.nlm.nih.gov/Omim/), or the *Seldin/Debry Human/Mouse Homology Map* (DeBry and Seldin 1996) which lists 1400 genes that have been mapped in human and mouse (http://www3.ncbi.nlm.-nih.gov//Homology/).

Testing candidate genes

If an appropriate candidate gene has been mapped within 10 cM of the apparent chromosomal location of the transgene insertion site, it is worthwhile to evaluate the candidate gene before initiating a cloning project. Insertional mutations often reduce the expression of the disrupted gene, which can be detected by Northern blotting with the cDNA probe, or by RT-PCR. Transgene insertions can also be identified by novel restriction fragments on Southern blots, compared with nontransgenic mice of the same strain.

Testing allelism with closely linked mutants

Half of the reported mutations in transgenic mice have been new alleles of previously described mutants (Meisler 1992). If a mutant with a similar phenotype has been mapped within 10 cM of the apparent location of the transgene insertion site, it is worthwhile to obtain mice carrying the earlier mutation to test for allelism. Many mutants can be obtained from the Jackson Laboratory or from individual investigators. Allelism is tested by crossing heterozygous carriers of the original mutation with heterozygous transgenic mice. For recessive mutations, the observation of affected offspring is evidence for allelism (e.g., Jones et al. 1993; Kohrman et al. 1995). For dominant mutations, a more severe phenotype among offspring suggests allelism. Identification of a previously described allele of your insertional mutation can provide a wealth of physiological and pathological information, since many known mutants have been studied in detail. The availability of a second allele for molecular analysis can also confirm the identity of the cloned gene.

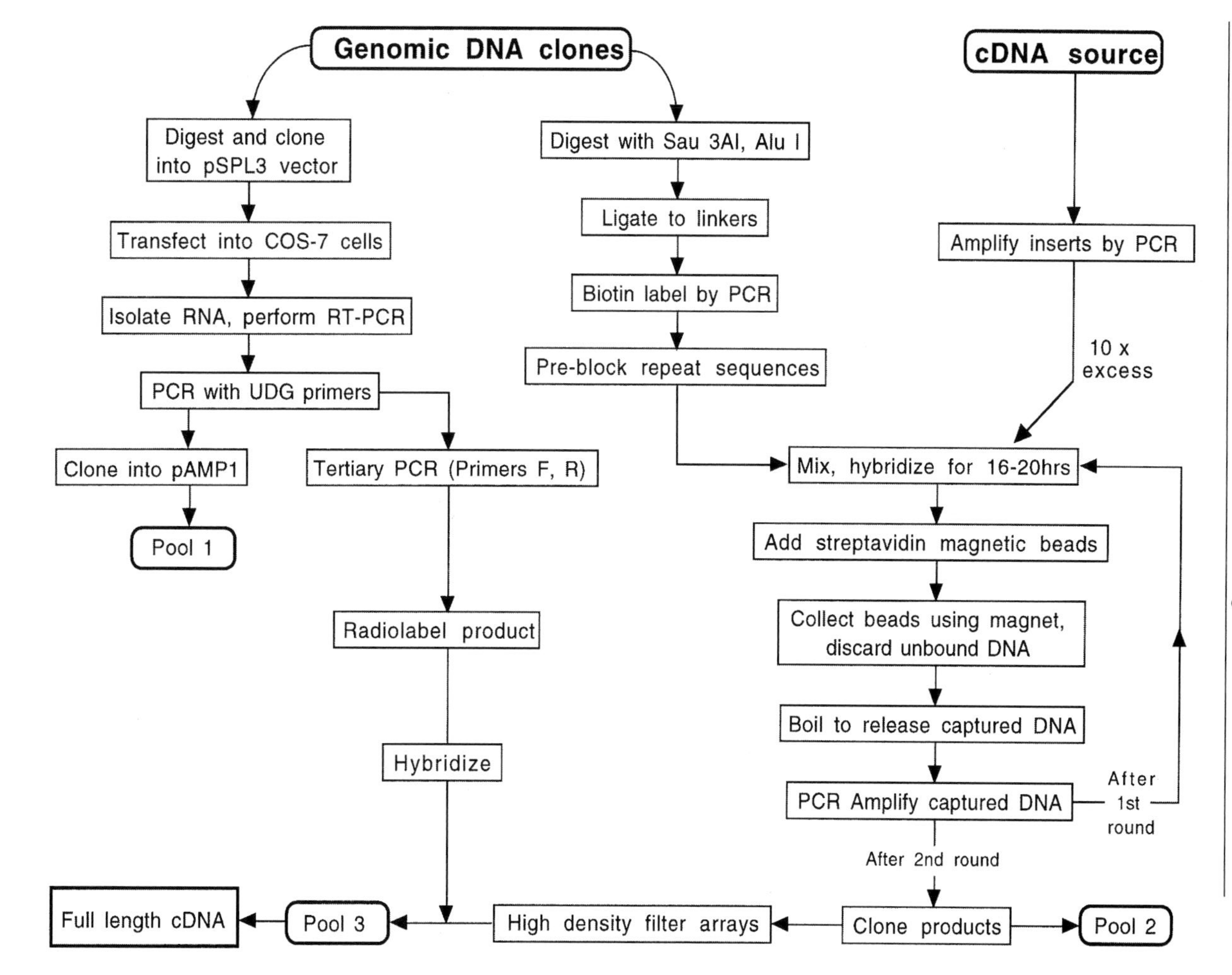

Fig. 27.2. Flow chart for mutant gene isolation by cDNA selection and exon trapping. *Pool 3* contains cDNA clones that cross-hybridize with exon-trapped products and is most likely to contain authentic fragments of the gene encoded at the transgene insertion site. Large scale sequencing of genomic DNA clones provides another important method for gene identification

27.3 Cloning the Mutated Gene

Isolation of Transgene Junction Clones and Wild-Type DNA Clones that Span the Insertion Site

DNA adjacent to the transgene insert can be isolated by constructing a library of genomic DNA from the transgenic mice and screening with a transgene-specific probe. Some of the positive clones will contain both transgene DNA and endogenous mouse DNA adjacent to the insertion site. We have chosen to use cosmid libraries for this purpose because of the relatively large insert size (~40 kb) and the stability of the clones. It is important to isolate flanking DNA from *both sides* of the insert for detection of deletions or rearrangements that may disrupt more than one gene. Flow charts describing the steps in cloning and gene isolation are presented in Fig. 27.1 and 2.

Preparation and Screening of a Cosmid Library of Transgenic DNA

Vector Many types of cosmid vectors and cloning strategies have been used to produce genomic libraries of high complexity. We have been successful with a modified sCOS vector (Stratagene) that contains a unique SalI cloning site and two *cos* packaging sites flanking a unique XbaI site. We use a cloning strategy based on partial fill-in of recessed 3' termini of vector and insert fragments. Vector arms are prepared by double digestion with SalI and XbaI, and the 4-bp overhangs of the arms are partially filled with dCTP and dTTP using the Klenow fragment of DNA polymerase as described (DiLella et al. 1987).

Genomic DNA High molecular weight DNA from heterozygous or homozygous transgenic mice is isolated from spleen or brain as described (Sambrook et al. 1989). To prevent shearing during extractions, aqueous and organic phases are mixed gently by hand. The size of the isolated DNA is estimated by pulsed-field gel electrophoresis (Burmeister and Ulanovsky 1992). The genomic DNA should be at least 200 kb in length to ensure that most molecules will possess two cohesive ends after partial digestion with MboI (Sambrook et al. 1989). Partial digestion with MboI is carried out to generate genomic DNA fragments between 30 and 45 kb in length, as monitored by gel electrophoresis. The partial fill-in of the MboI fragments in the presence of dGTP and dATP generates a 3' overhang of 2 bp that is complementary with the partially filled in SalI ends of the vector.

Ligation

Vector and insert DNA are ligated together at 15 °C under standard conditions (DiLella et al. 1987). Since the 2-bp overhangs are not self-complementary, generation of vector concatamers and noncontiguous chimeric inserts is minimized. Control ligations containing vector alone or insert alone are processed in parallel. The success of the fillin reactions can be evaluated by analyzing the control ligations by gel electrophoresis. The mobility of partially filled in DNA in the control ligations will not be altered, while DNA that escaped fillin will be ligated into concatameric fragments of higher molecular weight that can be seen on the gels.

Library construction

Ligations are packaged using commercially available extracts according to manufacturer's instructions. The resulting phage mixtures are used to infect bacterial cells, plated onto 137-mm nylon filters on LB-ampicillin plates (30,000 to 50,000 colonies per filter), and replicated to additional filters as described (DiLella et al. 1987). An adequate ligation should produce approximately 1×10^6 colonies per ug of packaged insert DNA. This efficiency should provide a library with $\geq$ four-fold redundancy (approximately 4×10^5 clones).

Screening the cosmid library with transgene probes

Duplicate filters are hybridized with a radiolabeled transgene-specific probe by standard procedures (Ausubel et al. 1995). Depending upon the original density of colonies on the master plate, two or three rounds of screening may be required to purify positive clones. The final screening may be carried out by colony hybridization or by PCR with transgene specific primers.

Characterization of Flanking DNA in Transgene Junction Clones and Isolation of Single-Copy DNA

Restriction fragment content of cosmids

One to five μg of DNA from each purified transgene-positive cosmid are digested with several 6-bp recognition site restriction enzymes. The restriction fragment content of different clones is compared by carrying out a fill-in reaction with the appropriate radiolabeled nucleotide to label each fragment, followed by agarose gel electrophoresis and autoradiography of the dried gel.

Southern blot analysis of cosmids

Gels containing 50 ng of digested cosmid DNA per lane are Southern blotted and probed consecutively with radiolabeled cosmid vector, transgene, and total mouse DNA for detection of repetitive sequences. The sizes and hybridization of common restriction fragments are used to assign cosmid clones to three classes: clones internal to the transgene array that contain only transgene- and vector-positive fragments, and junction clones from each side of the insertion site that contain transgene-positive fragments, fragments with repetitive mouse DNA, and some nonhybridizing fragments.

Single-copy fragments

Fragments that do not hybridize with any of the probes above are likely to contain nonrepetitive mouse DNA. These fragments are subcloned and their single copy status is confirmed by labeling each fragment and probing a Southern blot of total mouse genomic DNA. Probes that hybridize with one or a few genomic restriction fragments on the Southern blots can be used for genetic mapping of the insertion site (Sec. 27.2) and for isolation of wild-type DNA spanning the insertion site (Sec. 27.3). Sequencing these subclones is worthwhile, since they may contain exons from the disrupted gene. Comparison with sequence databases may identify related genes or ESTs. Analysis with gene identification programs such as GRAIL (Uberbacher and Mural 1994; http://avalon.epm.ornl.gov/Grail-bin/EmptyGrailForm) and Gene Finder (Solovyev et al. 1994; http://dot.imgen.bcm.tmc.edu:9331/gene-finder/gf.html) may also identify potential exons.

Isolation of Wild-Type DNA Clones Spanning the Insertion Site

Our strategy is to isolate P1 clones (Pierce et al. 1992) that contain DNA from both sides of the transgene insert and therefore should contain the mutated gene (Fig. 27.1). The 70-kb average insert of P1 clones is large enough to contain a major portion of the disrupted gene, but small enough to be limited to one or two genes closest to the transgene insertion site. Other advantages of P1 clones include the stability of inserts during propagation, the low incidence of chimeras (<1 %), and the ease of preparation of purified DNA. Thirty ml overnight cultures of bacteria are induced with 0.5 mM IPTG for 5 h to increase P1 copy number from 1 to 20 per cell. Standard alkaline lysis protocols then yield 50 to 150 μg of supercoiled P1 DNA.

Screening P1 libraries

P1 libraries can be screened by hybridization with unique sequence probes from the junction clones, or by PCR using primers from non-repetitive sequences. Commercial screening services are available.

Inability to isolate a P1 clone spanning the insertion site suggests that a deletion or other rearrangement occurred at the time of transgene insertion (Keller et al. 1994). This can be confirmed by comparing the genetic map positions of markers from the two transgene flanks (Sect. 27.2) or by physically mapping the two flanks by pulsed field gel electrophoresis of wild-type DNA. Large deletions or other rearrangements at the insertion site may alter the expression of multiple genes, making it impossible to isolate a singe gene that can account for the mutant phenotype. The gene identification strategy described below is appropriate for "simple" insertion sites, which we operationally define by the successful isolation of a P1 clone containing *both* flanks of the transgene. Sequencing of the P1 clone is another option.

Gene Identification in Wild-Type P1 Clones that Span the Insertion Site

A practical strategy for identification of the mutated gene is to apply gene isolation methods to the entire P1 clone, which is small enough to contain exons from the mutated gene and very few others. Exon amplification (Church et al. 1994) and cDNA selection (Tagle et al. 1993, 1994; Couch et al. 1994) are complementary methods that isolate candidate gene fragments by in vivo splicing of subclones containing exons, or by hybridization between transcripts and exons within the genomic DNA. Since both methods frequently generate many false positives, it is useful to identify those products that are independently isolated by both methods (Fig. 27.2).

Exon amplification

Exon amplification is carried out on individual P1 clones using the pSPL3 vector and protocols described by Church et al. 1994. The vector and primers are available commercially (Exon Trapping System; Gibco BRL). We have added a third PCR step using the new primers F and R that are complementary to pSPL3:

F: $^{5'}$GTGCTCTAGAGTCGACCCAGC$^{3'}$ R: $^{5'}$CCCCTCGGGAGATCTCCAG$^{3'}$
677 *Sal I* 697 3115 *Bgl II* 3097

Primers F and R form a tertiary nested primer set within primers dUSD2 and dUSA4 (Gibco BRL). The tertiary amplification step with primers F and R increases the specificity of amplification and reduces the amount of vector sequence in the final product from 153 bp to 43 bp. This significantly reduces background hybridization due to vector DNA when the amplified exons are used as a hybridization probe (next step). After amplification with primers F and R, the PCR product is purified using the Wizard PCR Preps DNA Purification System (Promega) and radiolabeled using a random priming mix containing primers F and R (0.5 μM, final concentration) to increase the specific activity of the product. The radiolabeled product is used as a hybridization probe on high density filters containing the cDNA selection products (Fig. 27.2).

cDNA source

The phenotype of the mutant may indicate which tissues are likely to express the mutated gene. When the primary affected tissue is not evident, cDNA from several tissues with mRNA populations of high complexity, such as brain, testis, and HeLa cells, can be combined.

Direct cDNA selection

Direct cDNA selection is carried out using a biotin-streptavidin magnetic bead system as described by Lovett (1994) with the following modifications.

(1) Input genomic DNA is ligated to linkers and biotinylated by PCR using a 5' biotin endlabeled oligonucleotide complementary to the linkers.

(2) End products of the selection system are cloned with the CloneAmp system (GIBCO/BRL).

(3) Recombinant colonies are picked and gridded into 96 well microtiter dishes from which gridded filters are prepared by hand or with a Biomek (Beckman) robot.

Use of the robot makes it possible to prepare 8×12 cm high density filters that contain 1472 clones from 16 92-well microtiter dishes. Multiple copies of the high density filters can be readily prepared for hybridization with different probes.

Identification of clones, independently isolated by exon amplification and cDNA selection

cDNA colonies that hybridize with the mixed exon probe are purified and sequenced with vector primers (Pool 3, Fig. 27.2). These products are between 200 and 400 bp in length. In our experience, all clones from pool 3 have been authentic gene fragments derived from the P1 clone. Thus, although the combined protocol requires setting up two methods of gene identification, the small size of the P1 target and the efficient elimination of false positive clones results in the isolation of a small number of authentic gene fragments.

Isolation of the full-length cDNA

After the clones from pool 3 are sequenced, PCR-based methods can be used to obtain the full length cDNA (Burgess et al. 1995). Efficient methods for extending the cDNA include vector-insert PCR using a cDNA library as template, 5' and 3' RACE using first-strand cDNA as template, and inter-exon PCR with either type of template. Screening cDNA libraries by hybridization may be useful if the cDNA fragment is long enough to label efficiently, and if the transcript is of sufficient abundance to be represented in available libraries. ESTs identified from public databases can also be very useful for generating the full-length cDNA.

Genomic DNA sequencing alternative

When a high throughput DNA sequencing service is available, direct sequencing of the cosmids flanking the transgene insert, or the P1 clone spanning the insertion site, can substitute for the exon amplification and cDNA selection steps in this protocol. In our hands, *partial* random sequencing of these genomic clones has not been an efficient method for finding exons.

Identification of the Insertionally Mutated Gene

The cDNA corresponding to the insertionally mutated gene can be recognized by comparison of DNA and RNA from mutant and wild-type tissues. Transgene insertion usually causes complete loss of expression or truncation of the trancript. If another allele has been identified (Sect. 27.2), then detection of a mutation in the second allele can confirm the identification of the mutated gene. In vivo correction of the mutant phenotype by the wild-type gene is necessary only when a second allele is not available and the data do not clearly demonstrate that a single gene has been altered by the insertional mutation.

Southern blots

Many gene disruptions can be detected on Southern blots using the cDNA as probe (Hodgkinson et al. 1993; Hughes et al. 1993; Burgess et al. 1995). Genomic DNA from homozygous transgenic mice, heterozygotes, and wild-type mice of the same strain(s) as the microinjected eggs is digested with several restriction enzymes and probed with a radiolabeled full-length cDNA.

Northern blots and RT-PCR

RNA is prepared from the appropriate tissue(s) of homozygous transgenic mice and wildtype controls. Complete loss of expression in transgenic tissues is often observed in insertional mutants and provides strong evidence that the mutant gene has been correctly identified.

27.4 Conclusions

Determining the function of the estimated 100,000 genes in the mammalian genome is a challenging goal for biomedical research. The random insertion of transgenes into functional mouse genes provides an opportunity for direct association of complex phenotypes with genetic defects in novel genes. Since 3 % of transgenic mice carry visible mutations, it is worthwhile for facilities that generate more than 100 transgenic lines per year to systematically monitor their lines for insertional mutations by analysis of homozygous transgenic mice from each line.

Acknowledgment. Preparation of this chapter was supported by NIH GM24872.

References

Anonymous (1994) Towards high resolution maps of the mouse and human genomes – a facility for ordering markers to 0.1 cM resolution. European Backcross Collaborative Group. Hum Mol Genet 3:621–627

Ausubel FM (1995) Current Protocols in Molecular Biology. In: Ausubel FM, Brent R, Kingston RE, Moore DD, Seidman JG, Smith JA, Struhl K (eds) Short protocols in molecular biology. John Wiley, New York, Harvard Medical School, pp 6.30.3–6.3.4

Brown A, Bernier G, Mathieu M, Rossant J, Kothary R (1995) The mouse dystonia musculorum gene is a neural isoform of bullous pemphigoid antigen 1. Nat Gen 10:301–306

Burgess DL, Kohrman DC, Galt J, Plummer NW, Jones JM, Spear B, Meisler MH (1995) Mutation of a new sodium channel gene, Scn8a, in the mouse mutant " motor endplate disease." Nat Genet 10:461–465

Burmeister M, Ulanovsky L (eds) (1992) Pulsed-field gel electrophoresis. In: Methods in molecular biology. Humana Press, Totowa

Church DM, Stotler CJ, Rutter JL, Murrell JR, Trofatter JA, Buckler AJ (1994) Isolation of genes from complex sources of mammalian genomic DNA using exon amplification. Nat Genet 6:98–105

Couch FJ, Weber BL, Collins FS, Tagle DA (1994) Isolation of expressed sequences from the chromosome 17q21 BRCA1 region by magnetic bead capture. In: Hochgeschwender U, Gardiner K (eds) Identification of transcribed sequences. Plenum, New York, pp 51–64

D'Arcangelo G, Miao GG, Chen SC, Soares HD, Morgan JI, Curran T (1995) A protein related to extracellular matrix proteins deleted in the mouse mutant reeler. Nature 374:719–723

DeBry RW, Seldin MF (1996) Human/mouse homology relationships. Genomics 33:337–351

DiLella AG, Woo SLC (1987) Cloning large segments of genomic DNA using cosmid vectors. Methods Enzymol 152:199–212

Doolittle DP, Davisson MT, Guidi JN, Green MC (1996) Catalog of mouse loci. In: Lyon MF, Rastan S, Brown SDM (eds) Genetic variants of strains of the laboratory mouse, 3rd end. Oxford University Press, New York, pp 17–854

Gendron-Macguire M, Gridley T (1993) Identification of transgenic mice, in guide to techniques in mouse development. Methods Enzymol 225:794–799

Hodgkinson CA, Moore KJ, Nakayama A, Steingrimsson E, Copeland NG, Jenkins NA, Arnheiter H (1993) Mutations at the mouse microphthalmia locus are associated with defects in a gene encoding a novel basic-helix-loop-helix-zipper protein. Cell 74:395–404

Hughes MJ, Lingrel JB, Krakowsky JM, Anderson KP (1993) A helix-loop-helix transcription factor gene is located at the *mi* locus. J Biol Chem 268:20687–20690

Jones JM, Elder JT, Simin K, Keller SA, Meisler MH (1993) Insertional mutation of the hairless locus on mouse chromosome 14. Mamm Genome 4:639–643

Keller SA, Jones JM, Boyle A, Barrow LL, Killen PD, Green DG, Kapousta NV, Hitchcock PF, Swank RT, Meisler MH (1994) Kidney and retinal defects (Krd), a transgene-induced mutation with a deletion of mouse chromosome 19 that includes the Pax2 locus. Genomics 23:309–320

Kohrman DC, Plummer NW, Schuster T, Jones JM, Jang W, Burgess DL, Galt J, Spear BT, Meisler MH (1995) Insertional mutation of the motor endplate disease (*med*) locus on mouse chromosome 15. Genomics 26:171–177

Lovett M (1994) Fishing for complements: finding genes by direct selection. Trends Genet 10:352–357

Meisler MH (1992) Insertional mutation of 'classical' and novel genes in transgenic mice. Trends Genet 8:341–344

Perry WL 3rd, Vasicek TJ, Lee JJ, Rossi JM, Zeng L, Zhang T, Tilghman SM,
Costantini F (1995) Phenotypic and molecular analysis of a transgenic insertional allele of the mouse Fused locus. Genetics 141:321–332

Pierce JC, Sternberg N, Sauer B (1992) A mouse genomic library in the bacteriophage P1 cloning system: organization and characterization. Mamm Genome 3:550–558

Rowe LB, Nadeau JH, Turner R, Frankel WN, Letts VA, Eppig JT, Ko MSH, Thurston SJ, Birkenmeier EH (1994) Maps from two interspecific backcross DNA panels available as a community genetic mapping resource. Mamm Genome 5:253–274

Sambrook J, Fritsch EF, Maniatis T (1989) Molecular cloning, a laboratory manual. 2nd edn. Cold Spring Harbor Laboratory Press, Cold Spring Harbor, 9.14–9.19 (a) and 9.24–9.26 (b)

Shi Y-P, Huang TT, Carlson EJ, Epstein CJ (1994) The mapping of transgenes by fluorescence in situ hybridization on G-banded mouse chromosomes. Mammal Genome 5:337–341

Solovyev VV, Salamov AA, Lawrence CB (1994) Predicting internal exons by oligonucleotide composition and discriminant analysis of spliceable open reading frames. Nucleic Acids Res 22:5156–5163

Tagle DA, Swaroop M, Lovett M, Collins FS (1993) Magnetic bead capture of expressed sequences encoded within large genomic segments. Nature 361:751–753

Tagle DA, Swaroop M, Elmer L, Valdes J, Blanchard-McQuate K, Bates G, Baxendale S, Snell R, MacDonald M, Gusella J, Lehrach H, Collins FS (1994) Magnetic bead capture of cDNAs: A strategy for isolating expressed sequences encoded within large genomic segments. In: Uhlén M, Hornes E, Olsvik O (eds) Advances in biomagnetic separation. Eaton Publ, London, pp 91–106

Ting C-N, Kohrman D, Burgess DL, Boyle A, Altschuler RA, Gholizadeh G, Samuelson
LC, Jang W, Meisler MH (1994) Insertional mutation on mouse chromosome *18* with vestibular and craniofacial abnormalities. Genetics 136:247–254
Uberbacher EC, Mural RJ (1991) Locating protein-coding regions in human DNA sequences by a multiple sensor-neural network approach. Proc Natl Acad Sci USA 88:11261–11265
Woychik RP, Maas RL, Zeller R, Vogt TF, Leder P (1990) "Formins": proteins deduced from the alternative transcripts of the *limb deformity* gene. Nature 346:850–853
Zhou X, Benson KF, Ashar HR, Chada K (1995) Mutation responsible for the mouse pygmy phenotype in the developmentally regulated factor HMGI-C. Nature 376:771–774

Chapter 28

Contributions of Transgenic and Knockout Mice to Immunological Knowledge

José Moreno[1]*, Laura C. Bonifaz[1], and Jesús Martínez-Barnetche[1]

The advent of transgenic mouse technology in the early 1980 s provided a major tool for exploring immunological phenomena which had remained obscure for several decades. Many aspects of B cell and T cell ontogeny were finally understood by the use of a number of transgenic mouse models. By the late 1980s, this was further improved by the availability of endogenous gene inactivation by homologous recombination in embryonic stem cells grown in culture. In this chapter, we summarize what we consider the major contributions of this technology to our current views about the function of the immune system. Given the enormous amount of information available, instead of an exhaustive review, we have focused on data supported by the most convincing evidence, trying to avoid confusing or inconclusive data. For further insight the reader is referred to related recent reviews (Pfeffer and Mak 1994; Ghosh 1995; Müller 1995).

28.1 Lymphoid Cell Development

The identification of bone marrow stem cell populations has been greatly improved by the ability to inactivate the genes coding for a number of transcription factors (Dorshkind 1994). Targeted disruption of the transcription factor Pu.1 abrogates the development of both lymphoid and myeloid cells but has no effect on erythroid lineages (Scott et al. 1994), indicating the existence of a common precursor for both these cell lineages. The next control point of lymphopoiesis is the commitment of the

* corresponding author: phone: 525 627-6900 ext 1608; fax: 525–761 17 25; e-mail: morenoib@mail.internet.com.mx
[1] Research Unit in Immunobiology and Rheumatology, Hospital de Especialidades, Centro Medico Nacional Siglo XXI, IMSS, Mexico DF, Mexico

Table 28.1 Endogenous gene disruption by homologous recombination. Effects on T lymphocytes

Inactivated gene	Comments
RAG-1 and/or RAG-2	Absence of T cells due to lack TcR gene rearrangement
TcRβ	Accumulation of double negative, thymocytes, few double positives, normal TcRα rearangement
TcRα	Double negatives and double positives present, no single positives
$p56^{lck}$	Arrest in double positive maturation in early phases. no γδ T cells
CD3-ζ	Arrest in double positive maturation in early phases
CD45	Arrest in double positive maturation
Class II, invariant chain	Absence of $CD4^+$ T cells, few $CD4^+CD8^{-/+}$, normal NK 1.1^+CD4^+ normal CD8
CD4	Normal CD8 number and function
Class I, β2 microglobulin, TAP1/TAP2	Absence of $CD8^+$, normal CD4
RLT-δ	Absence of γδ T cells, increased susceptibility *Mycobacterium tuberculosis*
$p59^{fyn}$	Normal ontogeny, deficient responses of mature T cells

lymphoid/myeloid stem cell into a common lymphoid precursor, as indicated by mice with endogenous inactivation of the transcription factor *Ikaros*, which have a complete absence of T cells, B cells, and natural killer (NK) cells (Georgopoulos et al. 1994). *Ikaros–/–* mice have normal numbers of both myeloid and erythroid cells. The *Ikaros* transcription factor belongs to the zinc finger family of proteins.

Rearrangement of antigen-receptor genes depends on the participation of at least two recombination-activating gene (RAG) products, known as RAG-1 and RAG-2, as well as on the activity of terminal-deoxynuclotidyl transferase which adds nucleotides to the open DNA ends, resulting in N-segment diversity (Schatz et al. 1989; Oettinger et al. 1990). These products have been shown to interact with double-stranded DNA and RNA and, therefore, appear to be the lymphoid-specific recombinases (Spanopoulou et al. 1995). Inactivation by homologous recombination of either RAG-1 or RAG-2 results in combined immunodeficiency due to an absolute absence of both T cells and B cells (Mombaerts et al. 1992; Shinkai et al. 1992). In addition to this contribution, RAG-1 and RAG-2 knockout mice have also been of great value in exploring immunological phenomena mice due to a lack of endogenous immunocompetent cells.

28.2
T Cell Ontogeny

T lymphocytes derive from precursors that, upon entry into the thymus, undergo a number of changes that begin with the rearrangement of the T cell receptor (TcR) α- and β-genes to end with the differentiation into mature $CD4^+$ and $CD8^+$ T cell subsets. A great deal of what we know today about T cell ontogeny is based on studies accomplished first in transgenic and later in knockout mice. Early studies (Uematsu et al. 1988) demonstrated that introduction of a rearranged β chain, as a transgene, prevented (VDJ) rearrangement of endogenous β chain genes (the phenomenon of allelic exclusion). All T cells in these mice bear the same TcRβ chain, regardless of specificity and phenotype (CD4 or CD8).

Additional studies by von Boehmer and his coworkers with double transgenic mice for rearranged TcR α and β genes, demonstrated that unlike the β chain, the α chain genes were not subject to allelic exclusion (Kisielow et al. 1988a, b; Sia Teh et al. 1988). This posed a challenge, because to analyze in detail T cell phenomena, it was desirable to use mice in which the majority (if not all) T cells expressed the same TcR. This problem was solved by backcrossing αβ TcR double transgenic mice into severe combined immunodeficient (SCID) mice which fail to rearrange endogenous (T and B cell) antigen receptor genes (Scott et al. 1989). These mice allowed to make a number of capital observations that stand today as our major basis of knowledge of T cell development.

The TcR genes used to generate these mice derived from a $CD8^+$ cytotoxic T cell clone generated in an $H\text{-}2^b$ female mouse against the male (H-Y) antigen. The restricting MHC class I molecule was $H\text{-}2D^b$. Although the number of T cells (all with the same TcR) in the lymphoid tissue of $H\text{-}2^b$ females was normal, most of them were $CD8^+$, all were H-Y-specific, and were restricted by $H\text{-}2D^b$ (Sia Teh et al. 1988; Scott et al. 1989). These initial observations confirmed previous in vitro studies indicating that the same TcR is responsible for both peptide specificity and the MHC restriction (Yagüe et al. 1985). Moreover, the results indicated that the specificity of the TcR dictates the phenotype (CD4 or CD8) of a mature T cell.

In $H\text{-}2^b$ males the scenario was quite different. Although they had very few T cells in peripheral lymphoid organs, the number of CD4/CD8 double positive thymocytes in the thymuses was normal. However, CD8 single positive cells were drastically reduced in the thymus and in the periphery. This result indicated that T cells bearing self-reactive (males are H-Y$^+$) TcRs are deleted in the thymus at some stage between the transition from double positive to single positive thymocytes (Kisielow et al. 1988a).

Of additional interest, when these mice were backcrossed to an H-2 haplotype other than H-2^b, no T cells were found in female (and male) mice. Negative selection could be easily ruled out as responsible for the absence of T cells because females are H-Y negative. Therefore, the only viable explanation was that the TcR expressed in these mice had to be positively selected by an H-2^b thymus (Kisielow et al. 1988b). Positive selection thus appears to be a process where TcR binding to MHC molecules in the thymus rescues double positive thymocytes from programmed cell death (Huesmann et al. 1991). Studies from different groups confirmed these findings with TcRs of different specificities and MHC restrictions (Sha et al. 1988a, b). Similar observations were made for CD4$^+$ cells which are restricted by class II molecules (Berg et al. 1989, 1990). More recent studies (Chan et al. 1994) strongly suggest that coreceptor engagement during positive selection is necessary for commitment of double positive thymocyte to either CD8 or CD4 single positive T cells.

A population of T cells express a different kind of TcR formed by the γ and δ chains (Haas et al. 1993). Disruption of the δ chain gene by homologous recombination yields mice lacking $\gamma\delta$ T lymphocytes (Itohara et al. 1993). Although the number and functions of $\alpha\beta$ T cells in these mice appears normal, they have an increased susceptibility to infections by *Mycobacteriae* and by *Lysteria monocytogenes*. The role of $\gamma\delta$ T cells in these infections is not clear, however, because other studies indicate that CD4$^+$ and CD8$^+$ $\alpha\beta$ TcR-expressing T cells play an important protective role in these diseases (reviewed by Haas et al. 1993).

Studies accomplished in mice transgenic for MHC genes under promoters with selective tissue expression yielded additional insight into the mechanisms of T cell ontogeny. Thus, it was found that positive selection of T cells is accomplished by MHC molecules expressed by thymic cortical epithelial cells (Marrack et al. 1988; van Ewjik et al. 1988; Berg et al. 1989, 1990). In contrast, negative selection can be mediated by MHC molecules expressed by different cell types within the thymus (Marrack et al. 1988; Matzinger and Guerder 1989; Hoffmann et al. 1992; Speiser et al. 1992; Vasquez et al. 1992). It appears that for the latter process to occur, the key issue is the maturational status of the T cell precursors (Finkel et al. 1992; Speiser et al. 1992; Vasquez et al. 1992).

The availability of mice with targeted disruption of MHC class I or class II genes indicated an absolute need for these molecules for the development of CD8$^+$ or CD4$^+$ T cells respectively. Class II negative mice were generated by endogenous inactivation of IAβ-chain gene (Cosgrove et al. 1991; Grusby et al. 1991), whereas class I negative mice were accomplished by targeted disruption of the β_2 microglobulin (β_2M) gene (Zijlstra et al. 1990; Koller et al. 1990). β_2M–/– mice lack class I expression and have no CD8$^+$ cells. Grown in a germ-free environment, these mice appear normal. However, β_2M–/– mice are highly susceptible to certain types of infectious agents such as *Trypanosoma cruzi* (Tarleton et al. 1992) and influenza virus (Bender et al. 1992). On the other hand, class II negative mice lack CD4$^+$ T cells and are profoundly immunodeficient, although certain types of B cells responses remain functional (Cosgrove et al. 1991).

The class II and class I negative phenotypes, with identical effects on T cell development, could be reproduced in mice with targeted disruption of the class II-associated invariant chain gene (Viville et al. 1993) and the peptide transporter TAP-1 (van Kaer et al. 1992) gene, respectively. However, CD4 is not essential for class II-restricted T cell responses as mice with targeted disruption of the CD4 gene, have class II restricted CD4$^-$/CD8$^-$ (Rahemtulla et al. 1994) or CD8$^+$ (Matechak et al. 1996) $\alpha\beta^+$ T cells.

Transgenic mouse with a hybrid gene containing the extracellular domains of CD8 with the CD4 transmembrane and cytoplasmic domains, backcrossed to transgenic mice for a class I-restricted TcR $\alpha\beta$ develop class I-restricted CD4$^+$ T cells, which also express the transgenic chimeric coreceptor (Seong et al. 1992). This suggests that the cytoplasmic and/or transmembrane domains of CD4 deliver a signal during thymocyte development that allows the differentiation into CD4 single positive T cells. Further studies have shown that differentiation into CD4$^+$ T cell subset is mediated by regulatory elements located 6.7 kb and 13 kb upstream of the human and mouse CD4 genes (Blum et al. 1993).

Besides the TcR, many other molecules are associated with T cell development, including: surface accessory molecules, coreceptors and intracellular signal-transducing molecules (Levelt and Eichmann 1995). The manner in which most of them actually function remains unknown. However, studies carried out in transgenic and knockout mice have made it possible to trace the time points during T lymphocyte ontogeny in which some of them act.

Signal transduction by the TcR requires the participation of the CD3 complex which is a pentamer noncovalently associated on the cell membrane to the TcR (Romeo et al. 1992). The CD3 subunits are the γ, δ, ε (2×) and ζ (2×) chains. Some T cells, instead of a $\zeta\zeta$ dimer, express the dimer $\zeta\eta$ as a result of alternative splicing of ζ chain mRNA (Wegener et al. 1992). In CD3 ζη knockouts (Malissen et al. 1993), T cell maturation is arrested at the same stage as in p56^{lck} knockout mice (see below). In contrast, mice deficient in p59^{fyn} protein tyrosine kinase (PTK), which participates in TcR signal transduction, have normal T cell numbers but T cell responses are impaired with a decreased Ca^{2+} mobilization and IL-2 release after TcR-stimulation (Appleby et al. 1992; Stein et al. 1992).

Among the surface molecules known to participate in T cell activation is the leukocyte common antigen CD45 (Thomas 1989). This molecule with several isoforms expressed by different cell types has protein tyrosine phosphatase (PTPase) activity in the cytoplasmic domain (Hurley et al. 1993). CD45 associated PTPase dephosphorylates TcR associated signal transduction PTKs p56^{lck} and p59^{fyn} which, hence, become activated. Disruption of CD45 exon 6 by homologous recombination results in an arrest of T cell development just before CD4/CD8 double positive to single positive transition (Kishihara et al. 1993).

The PTK p56^{lck} was initially discovered as a signal transducing molecule associated to coreceptors CD4 and CD8 in the respective T cell subsets (Weiss and Littman, 1994). Inactivation of the PTK p56^{lck} gene yielded mice with drastically reduced numbers of peripheral T cells (Molina et al. 1992). Intrathymically, there is an arrest in thymocyte maturation at the double positive, TcR^{int} stage (the earliest stage being TcR^{low} and the latest TcR^{high}) with only a few double positive thymocytes and an incomplete TcRβ allelic exclusion (Wallace et al. 1995). The few T cells present in these mice fail to develop an efficient immune response by both CD4$^+$ and CD8$^+$ T cells (Liao and Littman 1995). However, p56^{lck} is not required for clonal deletion on T lymphocytes (Nakayama and Loh 1992). Thymocyte development regulated by p56^{lck} does not depend on its interaction with CD4 or CD8 (Levin et al. 1993). Finally, these mice also lack γδ T cells (Pfeffer and Mak 1994).

Recent studies (Amakawa et al. 1996) demonstrate that the tumor necrosis factor (TNF) family molecule, CD30, also plays a role in thymic selection. Expression of CD30 is restricted to lymphoid tissue, apparently cells of dendritic lineage. CD30-CD30 ligand interactions appear to provide a death signal. Thus, CD30-deficient mice have a profound defect in T cell negative selection although positive selection remains normal.

28.3
T Lymphocyte Function and Peripheral Tolerance

Perhaps one of the areas most favored by the arrival of transgenic mice was the understanding of the mechanisms underlying T cell tolerance. The initial observations were accomplished in MHC transgenic mice expressing class I (Allison et al. 1988) or class II (Lo et al. 1988; Savertnick et al. 1988; Böhme et al. 1989) molecules by Langerhans' islet β cells driven by the rat insulin promoter (RIP). These mice were developed on the basis of the hypothesis that ectopic expression of MHC molecules could be one of the mechanisms underlying organ-specific autoimmune diseases (reviewed by Arnold and Hämmerling 1991). Although in some cases these mice were diabetic (Lo et al. 1988; Savertnick et al. 1988), there was no evidence of autoimmune or inflammatory-mediated destruction of the Langerhans islets. Indeed, instead of an immune response to the aberrantly expressed allogeneic MHC molecules, many of these mice had peripheral tolerance to them due to T cell clonal anergy. Additional studies carried out in independently generated MHC-transgenic mice indicated that the lack of T cell-mediated response to the allogeneic pancreatic MHC molecules was not in all cases mediated by clonal anergy (Lo et al. 1988; Savertnick et al. 1988; Burkly et al. 1989; Morahan et al. 1989). Rather, in some cases, despite an apparent *in vivo* tolerance, there were *in vitro* responses to the same allogeneic MHC molecules (Böhme et al. 1989).

These initial studies led later to the full establishment of the need for costimulatory signals, besides peptide/MHC recognition, for T cell activation. A number of elegant studies established that expression of costimulatory molecules by an antigen-presenting cell (APC) distinguishes professional (capable of fully activating T cells) from nonprofessional (tolerance-inducing) APC (reviewed by Rudd 1996). The best-characterized costimulatory signal is the interaction of the T cell surface molecule CD28 with its ligands on the APC, known as B7.1 (CD80) and B7.2 (CD86). A number of transgenic mouse models expressing costimulatory molecules in a tissue-specific fashion allowed the establishment of transgenic models of autoimmune diseases (see below). Surprisingly, disruption of CD28 genes had little effect on T cell function in mice, suggesting that other costimulatory signals participate (alternatively?) in T cell activation (Shahinian et al. 1993). A potential candidate for a second costimulatory receptor on T cells has been recently described as a 70-kDa surface molecule known as SLAM (Cocks et al. 1995).

A second ligand for the costimulatory molecules B7.1 and B7.2 is CTLA-4 (Linsley et al. 1991). Binding studies indicate that the affinity of CTLA-4 for B7.1 and B7.2 is 10 to 20 times higher than affinity of CD28 (Linsley et al. 1994). As opposed to CD28, that is constitutively expressed by T cells, CTLA-4 is induced on T cells upon activation (Freeman et al. 1992). The function of CTLA-4 remained controversial until the availability of mice with targeted disruption of the CTLA-4 gene (Tivol et al. 1995; Waterhouse et al. 1995). These mice had large lymph nodes with splenomegaly, widespread lymphoid infiltrates, tissue destruction, and greatly reduced lifespan. These findings indicate that CTLA-4 plays a regulatory role in T cell function that serves to prevent an uncontrolled expansion of antigen-specific T lymphocytes after the antigenic challenge.

$CD8^+$ T cells are the major antigen-specific cytotoxic effector cells (CTLs). One of the mechanisms of T cell-mediated cytotoxicity is the release of perforin, a protein related to the membrane attack complex of the complement system (Shinkai et al. 1988). Endogenous inactivation of perforin genes results in mice with T cells capable of mounting an immune response but unable to clear infectious agents such as lymphochoriomeningitis virus (LCMV, Walsh et al. 1994). However, CTLs in these mice still retain cell-mediated cytotoxicity to cells expressing on the surface the Fas molecule, which is a receptor for apoptotic signals. In independently generated perforin knockout mice, CTL and natural killer cell activity are decreased but not absent (Lowin et al. 1994).

Tak Mak and collaborators (Penninger et al. 1995) demonstrated the role of some $\gamma\delta$ T cells in immunosurveillance and protection against hematopoietic tumors. They developed transgenic mice with a rearranged TcRγ chain gene derived from leukemia-reactive $\gamma\delta$ T cell lines. These mice were resistant to acute T cell leukemias but not to nonhematopoietic tumors.

28.4 B Cell Ontogeny

A number of transcription factors involved in B cell differentiation have been identified by targeted gene disruption. Among such factors are the products of the EA2 gene that encodes a transcription factor that belongs to the helix-loop-helix (HLH) family of proteins. In the immune system, the only cell type affected by the lack of E2A is the B lymphocyte (Bain

Table 28.2. Endogenous gene disruption by homologous recombination. Effects on B lymphocytes.

Gene	Comments
RAG-1, RAG-2	Lack of Ig gene rearrangement, accumulation of $CD43^+$ cells
IgM transmembrane domain	Accumulation of late $CD43^+$ cells, no allelic exclusion
λ5	Accumulation of $CD43^+$ cells, slow late appearance of some mature B cells
IgD	Absence of B cell affinity maturation (secondary response)
IL-2	Increase of IgG1, IgG2a e IgG2b, autoantibodies
CD40L	Hyper-IgM, absence of germinal centers, low IL-12 and IFNγ, impaired macrophage activation
CD28	Diminished IgG1 and IgG2b

et al. 1994; Zhuang et al. 1994). Two products resulting from alternate mRNA splicing of the E2A gene, E12 and E47, are involved in early B cell differentiation (Murre et al. 1989). Expression of E47 promotes immunoglobulin gene rearrangement in a pre-T cell line, although it has no effect on T cell receptor genes (Schlissel et al. 1991). HLH proteins bind to a consensus DNA sequence (GCAGXTG) in the E box elements of the immunoglobulin gene enhancers (Murre et al. 1994). Although, in addition to EA2 products, other E box-binding products have been identified, EA2−/− mice have an absolute absence of B cells as a result of an arrest prior to D to J heavy chain gene rearrangement. Moreover, these mice have a dramatically reduced transcription of RAG-1 and the B cell-specific products *mb-1*, CD19 and λ5 (Bain et al. 1994; Zhuang et al. 1994), indicating that the E2A products are essential for B cell development. Early B cell precursors have two types of germline heavy chain transcripts: μ^o and Iμ. Transcription of μ^o starts downstream from the Ig heavy chain J region, whereas Iμ is transcribed from the heavy chain enhancer, located between the V and D regions. In EA2−/− mice only Iμ transcription is affected. The actual meaning of this latter difference is currently unknown.

Gene targeting by homologous recombination also led to the identification of an additional factor implicated in B cell development. The product of the Pax5 gene, known as BSAP, is apparently induced by HLH proteins (Urbànek et al. 1994). Finally, studies carried out in transgenic mice with constitutive expression of an inhibitor of the activity of HLH transcription factors, the *Id1* protein, yielded a phenotype similar to that observed in EA2−/− mice (Sun 1994).

Prior to immunoglobulin gene rearrangements, B cell precursors express a transcription factor known as early B cell factor (EBF). Expression of EBF is constant during all stages of antigen-independent B cell differentiation and it plays a role in inducing expression of the B cell receptor associated protein Ig-α. Targeted disruption of EBF results in an absence of B cells and accumulation of committed B cell precursors before the onset of immunoglobulin gene rearrangement (Lin and Grosschedl 1995). This indicates that EBF plays a role in the induction of immunoglobulin gene rearrangement.

Early in ontogeny, B cell precursors activate their gene recombination machinery (RAG-1, RAG-2, and TdT) to rearrange immunoglobulin genes (Rajewsky 1996). The majority of B cell precursors initially rearrange the heavy chain gene to constitute a pre-B cell. Productive heavy chain rearrangement yields a pre-B cell in which the heavy chain assembles with a surrogate light chain (λ5) to constitute the pre-BCR). λ5 knockout mice and mice with disruption of the μ chain transmembrane domain (Kitamura and Rajewsky 1992; Löffert et al. 1996), both of which fail to express the pre-BCR, have an arrest in early B cell development, just before B cell precursors become small resting pre-B cells. This indicates a role for pre-BCR in B cell development. Apparently, the pre-BCR participates in a selective process which allows only cells with a productive heavy chain gene rearrangement to go on to rearrange the light chains. The putative ligand of the pre-BCR remains unknown.

B cell receptors (BCR) consist of surface immunoglobulin non-covalently-linked to signal-transducing proteins Ig-α and Ig-β (reviewed by Rajewsky 1996). A recent study in mice with targeted disruption of Ig-β (Gong and Nussenzweig 1996) indicated that an Ig-β-mediated event is necessary to complete DJ to VDJ rearrangement. Thus, Ig-β signaling appears to precede the existence of a functional heavy chain and, hence, its association with an antigen receptor chain, including the pre-B cell receptor.

Initial studies demonstrated that the transgenic introduction of a rearranged immunoglobulin heavy chain gene prevented rearrangements of the endogenous heavy chain genes (Weaver et al. 1995). This provided the first molecular evidence for the phenomenon of allelic exclusion, which prevents the expression of the second allele once the first heavy chain gene has been productively rearranged. Although the intrinsic mechanisms of allelic exclusion have not been completely elucidated, several lines of evidence support a role of the pre-BCR on it. First, as opposed to the complete, rearranged heavy chain gene, transgenic expression of only soluble μ chains, by disruption of the *trans-*

membrane domain, fails to induce allelic exclusion (Kitamura and Rajewsky 1992); second, targeted disruption of the surrogate light chain also prevents allelic exclusion (Löffert et al. 1996).

Cells with a productive heavy chain gene rearrangement are positively selected by binding to an, as yet, unidentified ligand. These cells down-regulate the surrogate light chain and go on to rearrange a (ϰ or λ) light chain gene to become small resting pre-B cells (Grawunder et al. 1995). Cells with functional BCR express initially surface IgM (immature B cells) followed by expression of IgD plus IgM which are the mature virgin B cells present in peripheral lymphoid organs. Upon antigen selection and differentiation in the presence of the appropriate T cell help, B cells develop into secondary response B cells which express on their surface other isotypes such as IgG, IgA, or IgE, almost always together with IgM. Mice transgenic for a rearranged γ2b heavy chain (Roth et al. 1993) have an early defect in B cell maturation. However, the number of B cells in older mice is near normal. All B cells express surface μ besides γ2b. Similar to rearranged μ genes, the γ2b transgene prevents the rearrangement of endogenous heavy chain genes. Finally, these mice were crossed with mice with endogenous inactivation of μ constant regions, resulting in a lack of B cell development. This indicates that transmembrane IgG2b can form a functional pre-BCR for allelic exclusion but it cannot replace surface IgM for B cell development.

28.5
B Cell Function

As mentioned, the BCR consists of clonally variable immunoglobulin heavy and light chains, which confer antigen specificity, noncovalently linked to invariant signal transducing units Ig-α and Ig-β. Intracellularly, signal transduction in B lymphocytes is mediated by several molecules, including the Src-related PTKs *Lyn*, *Fyn*, *Blk*, and *Btk*. Targeted disruption of *lyn* in mice results in a phenotype with a moderate decrease in the number of peripheral B cells (Nishizumi et al. 1995). Signal transduction is altered and several intracellular proteins including *vav*, *Cbl*, and HS-1 fail to be phosphorylated after IgM cross-linking. Interestingly, these mice have elevated levels of IgM, IgA, anti-double-stranded (ds) DNA autoantibodies, mostly IgM, and evidence of glomerulonephritis. However, further studies are required do define the nature of the autoimmune response and possible relation to anti-dsDNA antibodies present in systemic lupus erythematosus (SLE).

Similar to T cells, B cells express a number of lineage-specific molecules that may play different roles on B cell activation and differentiation. One such molecule is CD19, which is expressed from early B cell precursors to mature B cells, where CD19 participate on B cell activation together with surface immunoglobulin (Rajewsky 1996). Targeted disruption of CD19 genes does not affect B cell development, but results in a profound impairment of B cell responses to protein (T cell-dependent) antigens (Rickert et al. 1995) without affinity maturation.

B-1 cells are a second, distinct, population of B lymphocytes which appear to segregate from regular B cells during ontogeny (Arnold et al. 1994). The role of B-1 cells in immunity is to mediate the majority of T cell independent antibody responses (Rajewsky 1996). CD19-/- mice have an absence of B-1 B cells, which are the major producers of serum IgM (Rajewsky 1996).

Among the surface molecules thought to play a role in B cell responses to antigen are the complement receptors. Two receptors CR1 (CD35) and CR2 (CD21) bind activated products of the C3 fraction of serum complement. Both receptors are encoded by a single locus in mouse chromosome 1 (Kurtz et al. 1990). In an interesting study, Ahearn et al. (1996) produced mice deficient in CR1 and CR2 by targeted disruption of the *Cr2* locus. These mice have reduced numbers of B-1 type B cells, with normal serum IgM levels. Nevertheless, antibody response to T-dependent antigens is dramatically decreased; the germinal centers are sparse and small. The immune defect was corrected by transfer of *Cr2+/+* B cells. This study indicates that antibody responses are strongly influenced by membrane complement receptors on B cells.

B cells constitutively express on their surface the CD40 molecule. Engagement of CD40 by its ligand, which is induced on T cells by activation (Banchereau et al. 1994), is an event required for class switch of immunoglobulin genes. This phenomenon is best exemplified by patients with X-linked hyper-IgM immunodeficiency in which isotypes other than IgM are absent (reviewed by Banchereau et al. 1994).

The hyper-IgM syndrome was reproduced in mice with targeted disruption of the CD40L gene (Xu et al. 1994). Lymphoid tissue in these mice shows an absence of germinal centers. In additional studies (Grewal et al. 1995), it was found that CD40L-deficient mice have impaired T cell responses to exogenous antigens. Moreover, in these mice (Campbell et al. 1996; Soong et al. 1996) and in mice with endogenous inactivation of CD40 (Kamanaka et al. 1996), it was found that besides its known role on B cell activation, CD40-CD40L interactions participate in cellular immune functions. Thus, it appears that a signal received through CD40 allows macrophages to release IL-12, which will eventually turn into INFγ and TNFα secretion. CD40L –/– as well as CD40 –/– mice have also a defect in cellular immunity that results in an increased susceptibility to infections by intracellular pathogens such as *Leishmania major* (Campbell et al. 1996; Kamanaka et al. 1996) and *Leishmania amazonensis* (Soong et al. 1996).

28.6 B Cell Tolerance

In normal mice, the reduced number of B cells specific for a given antigen makes it difficult to study their fate during development and maturation. The availability of mice transgenic for rearranged heavy and light chains of known specificity has been of great value in studies of B cell tolerance. Goodnow and colleagues initiated a series of studies in transgenic mice expressing the heavy and light chains of an anti-hen egg-white lysozyme (HEL) antibody (Goodnow et al. 1988). In these mice the vast majority of B lymphocytes bear the same antigen receptor which is, in addition, functional. When anti-HEL transgenic mice were crossed to HEL-transgenic mice, the result was a functional inactivation of B cells which was reversible upon transfer into normal mice (Goodnow et al. 1991), indicating the reversibility of the phenomenon of clonal anergy. To maintain B cell clonal anergy, it is necessary to be in the continuous presence of the relevant antigen. Thus, this form of tolerance could be a reversible status that carries the possibility of contributing to autoimmune disease.

It was later found in other models that, depending on the nature of the antigen, B cell tolerance can be mediated by functional inactivation or clonal deletion. It appears now that for deletion, immobilized antigens, capable of cross-linking B cell surface immunoglobulin receptors, are optimal (Nemazee and Bürki 1989; Erickson et al. 1991; Hartley et al. 1991; Russell et al. 1991; Chen et al. 1995), whereas soluble antigens induce preferentially clonal inactivation in the periphery (Goodnow et al. 1989; Hartley et al. 1991).

Further studies demonstrated that anergic B cells are short-lived if the antigen they recognize is present (Cyster et al. 1994; Fulcher and Basten 1994; Cyster and Goodnow 1995). This is reversible upon transfer into mice lacking HEL. These results indicate an exclusion in the periphery of self antigen-reactive B cells by an, as yet, unknown mechanism. Elimination of these cells is through an apoptotic mechanism mediated by a CD95 (Fas)-Fas ligand interaction (Rathmell et al. 1995). When HEL-anti-HEL double transgenic mice were rendered CD45 deficient by gene targeting there was no tolerance, indicating that CD45 expression is required for tolerance induction (Cyster et al., 1996). In CD45+/+ anti-HEL mice, circulating HEL induces clonal anergy, whereas in the absence of CD45 the presence of antigen promoted positive selection of HEL-specific B cells.

In the same transgenic mice it was found that anergic B cells have impaired a number of additional functions including signal transduction through the (surface immunoglobulin) BCR (Cooke et al. 1994) and through the surface molecule CD38 (Lund et al. 1995), as well as an inability to provide CD28-mediated costimulation of naive T lymphocytes (Ho et al. 1994).

In contrast to the effects of soluble (non cross-linking) antigens which induce peripheral clonal inactivation of B cells, membrane antigens (Hartley et al. 1991; Russell et al. 1991), or other cross-linking antigens, such as dsDNA (Chen et al. 1995), induce either peripheral or central clonal deletion of self-reactive B cells. These models provide evidence that, similar to T cells, B cell tolerance can be induced at two different levels (central and peripheral) and by at least two different mechanisms.

28.7 Cytokine and Cytokine Receptor Genes

The redundancy in cytokine actions prevented for a long period ascertaining the actual role of the individual molecules in homeostasis and disease. The availability of transgenic mice overexpressing cytokines allowed some of the pathogenic effects of their excess to be established. It was, however, the availability of mice with targeted disruption of individual cytokine genes or their receptors what has made it possible to distinguish shared and selective functions of many of these molecules.

The major growth factor for T lymphocytes is interleukin (IL)-2 (Smith 1984). This cytokine has additional roles in the maturation of NK cells and B cell differentiation (Miyajima et al. 1992). IL-2 exerts its effects through a receptor (IL-2R) constituted by three transmembrane glycoproteins. The β and the γ chains together form the intermediate affinity IL-2R which is constitutively expressed by T and NK cells. Upon

Table 28.3 Endogenous gene disruption by homologous recombination. Cytokines and cytokine receptors

Gene	Comments
IL-2	Late B cell immunodeficiency, autoimmune phenomena
IL-2Rγ (common γ chain)	Severe combined immunodeficiency
IL-4	IgE absent, decreased IgG1, lack of Th2 cell development
IFNγ, IFNγR	Defects in macrophage activation, infections by BCG, *Lysteria*, vaccinia, etc., lethal
LIF	Failure in blastocyst implantation, defect in multipotential hemopoietic stem cells
TGF-β	Multiple defects, waste syndrome, early death, increased IFNγ and TNFα, multiple inflammatory lesions
TNFαR (p55)	Resistance to LPS, death by *Lysteria, Mycobacterium tuberculosis,* absence of germinal centers
LTXα	Absence of peripheral lymph nodes and epithelial lymphoid tissue, lack of germinal center formation
IL-10	Ulcerative colitis-like syndrome
IL-12	Lack of Th1 cells and low INFγ

activation, T cells express the α chain, which by itself is a low affinity IL-2R, but together with the βγ complex yields a high affinity IL-2R (reviewed by Sato and Miyajima 1994).

Surprisingly, targeted disruption of the IL-2 gene yielded what appeared to be an essentially normal mouse (Schorle et al. 1991) with only a moderate B cell function impairment by old age. This unexpected phenotype could be explained in part by the ability of some of the IL-2R components to bind other cytokines, most noticeably IL-15, which can replace some functions of IL-2. Later studies, however, have shown that IL-2-deficient mice develop T cell-dependent autoimmune phenomena (Kündig et al. 1993; Sadlak et al. 1993; Schorle et al. 1993). Moreover, disruption of the IL-2Rβ chain yields mice with T cell hyperesponsiveness, an excess of plasma cells and granulocytes, and similar autoimmune features (Suzuki et al. 1995). This indicates that IL-2, perhaps through the IL-2Rβ chain, also mediates regulatory signals to prevent an uncontrolled immune response. IL-2-deficient mice grown in a nongerm-free environment develop an ulcerative colitis-like syndrome (Sadlack et al. 1993).

Absence of the IL-2Rγ chain results in severe combined immunodeficiency with impaired T and B cell development and absence of NK cells and γδ T cells (Cao et al. 1995). This is explained because, in addition to mediating the response to IL-2, the γ chain is a common component of several cytokine receptors, including IL-4, IL-7, IL-9, and IL-15. Indeed, the chain formerly called IL-2Rγ is now known as the common cytokine receptor γ chain (cRγC). In humans, the majority of X-linked severe combined immunodeficiencies are due to mutations in the cRγC which affect cytokine functions to a variable extent (Sato and Miyajima 1994; Fischer and Leonard 1995). A similar immune phenotype was accomplished in mice deficient in the *Jak3* kinase, which is part of the signaling pathway departing from the cRγC (Park et al. 1995).

Transgenic mice overexpressing the IL-4 gene develop an hyper-IgE syndrome and allergic phenomena with a paucity of Th1 type $CD4^+$ T cells and an increase in $CD8^+$ T cells (Tepper et al. 1990). On the other hand, IL-4-deficient mice lack IgE, have very low levels of IgG1 (Kühn et al. 1991), and fail to develop Th2 cells (Kopf et al. 1993), confirming the need of IL-4 for the generation of this helper T cell subset. Finally, despite IL-4 –/– mice have impaired mucosal immune responses in the digestive tract; serum IgA levels are normal (Vajdy et al. 1995).

These models confirmed the roles assigned to IL-4 by many elegant in vitro and in vivo studies achieved by several groups of immunologists over the past two decades (reviewed by Seder and Paul 1994). On the other hand, disruption of the gene encoding the signal transducer and activator of transcription Stat6 (Kaplan et al. 1996), which is activated after IL-4R ligation, also resulted in an absence of Th2 cells and lack of B cell responses to the Th2 cytokines IL-4 and IL-13. These mice had no other detectable abnormalities, indicating that Stat6 activity is restricted to the immune system.

One of the most dramatic examples of what lack a cytokine might mean to an animal is exemplified by mice with targeted disruption of transforming growth factor (TGF)β1 gene (Shull et al. 1992; Kulkarni et al. 1993). These mice have a wasting disease with widespread mixed cell infiltrates, necrotic lesions, and early death.

Mice with inactivation of interferon-(IFN)γ genes (Dalton et al. 1993, Flynn et al. 1993) or components of its receptor (Huang et al. 1993) are immunodeficient with multiple defects. These mice cannot activate the respiratory burst necessary to kill many intracellular pathogens by means of reactive oxygen intermediates. These animals die from many different infections, including tuberculosis, have a decreased expression of MHC class II molecules, and also have diminished NK cell activity.

Finally, these mice have low levels of IgG2a, confirming the role of IFNγ in the isotype switch to IgG2a.

Tumor necrosis factor (TNF)α is a major inflammatory cytokine produced by mononuclear phagocytes. It exerts its effects through two different receptors, p55 and p75, which deliver different types of signals to target cells (Miyajima et al. 1992). p55 TNFR deficient mice are highly susceptible to infections by *Mycobacterium tuberculosis* while retaining response to BCG vaccination (Flynn et al. 1995).

A second type of TNF, known as TNFβ has two forms: lymphotoxin α (LTX α), a secreted protein derived mainly from TH1-type T cells, or a membrane form (LTXβ). These proteins bind to the same receptors as TNFα and were thought to have identical functional effects on target cells (Vassalli 1992). However, disruption of LTX genes results in an absence of lymph nodes and epithelial associated lymphoid tissue (De Togni et al. 1994), and a failure to form germinal centers in the spleen (Matsumoto et al. 1996). The latter phenomenon does not appear to be explained by a different receptor for LTX because p55 TNFR deficient mice also show failure in germinal center formation. As these mice have normal TNFα production, this indicates that although subtle, there are functional differences between both types of TNFs through an, as yet, unknown mechanism.

IL-5-deficient mice yielded an unexpected finding: eosinophil development and basal eosinphil levels are normal. However, IL-5–/– mice do not develop helminth-induced eosinophilia (Kopf et al. 1996), and the development of B-1 cells is severely impaired, being reduced by up to 80 % (Kopf et al. 1996). Identical findings were obtained in mice with disruption of the IL-5Rα chain gene (Yoshida et al. 1996).

Other phenotypes observed in mice with targeted disruption of cytokine genes include severely impaired T and B cell development in IL-7-disrupted mice (van Freeden-Jeffry et al. 1995), impaired development of cortical bone (Poli et al. 1994) and lack of mucosal IgA secretion (Ramsay et al. 1994) in IL-6-deficient mice; a diminished Th1-type T cell response with marginal secretion of IFNγ and poor delayed type hypersensitivity in IL-12-deficient mice (Magram et al. 1996); anemia and inflammatory bowel disease in mice with targeted disruption of the IL-10 gene (Kühn et al. 1993).

28.8 Studies on Autoimmunity

The information about central and peripheral T and B cell tolerance generated in transgenic mice has allowed the development of animal models aiming at understanding the molecular mechanisms of autoimmunity and autoimmune disease. For practical purposes, we focus mainly on insulin-dependent diabetes mellitus and experimental allergic (autoimmune) encephalomyelitis (EAE). These models have provided a great deal of information regarding predisposition and triggering of immune response toward self antigens.

28.9 Insulin-Dependent Diabetes Mellitus (IDDM)

This disease is characterized by autoimmune destruction of the islets of Langerhans, accompanied by lymphocytic infiltrate within (insulitis) or around (peri-insulitis) the islets. Although there is a clear strong genetic linkage in the predisposition to IDDM (i.e., HLA-DQβ and DQα alleles), its incidence in monocygotic twins is only 20 to 30%. (Teofilopoulos 1995; Vyse and Todd 1996). Therefore, additional genetic or environmental factors (or both) might be contributing to the development of disease (Tisch and McDevitt 1996).

Nonobese diabetic (NOD) mice spontaneously develop a disease closely resembling human IDDM. By 3 weeks of age, a significant proportion of these mice develop peri-insulitis that evolves to insulitis and overt diabetes by 10 weeks of age. NOD mice have an unusual MHC class II allele, similar to IA^{b} but with a serine 57 in the β chain (Morell et al. 1988 and references therein). The human equivalents for IA^{bnod} are the HLA class II alleles DQB1*0201 and DQB1*0302, both with a conserved small noncharged amino acid at position 57. Class II β alleles with positive residues (such as aspartate) at position 57 have a dominant protective role for the development of IDDM. Although other recessive non-MHC genes have been linked to diabetic predisposition, none of them is absolutely necessary for the development of the disease. (Tisch and McDevitt 1996).

The role of MHC class II products in susceptibility and protection from IDDM is rather complex. The introduction of an IEα chain or non-NOD IA alleles as transgenes into NOD mice prevent insulitis and diabetes (Nishimoto et al. 1987; Böhme et al. 1990; Slattery et al. 1990; Miyazaki et al. 1990). The mechanisms involved in protection are not clear. Passive transfer of NOD-IA^k transgenic T cells into NOD-SCID or into thymectomized-lethally irradiated NOD mice induces IDDM, suggesting that protection is not due to central or peripheral T cell tolerance (Singer et al. 1993).

As mentioned, the clinical expression of IDDM appears to be the end stage of a complex multistep process, with environmental factors playing an important role. Infectious agents sharing epitopes with self antigens could trigger the disease in individuals with the appropriate genetic background. Alternatively, a chronic inflammatory response to a foreign agent could enhance (or modify) self antigen presentation and the arrival of potentially autorreactive cells in the islets, leading to islet destruction. Insulitis is necessary but not sufficient for the progression of the disease (Flavell et al. 1995).

In a series of experiments accomplished over the past 6 years, two independent groups (Ohashi et al. 1991; Oldstone et al. 1991; von Herrath et al. 1995) defined many of the mechanisms involved in the development of IDDM. Mice transgenic for the viral antigen lymphochoriomeningitis (LCMV) glycoprotein (gp), driven by the RIP have LCMVgp expressed only by pancreatic β cells but do not develop spontaneous insulitis (Ohashi et al. 1991). However, upon LCMV infection, RIP-LCMVgp mice rapidly develop autoimmune insulitis and diabetes. In parallel studies, Oldstone et al. reported identical findings in mice transgenic for either LCMVgp or LCMV nucleoprotein. These findings demonstrate that, in normal individuals, self-reactive T cells not subject to tolerance induction (*ignorant* – not aware of the presence of the antigen they recognize) can, upon appropriate antigen presentation, be activated and induce autoimmune disease.

Interestingly, mice double transgenic for RIP-LCMVgp and a rearranged TcR from a GP peptide 33–41-specific T cell clone do not develop insulitis nor diabetes, despite the fact that 80 % of their T cells bear the potentially autoreactive TcR (Ohashi et al. 1991). This peripheral ignorance can be reverted in triple transgenic mice expressing RIP-LCMVgp and LCMV 33–41 TcR together with the RIP-B7.1 gene (Harlan et al. 1994). These mice develop early severe IDDM. Double transgenics expressing LCMV GP plus B7.1 also develop IDDM, although in a delayed fashion.

Another approach to understanding the role of inflammation in IDDM is the use of transgenic mice expressing cytokines by islet β cells under the control of RIP. RIP-IFNγ (Savertnick et al. 1988), RIP-tumor necrosis factor (TNF)α (Higuchi et al. 1992; Picarella et al. 1993), RIP-lymphotoxin (LT)α (Picarella et al. 1992), RIP-IFNα (Stewart et al. 1993), RIP-IL-10 (Wogensen et al. 1993), and RIP-IL-2 (Heath et al. 1992) transgenic mice develop nonantigen-specific inflammatory infiltrates in the islets. Some of them even develop diabetes without additional stimulus. Islet infiltration appears to be, at least in part, the result of cytokine-induced expression of adhesion molecules, but without antigen-specific T cell activation.

The lack of activation of self-reactive T cells in these models is explained by the absence of co-stimulatory molecules on β cells. The isolated transgenic expression of the co-stimulatory molecule B7.1 (CD28 ligand) or TNFα by β cells does not induce disease by itself. However, 100 % of double transgenic mice expressing simultaneously RIP-B7.1 plus RIP-TNFα develop autoimmune IDDM (Guerder et al. 1994). Thus, the expression of costimulatory signals allows the inflammatory process to induce the activation self antigen-reactive T cells that would otherwise remain ignorant.

28.10 Experimental Autoimmune Encephalomyelitis

The observation that some humans developed a disease resembling multiple sclerosis (MS) after immunization with impure rabies vaccines led to the establishment of experimental models of demyelinating diseases (Hemachudha et al. 1987). Genetic factors play a role in the predisposition to MS. As in other autoimmune diseases, certain HLA class II alleles are associated with an increased risk for MS (reviewed by Steinman 1996). Although there is some familiar clustering, the low concordance rate between monocygotic twins suggests that environmental factors also play a role.

Immunization of SJL/J mice with pertussis vaccine plus myelin basic protein (MBP) emulsified in complete Freund's adjuvant induces encephalomyelitis and demyelination resembling human multiple sclerosis (MS) (Bernard and Carnegie 1975). Although there is no formal proof that MBP-specific T cells cause MS, EAE is widely accepted as a model for MS. As in the human disease, the genetic background of different mice strains has an influence on the susceptibility and outcome of EAE. Some H-2 alleles have been linked to EAE susceptibility (Fritz et al. 1985; Kono et al. 1988) but it is widely accepted that other loci besides H-2 are crucial.

The brains of MS patients have an increased expression of MHC molecules (McFarlin and McFarland 1982). H-2^k mice transgenic for H-$2K^b$ under the control of the MBP promoter (Turnley et al. 1991) resemble shivering mice which have a null mutation of the MBP gene. The transgene is expressed only by oligodendrocytes, resulting in profound demyelinization without lymphocytic infiltrates. These findings are consistent with studies accomplished in mice expressing Class I molecules by islet β cells (Allison et al. 1988), which develop IDDM without immune destruction of Langerhans islets. In both models, β cell and oligodendrocyte dysfunction is a consequence of transgene expression which interferes with normal cell function, and not an immune-mediated phenomenon.

$CD4^+$ T lymphocytes are the main effector cells in EAE. Adoptive transfer of T lymphocytes activated in vitro with MBP- or proteolipid apoprotein (PLP) into susceptible SJL/J recipients induces disease (Pettinelli and McFarlin 1981; Mokhtarian et al. 1984). Mice pretreated with anti-CD4 monoclonal antobodies (mAb) fail to develop EAE upon transfer (Waldor et al. 1985; Sriram and Roberts 1986). Analysis of TcR variable regions of MBP-specific T cell lines demonstrated a high usage of Vβ8.2. Accordingly, transfer of these cells into healthy recipients reproduces disease whereas anti-Vβ8.2 mAb treatment blocks the induction (Acha-Orbea et al. 1988; Owhashi and Heber-Katz 1988; Zaller et al. 1990).

Encephalitogenic T cells in IA^u mice (susceptible to EAE) are directed at MBP peptide 1–11. In MBP 1–11 TcR transgenic mice (Goverman et al. 1993) all mature T cells are $CD4^+$(the TcR was obtained from a class II-restricted $CD4^+$ T cell clone). In these mice, autoreactive T cells are not negatively selected and remain healthy when bred in a pathogen-free facility. However, animals bred in a nonpathogen-free facility frequently suffer from EAE. The nonresponsive status is not due to clonal anergy because T cells from these mice proliferate and secrete IL-2 in response to purified MBP in vitro. Treatment with pertussis toxin, which temporarily breaks the blood-brain barrier, induces severe EAE in TcR transgenic mice but not in their control littermates. Thus, in immunoprivileged sites, such as the brain, temporary opening of physiologic barriers may be sufficient for the development of autoimmune disease as long as ignorant autoreactive T cells are present.

The clinical courses of MS and EAE are variable, sometimes being progressive, and others manifested by episodes of relapse followed by remission (Steinman 1996). It has been postulated that a T lymphocyte subset other than CD4 could be responsible for mediating the recovery stage. In MBP 1–9 TcR transgenic mice backcross into RAG-1 knockouts all T cells express the transgenic TcR and all mice spontaneously develop EAE (Lafaille et al. 1994). Thus, in this model, autoreactive $CD4^+$ T cells are the only encephalitogenic effector cells needed for the development of EAE. The low EAE incidence in RAG-1 competent mice suggests that in normal mice an as yet unidentified cell with rearranged antigen receptor genes prevents the triggering and progression of the disease.

Although other models of autoimmune disease have been developed in transgenic mice, the underlying mechanisms are somewhat less well understood. Nevertheless, we will refer to a transgenic mice model of autoimmune hemolytic anemia developed with the rearranged heavy and light chain genes of an anti-red cell antibody (Okamoto et al. 1992). These mice did not develop tolerance to the self antigen. A variable number of animals had hemolytic anemia with varying degrees of severity. All autoantibodies were IgM and were secreted by intraperitoneal B-1 cells. As autoantibody-secreting B-1 cells that leave the peritoneal cavity undergo self antigen-induced apoptosis (Murakami et al. 1992), it appears that cells remaining in the peritoneal cavity are sequestered from self antigen therefore preventing deletion. This model provides evidence that in some cases B-1 cells which are thought to produce mainly natural (nonpathogenic) autoantibodies are capable of generating autoimmune diseases. It is possible that autoantibody secretion by these cells is induced by an antigen-independent mechanism.

28.11 Studies in Animals with Human MHC Molecules

Soon after transgenic mouse technology was available, several groups developed mice expressing human HLA genes. One of the main purposes was to develop animal models expressing disease-associated HLA alleles to study their precise role in disease. Mice double transgenic for the human class I molecule HLA-B27 and β_2M (Krimpenfort et al. 1987) expressed the transgenes which functioned as restriction elements (Kievits et al. 1987), but did not result in the expected disease: ankylosing spondylitis (AS). It was necessary to introduce the B27 and human β_2M genes in rats (Hammer et al. 1990) to induce a disease resembling Reiter's disease, a form of spondyloarthritis related to AS. However, up to date, the manner in which B27 favors the development of AS and other seronegative spondyloarthritis is still unclear.

A number of other human class I and class II MHC genes have been expressed in transgenic mice. Generally, the main observation has been that they function as restriction elements for mouse T cells, although in certain instances a species barrier exists that is partially explained by a failure of the mouse coreceptors CD4 and CD8 to interact with human MHC molecules.

There are many additional models of both transgenic and knockout mice that have improved our knowledge about the immune response, including those of genes involved in either induction or prevention of apoptosis. However, due to space limitations, we refer the reader to recent reviews (Cohen et al. 1992, Fraser and Evan 1996).

Transgenic and knockout experiments have contributed importantly to our current knowledge in immunology. In the forthcoming years, targeted disruption of additional genes should be of great value to continue expanding our understanding of the immune response. Currently, mice with tissue-directed deletion of endogenous genes are beginning to provide additional information to immunologists.

28.12 Some Technical Hints to Study Transgenic and Knockout Mice of Immunological Interest

The detailed protocols to develop transgenic and knockout mice are described throughout this manual. Although an enormous number of techniques are used to study the expression and some functions of the transgenic (or missing) products, we will here describe only how to prepare cells from lymphoid tissue for different purposes and how to examine them by flow cytometry, which nowadays constitutes one of the most powerful tools to study immunologically relevant cells.

Preparation of Lymphoid Cells from Different Tissues

Many of the immunologically relevant proteins are expressed on the cell surface. Depending on the nature of the mice under study and the transgenes employed, one may need to obtain cells from different lymphoid organs. The first step in all cases is to kill the mouse by cervical dislocation or, if preferred, by ether anesthesia. Then, if cells are needed for culture, it is necessary to perform asepsis of the skin. For this purpose, 70 % ethanol is usually appropriate.

Spleen For peripheral lymphoid cells, the spleen is usually an easily accessible source of cells for analysis. With scissors, make a small cut (about 5 mm) in the center of the abdominal skin, being careful not to perforate the muscle wall. With both hands pull the skin to expose the whole abdominal wall. Place the mouse on a board lying on its right side. Use the scissors again to open the muscular wall (a 4-mm cut is sufficient) in the upper left quadrant where the spleen can be seen as a purple shadow. With pliers, take the spleen carefully and cut the pediculus.

Thymus The thymus is located in the upper mediastinum, right above the heart. To obtain the thymus, a dead mouse is placed, face up, on a board. The skin should have been opened as described for the spleen. Cut the sternal bone from bottom to top, avoiding to cut the heart and vessels. The thymus is seen as a white organ of variable size, depending on the age of the animal. Take it with pliers and use small scissors to release carefully the whole organ.

Lymph nodes This is the preferred tissue to obtain T cells from immunized mice. Although this is used essentially to achieve functional assays of specific T cells, not normally for flow cytometry, here we describe how to obtain them. There are several possibilities. However, we describe here the obtention of para-aortic lymph nodes that we have employed with good results. Mice are immunized at the base of the tail with the desired antigen (usually 20 to 50 μg is sufficient) emulsified in complete Freund's adjuvant (CFA). Immunize on both sides of the tail in a maximum volume of 50 μl on each side. Eight to ten days latter, mice are killed and the abdominal wall is exposed by opening the skin as described. Open the abdominal cavity, put the bowels aside to visualize the aorta. If the mice was appropriately immunized, lymph nodes are readily seen as 2- to 3-mm round white masses along the abdominal aorta. Take the pediculus (not the lymph nodes) with pliers, pull, and place in a petri dish. Take all possible lymph nodes. Try not to cut the vessels because bleeding impairs the visualization of the lymph nodes.

Lymph nodes from nonimmunized mice are more difficult to visualize. However, for flow cytometry of nonimmunized lymph node lymphocytes, it is easier to locate inguinal and axillary lymph nodes that are usually larger than para-aortic lymph nodes.

Once the lymphoid organ under study is released, it can be placed in a petri dish, in culture medium, or isotonic saline solution to keep it until the next step. Then, depending on the type of analysis to be accomplished, you may want to obtain suspensions of cells or to make cuts for histological sections. For flow cytometry, cells must be in suspension and free of clumps and debris.

Place the tissue in sieves with mesh 100 screens, use a syringe embolous to press the tissue. To collect the cells, the sieves are placed on an appropriate size beaker. Add medium in small amounts to facilitate the procedure and until a small membrane (capsule) is the only tissue left above the screen. Spin the cell suspension and resuspend at the desired concentration. To eliminate clumps and connective tissue, it is desirable to pass the cell suspension through a piece of 100-mesh nylon cloth.

Staining with Antibodies for Analysis by Flow Cytometry

It is important to choose the appropriate antibody or antibodies (usually monoclonal) to study the mice and to decide whether it is necessary to perform analysis with one, two, or more dyes. For one dye, either direct or indirect immunofluorescence can be employed, whereas for two or more dyes the method of choice is direct immunofluorescence, although one of the antibodies may be biotin-labeled.

Important considerations

The type of isotonic solution to stain the cells depends on the cell type under study. We prefer Dulbeco's PBS without calcium and magnesium. The addition of EDTA at a final concentration of 5 mM may be useful to prevent clumping. Addition of a source of proteins helps to keep the cells in better shape. This can be accomplished with serum (1 to 2 %) or BSA (1 %). We prefer serum. For cells with Fc receptors, it is important to add a source of **nonmouse** IgG antibodies that outnumber and prevent the cytophylic binding of our mAb to the cells, which can give spurious results. The best Fc receptor blocking antibodies are found in rabbit, horse, and human serum. We normally use horse serum. Finally, to prevent capping, the solution is kept continuously cold (on ice) in addition to being supplemented with sodium azide (0.05 to 0.1 % final concentration). Here we refer to this solution as working solution (WS).

The source of the antibody is also important. Monoclonal antibodies can be used purified (the ideal situation), as culture medium or as ascites (although animal protection laws in many countries prohibit the production of mAb as ascites in mice, there may still be stocks of ascites remaining in many freezers from different laboratories). Whatever the source, the antibody must be titrated to establish the optimal dilution. Depending on the antibody affinity, purified antibodies are used at ~5 μg/ml. For supernatants, it is usually necessary to add 50 to 100 μl undiluted (sometimes the antibody titer is so low that it needs to be concentrated). Ascites contain large amounts of antibody (in the range of mg/ml). Therefore, the highest dilution giving optimal staining should be used. In must be kept in mind that even at the optimal dilution, ascites can give high backgrounds.

Direct immunofluorescence

For this method, one must use antibodies (usually mAb) labeled with a fluorescent dye. The most commonly used dyes for flow cytometry are fluorescein isothiocyanate (FITC), which gives green fluorescence (fluorescence-1) at 488 nM (the standard wavelength of flow cytometers), phycoerythrin (PE), yellow fluorescence (fluorescence-2), and peridinin chlorophyll protein (PerCP), red fluorescence (fluorescence-3).

One-color analysis

An appropriate number of cells (around 10^6) suspended in WS is placed either in a 1.5-ml Eppendorf tube or in a round-bottomed 96-well plate. The latter is the choice when the number of samples is high (usually more than 24 samples). Suspend the cells in a maximum volume of 20 to 50 µl; using a small volume avoids overdilution of the antibody which is added to the cell suspension at saturating concentrations. The mixture is incubated on ice for 30 to 60 min. At the end of the incubation time, remove excess unbound mAb by filling the tubes (or wells) with WS and centrifuge at 400 *g* for 5 min. Aspirate the supernatant, resuspend the cells, fill the tubes again with WS and spin again. Usually, two washes are sufficient for removal of the unbound antibody. Aspirate the supernatant and resuspend the cells in ~400 µl WS for analysis in the flow cytometer.

Two-color analysis

When both antibodies are directly labeled with different dyes, both can be added at the same time and all the procedures are as described for one-color analysis. When one of the antibodies is biotinylated, they can also be added at the same time. However, a second reagent (dye-labeled streptavidin) must be added at the end of the incubation with the first two reagents (see indirect immunoflourescence below). If three-color immunofloursecence is to be performed, the procedure is the same.

It is not advisable to perform two- or more-color analysis when one of the reagents is not labeled because it takes more steps (at least three) and normally will have a high background.

Indirect immunofluorescence

This method can be used with two variants. In the first one, an unlabeled antibody (for mouse usually rat mAb is used) is added to cells and its reactivity to the cell surface is determined with a second antibody which consists of a fluorescent antiserum (usually goat or rabbit anti-rat or the species of origin of the first antibody) directed against the original antibody. In the second method, the first antibody is biotinylated and as a second reagent avidin or streptavidin labeled with a fluorescent dye (usually FITC) is added.

First antibody

An appropriate number of cells (around 10^6) suspended in WS is placed either in a 1.5-ml Eppendorf tube or in a round-bottomed 96-well plate. The latter is the choice when the number of samples is high (usually more than 24 samples). Do not overdilute the antibodies, suspend the cells in a maximum volume of 20 to 50 μl. Add the antibody at saturating concentrations to the suspension and incubate on ice for 30 to 60 min. At the end of the incubation time, remove excess unbound mAb by filling the tubes (or wells) with WS and centrifuge at 400 *g* for 5 min. Aspirate the supernatant, resuspend the cells, fill the tubes again with WS, and spin again. Usually, two washes are sufficient for removal of the unbound antibody. Aspirate the supernatant leaving the cells in 20 to 50 μl WS.

Second antibody

This is, usually, an FITC-labeled goat or rabbit polyclonal antiserum directed against the first antibody which usually derives from mouse, rat, or hamster. We recommend antisera from Jackson or Southern which have a wide variety of choices, high activity and low background. The antiserum must be titrated and added at the appropriate dilution (which is highly variable) to the cell suspension. Cells are left on ice at 4 °C for 30 to 60 min (30 min is usually long enough) in the dark (they can be covered with aluminum foil). This is followed by two washes as described for the first antibody and finally resuspended in ~400 μl WS.

When B cells are stained with mouse antibodies as the first antibody, this must be biotinylated because use of a goat or rabbit antimouse antiserum as the second reagent (see below) will stain surface immunoglobulin, which will raise the baseline at levels that make difficult to examine the results.

Never add first and second reagents simultaneously to the cells, as doing so decreases the efficiency and accuracy of the assay.

Controls

As fluorescence intensity is a relative measure, it is absolutely necessary to always include negative controls which should contain an irrelevant first antibody of the same isotype as the mAb used for analysis, followed by the appropriate second reagent (in the case of indirect immunoflourescence). If two or more antibodies of different isotypes are used, the same number of irrelevant antibodies and isotypes should be used. These samples are handled in parallel during the staining procedure.

In certain instances, the stained cells can be fixed with 3% formaldehyde and left in the dark and in a cold place for several days before analysis or if a reappraisal is desired. To fix the cells, add 500 μl formaldehyde in PBS after the last wash, leave the cells on it 5 min, stop the reaction with 500 μl glycine 200 mM in PBS.

Just before analysis

For one- and two-color immunofluorescence we suggest excluding dead cells from analysis (they increase the background). This can be accomplished by the addition of propidium iodide (PI). This compound dyes dead cells which then emit red fluorescence (channel 3). PI is prepared as a 100× stock solution of 50 μg/ml and it is added to the cells immediately before analysis. It should be maintained in the dark at 4 °C. Cells positive for red fluorescence can be excluded from analysis by setting a gate that allows examination only of cells negative for red fluorescence. Do not use PI with fixed cells because they are stained with this dye.

This section of the chapter is intended to provide some hints for only a few of thousands of immunological methods that are used in the study of transgenic mice. It does not pretend by any means to replace a number of outstanding practical guides of immunological methods. For the procedure of flow cytometry we refer the reader to the appropriate sources (Coon and Weinstein 1991; Coligan et al. 1994).

This work was supported in part by grant CI1*-CT92-0027 from the Commission of the European Communities and grant 0221P-N9506 from CONACyT to Dr. José Moreno.

References

Acha-Orbea H, Mitchell DJ, Timmermann L, Wraith DC, Tausch GS, Waldor MK, Zamvil SS, McDevitt HO (1988) Limited heterogeneity of T cell receptors from lymphocytes mediating autoimmune encephalomyelitis allows specific immune intervention. Cell 54:263–273

Ahearn JM, Fischer MB, Croix D, Goerg S, Ma M, Xia J, Zhou X, Howard RG, Rothstein TL, Carroll MC (1996) Disruption of the *Cr2* locus results in a reduction of B-1a cells and an impaired B cell response to T-dependent antigen. Immunity 4:251–262

Allison J, Campbell IL, Morahan G, Mandel TE, Harrison L, Miller JFAP (1988) Diabetes in transgenic mice resulting from overexpression of class I histocompatibility molecules in pancreatic β cells. Nature 333:529–533

Amakawa R, Hakem A, Kundig TM, Matsuyama T, Simard JJL, Timms E, Wakeham A, Mittruecker H-W, Griesser H, Takimoto H, Schmits R, Shahinian A, Ohashi PS, Penninger JM, Mak TW (1996) Impaired negative selection of T cells in Hodgkin's disease antigen CD30-deficient mice. Cell 84:551–562

Appleby MW, Gross JA, Cooke MP, Levin SD, Quin X, Perlmutter RM (1992) Defective T cell receptor signaling in mice lacking the thymic isoform of $p59^{fyn}$. Cell 70:751–763

Arnold B, Hämmerling GJ (1991) MHC class-I transgenic mice. Annu Rev Immunol 9:297–322

Arnold LW, Pennell CA, McCray SK, Clarke SH (1994) Development of B-1 cells: segregation of phosphatidyl choline-specific B cells to the B-1 population occurs after immunoglobulin gene expression. J Exp Med 179:1585–1595

Bain G, Robanus-Maandag EC, Izon DJ, Amsen D, Kruisbeek AM, Weintraub BC, Krop I, Schlissel MS, Feeney AJ, van Roon M, van der Valk M, te Riele HPJ, Berns A, Murre C (1994) E2A proteins are required for proper B cell development and initiation of immunoglobulin gene rearrangements. Cell 79:885–892

Banchereau J, Bazan F, Blanchard D, Brière F, Galizzi JP, van Kooten C, Liu YJ, Rousset F, Saeland S (1994) The CD40 antigen and its ligand. Annu Rev Immunol 12:881–922

Bender BS, Croghan T, Zhang L, Small PA (1992) Transgenic mice lacking class I major histocompatibility complex-restricted T cells have delayed viral clearance and increased mortality after influenza virus challenge. J Exp Med 175:1143–1145

Berg L, J, Pullen AM, Fazekas de St. Groth B, Mathis D, Benoist C, Davis MM (1989) Antigen/MHC-specific T cells are preferentially exported from the thymus in the presence of their MHC ligand. Cell 58:1035–1046

Berg LJ, Frank GD, Davis MM (1990) The effects of MHC gene dosage and allelic variation on T cell receptor selection. Cell 60:1043–1053

Bernard CCA, Carnegie PR (1975) Experimental allergic encephalomyelitis in mice: immunologic response to mouse spinal cord and myelin basic proteins. J Immunol 114:1537–1540

Blum MD, Wong GT, Higgins KM, Sunshine MJ, Lacy E (1993) Reconstitution of the subclass-specific expression of CD4 in thymocytes and peripheral T cells of transgenic mice: identification of a human CD4 enhancer. J Exp Med 177:1343–1358

Böhme J, Haskins K, Stecha P, van Ewjik W, LeMeur M, Grelinger P, Benoist C, Mathis D (1989) Transgenic mice with I-A on islet cells are normoglycemic but immunologically intolerant. Science 244:1179–1183

Böhme J, Schuhbaur B, Kanagawa O, Benoist C. and Mathis D (1990) MHC-linked protection from diabetes dissociated from clonal deletion of T cells. Science 249:293–295

Burkly LC, Lo D, Kanagawa O, Brinster RL, Flavell RA (1989) T-cell tolerance by clonal anergy in transgenic mice with nonlymphoid expression of MHC class II I-E. Nature 342:564–566

Campbell KA, Ovendale PJ, Kennedy MK, Fanslow WC, Reed SG, Malizewski CR (1996) CD40 ligand is required for protective cell mediated immunity to *Leishmania major*. Immunity 4:283–289

Cao X, Shores EW, Hu-Li J, Anver MR, Kelsall BL, Russell SM, Drago J, Noguch M, Grinberg A, Bloom ET, Paul WE, Katz SI, Love PE, Leonard WJ (1995) Defective lymphoid development in mice the expression of the common cytokine receptor γ chain. Immunity 2:223–238

Chan SH, Waltzinger C, Baron A, Benoist C, Mathis D (1994) Role of coreceptors in potitive selection and lineage commitment. EMBO J. 13:4482–4489

Chen C, Nagy Z, Radic MZ, Hardy RR, Huszar D, Camper SA, Weigert M (1995) The site and stage of anti-DNA B-cell deletion. Nature 373:252–255

Cocks BJ, Chang C-CJ, Carballido JM, Yssel H, de Vries JE, Aversa G (1995) A novel receptor involved in T-cell activation. Nature 376:260–263
Cohen JJ, Duke RC, Fadok VA, Sellins KS (1992) Apoptosis and programed cell death in immunity. Annu Rev Immunol 10:267–293
Coligan JE, Kruisbeck AM, Margulies DM, Shevach EM, Strober W (eds) (1994) Current protocols in immunology. John Wiley, New York
Cooke MP, Heath AW, Shokat KM, Zeng Y, Finkelman FD, Linsley PS, Howard M, Goodnow CC (1994) Immunoglobulin signal transduction guides the specificity of B cell-T cell interactions and is blocked in tolerant self-reactive B cells. J Exp Med 179:425–438
Coon JS, Weinstein RS (eds) (1991) Diagnostic flow cytometry. Williams and Wilkins, Baltimore
Cosgrove D, Gray D, Dierich A, Kaufman J, LeMeur M, Benoist C, Mathis D (1991) Mice lacking MHC class II molecules. Cell 66:1051–1066
Cyster JG, Goodnow CC (1995) Antigen-induced exclusion from follicles and anergy are separate and complementary processes that influence peripheral B cell fate. Immunity:3:691–701
Cyster JG, Hartley SB, Goodnow CC (1994) Competition for follicular niches excludes self-reactive cells from the recirculating B-cell repertoire. Nature 371:389–395
Cyster JG, Healy JI, Kishihara K, Mak TW, Thomas ML, Goodnow CC (1996) Regulation of B-lymphocyte negative and positive selection by tyrosine phosphatase CD45. Nature 381:325–328
Dalton DK, Pitts-Meek S, Keshav S, Figari IS, Bradley A, Stewart TA (1993) Multiple defects of immune function in mice with disrupted interferon-gamma genes. Science 259:1739–1742
De Togni P, Goellner J, Ruddle NH, Streeter PR, Fick A, Mariathasan S, Smith SC, Carlson R, Shormnick LP, Strauss-Schoenberger J, Russell JN, Kar R, Chaplin D (1994) Abnormal development of peripheral lymphoid organs in mice deficient in lymphotoxin. Science 264:703–707
Dorshkind K (1994) Transcriptional control points during lymphopoiesis. Cell 79:751–753
Erickson J, Radik MZ, Camper SA, Hardy RR, Carmack C, Weigert M (1991) Expression of anti-DNA transgenes in non-autoimmune mice. Nature 349:331–334
Finkel TH, Kappler JW, Marrack PC (1992) Immature thymocytes are protected from deletion early in ontogeny. Proc Natl Acad Sci USA 89:3372–3374
Fischer A, Leonard WJ (1995) Inherited immunodeficiencies. Immunologist 3:237–240
Flavell RA, Kratz NM, Ruddle TH (1995) Contribution of insulitis to diabetes development in TNF transgenic mice. Curr. Top Microbiol Immunol 206:33–50
Flynn JL, Chan J, Triebold KJ, Dalton DK, Stewart TA, Schreiber R, Bloom BR (1993) An essential role for interferon γ in resistance to *Mycobacterium tuberculosis* infection. J Exp Med 178:2249–2254
Flynn JL, Goldstein MM, Chan J, Triebold KJ, Pfeffer K, Lowenstein CJ, Schreiber R, Mak TW, Bloom BR (1995) Tumor necrosis factor-α is required for protective immune response against Mycobacterium tuberculosis in mice. Immunity 2:561–572
Fraser A, Evan G (1996) A license to kill. Cell 85:781–784

Freeman GJ, Lombard DB, Gimmi CD, Brod SA, Lee K, Laning JC, Hafler DA, Dorf ME, Gray G, Reiser H, June CH, Thompson CB, Nadler LM (1992) CTLA-4 and CD28 mRNAs are coexpressed in most activated T cells after activation. Expression of CTLA-4 and CD28 messenger RNA does not correlate with the pattern of lymphokine production. J Immunol 149:3795–3801

Fritz RB, Skeen MJ, Chou C-HJ, Garcia M, Egoov IK (1985) Major histocompatibility complex-linked control of the murine response to myelin basic protein. J Immunol 134:2328–2332

Fulcher DA, Basten A (1994) Reduced life span of anergic self-reactive B cells in a double-transgenic model. J Exp Med 179:125–134

Georgopoulos K, Bigby M, Wong J-M, Molnar A, Wu P, Winandy S, Sharpe A (1994) The Ikaros gene is required for the development of all lymphoid lineages. Cell 79:143–156

Ghosh S (1995) Transcriptional regulation in lymphocyte differentiation. Immunologist 3:168–171

Gong S, Nussenzweig MC (1996) Regulation of an early developmental checkpoint in the B cell pathway by Ig-β. Science 272:411–413

Goodnow CC, Crosbie J, Addelstein S, Lavoie TB, Smith-Gill SJ, Brink RA, Pritchard-Briscoe H, Whoterspoon JS, Loblay SH, Raphael K, Trent RJ, Basten A (1988) Altered immunoglobulin expression and functional silencing of self-reactive B-lymphocytes in transgenic mice. Nature 334:676–682

Goodnow CC, Crosbie J, Jorgensen H, Brink RA, Basten A (1989) Induction of self tolerance in peripheral mature B lymphocytes. Nature 342:385–391

Goodnow CC, Brink RA, Adams E (1991) Breakdown of self-tolerance in anergic B lymphocytes. Nature 352:532–536

Goverman J, Woods A, Larson L, Weiner LP, Hood L, Zaller DM (1993) Transgenic mice that express a myelin basic protein-specific T cell receptor develop spontaneous autoimmmunity. Cell 72:551–560

Grawunder U, Leu TMJ, Schatz DG, Werner A, Rolink AG, Melchers F, Winkler TH (1995) Down-regulation of *RAG1* and *RAG2* gene expression in pre-B cells after functional immunoglobulin heavy chain rearrangement. Immunity 3:601–608

Grewal IS, Xu J, Flavell R (1995) Impairment of antigen-specific T-cell priming in mice lacking CD40 ligand. Nature 378:617–620

Grusby MJ, Johnson RS, Papaioannou VE, Glimcher LH (1991) Depletion of $CD4^+$ T cells in major histocompatibility complex II-deficient mice. Science 259:1417–1420

Guerder S, Picarella DE, Linsley P, Flavell RA (1994) Costimulator B7–1 confers antigen presenting-cell function to parenchymal tissue and in conjunction with tumor necrosis factor α leads to autoimmunity in transgenic mice. Proc Natl Acad Sci USA 91:5138–5142

Haas W, Pereira P, Tonegawa S (1993) Gamma/delta cells. Annu Rev Immunol 11:637–685

Hammer RE, Maika SD, Richardson JA, Tang J-P, Taurog J (1990) Spontaneous inflammatory disease in transgenic rats expressing HLA-B27 and human β_2M: an animal model of HLA-B27-associated human disorders. Cell 63:1099–1112

Harlan DM, Hengartner H, Huang ML, Kang Y-H, Abe R, Moreadith RW, Pircher H, Gray GS, Ohashi PS, Freeman GJ, Nadler LM, June CH, Aichele P (1994) Mice expressing both B7–1 and viral glycoprotein on pancreatic beta cells along with glycoprotein-specific transgenic T cells develop diabetes due to a breakdown of T-lymphocyte unresponsiveness. Proc Natl Acad Sci USA 91:3137–3141

Hartley SB, Crosbie J, Brink R, Kantor AB, Basten A, Goodnow CC (1991) Elimination from peripheral lymphoid tissues of self-reactive B lymphocytes recognizing membrane-bound antigens. Nature 353:765–769

Heath WR, Allison J, Hoffmann MW, Schönrich G, Hämmerling GJ, Arnold B, Miller JFAP. (1992) Autoimmune diabetes as a consequence of locally produced interleukin-2. Nature 359:547–549

Hemachudha T, Griffin DE, Giffels JJ, Johnson RT, Moser AB, Phanuphak P (1987) Myelin basic protein as an encephalitogen in encephalomyelitis and polyneuritis following rabies vaccination. N Engl J Med 316:369–374

Higuchi Y, Herrera P, Muniesa P, Huarte J, Belin D, Ohashi P, Aichele P, Orci L, Vassalli J-D, Vassalli P (1992) Expression of a tumor necrosis factor a transgene in murine pancreatic β cells results in severe and permanent insulitis without evolution towards diabetes. J Exp Med 176:1719–1731

Ho WY, Cooke MP, Goodnow CC, Davis MM (1994) Resting and anergic B cells are defective in CD28-dependent costimulation of naive $CD4^+$ T cells. J Exp Med 179:1539–1549

Hoffmann MW, Allison J, Miller JFAP (1992) Tolerance induction by thymic medullary epithelium. Proc Natl Acad Sci USA 89:2526–2530

Huang S, Hendricks W, Althage A, Hemmi S, Bluethmann H, Kamijo R, Vilcek J, Zinkernagel RM, Auget M (1993) Immune response in mice that lack the interferon-gamma receptor. Science 259:1742–1745

Huesmann M, Scott B, Kisielow P, von Boehmer H (1991) Kinetics and efficacy of positive selection in the thymus of normal and T cell receptor transgenic mice. Cell 66:533–540

Hurley TR, Hymann R, Sefton BM (1993) Differential effects of the expression of CD45 tyrosine protein phosphatase on the tyrosine phosphorylation of lck, fyn, c-src tyrosine protein kinases. Mol Cell Biol 13:1651–1656

Itohara S, Mombaerts P, Lafaille J, Iacomini J, Nelson A, Clarke AR, Hooper ML, Farr A, Tonegawa S (1993) T cell receptor delta gene mutant mice: independent generation of alpha beta T cells and programmed rearrangements of gamma delta TCR genes. Cell 72:337–348

Kamanaka N, Yu P, Yasui T, Yoshida K, Kawanabe T, Horii T, Kishimoto T, Kikutani H (1996) Protective role of CD40 in *Leishmania major* infection at two distinct phases of cell-mediated immunity. Immunity 4:275–281

Kaplan MH, Schindler U, Smiley ST, Grusby MJ (1996) Stat6 is required for mediating responses to IL-4 and for the development of Th2 cells. Immunity 4:313–319

Kievits F, Ivanyi P, Krimpenfort P, Berns A, Ploegh HL (1987) HLA-restricted recognition of viral antigens in HLA transgenic mice. Nature 329:447–449

Kishihara K, Penninger J, Wallace VA, Kündig TM, Kawai K, Wakeham A, Timms E, Pfeffer K, Ohashi PS, Thomas ML, Furlonger C, Paige CJ, Mak TW (1993) Normal B lymphocyte development but impaired T cell maturation in CD45-exon 6 protein tyrosine phosphatase -deficient mice. Cell 74:143–156

Kisielow P, Blüthmann H, Staerz U.D, Steinmetz M, von Boehmer H (1988a) Tolerance in T-cell-receptor transgenic mice involves deletion of nonmature $CD4^+8^+$ thymocytes. Nature 333:742–746

Kisielow P, Sia Teh H, Blüthmann H, von Boehmer H (1988b) Positive selection of antigen-specific T cells in thymus by restricting MHC molecules. Nature 335:730–733

Kitamura D, Rajewsky K (1992) Targeted disruption of m chain membrane exon causes loss of heavy chain allelic exclusion. Nature 356:154–156

Koller BH, Marrack P, Kappler JW, Smithies O (1990) Normal development of mice deficient in β_2M, MHC class I proteins, $CD8^+$ T cells. Science 248:1227–1230

Kono DH, Urban JL, Horvath SJ, Ando DG, Saavedra RA, Hood L (1988) Two minor determinants of myelin basic protein induce experimental allergic encephalomyelitis in SJL/J mice. J Exp Med 168:213–227

Kopf M, Le Gros G, Bachmann M, Lammers MC, Bluethmann H, Köhler G (1993) Disruption of the murine IL-4 gene blocks Th2 cytokine responses. Nature 362:245–248

Kopf M, Brombacher F, Hodgkin PD, Ramsay AJ, Milbourne EA, Dai WJ, Ovington KS, Behm CA, Köhler G, Young IG, Matthaei KI (1996) IL-5-deficient mice have a developmental defect in $CD5^+$ B-1 cells and lack eosinophilia but have normal antibody and cytotoxic T cell responses. Immunity 4:15–24

Krimpenfort P, Rudenko G, Hoch-Steinbach F, Guessow D, Berns A, Ploegh H (1987) Crosses of two independently derived transgenic mice demonstrate functional complementation of the genes encoding heavy (HLA-B27) and light (β2-microglobulin) chains of HLA class I antigens. EMBO J 6:1673–1676

Kühn R, Rajewsky K, Müller W (1991) Generation and analysis of interleukin-4 deficient mice. Science 254:707–710

Kühn R, Löhler J, Rennick D, Rajewsky K, Müller W (1993) Interleukin-10-deficient mice develop chronic enterocolitis. Cell 75:263–274

Kulkarni AB, Huh CG, Becker D, Geyser A, Lyght M, Flanders KC, Roberts AB, Sporn MB, Ward JM, Karlsson S (1993) Transforming growth factor β 1 null mutation in mice causes excessive inflammatory response and early death. Proc Natl Acad Sci USA 90:770–774

Kündig TM, Schorle H, Bachmann MF, Hengartner H, Zinkernagel R, Horak I (1993) Immune responses in interleukin-2 deficient mice. Science 262:1059–1061

Kurtz CB, O'Toole E, Christensen SM, Weiss JH (1990) The murine complement receptor gene family. IV. Alternative splicing of *Cr2* gene transcripts predicts two distinct gene products that share homologous domains with both human CR2 and CR1. J Immunol 144:3581–3591

Lafaille J, Nagashima K, Katsuki M, Tonegawa S (1994) High incidence of spontaneous autoimmune encephalomyelitis in immunodeficient anti-myelin basic protein T cell receptor transgenic mice. Cell 78:399–408

Levelt CN, Eichmann K (1995) Receptors and signals in early thymic selection. Immunity 3:667–672

Levin SD, Abraham KM, Anderson SJ, Forbush KA, Perlmutter RM (1993) The protein tyrosine $p56^{lck}$ regulates thymocyte development independently of its interaction with CD4 and CD8 coreceptors. J Exp Med 178:245–255

Liao XC, Littman DR (1995) Altered T cell receptor signaling and disrupted T cell development in mice lacking *ltk*. Immunity 3:757–769.

Lin H, Grosschedl R (1995) Failure of B-cell differentiation in mice lacking the transcription factor EBF. Nature 376:263–267

Linsley PS, Brady W, Umes M, Grosmaire GS, Damle NK, Ledbetter JA (1991) CTLA-4 is a second receptor for the B cell activation antigen B7. J Exp Med 174:561–569

Linsley PS, Greene JL, Brady W, Bayorath J, Ledbetter JA, Peach R (1994) Human B7.1 (CD80) and B7.2 (CD86) bind with similar avidities but different kinetics to CD28 and CTLA-4 receptors. Immunity 1:793–801

Lo D, Burkly LC, Widera G, Cowing C, Flavell RA, Palmiter RD, Brinster RL (1988) Diabetes and tolerance in transgenic mice expressing class II MHC molecules in pancreatic beta cells. Cell 53:159–168

Löffert D, Ehlich A, Müller A, Rajewsky K (1996) Surrogate light chain expression is required to establish immunoglobulin heavy chain allelic exclusion during early B cell development. Immunity 4:133–144

Lowin B, Beermann F, Schmidt A, Tschopp J (1994) A null mutation in the perforin gene impairs cytolytic T lymphocyte- and natural killer cell-mediated cytotoxicity. Proc Natl Acad Sci USA 91:11571–11575

Lund FE, Solvason NW, Cooke MP, Heath AW, Grimaldi JC, Parkhouse RME, Goodnow CC, Howard MC (1995) Siganling through murine CD38 is impaired in antigen receptor-unresponsive B cells. Eur J Immunol 25:1338–1345

Magram J, Connaughton SE, Warrier RR, Carvajal DM, Wu C, Ferrante J, Stewart C, Sarmiento U, Faherty DA, Gately MK (1996) IL-12-deficient mice are defective in INF-γ production and type 1 cytokine responses. Immunity 4:471–481

Malissen M, Gillet A, Rocha B, Trucy J, Vivier E, Boyer C, Köntgen F, Brun N, Mazza G, Spanopoulou E, Guy-Grand D, Malissen B (1993) T cell development in mice lacking the CD3-ζ/η gene. EMBO J 12:4357–4366

Marrack P, Lo D, Brinster R, Palmiter R, Burhly L, Flavell RA, Kappler J (1988) The effect of thymus environment on T cell development and tolerance. Cell 53:627–634

Matechak EO, Killeen N, Hedrick SM, Fowlkes BJ (1996) MHC class II-specific T cells can develop in the CD8 lineage when CD4 is absent. Immunity 4:337–347

Matsumoto M, Mariathasan S, Nahm MH, Baranyay F, Peschon J, Chaplin DD (1996) Role of lymphotoxin and the type I TNF receptor in the formation of germinal centers. Science 272:1289–1291

Matzinger P, Guerder S (1989) Does T-cell tolerance require a dedicated antigen-presenting cell? Nature 338:74–76

McFarlin DE, McFarland HF (1982) Multiple sclerosis. N Engl J Med 307:1183–1188

Miyajima A, Kitamura T, Harada N, Yokota T, Arai K-I (1992) Cytokine receptors and signal transduction. Annu Rev Immunol 10:295–331

Miyazaki T, Uno M, Uehira M, Kikutani H, Kishimoto T, Kimoto M, Nishimoto H (1990) Direct evidence for the contribution of the unique I-A NOD to the development of insulitis in non-obese diabetic mice. Nature 345:722–724

Mokhtarian F, McFarlin DE, Raine CS (1984) Adoptive transfer of myelin basic protein-sensitized T cells produces chronic relapsing demyelinating disease in mice. Nature 309:356–358

Molina TJ, Kishihara K, Siderovski DP, van Ewijk W, Narendran A, Timms E, Wakeham A, Paige CJ, Hartmann K-U, Veilette A, Davidson D, Mak TW (1992) Profound block in thymocyte development in mice lacking $p56^{lck}$. Nature 357:161–164

Mombaerts P, Iacomini J, Johnson RS, Herrup K, Tonegawa S, Papaioannou V (1992) RAG-1 deficient mice have no mature B and T lymphocytes. Cell 68:869–878

Morahan G, Allison J, Miller JFAP (1989) Tolerance of class I histocompatibility antigens expressed extrathymically. Nature 339:622–624

Morell PA, Dorman JS, Todd JA, McDevitt HO, Trucco M (1988) Aspartic acid at position 57 of the HLA-DQβ chain protects against type I diabetes: a family study. Proc Natl Acad Sci USA 85:8111–8115

Müller W (1995) Dissecting the cytokine network. Immunologist 3:216–218

Murakami M, Tsubata T, Okamoto M, Shimizu A, Kumagai S, Imura H, Honjo T (1992) Antigen-induced apoptotic death of Ly-1 B cells responsible for autoimmune disease in transgenic mice. Nature 357:77–80

Murre C, McCaw PS, Baltimore D (1989) A new DNA binding and dimerization motif in immunoglobulin enhancer binding, *daughterless MyoD* and *myc* proteins. Cell 56:777–783

Murre C, Bain G, van Djik MA, Engel I, Fumari BA, Massari ME, Mathews JR, Quong MW, Rivera RR, Stuiver MH (1994) Structure and function of helix-loop-helix proteins. Biochim Biophys Acta 1218:129–135

Nakayama K, Loh DY (1992) No requirement for $p56^{lck}$ in the antigen-stimulated clonal deletion of T lymphocytes. Science 257:94–96

Nemazee DA, Bürki K (1989) Clonal deletion of B lymphocytes in a transgenic mouse bearing anti-MHC class I antibody genes. Nature 337:562–566

Nishimoto H, Kikutani H, Yamamura K, Kishimoto T (1987) Prevention of autoimmune insulitis by expression of I-E molecules in NOD mice. Nature 328:432–434

Nishizumi H, Tainuchi I, Yamanashi Y, Kitamura D, Ilic D, Mori S, Watanabe T, Yamamoto Y (1995) Impaired proliferation of peripheral B cells and indication of autoimmune disease in *lyn*-deficient mice. Immunity 3:549–560

Oettinger MA, Schatz DG, Gorka C, Baltimore D (1990) RAG-1 and RAG-2, adjacent genes that synergistically activate V(D)J recombination. Science 248:1517–1523

Ohashi P, Oehen S, Buerki K, Pircher H, Ohashi CT, Odermatt B, Malissen B, Zinkernagel RM, Hengartner H (1991) Ablation of "tolerance" and induction of diabetes by virus infection in viral antigen transgenic mice. Cell 65:305–317

Okamoto M, Murakami M, Shimizu A, Ozaki S, Tsubata T, Kumagai S, Honjo T (1992) A transgenic model of autoimmune hemolytic anemia. J Exp Med 175:71–79

Oldstone MBA, Nerenberg M, Southern P, Proce J, Lewicki H (1991) Vitus infection triggers insulin-dependent diabetes mellitus in a transgenic model: role of anti-self (virus) immune response. Cell 65:319–331

Owhashi M, Heber-Katz E (1988) Protection from experimental allergic encephalomyelitis conferred by a monoclonal antibody directed against a shared idiotype on rate T cell receptors specific for myelin basic protein. J Exp Med 168:2153–2164

Park SY, Saijo K, Takahashi T, Osawa M, Arase H, Hirayama N, Miyake K, Nakauchi H, Shirasawa T, Saito T (1995) Developmental defects of lymphoid cells in Jak3 kinase-deficient mice. Immunity 3:771–782

Penninger JM, Wen T, Timms E, Potter J, Wallace V, Matsuyama T, Ferrick D, Sydora B, Kronenberg M, Mak TW (1995) Spontaneous resistance to acute T-cell leukemias in TCRVγ1.1γ4Cγ4 transgenic mice. Nature 375:241–244

Pettinelli CD, McFarlin DE (1981) Adoptive transfer of experimental allergic encephalomyelitis in SJL/J mice after in vitro activation of lymph node cells by myelin basic protein: requirement for Lyt $1^{+}2^{-}$ lymphocytes. J Immunol 127:1420–1423

Pfeffer K, Mak TW (1994) Lymphocyte ontogeny and activation in gene targeted mutant mice. Annu Rev Immunol 12:367–411

Picarella DE, Kratz A, Ruddle NH, Flavell RA (1992) Insulitis in transgenic mice expressing TNF-β (Lymphotoxin) in the pancreas. Proc Natl Acad Sci USA 89:10036–10040

Picarella DE, Kratz A, Li C-B, Ruddle NH, Flavell RA (1993) Transgenic TNF-α production in islets leads to insulitis, not diabetes: distinct patterns of inflammation in TNF-α and TNF-β transgenic mice. J Immunol 149:4136–4150

Poli V, Balena R, Fattori E, Markatos A, Yamamoto M, Tanaka H, Ciliberto G, Rodan GA, Constantini F (1994) Interleukin-6 deficient mice are protected from bone loss caused by estrogen depletion. EMBO J 13:1189–1196

Rahemtulla A, Kündig TM, Narendran A, Bachmann MF, Julius M, Paige C, Ohashi PS, Zinkernagel RM, Mak TW (1994) Class II major histocompatibility complex-restricted T cell function in CD4-deficient mice. Eur J Immunol 24:2213–2218

Rajewsky K (1996) Clonal selection and learning in the antibody system. Nature 381:751–758

Ramsay AJ, Husband AJ, Ramshaw JA, Bao S, Matthaei KI, Koehler G, Kopf M (1994) The role of interleukin-6 in mucosal IgA antibody responses in vivo. Science 264:561–563

Rathmell JC, Cooke MP, Ho WY, Grein J, Townsend SE, Davis MM, Goodnow CC (1995) CD95 (Fas)-dependent elimination of self-reactive B cells upon interaction with $CD4^+$ T cells. Nature 376:181–183

Rickert RC, Rajewsky K, Roes J (1995) Impairment of T-cell dependent B-cell responses and B-1 cell development in CD19-deficient mice. Nature 376:352–355

Romeo C, Amiot M, Seed B (1992) Sequence requirements for the induction of cytolysis by the T cell antigen/Fc receptor ζ chain. Cell 68:889–897

Roth PE, Doglio L, Manz JT, Kim JY, Loh D, Storb U (1993) Immunoglobulin γ2b transgenes inhibit heavy chain gene rearrangement, but cannot promote B cell development. J Exp Med 178:2007–2021

Rudd CE (1996) Upstream-downstream: CD28 cosignaling pathways and T cell function. Immunity 4:527–534

Russell DM, Dembic Z, Morahan G, Miller JFAP, Burki K, Nemazee D (1991) Peripheral deletion of self-reactive B cells. Nature 354:308–311

Sadlack B, Merz H, Schorle H, Schimpl A, Feller AC, Horak I (1993) Ulcerative colitis-like disease in mice with a disrupted interleukin-2 gene. Cell 75:253–261

Sato M, Miyajima A (1994) Multimeric cytokine receptors: common versus specific functions. Curr Op Cell Biol 6:174–179

Savertnik N, Liggitt D, Pitts S, Hansen SE, Stewart TA (1988) Insilin-dependent diabetes mellitus induced in transgenic mice by ectopic expression of class II MHC and interferon-gamma. Cell 52:773–782

Schatz D, Oettinger MA, Baltimore D (1989) The V(D)J recombination activating gene RAG-1. Cell 59:1035–1048

Schlissel M, Voronova A, Baltimore D (1991) Helix-loop-helix transcription factor E47 activates germ-line immunoglobulin heavy chain gene transcription and rearrangement in a pre-T cell line. Genes Dev 5:1367–1376

Schorle H, Holtshcke T, Hünig T, Schimpl A, Horak I (1991) Development and function of T cells in mice rendered interleukin-2 deficient by gene targeting. Nature 352:621–624

Schorle H, Schimpl A, Merz H, Feller AC, Horak I (1993) Immunodeficiency caused by a targeted disruption of interleukin-2 gene in mice. J Cell Biochem 17(B):54

Scott B, Blüthmann H, Sia Teh H, von Boehmer H (1989) The generation of mature T cells requires interaction of the αβ T-cell receptor with major histocompatibility antigens. Nature 338:591–593

Scott EW, Simon MC, Anastasi J, Singh M (1994) Requirement of transcription factor Pu.1 in the development of multiple hematopoietic lineages. Science 265:1573–1577

Seder RA, Paul WE (1994) Acquisition of lymphokine-producing phenotype by $CD4^+$ T cells. Annu Rev Immunol 12:635–673

Seong RH, Chamberlain JW, Parnes JR (1992) Signal for T-cell differentiation to a CD4 cell lineage is delivered by CD4 transmembrane region and/or cytoplasmic tail. Nature 356:718–720

Sha WC, Nelson CA, Newberry RD, Kranz DM, Russell JH, Loh DY (1988a) Selective expression of an antigen receptor on CD8-bearing T lymphocytes in transgenic mice. Nature 335:271–274

Sha WC, Nelson CA, Newberry RD, Kranz DM, Russell JH, Loh DY (1988b) Positive and negative selection of an antigen receptor on T cells in transgenic mice. Nature 336:73–76

Shahinian A, Pfeffer K, Lee KP, Kuendig TM, Kishihara K, Wakeham A, Kawai K, Ohashi PS, Thompson CB, Mak TW (1993) Differential costimulatory requirements in CD28-deficient mice. Science 261:609–612

Shinkai Y, Takio K, Okamura K (1988) Homology of perforin to the ninth component of complement (C9) Nature 334:525–527

Shinkai Y, Rathburn G, Lam KP, Oltz EM, Stewart V, Mendelsohn M, Charron J, Datta M, Young F, Stall AM, Alt FW (1992) RAG-2 deficient mice lack mature lymphocytes owing to inability to initiate V(D)J rearrangement. Cell 68,855–868

Shull MM, Ormsby I, Kier AB, Pawloski S, Diebold RJ, Yin M, Allen R, Sidman C, Proetzel G, Calvin D, Annunziata N, Doetschman T (1992) Targeted disruption of the mouse transforming growth factor-β 1 gene results in multifocal inflammatory disease. Nature 359:693–699

Sia Teh H, Kisielow P, Scott B, Kishi H, Uematsu Y, Blüthmann H, von Boehmer H (1988) Thymic major histocompatibility complex antigens and the αβ T-cell receptor determine the CD4/CD8 phenotype of T cells. Nature 355:229–233

Singer SM, Tisch R, Yang X-D, McDevitt HO (1993) An $A\beta^{d}$ transgene prevents diabetes in nonobese diabetic mice by inducing regulatory T cells. Proc Natl Acad Sci USA 90:9566–9570

Slattery RM, Kjer NL, Allison J, Charlton B, Mandel TE, Miller JFAP (1990) Prevention of diabetes in non-obese diabetic $I\text{-}A^{k}$ transgenic mice. Nature 345:724–743

Slattery RM, Miller JFAP, Heath WR, Charlton B (1993) Failure of a protective major histocompatibility complex class II molecule to delete autoreactive T cells in autoimmune diabetes. Proc Natl Acad Sci USA 90:10808–10810

Smith KA (1984) Interleukin 2. Annu Rev Immunol 2:319–333

Soong L, Xu J-C, Grewal IS, Kima P, Sun J, Longley K, Ruddle NH, McMahon D, Flavell RA (1996) Disruption of CD40-CD40 ligand interactions results in enhanced susceptibility to *Leishmania amazoniensis* infection. Immunity 4:263–273

Spanopoulou E, Cortes P, Shih C, Huang C, Silver D. P, Svec P, Baltimore D (1995) Localization, interaction, RNA binding properties of the V(D)J recombination-activating proteins RAG1 and RAG2. Immunity 3:715–726

Speiser DE, Pircher H, Ohashi P, Kyburz D, Hengartner H, Zinkernagel RM (1992) Clonal deletion induced by either radioresistant tymic host cells or lymphohemopoietic donor cells at different stages of class I-restricted T cell ontogeny. J Exp Med 175:1277–1283

Sriram S, Roberts C.A (1986) Treatment of established chronic relapsing experimental allergic encephalomyelitis with anti-L3T4 antibodies. J Immunol 136:4464–4469

Stein PL, Lee HM, Rich RS, Soriano P (1992) $pp59^{fyn}$ mutant mice display differential signaling in thymocytes and peripheral T cells. Cell 70:741–750

Steinman L (1996) Multiple sclerosis: a coordinated attack against myelin in the central nervous system. Cell 85:299–302

Stewart TA, Hultgren B, Huang X, Pitts-Meek S, Hully J, MacLachlan NJ (1993) Induction of type I diabetes by interferon-α in transgenic mice. Science 260:1942–1946

Suzuki H, Kündig T, Furlonger C, Wakeham A, Timms E, Matsuyama T, Schmits R, Simard JJL, Ohashi P, Griesser H, Taniguchi T, Paige C, Mak TW (1995) Deregulated T-cell activation and autoimmunity in mice lacking interleukin-2 receptor β. Science 268:1472–1476

Tarleton RL, Koller BH, Latour A, Postan M (1992) Susceptibility of β_2 microglobulin deficient mice to trypanosoma cruzi infection. Nature 356:338–340

Teofilopoulos AN (1995) The basis of autoimmunity: Part II Genetic predisposition. Immunol Today 16:150–158

Tepper RI, Levinson DA, Stanger BZ, Campos-Torres J, Abbas A, Leder P (1990) IL-4 induces allergic-like inflammatory disease and alters T cell development in transgenic mice. Cell 62:457–467

Thomas ML (1989) The leukocyte common antigen family. Annu Rev Immunol 7:339–369

Tisch R, McDevitt HO (1996) Insulin-dependent diabetes mellitus. Cell 85:291–297

Tivol EA, Borrielo F, Schweitzer AN, Lynch WP, Bluestone JA, Sharpe AN (1995) Loss of CTLA-4 leads to massive lymphoproliferation and fatal multiorgan tissue destruction, revealing a critical negative regulatory role of CTLA-4. Immunity 3:541–547

Turnley A.M, Morahan G, Okano H, Bernard O, Mikoshiba K, Allison J, Bartlett P.F, Miller J.F.A.P (1991) Dysmyelination in transgenic mice resulting from expression of class I histocompatibility molecules in oligodendrocytes. Nature 353:566–569

Uematsu Y, Ryser S, Dembic Z, Borgulya P, Krimpenfort P, Berns A, von Boehmer H, Steimetz M (1988) In transgenic mice the introduced functional T cell receptor β gene prevents expression of endogenous β genes. Cell 53:831–841

Urbànek P, Wang Z-Q, Fetka Y, Wagner EF, Busslinger M (1994) Complete block of early B cell differentiation and altered patterning of the posterior midbrain in mice lacking Pax5/BSAP. Cell 79:901–912

Vajdy M, Kosko-Vilobis MH, Kopf M, Köhler G, Lycke N (1995) Impaired mucosal immune responses in interleukin 4-targeted mice. J Exp Med 181:41–53

van Ewjik W, Ron Y, Monaco J, Kappler J, Marrack P, Le Meur M, Gerlinger P, Durand B, Benoist C, Mathis D (1988) Compartmentalization of MHC class II gene expression in transgenic mice. Cell 53:357–370

van Freeden-Jeffry U, Vieira P, Lucian LA, McNeil T, Bordach SEG, Murray R (1995) Lymphopenia in interleukin (IL)-7 gene deleted mice identifies IL-7 as a non redundant cytokine. J Exp Med 181:1519–1526

van Kaer L, Ashton-Rickardt PG, Ploegh HL, Tonegawa S. (1992) TAP-1 mutant mice are deficient in antigen presentation, surface class I molecules, $CD4^-8^+$ T cells. Cell 71:1205–1214

Vasquez N, Kaye J, Hedrick SM (1992) In vivo and in vitro clonal deletion of double positive thymocytes. J Exp Med 175:1307–1316

Vassally P (1992) The pathophysiology of tumor necrosis factors. Annu Rev Immunol 10:411–452

Viville S, Neefjes J, Lotteau V, Dierich A, LeMeur M, Ploegh HL, Benoist C, Mathis D (1993) Mice lacking the MHC class II-associated invariant chain. Cell 72:635–648 von Herrath MG, Guerder S, Lewicki H, Flavell RA, Oldstone MBA (1995) Coexpression of B7-1 and viral ("self") transgenes in pancreatic β cells can break peripheral ignorance and lead to spontaneous autoimmune diabetes. Immunity 3:727-738

Vyse TJ, Todd JA (1996) Genetic analysis of autoimmune disease. Cell 85:311-318

Waldor MK, Sriram S, Hardy R, Herzenberg LA, Lanier L, Lim M, Steinman L (1985) Reversal of experimental allergic encephalomyelitis with monoclonal antibody to a T-cell subset marker. Science 227:41-417

Wallace VA, Kawai K, Levelt CN, Kishihara K, Molina T, Timms E, Pircher H, Penninger J, Ohashi PS, Eichmann K, Mak TW (1995) T lymphocyte development in $p56^{lck}$ deficient mice: allelic exclusion of TcR β locus is incomplete but thymocyte development is not restored by TcR β or TcR αβ transgenes. Eur. J Immunol 25:1312-1318

Walsh CM, Matloubian M, Liu C.-C, Ueda R, Kurahara CG, Christensen JL, Huang MTF, Young JD-E, Ahmed R, Clark WR (1994) Immune function in mice lacking the perforin gene. Proc Natl Acad Sci USA 91:10854-10858

Waterhouse P, Penninger JM, Timms E, Wakeham A, Shahinian A, Lee KP, Thompson CB, Griesser H, Mak TW (1995) Lymphoproliferative disorders with early lethality in mice deficient in CTLA-4. Science 270:985-988

Weaver D, Constatini F, Imanishi-Kari T, Baltimore D (1995) A transgenic immunoglobulin Mu gene prevents rearrangement of endogenous genes. Cell 42:117-127

Wegener A-MK, Letourneur F, Hoeveler A, Brocker T, Lutton F, Malissen B (1992) The T cell receptor/CD3 complex is composed of at least two autonomous transduction modules. Cell 68:83-95

Weiss A, Littman DR (1994) Signal transduction by lymphocyte antigen receptors. Cell 76:263-274

Wogensen L, Huang X, Sarvetnick N (1993) Leukocyte extravasation into the pancreatic tissue in transgenic mice expressing interleukin-10 in the islets of Langerhans. J Exp Med 178:175-185

Xu J, Foy TM, Laman JD, Elliot TA, Dunn JJ, Waldschmidt TJ, Elsemore J, Noelle RJ, Flavell RA (1994) Mice deficient for the CD40 ligand. Immunity 1:423-431

Yagüe J, White J, Coleclough C, Kappler J, Palmer E, Marrack P (1985) The T cell receptor: the α and β chains define idiotype, antigen and MHC specificity. Cell 42:81-87

Yoshida T, Ikuta K, Sugaya H, Maki K, Takagi M, Kanazawa H, Sunaga S, Kinashi T, Yoshimura K, Miyazaki J, Takaki S, Takatsu K (1996) Defective B-1 cell development and impaired immunity against Angiostrongilus cantoniensis in IL-5Rα deficient mice. Immunity 4:483-494

Zaller DM, Osman G, Kanagawa O, Hood L (1990) Prevention and treatment of murine experimental allergic encephalomyelitis with T cell receptor V β specific antibodies. J Exp Med 171:1943-1955

Zhuang Y, Soriano P, Weintraub H (1994) The helix-loop-helix gene E2A is required for B cell formation. Cell 79:875-884

Zijlstra M, Bix M, Simister NE, Loring JM, Raulet DH, Jaenisch R (1990) Beta-2-microglobulin deficient mice lack $CD4^-8^+$ cytolytic T cells. Nature 344:742-746

Chapter 29

Generation and Application of Transgenic Rabbits

Urban Besenfelder[1], Bernhard Aigner[1], Mathias Müller[1*], and Gottfried Brem[1*]

The generation of transgenic farm animals was first reported more than a decade ago (Brem et al. 1985; Hammer et al. 1985). The gene transfers were carried out by microinjection of DNA constructs into pronuclei of fertilized oocytes. This technique still represents the method of choice for generating transgenic farm animals (for review see Brem and Müller 1994). Other methods, such as using sperm cells or liposomes as a DNA vehicle, have not yet been established for practical use (review by Brem 1993b). Depending on the species used for gene transfer, the efficiency (transgenic newborns/microinjected zygotes) is about 1–3 % (see Table 29.1). In mice, the handling of totipotential embryonic stem (ES) cell lines has become a routine method for altering the genome. The establishment of ES cell lines in farm animals is eagerly awaited but cannot be guaranteed for these species in the near future (see below).

Rabbits are one of the latest domesticated species. Originating from Spain (hebrew: i-shephan-im, latinized: Hispania stands for land of rabbits), wild rabbits were kept in rabbit gardens or hunting-grounds in ancient Rome. Further domestication took place in late antiquity and during the Middle Ages in monesaries in France. Rabbit breeds and hybrid strains were developed during the 19th century based on different mutations of coat color and other visible traits.

For many centuries rabbits have been used for both livestock production and animal experimental studies. Classical experimental purposes utilized them for antibody production, development of new surgical techniques, studies of physiology, e.g., circulation and blood pressure, and toxicity tests of new drugs. On the other hand rabbits are also important in livestock production especially in Mediterranean and

* corresponding author: phone: 43-1-25077 5601; fax: 43-1-25077 5692; e-mail: muller@ifa1.boku.ac.at

[1] Institut für Tierzucht und Genetic, Veterinärmedizinische Universität Wien, Josef-Baumann-Gasse 1, Vienna, Austria

Table 29.1 Percent efficiencies of gene transfer into mammals by DNA microinjection[a]

	Pig	Mouse	Rat	Rabbit	Sheep	Goat	Cattle
Animals born/injected embryos transferred	5–10	10–20	15–25	10–15	10–15	15	10–15
Transgenic animals/offspring (integration frequency)	10–15	15–25	18	10–12	5–15	7	2– 5
Transgenic/injected embryos transferred (gene transfer efficiency)	0.5–1 (–3)	2–3 (–5)	4–5	1–2	1–2	1	0.2

[a] For references see Brem (1993b); Brem and Müller (1994); Wall (1996).

developing countries for meat, fur, and angora wool production. Thus, gene transfer into rabbits is a very interesting technique for improving their performance and for application in research and livestock production.

29.1 Generation of Transgenic Rabbits

The production of transgenic rabbits is carried out according to the following steps:

- Cloning of the gene construct
- Preparation of the DNA solution for microinjection
- Collection of rabbit zygotes
- Microinjection into the pronuclei of fertilized oocytes
- Transfer of the injected embryos
- Detection of the transgene in the produced rabbits
- Establishment of transgenic lines carrying the hereditary transgene
- Examination of expression and biological activity of the transgene.

Cloning of the Gene Construct

A gene construct ideally provides all elements controlling temporal-spatial and tissue specific expression. The coding portion (structural gene or cDNA) is linked to 5' *cis*-control elements, i.e., promoter and enhancer/silencer regions, which specify tissue, time, and quantity of transcription. 3' control regions provide correct RNA processing signals and are thought to be implicated in RNA stability. Although there are no golden rules for an optimal transgene design, it is currently envisaged to combine the coding sequences with 5' and 3' flanking regions as large as possible in order to shield the inserted DNA constructs from influences of the surrounding chromatin.

The integration frequency of the gene construct is influenced by the DNA conformation (linear or circular). The linear form integrates five times better than the circular form (Brinster et al. 1985). Under ideal conditions, it is possible to reach a 25 % integration frequency (i.e., number of transgenics/number of newborn animals) with the injection of linearized DNA fragments into fertilized mouse ova (see Table 29.1).

For the successful expression of the transgenes, it is necessary to remove prokaryotic sequences (i.e., vector sequences) because they have been shown to have inhibitory effects. In order to increase gene expression, genes should be used in their original genomic structure rather than in the form of their cDNA. The exon-intron structure seems to raise the transcription efficiency of transgenes (Brinster et al. 1988; Choi et al. 1991; Palmiter et al. 1991). If the genomic clones of the gene of interest are not available or exceed the cloning capacity of the vectors, it is advisable to include heterologous introns in the gene construct.

Preparation of the DNA Solution for Microinjection

The gene construct is removed from the vector sequences (usually plasmids or cosmids) by restriction endonucleases and subsequent agarose gel electrophoresis. The transgene fragments are purified from the gel and diluted in injection buffer containing Tris-Cl/EDTA (10 mM Tris pH 7.5, 0.2 mM EDTA pH 8) or MSOF (MOPS buffered synthetic oviduct fluid: 107.7 mM NaCl, 7.16 mM KCl, 1.71 mM CaCl2, 1.19 mM KH2PO4, 0.49 mM MgCl2, 5 mM NaHCO3, 0.33 mM Na-pyruvate, 3.3 mM Na-Lactate, 1.5 mM glucose, 1 mM L-glutamine, 1× concentration of essential and nonessential amino acids, 0.8 % bovine serum albumin, 20 mM MOPS; osmolarity 265–275, pH 7.2–7.4). The use of a physiological medium (e.g., MSOF) for microinjection resulted in our experiments in significantly greater efficiency of transgene production (G. Brem, N. Zinovieva, and U. Besenfelder, unpubl.). Routinely, 1 pl DNA solution containing about 500 to 1000 copies of the gene constuct are microinjected. This concentration of the injection solution has been shown to result in high integration frequency (Brinster et al. 1985). The DNA microinjection solution has to be free from particles or other contaminants in order to prevent clogging of the injection needle (diameter 1 mm) or damaging of the injected cells.

Preparation of Donors and Collection of Embryos

For gene transfer experiments rabbit strains, cross-breedings or hybrid strains (e.g., ZIKA) are used. Depending on the genetic background, the embryo donors and recipients should be at least 4 months of age. The use of animals experienced in breeding as recipients avoids some parturition problems and losses of offspring.

Adult rabbits are housed in cages (0.5 and 0.8 m^2 per experimental and breeding animal, respectively) and kept under conditions of 14 h light and 10 h dark in air-conditioned stables. Feeding rabbits a combination of nutrient concentrate (80 g per day) in addition to water and hay ad libitum results in good breeding conditions for the animals. At least 17 days prior to superovulation or synchronization, the females are separated to avoid pseudopregnancy status, which interferes with the hormonal treatment.

Four days prior to the collection of the embryos the donor rabbits receive a single intramuscular (im) injection of 20–30 IU PMSG (pregnant mares serum gonadotropin) per kg body weight. Ovulation is induced by intravenous (iv) injection of 180 IU HCG (human chorionic gonadotropin) per animal 3 days after PMSG administration. Immediately after the HCG application, the donor animals are mated naturally or artificially inseminated twice within 1 h to ensure fertilization of a maximum number of oocytes. The activity of PMSG can be neutralized after fertilization has taken place by application of anti-PMSG antibodies.

For embryo collection, the animals are slaughtered 19 to 20 h after insemination to obtain the reproductive organs (ovaries, oviducts, and cranial parts of the uterus horns). After removal of mesosalpinx and fat tissue, the oviducts are flushed through the infundibulum with PBS (phosphate buffered saline). Storage and short-term in vitro culture of embryos before and after manipulation is performed in PBS supplemented with 20 % fetal calf serum. Zygotes (fertilized oocytes) of good quality are free of cumulus cells and show a bright cytoplasma with two easily detectable large pronuclei mostly placed in the center of the cell. Normally, 20 to 30 oocytes can be collected per donor and about 80 to 90 % of them are fertilized and suitable for microinjection. Surgical embryo collection as alternative method is laborious and time-consuming, thus limiting the use of this technique for standard experiments.

Microinjection of Zygotes

Microinjection is carried out under 400× magnification using an inverted microscope (e.g., Zeiss ICM 405) and Nomarski interference contrast optics (see Fig. 29.1). Zygotes are placed on a depression slide in a drop of medium with a top layer of paraffin oil. For injection, the zygote is fixed on the holding pipette by suction, and the injection pipette is inserted carefully through zona pellucida, cell, and nuclear membrane until the tip is positioned within the larger (male) pronucleus. Successful microinjection of DNA solution by air pressure is indicated by visible swelling of the pronucleus. An experienced person is able to successfully inject between 100 and 200 zygotes per hour.

After injection, the zygotes are stored for a short time before being transferred to the oviducts of recipients. Under optimal conditions, less than 10 % of the injected rabbit embryos show morphological damage and lysis. Surviving embryos are washed, pooled in batches of ten zygotes, and transferred into the oviduct. Long-term in vitro culture of rabbit embryos (one-cell to blastocyst stage) is possible in principle (Carney and Foote 1991). However, even the embryos developed in vitro to advanced stages will not implant after transfer to the uterus. It is suggested that, due to the lack of transport through the oviduct, the embryos miss the mucin layer, which is put on the zona pellucida, and seems to be necessary for implantation (Adams 1970).

Transfer of Injected Embryos and Production of Offspring

Recipient rabbits are caged individually 3 weeks before the transfer and synchronization is performed by induction of ovulation with 120 IU HCG. Embryo transfer to the oviducts can be done surgically or endo-

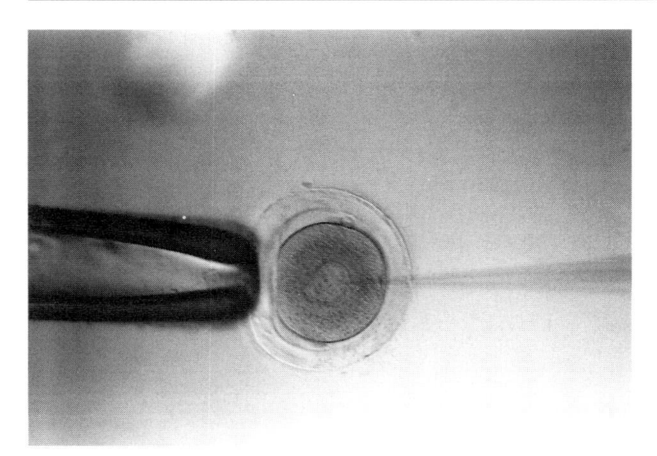

Fig. 29.1. Microinjection into pronuclei of fertilized rabbit oocyte

scopically. The recipients are anesthezised by iv injection of xylacine-ketamine (1.8 mg 2 % xylacine Rompun Bayer Leverkusen and 15 mg 10 % ketamine hydrochloride WDT Garbsen per kg body weight). After taking hygienic precautions, surgical transfer is done by midline incision (Ross et al. 1988) and subsequent transfer of 8 to 12 injected embryos to each oviduct. Examination for pregnancy by palpation takes place on day 10 to 12 after transfer.

Recently, we have developed laparoscopic embryo transfer techniques into the oviduct of rabbits (Besenfelder and Brem 1993). After making a small incision (<1 cm), an endoscope trocar is introduced through the abdominal wall. The trocar is removed, the abdomen inflated with air and the endoscope is inserted. A transfer capillary loaded with the microinjected embryos is then inserted through a vein catheter into the oviduct and the embryos are transferred into the ampulla via the infundibulum. Because this technique requires minimal operative procedure and manipulation of the reproductive organs, the pregnancy rate after transfer of untreated control embryos is up to 90 %. Pregnancy rates after transfer of microinjected embryos is between 60 and 80 %, depending on the construct. In addition, laparoscopic transfer is less time-consuming (5–10 min per recipient) than surgery (20–30 min per recipient). Thus, laparoscopic transfer is the method of choice for efficient routine gene transfer experiments.

Pregnancy is detected by palpation of the females at day 12 after synchronization, when the implantation sites are clearly distinguishable as tight round spheres (Hassan et al. 1992). In young does, manual detection is already possible at day 9. During the last 10 days of pregnancy, the fetuses can be palpated inside the fetal membranes. Normally, synchronization and induction of parturition is not necessary. The application of $PGF_{2\alpha}$ 2 days ante partum reduces the risk of prolonged pregnancy (>day 32) and birth complications. The use of oxytocin (2–3 IU per animal) is only advised after the birth of at least one pup to ensure that uterus contraction will direct the pups to the birth canal.

Does coming to birth for the first time show a greater nervousness than animals having already born several litters. This behavior is characterized by incomplete nest-building, birth of pups over the whole cage area, and cannibalism. To reduce the loss of offspring, it is necessary to offer these does a dark, quiet place, to put all pups inside the nest, and to remove the dead offspring. Rabbits nurse only briefly once every night or early in the morning. Therefore, aggressions against the pups during the first days postpartum can be avoided by separating the pups from the does except for the time necessary for sucking. Pups of poor condition can be fed by fixing the doe in dorsal recumbancy, injecting 2 to 3 IU oxytocin and assisting sucking. Around the 3rd week postparturition, the pups begin to consume solid food, e.g., pellets. At this time, attention should be paid that the pups get enough milk to overcome this critical period in the changing of the enteral metabolism.

Detection of Transgenic Founders and Establishing Transgenic Lines

Offspring derived from gene transfer experiments are examined by DNA analyses. Tissue samples for DNA isolation are routinely obtained from tail or ear biopsies or blood, and DNA can be extracted according to various established methods (Kawasaki 1990; Ausubel et al. 1994). The integration of the transgene into the host genome is analyzed by Southern blotting, dot/slot blotting, or modifications thereof (Southern 1975; Brenig et al. 1989). The use of the polymerase chain reaction (PCR; Saiki et al. 1988; Attal et al. 1995) enables the detection of transgenes even in a small number of cells. Depending on the strain, transgenic founders will be mated earliest on month 4 postparturition to nontransgenic rabbits. The offspring will be investigated for integration of gene constructs. Semiquantitative PCR can be used to detect homozygous transgenic animals or to estimate the number of gene constructs integrated per cell (e.g., Aigner and Brem 1995).

An alternative method to detect transgenic founders and establish transgenic lines is the co-injection and cointegration of marker transgenes. The exact mechanism of transgene integration into the host genome remains to be elucidated (McFarlane and Wilson 1996). It has been observed, however, that the transgenes integrate as concatemers of several gene construct copies in a single integration site. Hence, coinjection of a marker gene with the DNA construct of interest in most cases results in cointegration and coexpression of the foreign DNA. A marker gene has to be readily detectable and its expression should not have any side effects on the transgenic animal. Suitable markers are provided by well-characterized color genes of mammals. The tyrosinase gene is known to be essential for the pigmentation of coat and retina in mammals. Lack of tyrosinase activity results in the albino phenotype. Transferring wild-type tyrosinase gene constructs to albino mice and rabbits (e.g., ZIKA) was shown to convert the albino into a melanized phenotype (Beermann et al. 1990; Aigner et al. 1996). In due course, tyrosinase gene constructs have been extensively studied as a marker for transmission and expression of (co)integrated transgenes (Beermann et al. 1991; Overbeek et al. 1991; Aigner and Brem 1994). In the case of nonalbino rabbits being preferred for gene transfer experiments another coat color gene, the agouti gene, could be used to induce the production of a phenotypically visible yellow band on otherwise black hair (for a review see Manne et al. 1995).

For establishing transgenic lines it is a prerequisite that germline cells of the primary transgenic rabbit (founder) carry transgene copies. Transgenic founders harboring cells with different genotypes are termed mosaics. A transgenic mosaic carries the transgene in only some of their cell populations. The results of previous experiments indicate that 30 % of the primary transgenic animals are mosaics (Wilkie et al. 1986) and do not pass the transgene to their offspring at the expected rate of 50 %.

Usually, the hereditary mode of a transgene follows the Mendelian rules. The transgenic animals of the F_1 and the following generations carry the transgene in all of their somatic and germline cells as long as they test positive for the transgene.

Examination of Expression and Biological Activity of the Transgene

Transgene expression depends on the regulatory (*cis*-acting) elements of the gene construct and is influenced by the chromatin surrounding of the integration site. It is monitored on RNA and protein level by standard molecular biology techniques. Commonly used RNA analyses include Northern analysis, RNase protection, or reverse transcription (RT) PCR for monitoring accuracy, tissue specificity, and levels of transgene expression (Ausubel et al. 1994). Translation of the transgene is analyzed by protein staining methods or, more specifically, by immuno (histo)chemical methods (e.g., Western blotting, immunoprecipitation, ELISA, RIA). One has to keep in mind that the available analytical methods have to be adapted to the side, time, and levels of transgene expression. In addition, the desired biological effects and possible side effects of the transgene expression have to be carefully monitored.

29.2 Application of Gene Transfer in Rabbits

The applications of transgenic rabbits are listed in Table 29.2 and discussed in greater detail in the following.

Production of Pharmaceutical Proteins by Transgenic Rabbits

Pharmaceutical proteins have a broad therapeutical spectrum in human and veterinary medicine. The classical method to isolate human proteins was by extracting them from human blood or human organs. In many cases, however, it is not possible to produce pharmaceutical proteins by conventional methods and reach high enough quantity and quality levels. Problems are, for example, the small amount of proteins harvested, the risk of contamination with pathogenic agents, and the ethical conscience on the supply of human blood and organs. The chemical synthesis of peptides with short amino acid chains as an alternative method is not sufficient.

Recombinant DNA technology has provided the possibility to produce specific proteins in high quantity at relatively low cost in bacteria or yeast expression systems. Using prokaryotic or primitive eukaryotic production systems also has a variety of disadvantages. They frequently lack the ability of performing the required post translational modifications of the recombinant protein, which in turn is necessary for biological activity. Intracellular inclusion bodies or protein degradation sometimes make the isolation and purification of the recombinant protein difficult.

To avoid some of these problems, higher eukaryotic production systems were established by using genetically transformed human and animal cells. Some proteins produced in these systems are already used in therapy, and others are still at the stage of clinical trial or at the level of

Table 29.2. Current applications of transgenic rabbits

1. Gene farming – bioreactor mammary gland
   - Production of pharmaceuticals for human or veterinary medicine
   - Production of enzymes
2. Animal models for human diseases
   - Atherosclerosis
   - Tumorgenesis
   - Viral infections
3. Improvement of efficiency and quality in rabbit production
   - Meat production and growth performance

basic research. In cell culture, optimal expression parameters have to be established for genetically engineered proteins. Transcription efficiency, messenger RNA turnover rate, translation efficiency and protein stability have to be at their optimal level. However, eukaryotic tissue culture systems are very costly and prone to interferences. Despite the obvious advantages in carrying out the required posttranslational modifications when compared to prokaryotes, some of these modifications are performed insufficiently or false. In other cases, the production rate and the protein purification have proven to be inefficient. Especially in human tissue culture systems, precautions have to be taken to avoid the possibility of contamination with DNA, cell proteins, virus, or other agents.

Thus, the possibility of establishing transgenic mice and farm animals has opened new ways for the production of pharmaceutical proteins by a method called gene farming. The idea is to produce pharmaceutical proteins in specified organs or body fluids of transgenic animals. The advantages of this method are that the production of proteins in vivo is more accurate and efficient than in vitro. The arising costs are by factor 5 to 10 lower than in tissue culture.

The mammary gland is the most interesting organ for the production of recombinant proteins (Clark et al. 1987; Mercier 1987; Van Brunt 1988; Hennighausen 1990; Brem 1993b). It has an enormous physiological potential for the daily production of proteins. Milk is easy to collect and usually has a high hygienic standard. Quite a few experiments with different species such as mice, rabbits, pigs, sheep, and goats show that it is basically possible to reach mammary specific expression for recombinant proteins in transgenic animals. For some recombinant proteins, other organs are of advantage if their posttranslational modification is only possible in specific somatic cells. One example would be recombinant antibodies produced in B-lymphocytes and extracted from the blood of transgenic animals.

The choice criterium for selecting the most suitable species for gene farming is usually based on the quantity of protein needed per year. A simplified rule is: the production of a protein in tons should be carried out by cows, in hundreds of kg by sheep or goats, and in kg per year by rabbits. Pigs are the species of choice for expression of foreign proteins in blood or somatic cells.

Expression in the mammary gland

Six major milk proteins make up 80 % of the protein contents in the milk of farm animals (reviewed by Bawden et al. 1994). The major proteins of cow milk are α_{S1}, α_{S2} and β-caseins (calcium-sensitive proteins) and ϰ-casein. Both whey proteins, β-lactoglobulin (BLG) and α-lactalbumin, are found in milk in considerable amounts. The whey acidic protein (WAP) is the major whey protein of mice, rats, and rabbits. Other milk proteins like serum albumin, lysozyme, lactoferrin, and immunoglobulins are usually detected at a concentration level of less than 1 g per 1 l milk.

The majority of milk proteins is secreted from epithelial cells of the mammary gland under multihormonal control. The promoters of all major milk protein genes have been used for studying the expression of transgenes in the mammary gland of mice, rats, rabbits, pigs, sheep, goats, and cattle (reviewed by Bawden et al. 1994). At present, the regulatory elements from BLG, WAP, α-lactalbumin, β-casein, and αs1-casein of several species have been utilized.

Rabbit appears more and more to be an intermediate animal well adapted for the production of a limited amount of proteins (see below). Rabbit husbandry can be carried out under specific pathogen-free conditions. Rabbits have a short generation time, and transgenic founders can be generated with a reasonable efficiency. The endogenous milk proteins are well characterized (Baranyi et al. 1995) and milking can be performed semiautomatically, resulting in a milk yield of 10 kg per rabbit and year (Duby et al. 1993). Thus, considering both economical and hygienic aspects, this species is suitable for gene farming (Brem 1993a; Houdebine 1994). Mammary gland-specific expression of recombinant proteins in rabbits has been achieved with WAP-, β-casein-, and αs1-casein-promoter driven gene constructs (Bühler et al. 1990; Brem et al. 1993, 1994, 1995; Riego et al. 1993; Limonta et al. 1995).

The whey acidic protein is produced at the highest concentration in rodent milk. Steroid and peptide hormones regulate the expression of the WAP gene and the stability of the mRNA. The WAP RNA makes up 50 % of the mRNA during lactation. WAP gene constructs were first used for directing the expression of nonmilk proteins into the mammary gland of transgenic animals (Andres et al. 1987; Gordon et al. 1987). Since then, various WAP promoter gene constructs tested in several species have resulted in good protein production levels in the mammary gland (Hennighausen 1992; Maga and Murray 1995). However, in some cases ectopic expression of the transgene has been observed (e.g., Günzburg et al. 1991). We have produced ten transgenic rabbit lines carrying the murine WAP promoter (2.4 kb) linked to genomic human growth hormone (hGH) coding sequences. hGH expression was monitored by RNA slot blot and Northern blotting, in situ hybridization, immunohistochemistry, and by radioimmuno assay (RIA) in serum of milk and blood of the transgenics. Expression levels in milk serum were up to 4 g/l. There was no hGH detectable in the blood serum of the transgenic rabbits. All animals were healthy and fertile and no changes in growth performance were observed (Brem 1993a). In transgenic mice the same gene construct unexpectedly resulted in high-level expression of hGH in the brain (Brem 1993a). By comparison with other gene constructs, it was suggested that the combination of the WAP promoter and the hGH structural gene resulted in the novel tissue specificity rather than the 2.4-kb WAP regulatory sequences per se. Since the brain-specific expression was observed in several transgenic lines, possible effects of the integration sites seemed to be unlikely.

A similar WAP-hGH gene construct (using 2.6 kb murine WAP regulatory sequences) was used in a separate gene transfer experiment. Five transgenic rabbits were obtained, one of which showed mammary gland-specific hGH expression (Limonta et al. 1995). hGH levels in the milk were found to be around 50 mg/l. As has been observed for WAP transgenic mice, there was an increased hGH level in the blood serum and ectopic transgene RNA expression in the ovaries. However, the transgene expression in the rabbit line showed no apparent detrimental effect.

A rabbit WAP promoter fragment was fused to the human erythropoietin cDNA (hEPO) and used for the generation of six transgenic rabbits (three females). One transgenic female expressed low levels (0.3 μg/ml) of hEPO in the milk (Castro et al. 1995). There was no report on assaying ectopic transgene expression and possible deleterious effects on increased EPO levels on blood cell formation.

β-Casein is present at high concentrations in the milk of cattle and sheep. Transgenic mice harboring the rat β-casein gene were first used to study this promoter in gene farming (Lee et al. 1988). The 14-kb clone carried the whole rat β-casein gene, 3.5 kb of the 5', and 3 kb of the 3' flanking regions. Expression of rat β-casein mRNA in the mammary gland made up about 0.01 to 1 % of the endogenous mouse β-casein mRNA. The original transcription site was used. One line showed expression of the transgene in the brain at a reduced level.

The use of the β-casein promoter for the expression and secretion of foreign proteins in the milk of rabbits was also examined. Bühler et al. (1990) have produced a gene construct carrying the rabbit β-casein promoter and the genomic sequence of the human interleukin-2 (hIL-2) gene. Four transgenic rabbit lines were tested. Their milk contained biologically active hIL-2. The concentration was around 50 to 340 μg/l milk.

A variety of transgenic rabbits carrying gene constructs controlled by the αS1-casein promoter have been generated. αS1-casein is a milk protein present at high concentration in milk (for review see Bawden et al. 1994). Recombinant DNA constructs based on αS1-casein regulatory elements should therefore have high expression potential in the mammary gland of transgenic animals. Meade et al. (1990) used a gene construct based on bovine αS1-casein and the human urokinase gene to produce transgenic mice. The transgenic mice produced urokinase at a concentration of 1 to 2 mg/ml milk. There was no expression in other organs. All transgenic mice were healthy and fertile.

We used gene constructs based on the αS1-casein gene to direct expression of proteins into the milk of rabbits. The mammary gland-specific expression cassette consists of bovine αS1-casein sequences providing the 5'-promoter elements and the 3'-elements necessary for mRNA processing. In addition, the construct includes intron/exon boundaries known to enhance transgene expression (see above). In order to achieve secretion of the protein into the mammary gland, the coding region of choice is fused to the αS1-casein signal peptide sequence.

Three different hybrid gene constructs based on the αS1-casein expression cassette and the bovine prochymosin gene were used to produce 14 transgenic rabbit lines (Brem et al. 1995). Chymosin (rennin) is the milk-coagulating enzyme of young mammals and is widely used in cheese production. Nine transgenic lines were analyzed for expression which ranged from 0.5 to 2 g/l milk. The highest expression was found in two transgenic animals at 10 g/l. Their lacation period, however, was very short. RNA analyses demonstrated that the fusion constructs were exclusively expressed in females in the mammary gland during lactation. Prochymosin in transgenic rabbit milk can be readily activated by lowering the pH (pH 2.5 for 1.5 h) and subsequent neutralization. One liter of cow's milk was clotted by 0.09 ml of activated transgenic rabbit's milk containing about 1 g/l chymosin.

Two αS1-casein expression cassettes comprising different length of 5'-regulatory sequences (2.9 and 11.0 kb) were used to produce human insulin-like growth factor (hIGF-1) (Brem et al. 1994 and Zinovieva et al., submitted). Eight hIGF-1 expressing transgenic lines were established. As expected, transgene expression was restricted to the mammary gland of lactating animals. hIGF-1 production varied from 0.5 to 2 g/l milk. hIGF-1 was correctly processed and biologically active and was purified from the milk to a nearly homogenous active form.

In transgenic mice and one transgenic rabbit founder, a different expression cassette composed of a 1.6-kb αS1-casein promoter fragment fused to cDNA encoding the human tissue plasminogen activator (htPA) resulted in transgene expression at very low levels (50 mg/l in mice and 8–50 μg/l in the rabbit; Riego et al. 1993). Thus, longer promoter fragments are needed for achieving high production levels in the mammary gland.

Expression in blood

The production of certain pharmaceutical proteins such as antibodies, hemoglobin, albumin, and other serum proteins would be ideally carried out in the blood of transgenic animals, since this is their natural production site. It can be expected that the required posttranslational modifications, the protein trafficking, and the stability of recombinant proteins in blood would be optimal.

The production of human antibodies in transgenic animals would be a breakthrough since they have a wide diagnostic and therapeutic potential, and the possibility of causing an immune reaction in the patient is practically not given. The genes or cDNAs coding for the light and heavy chains of human antibodies are generated from human hybridoma cells. Transgenic farm animals that are produced with these gene constructs produce human antibodies in their serum. The human antibodies can be extracted and purified from their blood. Even chimerized and humanized antibodies, which consists of the variable sequences of animals and constant human regions could also be expressed in variable isotypes in transgenic pigs and rabbits. Heterospecific antibodies have a wide therapeutic potential. They help to eliminate virus-infected cells by recognizing an epitope on the cell membrane representing the antigen and an epitope on the cytotoxic T cells of the recipient species. It should also be possible to produce antibodies composed of various isotypes in transgenic animals.

In a model experiment, we introduced genomic clones coding for light and heavy chains of mouse monoclonal antibodies (MAbs) into the germline cells of mice, rabbits and pigs (Weidle et al. 1991). In the primary transgenic mice, 20 to 40 μg/ml of antibody proteins were expressed. Their offspring produced up to 150 μg/ml. The two examined primary transgenic rabbits showed an expression of 150 to 300 μg antibodies in 1 ml serum, and their offspring showed an expression of 150 mg/ml. The antibody titer in the serum of one pig was at 1000 μg antibodies per ml serum. Its offspring showed expression at the same level. We were able to detect antibody expression even in the milk. The expression level was, as expected, very high in the colostral milk (first 3 days of lactation) and ranged at 100 μg/ml milk serum. The level went down to 1 μg/ml milk during further lactation. Transgenic Ab was purified from serum of transgenic rabbits and pigs and was shown to have two intact binding sites for the antigen when analyzed in ELISA. However, in isoelectric focusing only a small fraction of the transgenic product matched the mouse MAb. These findings could be due to heterologous Ab by association of endogenous L chains with the mouse transgene H chains (or vice versa). The electrophoretic differences could be also attributed to species- and cell-type-specific posttranslational modifications (e.g., glycosylation, deamidation), as has been observed for hybridoma cells cultured in differing media.

The feasibility of expressing antibody encoding genes in farm animals has been demonstrated by the experiment described above and another study (Lo et al. 1991). Since rabbits have on average 50–60 ml blood per kg body weight, for large-scale production of antibodies or other proteins in the blood, pigs would be the species of choice (Logan 1993).

For the concept of generating an animal that, upon challenge with an antigen, will produce a human Ig response, two objectives have to be achieved by transgene technologies:

- the endogenous immunoglobulin loci of the animal must be rendered inactive. This requires homologous recombination in totipotential cells;
- the human heavy and light chain loci have to be introduced into the germline of the animal. Since these loci are very large (up to 5 Mb in their entirety), either mini-loci containing a smaller number of variable regions and/or large gene constructs cloned in yeast or bacterial artificial chromosome (YAC or BAC) vectors have to be generated. The development of fully human monoclonal Abs with therapeutic potential in transgenic mice has been reported by several groups (e.g., Green et al. 1994; Lonberg et al. 1994; Zou et al. 1994). The transfer of this technology to farm animals is currently not possible because of the lack of totipotential cell lines, thus preventing gene targeting experiments.

Transgenic Rabbits as Disease Models

The effects of a given transgene in vivo are studied in transgenic rabbits in addition to transgenic mice and rats in order to evaluate the observed results for application in other species. The size of the animals makes it possible and easier to carry out analyses and further experiments with the transgenic animals. The physiological data of laboratory rabbit strains are well established and the majority of rabbit-specific reagents necessary for studying disease models are available. Furthermore, rabbit strains have a more diverse genetic background than in- and outbred mouse strains. This might be favorable when studying complex disease models or therapeutic applications, since it resembles more accurately the situation in humans.

Lipid metabolism imbalances and atherosclerosis

Atherosclerosis is the main affection of arteria in human. It is caused by multiple factors including genetic predisposition, hyperlipidemia, hypertonia, diabetes mellitus, stress, and others. Several transgenic rabbit models have been established to examine atherogenesis.

15-Lipoxygenase is expressed in foamy macrophages of atherosclerotic lesions and has been implicated in the oxidative modification of low density lipoprotein (LDL) during early stages of disease development. Transgenic rabbits overexpressing this enzyme have been generated and may be used for further mechanistic studies on the implication of lipoxygenase in atherogenesis; they are also an ideal model for testing the in vivo action of 15-lipoxygenase inhibitors (Shen et al. 1995).

Imbalances in the lipid metabolism, especially a high blood level of low density lipoproteins (LDL) and triglycerols with rather low level of high density lipoproteins (HDL), have been implicated in atherogenesis. Apolipoprotein B (apoB) is the main constituent of LDL, thus transgenic rabbits overexpressing apoB serve as a model for hyperlipidemia and artherosclerosis (Fan et al. 1995).

A similar disease pattern can be induced by imbalances in the plasma cholesterol homeostasis. Studies in transgenic rabbits overexpressing hepatic lipase (HL) showed that transgene expression resulted in a five-fold increase in total plasma cholesterol and in a dramatic decrease in HDL. This demonstrated the key role of this enzyme in plasma cholesterol metabolism and suggested a possible role in lipid imbalances (Fan et al. 1994).

ApoB mRNA undergoes mRNA editing, which results in the creation of a new stop codon and a truncated apoB polypeptide. A catalytic subunit (APOBEC-1) of the multiprotein editing complex has been characterized and overexpressed in transgenic mice and rabbits (Yamanaka et al. 1995). The transgenic animals had reduced apoB and LDL levels. However, the transgenics had liver dysplasia, and many transgenic mice developed hepatocellular carcinomas. In addition, it was shown that mRNAs other than apoB can be edited by the overexpressed catalytic subunit, which in turn is suggested to be the cause of carcinogenesis. Nevertheless, the findings compromise the potential use of APOBEC-1 for gene therapy to lower plasma levels of LDL. Lecithin:cholesterol acetyltransferase (LCAT) is an enzyme involved in the metabolism of HDLs. The impact of overexpressing LCAT on the serum concentrations of the different plasma lipoproteins was evaluated in an animal model. Rabbits were chosen since they express cholesteryl ester transfer protein, one target of the enzyme's activity. The results indicated that overexpression of LCAT leads to both hyper-alpha-lipoproteinemia and reduced concentrations of atherogenic lipoproteins (Brousseau et al. 1996; Hoeg et al. 1996).

Tumorgenesis

Transgenic rabbits were generated with the rabbit c-myc proto-oncogene fused to the immunoglobulin (Ig) heavy (Eμ) or light (Eϰ) chain enhancers (Knight et al. 1988; Sethupathi et al. 1994). Rabbits carrying the Eμ-myc gene construct developed leukemia at 17–21 days of age and subsequently died of acute lymphoblastic leukemia, resembling childhood leukemia in human (Knight et al. 1988). Of a total of 19 Eϰ-myc transgenic rabbits, 8 developed tumors. In four cases they were lymphomas of B-lymphoid lineage, the others were diagnosed as embryonic carcinoma, hepatoma, ovarian carcinoma and basal cell carcinoma. The unexpected development of nonlymphoid tumors was discussed as being due to the ability of the transcription factor NFϰB to activate the Eϰ chain enhancer in cells other than of B-lymphoid origin (Sethupathi et al. 1994).

A model for virus-induced tumorgenesis was established in rabbits carrying copies of the cottontail rabbit papillomavirus (CRPV) DNA alone or in combination with the proto-oncogene EJ-ras (Peng et al. 1993, 1995). Although CRPV transgenes were detectable in all tissues, the CRPV expression was restricted to skin, thus resembling the situation in virion-infected animals. Tumor development was also detected only in skin. The suppression of transgene expression in the other tissues was found to be correlated with hypermethylation of the gene construct. The expression of EJ-ras was indicated to be dependent on the expression of certain CRPV genes and therefore may be a crucial cofactor of the virus in the progression of carcinomas.

HIV-1 infection

A major obstacle to understanding HIV-1 infection and AIDS is the lack of a suitable laboratory animal model for studying disease progression and testing diagnostic, therapeutic, and preventive measures. Transgenic rabbits expressing human CD4 (hCD4) were generated to provide a model for HIV-1 entry into rabbit T cells. In vitro studies demonstrated that hCD4 transgenic lymphocytes are more susceptible to HIV-1 infection than those from control rabbits (Dunn et al. 1995; Snyder et al. 1995). In vivo studies revealed the infection, virus replication, and seroconversion of various HIV-1 proteins in the transgenics, although these rabbits are less sensitive to infection than humans. Thus, the rabbits may be a useful tool in studying HIV-1-induced pathogenesis, especially the mechanisms leading to lymphocyte apoptosis (Leno et al. 1995).

Gene therapy and increased disease resistance

A variety of strategies can be envisaged aiming at the improvement of disease resistance in farm animals by means of germline integration of transgenes (for review see Müller and Brem 1994, 1996). The use of antisense RNA (asRNA) to inhibit RNA function within cells or whole organisms has provided a valuable molecular genetic tool. AsRNA functions by binding in a highly specific manner to complementary sequences, thereby blocking the ability of the bound RNA to be processed and/or translated. An antisense approach for decreasing the susceptibility to viral infection has been tested in transgenic rabbits (Ernst et al. 1991). An asRNA gene construct complementary to adenovirus h5 RNA was used for gene transfer. Although transgenesis was accompanied with significant deletions and rearrangements of the asRNA transgenes, two transgenic lines were established carrying intact copies. Primary kidney cells expressing the transgene were found to be 90–98% more resistant to adenovirus infection than cells from nontransgenic animals.

Somatic gene transfer experiments do not aim at the stable integration of the gene construct in all cell types. Thus, there is no requirement of transferring the DNA in early embryonic stages. "Transient" transgenesis can be achieved by all passive or active transfection/transformation methods developed in tissue culture or animal models. The main routes for DNA delivery in somatic tissues are viral vector systems or noninfectious methods, including injection of free (naked) or carrier-bound DNA, particle bombardment (gene gun) or aerosols (for review see Mulligan 1993; Müller and Brem 1996 and refs. therein). The vast majority of experiments in this field aim at human gene therapy. For testing the delivery routes, the efficiency of delivery, the levels, duration, and tissue specificity of transgene expression and the side effects (especially immunoreactivity), as many different animal model as possible are required. Currently, in rabbits gene transfer mediated by adenoviral vectors (e.g., Kozarsky et al. 1994; Steg et al. 1994; Roessler et al. 1995), particle bombardment (e.g., Cheng et al. 1993), and liposomes (e.g., Takeshita et al. 1994) has been carried out successfully. In veterinary medicine and genetic engineering, somatic gene transfer might be particularly useful for DNA vaccination (genetic immunisation) and large-scale production of polyclonal antibodies (Waine and Mcmanus 1995).

Transgenic Rabbits in Livestock Production

In Mediterranean and Third World countries rabbits play an important role in meat production. The possibility of influencing growth promotion in mammals by transgenic means was first demonstrated in mice (Palmiter et al. 1982). Growth hormone (GH) transgenic mice showed an enhanced growth performance with a fourfold increase in growth rates and a twofold increase in final body weight. Subsequently, transgene expression of GH and other members of the growth hormone cascade controlled by various promoters in mice resulted in giantism, but also in a variety of pathological side effects (e.g., Brem et al. 1989; Wanke et al. 1992). Based on the experiments carried out in mice, many projects in farm animals concentrated on transgenes for the improvement of growth traits, i.e., daily weight gain, food conversion, and carcass composition (meat/fat ratio).

Gene transfer experiments into rabbits were performed with gene constructs encoding GH or growth hormone-releasing hormone (GHRH; Ross et al. 1988; Gol'dman et al. 1993; Rosokhatsky et al. 1994). In some cases, increased growth rates of the transgenic rabbits have been reported. It has to be mentioned, however, that the experiments published do not allow a judgement of the feasibility of using growth-enhancing genes for meat production in rabbits. So far, none of the approaches to influence growth performance of mammals by transferring GH cascade genes or muscle differentiation genes has resulted in all the desired effects, and all have been accompanied by pathological side effects. Accepting the complexity of exogenous factors influencing the growth of an individual and the fine interplay of endogenous growth promoting and inhibiting factors, this is not completely surprising (for review see Müller and Brem 1994). An additional important aspect is the acceptance of transgenic farm animals and the resulting transgenic food in the public. At the present time, public opinion would not permit the successful marketing of giant and/or turbo farm animals.

Fig. 29.2. YAC transgenic rabbit

Future transgenic projects in rabbits might address approaches to alter fur or wool quality and production. Experiments to improve wool growth have already been carried out in transgenic mice and sheep (Bawden et al. 1995; Damak et al. 1996).

29.3
Conclusion and Outlook

Despite the obvious benefits of transgenic animals for genetic engineering and disease models, several problems remain to be solved. The expression levels and sometimes the tissue specificity of the promoters used are usually unpredictable. One approach to achieve strict spatio-temporal pattern of expression from genes of interest is the use of yeast artificial chromosome (YAC) gene constructs providing extensive sequences flanking the coding unit of the gene in order to avoid unwanted side effects of transgene expression. Recently, the first YAC DNA transgenic farm animals (see Fig. 29.2) have been generated (Brem et al. 1996). Without doubt, an exciting development will be the in vitro establishment of ES cells from rabbits and their subsequent use in cloning and transgenesis experiments. The availability of this technique will give new impetus to gene transfer, because it will not only provide the possibility of additive gene transfer and homologous recombination but will also notably reduce problems such as low efficiency, nonexpression of transgenes, or insertional mutations. ES-like cells have been established from rabbits and used for the in vitro formation of embryoid bodies, differentiation and nuclear transfer experiments (Graves and Moreadith 1993; Du et al. 1995). However, in rabbits, as well as in other farm animals, attempts to establish ES cells did not fulfill all criteria of pluripotency, i.e., germline chimerism.

References

Adams C (1970) The development of rabbit eggs after culture in vitro for 1–4 days. J Embryol Exp Morphol 23:21–34

Aigner B, Brem G (1994) Tyrosinase gene as a marker gene for studying transmission and expression of transgenes in mice. Transgenics 1:417–429

Aigner B, Brem G (1995) Detection of homozygous individuals in gene transfer experiments by semiquantitative PCR. BioTechniques 18:754–758

Aigner B, Besenfelder U, Seregi J, Frenyo L, Sahin-Toth T, Brem G (1996) Expression of murine wild type tyrosinase gene in transgenic rabbits. Transgenic Res 5:405–411

Andres AC, Schönenberger CA, Groner B, Hennighausen L, LeMeur M, Gerlinger P (1987) Ha-ras oncogene expression directed by a milk protein promoter, tissue specificity, hormonal regulation, and tumor induction in transgenic mice. Proc Natl Acad Sci USA 84:1299–1303

Attal J, Cajero-Juarez M, Houdebine L (1995) A simple method of DNA extraction from whole tissues and blood using glass powder for detection of transgenic animals by PCR. Transgenic Res 4:149–150

Ausubel FM, Brent R, Kingston RE, Moore DD, Seidman JG, Smith JA, Struhl K (1994) Current protocols in molecular biology. Greene Publ and John Wiley, New York

Baranyi M, Brignon G, Anglade P, Ribadeau-Dumas B (1995) New data on the proteins of rabbit (*Oryctolagus cuniculus*) milk. Comp Biochem Physiol 111B:407–415

Bawden C, Sivaprasad A, Verma P, Walker S, Rogers G (1995) Expression of bacterial cysteine biosynthesis genes in transgenic mice and sheep: Toward a new in vivo amino acids biosynthesis pathway and improved wool growth. Transgenic Res 4:87–104

Bawden WS, Passey RJ, Mackinlay AG (1994) The genes encoding the major milk-specific proteins and their use in transgenic studies and protein engineering. Biotechnol Genet Engin Rev 12:89–137

Beerman F, Ruppert S, Hummler E, Bosch F, Müller G, Rüther U, Schütz G (1990) Rescue of the albino phenotype by introduction of a functional tyrosinase gene into mice. EMBO J 9:2819–2826

Beerman F, Ruppert S, Hummler E, Schütz G (1991) Tyrosinase as a marker for transgenic mice. Nucleic Acids Res 19:958

Besenfelder U, Brem G (1993) Laparoscopic embryo transfer in rabbits. J Reprod Fertil 99:53–56

Brem G (1993a) Inheritance and tissue-specific expression of transgenes in rabbits and pigs. Mol Reprod Dev 36:242–244

Brem G (1993b) Transgenic animals. In: Rehm H-J, Reed G, Pühler A, Stadler P (eds), Biotechnology. VCH, Weinheim, pp 745–832
Brem G, Müller M (1994) Large transgenic animals. In N. Maclean, (ed.) Animals with novel genes. Cambridge University Press, Cambridge, pp 179–244
Brem G, Brenig B, Goodman HM, Selden RC, Graf F, Kruff B, Springman K, Hondele J, Meyer J, Winnacker E-L (1985) Production of transgenic mice, rabbits and pigs by microinjection into pronuclei. Zuchthygiene 20:251–252
Brem G, Wanke R, Wolf E, Buchmüller T, Müller M, Brenig B, Hermanns W (1989) Multiple consequences of human growth hormone expression in transgenic mice. Mol Biol Med 6:531–547
Brem G, Besenfelder U, Hartl P (1993) Production of foreign proteins in the mammary gland of transgenic rabbits. Chim Oggi 11:21–25
Brem G, Hartl P, Besenfelder U, Wolf E, Zinovieva N, Pfaller R (1994) Expression of synthetic cDNA sequences encoding human insulin-like growth factor-1 (IGF-1) in the mammary gland of transgenic rabbits. Gene 149:351–355
Brem G, Besenfelder U, Zinovieva N, Seregi J, Solti L, Hartl P (1995) Mammary gland-specific expression of chymosin constructs in transgenic rabbits. Theriogenology 43:175
Brem G, Besenfelder U, Aigner B, Müller M, Liebl I, Schütz G, Montoliu L (1996) YAC transgenesis in farm animals: rescue of albinism in rabbits. Mol Reprod Dev 44:56–62
Brenig B, Müller M, Brem G (1989) A fast detection protocol for screening large numbers of transgenic animals. Nucleic Acids Res 17:6422
Brinster RL, Chen HY, Trumbauer ME, Yagle MK, Palmiter RK (1985) Factors affecting the efficiency of introducing foreign DNA into mice by microinjecting eggs. Proc Natl Acad Sci USA 82:4438–4442
Brinster RL, Allen JM, Behringer RR, Gelinas RE, Palmiter RD (1988) Introns increase transcriptional efficiency in transgenic mice. Proc Nat Acad Sci USA 85:836–840
Bühler T, Bruyer T, Went D, Stranzinger G, Bürki K (1990) Rabbit b-casein promoter directs secretion of human interleukin-2 into the milk of transgenic rabbits. Bio/Technology 8:140–143
Brousseau ME, Santamaria Fojo S, Zech LA, Berard AM, Vaisman BL, Meyn SM, Powell D, Brewer HB Jr, Hoeg JM (1996) Hyperalphalipoproteinemia in human lecitin cholesterol acyltransferase transgenic rabbits. In vivo apolipoprotein A-I catabolism is delayed in a gene dose-dependent manner. J Clin Invest 97:1844–1851
Carney EW, Foote RH (1991) Improved development of rabbit one-cell embryos to the hatching blastocyst by culture in a defined, protein-free culture medium. J Reprod Fertil 91:113–123
Castro F, Aguirre A, Fuentes P, Ramos B, Rodriguez A, De la Fuente J (1995) Secretion of human erythropoietin by mammary gland explants from lactating transgenic rabbits. Theriogenology 43:184
Cheng L, Ziegelhoffer P, Yang N (1993) In vivo promoter activity and transgene expression in mammalian somatic tissues evaluated by using particle bombardment. Proc Natl Acad Sci USA 90:4455–4459
Choi T, Huang M, Gorman C, Jaenisch R (1991) A generic intron incearses gene expression in transgenic mice. Mol Cell Biol 11:3070–3074

Clark AJ, Simons P, Wilmut I, Lathe R (1987) Pharmaceuticals from transgenic livestock. Trends Bio Technol 5:20–24

Damak S, Su H, Jay NP, Bullock DW (1996) Improved wool production in transgenic sheep expressing insulin-like growth factor 1. Bio/Technology 14:185–188

Du F, Giles J, Graves R, Yang X, Moreadith R (1995) Nuclear transfer of putative rabbit embryonic stem cells leads to normal blastocyst development. J Reprod Fertil 104:219–223

Duby R, Cunniff M, Belak J, Balise J, Robl J (1993) Effect of milking frequency on collection of milk from nursing New Zealand white rabbits. Anim Biotechnol 4:31–42

Dunn C, Mehtali M, Houdebine L, Gut J, Kirn A, Aubertin A (1995) Human immunodeficiency virus type 1 infection of human CD4-transgenic rabbits. J Gen Virol 76:1327–1336

Ernst L, Zakharchenko V, Suraeva N, Ponomareva T, Miroshnichenko O, Prokofiev M, Tikchonenko T (1991) Transgenic rabbits with antisense RNA gene targeted at adenovirus H5. Theriogenology 35:1257–1271

Fan J, Wang J, Bensadoun A, Lauer S Dang Q, Mahley R, Taylor J (1994) Overexpression of hepatic lipase in transgenic rabbits leads to a marked reduction of plasma high density lipoproteins and intermediate density lipoproteins. Proc Natl Acad Sci USA 91:8724–8728

Fan J, McCormick S, Krauss R, Taylor S, Quan R, Taylor J, Young S (1995) Overexpression of human apolipoprotein B-100 in transgenic rabbits results in increased levels of LDL and decreased levels of HDL. Arterioscler Thromb Vasc Biol 15:1889–1899

Gol'dman I, Ernst L, Gogolevskii P, Afanasyev V, Semenova V, Grashchuk M, Konovalov M, Klenovitskii P, Zhivalev I, Polezhaeva I (1993) Exploration of expression of the gene responsible for cattle growth hormone in rabbits transgenic by mMT 1/bGHatt construction containing MAR (matrix attachment region) element. Sov Agricult Sci 1:44–53

Gordon K, Lee E, Vitale JA, Smith AE, Westphal H, Hennighausen L (1987) Production of human plasminogen activator in transgenic mouse milk. Bio/Technology 5:1183–1187

Graves K, Moreadith RW (1993) Derivation and characterization of putative pluripotential ES cell lines from preimplantation rabbit embryos. Mol Reprod Dev 36:424–433

Green LL, Hardy MC, Maynard-Currie CE, Tsuda H, Louie DM, Mendez MJ, Abderrahim H, Noguchi M, Smith DH, Zeng Y et al. (1994) Antigen-specific human monoclonal antibodies from mice engineered with human Ig heavy and light chain YACs. Nat Genet 7:13–21

Günzburg WH, Salmons B, Zimmerman B, Müller M, Erfle V, Brem G (1991) A mammary-specific promoter directs expression of growth hormone not only to the mammary gland, but also to Bergman glia cells in transgenic mice. Mol Endocrinol 5:123–133

Hammer RE, Pursel VG, Rexroad CE Jr, Wall RJ, Palmiter RD, Brinster RL (1985) Production of transgenic rabbits, sheep and pigs by microinjection. Nature 315:680–683

Hassan H, El Feel F, Sallam M, Ahmed M (1992) Evaluation of milk production and litter milk efficiency in two- and three-way crosses of three breeds of rabbits. Egypt J Anim Prod 29:303–315

Hennighausen L (1990) The mammary gland as a biorecator: production of foreign proteins in milk. Protein Expression Purification 1:3–8
Hennighausen L (1992) The prospects of domesticating milk protein genes. J Cell Biochem 49:325–332
Hoeg JM, Vaisman BL, Demosky SJ Jr, Meyn SM, Talley GD, Hoyt RF Jr, Feldman S, Berard AM, Sakai AM, Wood D, Brousseau ME, Marcovina S, Brewer HB Jr, Santamaria Fojo S (1996) Lecithin: cholesterol acetyltransferase overexpression generates hyperalpha-lipoproteinemia and a nonatherogenic lipoprotein pattern in transgenic rabbits. J Biol Chem 271:4396–4402
Houdebine L (1994) Production of pharmaceutical proteins from transgenic animals. J Biotechnol 34:269–287
Kawasaki ES (1990) Sample preparation from blood, cells, and other fluids. In: Innis MA, Gefland DH, Sninsky JJ, White TJ (eds) PCR protocols: a guide to methods and applications. Academic Press, San Diego, pp 146–152
Knight K, Spieker Polet H, Kazdin D, Oi V (1988) Transgenic rabbits with lymphocytic leukemia induced by the c-myc oncogene fused with the immunoglobulin heavy chain enhancer. Proc Natl Acad Sci USA 85:3130–3134
Kozarsky K, McKinley D, Austin L, Raper S, Stratford Perricaudet L, Wilson J (1994) In vivo correction of low density lipoprotein receptor deficiency in the Watanabe heritable hyperlipidemic rabbit with recombinant adenoviruses. J Biol Chem 269:13695–13702
Lee KF, DeMayo FJ, Atiee SH, Rosen JM (1988) Tissue-specific expression of the rat b-casein gene in transgenic mice. Nucleic Acids Res 16:1027–1041
Leno M, Hague B, Teller R, Kindt T (1995) HIV-1 mediates rapid apoptosis of lymphocytes from human CD4 transgenic but not normal rabbits. Virology 213:450–454
Limonta J, Castro F, Martinez R, Puentes P, Aguilar A, Lleonart R, De la Fuente J (1995) Transgenic rabbits as bioreactors for the production of human growth hormone. J Biotechnol 15:49–58
Lo D, Pursel V, Linto PJ, Sandgren E, Behringer R, Rexroad C, Palmiter RD, Brinster RL (1991) Expression of mouse IgA by transgenic mice, pigs and sheep. Eur J Immunol 21:25–30
Logan JS (1993) Transgenic animals: beyond "funny milk". Curr Opin Biotechnol 4:591–595
Lonberg N, Taylor LD, Harding FA, Trounstine M, Higgins KM, Schramm SR, Kuo CC, Mashayekh R, Wymore K, McCabe JG et al. (1994) Antigen-specific human antibodies from mice comprising four distinct genetic modifications. Nature 368:856–859
Maga EA, Murray JD (1995) Mammary gland expression of transgenes and the potential for altering the properties of milk. Bio/Technology 13:1452–1456
Manne J, Argeson AC, Siracusa LD (1995) Mechanisms for the pleiotropic effects of the agouti gene. Proc Natl Acad Sci USA 92:4721–4724
McFarlane M, Wilson J (1996) A model for the mechanism of precise integration of a microinjected transgene. Transgenic Res 5:171–177
Meade H, Gates L, Lacy E, Lonberg N (1990) Bovine aS1-casein gene sequences direct high level expression of active human urokinase in mouse milk. Bio/Technology 8:443–446

Mercier JC (1987) Genetic engineering applied to milk producing animals: some expectations. In: Smith C, King JW, McKay JC (eds) Exploiting new technologies in animal breeding. Oxford University Press, Oxford, pp 122–131

Müller M, Brem G (1994) Transgenic strategies to increase disease resistance in livestock. Reprod Fertil Dev 6:605–613

Müller M, Brem G (1996) Approaches to influence growth promotion of farm animals by transgenic means. Scientific conference on growth promotion in meat production. European Comission Directorate-General VI Agriculture, Brussels, pp 213–232

Müller M, Brem G (1996) Intracellular, genetic or congenital immunisation – transgenic approaches to increase disease resistance of farm animals. J Biotechnol 44:233–242

Mulligan RC (1993) The basic science of gene therapy. Science 260:926–932

Overbeek P, Aguilar-Cordova E, Hanten G, Schaffner D, Patel P, Lebovitz R, Lieberman M (1991) Coinjection strategy for visual identification of transgenic mice. Transgenic Res 1:31–37

Palmiter RD, Brinster RL, Hammer RE, Trumbauer ME, Rosenfeld MG, Birnberg NC, Evans RM (1982) Dramatic growth of mice that develop from eggs microinjected with metallotionein-growth hormone fusion genes. Nature 360:611–615

Palmiter RD, Sandgren EP, Avarbock MR, Allen DD, Brinster RL (1991) Heterologous introns can enhance expression of transgenes in mice. Proc Natl Acad Sci USA 88:443–446

Peng X, Olson R, Christian C, Lang C, Kreider J (1993) Papillomas and carcinomas in transgenic rabbits carrying EJ-ras DNA and cottontail rabbit papillomavirus DNA. J Virol 67:1698–1701

Peng X, Lang C, Kreider J (1995) Methylation of cottontail rabbit papillomavirus DNA and tissue-specific expression in transgenic rabbits. Virus Res 35:101–108

Riego E, Limonta J, Aguilar A, Perez A, De Armas R, Solano R, Ramos B, Castro F, De la Fuente J (1993) Production of transgenic mice and rabbits that carry and express the human tissue plasminogen activator cDNA under the control of a bovine alpha S1 casein promoter. Theriogenology 39:1173–1185

Roessler B, Hartman J, Vallance D, Latta J, Janich S, Davidson B (1995) Inhibition of interleukin-1-induced effects in synoviocytes transduced with the human IL-1 receptor antagonist cDNA using an adenoviral vector. Hum Gen Ther 6:307–316

Rosokhatsky S, Smirnov A, Yefimov A, Zverkhovsky L, Kazakov V, Dvoryanchikov G, Smaragdov M (1994) Increased growth rate of rabbits transgenic by human GHRF gene. Russian Agricult Sci 4:1–4

Ross K, Brenig B, Meyer J, Brem G (1988) Attempts to produce transgenic rabbits carrying MTI-hGH recombinant gene. In: Beynen A, Solleveld HA (eds) New developments in biosciences: their implications for laboratory animal science. Martinus Nijhoff, Dordrecht, pp 337–341

Saiki RK, Gelfand DH, Stoffel S, Scharf SJ, Higuchi R, Horn GT, Mullis KB, Erlich HA (1988) Primer-directed enzymatic amplification of DNA with a thermostable DNA polymerase. Science 239:487–491

Sethupathi P, Spieker Polet H, Polet H, Yam P, Tunyaplin C, Knight K (1994) Lymphoid and non-lymphoid tumors in E kappa-myc transgenic rabbits. Leukemia 8:2144–2155

Shen J, Kuhn H, Petho-Schramm A, Chan L (1995) Transgenic rabbits with the integrated human 15-lipoxygenase gene driven by a lysozyme promoter: macrophage-specific expression and variable positional specificity of the transgenic enzyme. FASEB J 9:1623–1631

Snyder B, Vitale J, Milos P, Gosselin J, Gillespie F, Ebert K, Hague B, Kindt T, Wadsworth S, Leibowitz P (1995) Developmental and tissue-specific expression of human CD4 in transgenic rabbits. Mol Reprod Dev 40:419–428

Southern, E (1975) Detection of specific sequences among DNA fragments separated by gel electrophoresis. J Mol Biol 98:503–517

Steg P, Feldman L, Scoazec J, Tahlil O, Barry J, Boulechfar S, Ragot T, Isner J, Perricaudet M (1994) Arterial gene transfer to rabbit endothelial and smooth muscle cells using percutaneous delivery of an adenoviral vector. Circulation 90:1648–1656

Takeshita S, Losordo D, Kearney M, Rossow S, Isner J (1994) Time course of recombinant protein secretion after liposome-mediated gene transfer in a rabbit arterial organ culture model. Lab Invest, 71:387–391

Van Brunt J (1988) Molecular farming: transgenic animals as bioreactors. Bio/Technology 6:1149–1154

Waine GJ, Mcmanus DP (1995) Nucleic acids: vaccines for the future. Parasitol. Today 11:113–116

Wall RJ (1996) Transgenic livestock: progress and prospects for the future. Theriogenology 45:57–68

Wanke R, Wolf E, Hermanns W, Folger S, Buchmüller T, Brem G (1992) The GH-transgenic mouse as an experimental model for growth research: clinical and pathological studies. Horm Res 37:74–87

Weidle UH, Lenz H, Brem G (1991) Genes encoding a mouse monoclonal antibody are expressed in transgenic mice, rabbits and pigs. Gene 98:185–191

Wilkie TM, Brinster RL, Palmiter RD (1986) Germline and somatic mosaicism in transgenic mice. Dev Biol 118:9–18

Yamanaka S, Balestra M, Ferrell L, Fan J, Arnold K, Taylor S, Taylor J, Innerarity T (1995) Apolipoprotein B mRNA-editing protein induces hepatocellular carcinoma and dysplasia in transgenic animals. Proc Natl Acad Sci USA 92:8483–8487

Zou Y-R, Müller W, Gu H, Rajewsky K (1994) Cre-loxP-mediated gene replacement: a mouse strain producing humanized antibodies. Curr Biol 4:1099–1103

Chapter 30

Microinjection of *Drosophila* Eggs

Cristiana Mollinari[1] and Cayetano González[1]*

Introduction

The early *Drosophila* embryos are particularly well suited for microinjection purposes for three major reasons. Firstly, they are easy to collect. Well-fed, healthy females can lay between two and three eggs per hour. Therefore, the yield from a single bottle containing between one and two 100 females can exceed 200 eggs every 30 min. Secondly, they are easy to handle and prepare for microinjection. The eggs of *Drosophila* are shaped like an ellipsoid whose major and minor axis are about 420 and 150 μm and have a volume of around 10 nl. This relatively large size facilitates their handling and microinjection without the need for highly sophisticated and expensive setups. Surrounding the cell membrane there is a thin, stiff, vitelline membrane which is covered by a rigid proteinaceus chorion. The chorion is too hard to allow for the penetration of the fine needles required for microinjection, but it can easily be removed by mechanical methods or chemical treatment. Finally, the suitability of the *Drosophila* embryo for microinjection is due to the special characteristics of early embryogenesis in *Drosophila*. After fertilization and fusion of the male and female pronuclei, nuclear proliferation proceeds through a rapid series of synchronous mitotic divisions which occur without partitioning the nuclei by cell membranes. These nuclear cycles lack the G_1 and G_2 phases and are sustained and regulated by maternal products. The result is a syncytium which is maintained until around the 9th nuclear division, which occurs about 90 min after fertilization when the embryos are incubated at 25 °C. At this stage, the majority of the nuclei migrate to the periphery of the egg and some of those that migrate to the posterior end of the embryo are enveloped in cell membranes to form

* corresponding author: phone: +49-6221-387292; fax: +49-6221-387306/387512; e-mail: gonzalez@embl-heidelberg.de
[1] European Molecular Biology Laboratory, Meyerhofstrasse 1, 69012 Heidelberg

the pole cells that will give rise to the germline. More than 500 nuclei are left behind, within the syncytium, until cellularization finally occurs, after four more divisions, when the embryo reaches the cellular blastoderm stage. The syncytial nature and the rapid nuclear cycles present in the early *Drosophila* embryo allow the material injected into the early cleavage embryo to freely diffuse and access subcellular organelles at any stage of the nuclear cycle over a relatively short period.

Microinjection in *Drosophila* has a long history, as it was widely used for pioneer transplantation experiments which involved the microinjection into one embryo of cells, nuclei or cytoplasm which had been previously taken from a donor (Illmensee 1972). Nowadays, microinjection has a very wide range of applications depending on the nature of the biologically active compound which is injected. These include pure molecules like DNA, RNA, antibodies, bacterially expressed proteins, and drugs, as well as more or less crude preparations of ooplasm.

Applications of Microinjection

Microinjection of exogenous DNA into *Drosophila* embryos has been used to manipulate gene expression in many different ways. The best-known application of microinjection techniques in *Drosophila* is the generation of transgenic flies by P element-induced germline transformation. This technique was developed by the pioneer work of Rubin and Spradling (1982), who showed that the transposable P element can be used to drive the insertion of exogenous DNA into the fly genome. A detailed description of the protocol for injecting P-element vectors into *Drosophila* eggs can be found in Santamaria (1986). The techniques for P element-mediated germline transformation have been evolving very fast ever since, and there is now a wide range of applications, as well as of transformation vectors which can be used as DNA carriers. Transgenes which carry the genomic regions containing both regulatory and coding wild-type sequences are widely used for rescue experiments in which the link between the molecular information (cloned sequences) and the genetic data (mutants) can be established. Other applications include the identification of developmentally regulated enhancers by producing transgenes carrying a reporter gene such as LacZ (Silhavy and Beckwith 1985; Fasano and Kerridge 1988) or green fluorescent protein (GFP; Wang and Hazelrigg 1994), and the study of effects brought about by ectopic gene expression. This can be achieved by transgenes in which the gene of interest is under the control of any of the available regulatable promoters (Zuker et al. 1988). An interesting and new development in this regard is represented by the use of the *Saccharomyces cerevisiae* FRT/Flipase system. The yeast, FLP recombinase catalyses recombination between homologous sequences (FRTs). Recombination can occur between sites which are in the same chromosome, in different chromosomes, or between chromosomal and extrachromosomal (plasmid) sites, thus offering a wide range of potential applications. Two such applications are site-specific insertion of cloned sequences, and clonal analysis by the flipout cassette (Simpson 1993; Struhl and Basler 1993).

Germline transformation by P-element-based vectors is not restricted to *Drosophila melanogaster*. P elements can also transpose in other species, as shown by Brennan et al. (1984) using *Drosophila hawaiiensis*. Therefore, P-element-based germline transformation can be used for generating interspecific transgenic flies by transforming one *Drosophila* species with a transgene from a different one. This approach provides a way to study genetic differences affecting gene expression among related, but reproductively isolated, species. An example of such an application is provided by Scavarda and Hartl (1984), who were able to rescue the mutant phenotype of *Drosophila simulans ry* flies with a transgene carrying the wild-type *rosy* gene of *Drosophila melanogaster* cloned into a P-element-based transformation vector. This observation showed that the ry^{+} gene of *Drosophila melanogaster* is functionally compatible with its *D. simulans* homologue, despite the estimated 1 to 5 million years of reproductive isolation since the evolutionary divergence of these species.

Injection of DNA Without a Transposable Element

Other applications of DNA microinjection include the generation of stable transgenic flies without transformation vectors, transient expression of plasmid-borne genes, and mutagenesis. The first reports of transformation in *Drosophila* were carried in the 1960 s and 1970s, when several authors reported that it was possible to revert some mutant phenotypes by the injection of total genomic DNA from wild-type flies (Fox and Yoon 1966; Germeraad 1976). Although initially interpreted as the result of homologous transformation, where exogenous sequences replace existing sequences by homologous recombination, it is now clear that other mechanisms like the mobilization of transposable elements can account for some of these results. Real transformation of *Drosophila* embryos without transposable element-based vectors can nevertheless be achieved by microinjection of plasmid DNA. The fate of injected plasmid DNA in *Drosophila* has been studied by Steller and Pirrotta (1985), who found that the injected DNA is enclosed in nuclei-like structures, where it remained throughout embryonic development. A portion of the exogenous DNA in these nuclei is converted to a high-molecular-weight form consisting largely of tandem oligomers, some of which may be derived from homologous recombination between plasmids. Some exogenous DNA may also engage in nonhomologous recombination between plasmid and genomic DNA and be inserted into the chromosomes, although with a rather low efficiency.

Finally, DNA microinjection has also been used for mutagenic purposes. This has been done by the use of retroviral proviruses like RSV which are structurally similar to the transposable elements of eukaryotic genomes. Microinjection of whole RSV particles or their cloned DNA fragments into the polar plasm of *Drosophila* embryos was used by Gazaryan et al. (1984) to isolate a set of mutant lines. In some of these, the virus-specific sequence was detected by Southern blot hybridization. Like P-element-mediated mutagenesis, RSV is clearly selective, and certain loci are more prone to be mutagenized than others.

A Way to Study Gene and Protein Function

Microinjection of either specific antibodies raised against different gene products or purified preparations of these gene products themselves has been used for two main purposes: functional assays and in vivo labeling of subcellular structures. Functional studies can be carried out by microinjecting blocking antibodies in order to inactivate the endogenous product under study. Alternatively, the function of a given protein can be assayed by injecting the protein itself, either as a full-length version or as deletion derivatives with which to define the functional domains within the protein. With this aim, the gene of interest must be cloned into an appropriate expression vector and tagged with distinctive epitopes (like MYC or 6xHis) to facilitate its identification. The effects brought about by the microinjected compounds can be monitored either in fixed preparations or by applying real-time or time-lapse techniques.

Real-time or time-lapse observations are based on the ability of antibodies or purified gene products to specifically recognize and bind to different subcellular structures. These techniques allow for the direct observation of live specimens, thus circumventing the limitations associated to the use of fixed material, such as the difficulty of establishing the temporal order of events, or the artifacts produced by fixation treatment. Real- and nearly real-time observation of life specimens introduces the time dimension and makes it possible to determine the dynamic behavior of the organelle under study throughout the cell cycle. The method involves microinjection of a fluorescently labeled probe into the embryo and the use of any of the several techniques such as confocal laser scanning microscopy or optical imaging and 3-dimensional deconvolution, which allow the optical sectioning of living embryos. The combination of these techniques with microinjection of fluorescently labeled macromolecules enables the study of specific aspects of a living organism with high temporal and spatial resolution (Agard et al. 1989; Hiraoka et al. 1989; Pawley and Erlandsen 1989; Shotton and White 1989). A practical example of the application of microinjection techniques for functional studies is provided by the work of Buchenau et al. (1993) on the characterization of the topoisomerase II of *Drosophila melanogaster*. The functional characterization of this gene is particularly relevant, because no mutant alleles have been identified so far. These authors showed that topoisomerase II activity can be specifically disrupted both by a specific anti-topoisomerase II antibody and by treatment with epipodophyllotoxin VM26, a known *in vitro* inhibitor of topoisomerase II. The effect of both treatments was followed *in vivo* using confocal laser scanning microscopy following the labeling of the chromatin of live embryos with tetramethyl-rhodamine-coupled histones. Both antibodies and the drug prevented or hindered the segregation of chromatin daughter sets at the anaphase stage of mitosis. In addition, high concentrations of inhibitor interfered with the condensation of chromatin and its proper arrangement into the metaphase plate, which resulted in abnormal nuclei. Thanks to the direct visualization of the live embryos, these authors were able to show that the abnormal nuclei were drawn into the inner yolk mass of the embryo and that undamaged nuclei surrounding the affected region underwent extra rounds of replication, thus restoring the normal nuclear density.

More recently, Oegema et al. (1995) have used a similar approach to identify the functional domains present within the CP190 Centrosomal Protein (Frasch et al. 1986; Whitfield et al. 1988; Kellogg et al. 1989). The localization of the CP190 protein oscillates in a cell-cycle-specific manner between the nucleus during interphase, and the centrosome during mitosis. To characterize the regions of CP190 responsible for its dynamic behavior, Oegema et al. injected rhodamine-labeled fusion proteins spanning different regions of CP190 into early *Drosophila* embryos, and followed their localization using time-lapse fluorescence confocal microscopy. Deletion constructs containing only the centrosome localization sequence were found at centrosomes throughout the cell cycle, suggesting that during interphase, native CP190 is actively recruited away from the centrosome into the nucleus. Other examples of applications of in vivo labeling of subcellular structures in *Drosophila* embryos can be found in Kellogg et al. (1988) and Sullivan et al. (1990). Warn et al. (1987) reported the labeling of microtubules following microinjections of fluorescently labeled antibodies antitubulin. Drugs can also be used to perform functional assays and to label subcellular structures. One example of such drug treatment to study the cytoskeleton is represented by the studies to investigate the effects of perturbing the F-actin microfilament systems during early development. These include the microinjection of both actin-depolymerizing agents like cytochalasin, as well as phalloidin, a bicyclic heptapeptide derived from the mushroom *Amanita phalloides*, known to strongly promote polymerization (Planques et al. 1991).

Microinjection of RNA has also found many applications in functional studies, including knocking down the endogenous gene product by antisense technology, studying gain-of-function phenotypes produced by constitutively active derivatives, and rescuing of mutant phenotypes by sense mRNA. Antisense methods were first applied in *Drosophila* to inhibit the function of the *Kruppel* gene Preiss et al. (1985) and have been widely used ever since. LaBonne et al. (1989) were able to induce phenocopies of the severe hyperneuralization characteristic of the *pecanex* mutant phenotype following microinjection of antisense RNA. An example of the use of engineered versions of sense RNAs encoding gain-of-function derivatives is represented by the work of Smith and DeLotto (1994), to examine the effect of producing a preactivated form of the snake protease on the generation of dorsal-ventral polarity. For this purpose, the authors produced a construct encoding the snake catalytic chain fused in-frame to a signal peptide which directed the fusion product to the perivitelline space. Using this strategy, they could show a position-dependent activation of the signal transduction pathway involved in embryo polarity.

The range of substances that can be microinjected into the *Drosophila* embryo is by no means limited to the purified biologically active compounds referred to above. Microinjection of whole ooplasm between embryos of different genotypes as well as between different species has also been used to tackle some interesting biological problems. One instance of such applications are the studies on inherited rickettsial symbionts of the genus *Wolbachia*. These microorganisms are common in arthropods and have been implicated in processes like parthenogenesis, feminization, and cytoplasmic incompatibility. Cytoplasmic incompatibility is expressed when males that carry the bacterial endosymbiont are mated to females that lack the infection. Such a cross in *Drosophila* produces few viable offspring, while a cross between infected females and "clean" males yields normal progeny counts. Studies of cytoplasmic incompatibility have been hampered by the difficulty in culturing the symbiont outside its host. To circumvent this limitation, Boyle et al. (1993) transinfected *Wolbachia* into uninfected *D. simulans* and *D. melanogaster* hosts and monitored their cellular distributions and the extent of cytoplasmic incompatibility with confocal microscopy. Interspecific transinfection experiments have also been performed with cytoplasm from more distantly related species. For instance, it has been shown that *Wolbachia* found in mosquito can behave like a natural *Drosophila* infection with regards to its inheritance, cytoskeleton interactions, and ability to induce incompatibility when crossed with uninfected flies (Braig et al. 1994). These findings clearly indicate that these agents can be transferred over large phylogenetic distances. Altogether, these observations indicate that the *Wolbachia* symbionts of different insect hosts are very closely related and have probably moved horizontally between insect lineages.

Alternatives to Microinjection

Finally, and although they fall beyond the scope of this chapter, it is worth mentioning that microinjection is not the only choice to introduce biologically active compounds into the *Drosophila* embryos and that alternative techniques have been used in the past and are still being developed. The first of such techniques, if only for purely historical purposes, is straight-forward soaking. As reviewed in Walker (1989), exogenous DNA can be introduced into *Drosophila* embryos by soaking them in DNA solutions after dechorionation (reviewed by Hoy 1994). However, this method has not been pursued because of very low uptake.

A more workable alternative to microinjection is provided by permeabilization. Chemical treatments that permeabilize the vitellin membrane to a variety of drugs, without significantly disrupting embryo physiology, have been known for a long time. Permeabilization has been used to study the effect of drugs such as aphidicolin, for example, to show that the mitotic cycles of the syncitial embryo can take place without completion of DNA synthesis (Raff and Glover 1988). Drugs which alter microtubule dynamics such as colchicine or taxol, phosphatase inhibitors such as okadaic acid, kinase inhibitors, and many more have also been introduced into the embryo by permeabilization. The technique also permits the introduction of chemicals such as bromodeoxyuridine (BrdU) as a label for DNA synthesis. The incorporation of this drug into DNA can be visualized by anti BrdU antibodies. The main advantage of permeabilization is that, unlike microinjection, the delivery of drugs by this method is not invasive. Unfortunately, only compounds of moderate size can be introduced, thus limiting the use of the technique to the delivery of relatively small drugs.

This limitation does not apply to two relatively new techniques which allow the delivery of different compounds into the *Drosophila* embryos. The first is electroporation which consists of the delivery of short electric impulses which make membranes transiently permeable. Transient expression of plasmid-borne genes electroporated into *Drosophila* embryos has already been reported (Kamdar et al. 1992). The other recent development, known as biolistic, consists of shooting the embryo with microprojectiles which are coated with the molecule which has to be introduced into the embryo. This so-called gene gun was developed to transform major crop plants and yeast. Its use with *Drosophila* eggs is still very limited, but transient expression of DNA coated onto tungsten particles in *Drosophila* embryos has been shown to occur (Baldarelli and Lengyel 1990). The main advantages of these two systems is that they allow many more embryos to be treated at once than does microinjection into individual eggs. Nevertheless, the application of these techniques to *Drosophila* is still under development and whether or not they will replace microinjection techniques remains to be seen.

The following protocols give an extensive description of the different steps that have to be followed for microinjection, from egg collection to cytological observation of the microinjected material, including standard techniques for fluorescence labeling prior to microinjection. A protocol for permeabilization of embryos is also included. For consultation on these and related subjects the reader is referred to Santamaria (1986); Ashburner (1989); Gonzalez and Glover (1993); Goldstein and Fyrberg (1994).

Procedures

30.1
Embryo Collection and Handling Prior to Injection

1. Collect embryos from parents of the same age to ensure synchronous development; place around 200 females per bottle; 4–10 day-old females are ideal.
2. Eggs are collected at 23–25 °C for 30–40 min on plastic dishes filled with fruit juice agar and smeared with a thick paste of live yeast.
3. Keep the bottles in a quiet, dark place.
4. Discard the first 2-h collections to avoid contamination with old embryos.
5. Transfer the eggs from the collection plate to double-sided Scotch tape stuck to a microscope slide.
6. Dechorionate by rolling the eggs over the tape with a pair of forceps. Alternatively, dechorionation can be carried out by immersion for 3–5 min in 50 % Chlorox, followed by extensive washing in water.
7. Stage embryos following the morphological criteria described in Bownes (1975). For germline transformation experiments, stage 2 embryos, 10 to 70 min old, prior to the formation of the pole buds, are chosen. Other applications may require embryos at different stages.
8. Arrange the embryos over a line of glue close to the edge of a 24 mm^2 coverslip. The glue is prepared by dissolving double sided Scotch tape in n-heptane. A thin line of this solution is then laid with a pipette on the coverslip. The glue is ready after the heptane evaporates. Glue can also be prepared by extracting pieces of brown packing tape with heptane overnight.
9. Desiccate for 10 min to reduce internal pressure in the embryos Desiccation times can be shortened if the embryos are placed in a petri dish containing silica gel or Drierite.
10. When desiccation is complete, cover the embryos with a thin layer of 700 halocarbon oil (Voltalef, 10S grade or Series 95; Halocarbon Products, North Augusta, SC, USA).

The embryos are now ready for injection.

30.2 Needle Preparation

1. Needles are heat-pulled from 10-μl glass capillaries using any of the pullers which are commercially available. Alternatively, and much more conveniently, they can be bought, ready to use. Although some applications may require the use or needles to suit special requirements, commercial needles are suitable for most purposes (Femtotips, Eppendorf, Hamburg, Germany).
2. Normally the tip of the needle has to be broken to an opening of 2–4 μm. This can be achieved either with a fine tungsten needle or by pushing the needle against the edge of a microscope coverslip once the needle is mounted in the micromanipulator.
3. To fill the needle, apply a small drop of injection solution to a piece of parafilm fixed on a slide and fill the needle by suction. Alternatively, the needle can be loaded by introducing a sequencing pipette tip or a hand-pulled pipette in the wide open.
4. Filled needles may be stored by keeping the tips dipped in oil (3S Voltalef oil or paraffin oil).

30.3 Injection

1. Injection is normally carried out at 16–20 °C, in order to slow down the rate of development, and at a relative humidity higher than 70 %.
2. Place the slide on an inverted microscope and proceed to inject embryos using a needle held in a micromanipulator.
3. The solution to be injected must be cleaned by centrifugation and additional microfiltration.
4. The normal range of volume injected varies from 1–500 pl (500 pl is about 5 % of the total embryo volume).
5. The place of injection depends on the specific application. For germline transformation embryos are injected posteriorly. For most other applications injection is performed laterally at about 50 % of egg length.

6. The needle is introduced at a distance of about 5–10 % of egg length by moving either the microscope stage or the needle itself.
7. Injection is carried out steadily. Microinjection can be achieved with controlling devices (a 30 psi line is convenient) or hand-controlled with a syringe.
8. It has been reported that removing the cytoplasm that leaks out from the embryos after withdrawing the needle, increases viability.
9. Following injection, the coverslip containing the embryos is transferred to a small petri dish and allowed to develop as required.

30.4 Immunostaining of Embryos Following Microinjection

1. Cover the embryos with a few drops of heptane to dissolve the glue until they start floating and transfer them to an Eppendorf tube.
2. Remove the supernatant and add fresh heptane.
3. For methanol fixation place the embryos in a 1:1 mixture of heptane: 97 % methanol, 3 % 0.5 M Na_3EGTA pH 7.6. The devitellinized embryos fall to the bottom. Wash two times with 97 % methanol, 3 % 0.5 M Na_3EGTA pH 7.6 and incubate the embryos for at least 4 h at room temperature or overnight at 4 °C.
4. If formaldehyde fixation is preferred, transfer the embryos to a 1:1 mixture of heptane : 37 % formaldehyde, and incubate for 30 min on a spinning wheel, at room temperature. After fixation, devitellinize in a heptane/methanol interphase as described above or by hand-rolling the embryos on Scotch tape.
5. Before immunostaining is carried out, the embryos must be rehydrated by passing them through a series of 20, 40, 60, and 80 % PBS in methanol, and finally PBS alone.
6. Incubate the embryos in 10 % fetal calf serum, 0.3 % Tween in PBS for 1 h.
7. Add RNase 2 mg/ml if DNA is to be stained with propidium iodide.
8. Incubate with the primary antibody in 10 % fetal calf serum, 0.1 % Tween in PBS either at 4 °C overnight, or for 4 h at room temperature.

9. Wash 5× for 10 min with PBS.
10. Incubate with the secondary antibody for 2 h at room temperature or overnight at 4 °C.
11. Wash 5× for 10 min with PBS.
12. To mount the embryos, place 50 µl of 2.5 % propyl gallate in 85 % glycerol on a siliconized coverslip and transfer the embryos into the mounting medium with a brush. Lower a microscope slide onto the drop and seal the preparation with nail varnish.

30.5 Protein Labeling

Rhodamin labeling of histones

1. Core histones are reconstituted by adding calf thymus histones directly onto a DNA-cellulose column containing double-stranded DNA in a buffer at low ionic strength.
2. Incubate the core histones with N-hydroxy-succinimide tetramethyl-rhodamine (NHSR) (Molecular Probes IIIC., Junction City, OR) in a buffer containing 0.2 M NaCI, 20 mM potassium Hepes pH 7.4, and 1 mM Na_3EDTA.
3. Remove the unreacted fluorochrome from the resin.
4. Pack the NHSR-labeled nucleosome/DNA cellulose resin into a small column.
5. Elute the histones with a 0.2–2-M NaCI gradient containing 20 mM Tris-HCI, pH 7.4, and 1 mM Na_3 EDTA.
6. The fractions that contain histones H2A and H2B were identified by PAGE, pooled, and concentrated by centrifugation in Centricon tubes. (Centricon 10; Amicon, Danvers, MA).
7. Small aliquots can be frozen in liquid nitrogen and stored at −80 °C for over 1 year.

Rhodamine labeling of fusion proteins

8. Add 0.75 µl of 12.5 mg/ml tetramethyl-rhodamine NHS ester (Molecular Probes, Eugene OR), dissolved in either N,N-dimethylformamide or dimethylsulfoxide, to 75 µl of fusion protein (1–5 mg/ml) in FPLC (composition) buffer.

9. Incubate on ice for 5 min, and 7.5 μl of 2 M potassium glutamate, pH 8.

10. Stop the reaction by adding 0.75 μl of 0.5 M dithiothreitol.

11. Transfer into injection buffer using a small spin column of Bio-Gel P-6 resin. For the 6× His fusion proteins, the injection buffer is 50 mM Hepes, pH 7.6, 250 mM KCI and 50 mM Hepes pH 7.6. For the GST fusion proteins the buffer used is 100 mM KCl.

12. The extent of labeling can be assayed by spectroscopy. Proteins are considered overlabeled if the absorption peak at 522 is higher than the absorption at 556 nm. Labeling stoichiometries should be between 0.15 and 0.5 rhodamine/protein monomer.

FITC labeling of antibodies and other proteins

13. IgG or any other protein is dissolved in 0.2 M Na_2HPO4.

14. Fluorescein isothiocynate (FITC) is dissolved prior to use in 0.1 M Na_2HPO_4.

15. Use 12.5 μg FITC per mg of protein aiming at a ratio of about three fluorescent molecules per molecule of protein.

16. Add FITC solution very slowly while stirring the protein solution for 2–3 min.

17. Incubate for 30 min to 1>h at room temperature and adjust the pH 9.5 during the coupling using Na_3PO4 (0.1 M).

18. Purify the conjugate by Sephadex G-25 chromatography to remove all free dye.

19. Concentrate by centrifugation in Centricon 10 tubes.

30.6 RNA Microinjection

1. Generate RNA from any template plasmid containing specific RNA polymerase promoters.

2. Purify the plasmid using the Plasmid Midi Kit from Qiagen (Chatsworth, CA, USA) extract with phenol and precipitate with 0.1 vol of 3 M sodium acetate and 1.5 vol of ethanol.

3. Dissolve the DNA in diethylpyrocarbonate (DEPC)-treated distilled water.
4. RNAs are generated using any commercial RNA polymerase kit and dissolved in nuclease-free distilled water at 1 mg/ml.
5. RNA concentration should vary according to the specific requirements as too much RNA can produce unspecific effects. A concentration of about 200 μg/ml is generally suggested.

30.7 Permeabilization of Embryos With Octane for Drug Treatment

1. Collect and dechorionate embryos as described before.
2. Place embryos on a small piece of wet, black blotting paper on a petri dish.
3. Dip the blotting paper into the organic phase of a mixture of 1 ml TCM.
   (5 vol Schneider's *Drosophila* medium revised: 4 vol MEM 25 mM) 1 ml octane (previously saturated in TCM) in a 10 ml glass vial. The embryos will fall onto the interphase, where they will clump together, forming a circle.
4. Place the vial on a flat horizontal surface and gently shake with a rotatory movement for 2 min.
5. Aspirate out the embryos with a Pasteur pipette. They will form a clump at the interphase.
6. Discard as much TCM as possible and drop the embryos into a vial containing 1 ml of TCM and the appropriate amount of the drug. The embryos will stay on the surface covered by some octane.
7. Blow air upon the embryos, observing them under the dissecting microscope until the octane has evaporated and they become reflectant.
8. After incubation with the drug, add a few drops of heptane saturated in TCM in order to re-form an aqueous/organic interphase. Collect the embryos with a Pasteur pipette as before and drop them into vial for fixation.

Comments

Like any other experimental technique, microinjection can lead to artifactual results. These can come from many sources, but the most common are due to the traumatic effects of microinjection itself and to the presence of biologically active contaminants in the injection mixture. For instance, puncture damage on its own has been reported to result in neurogenic-like cuticle defects. A frequent example of artifactual effects due to contaminants is the presence of transcript produced by leaky transcription from the "wrong" promoter in RNA microinjection experiments. It is therefore essential to ensure that appropriate controls are always carried out.

References

Agard DA, Hiraoka Y, Shaw P, Sedat JW (1989) Fluorescence microscopy in three dimensions. Methods Cell Biol. 30:353–377

Ashburner M (1989) *Drosophila*, a laboratory handbook. Cold Spring Harbor Laboratory Press, Cold Spring Harbor

Baldarelli RM, Lengyel JA (1990) Transient expression of DNA after ballistic introduction into *Drosophila* embryos. Nucleic Acids Res 18:5903–5904

Bownes M (1975) A photographic study of development in the living embryo of *Drosophila melanogaster*. J Embryol Exp Morphol 33:789–801

Boyle L, O'Neil S, Robertson HM, Karr HM (1993) Interspecific and intraspecific horizontal transfer of Wolbachia in *Drosophila*. Science 260:1796–1799

Braig HR, Guzman H, Tesh RB, O'Neil SL (1994) Replacement of the natural Wolbachia symbiont of *Drosophila simulans* with mosquito counterpart. Nature 367:453–455

Brennan MD, Rowan RG, Dickinson WJ (1984) Introduction of a functional P-element into germline of *Drosophila hawaiiensis*. Cell 38:147–151

Buchenau P, Arndt-Jovin DJ, Saumweber H (1993) In vivo observation of the puff-specific protein no-on transient A (NONA) in nuclei of *Drosophila* embryos. J Cell Sci 106:189–199

Fasano L, Kerridge S (1988) Monitoring positional information during oogenesis in adult *Drosophila*. Development 104:245–253

Fox AS, Yoon SB (1966) Specific effects of DNA in *Drosophila melanogaster*. Genetics 53:897–911

Frasch VE, Glover DM, Saumweber H (1986) Nuclear antigens follow different pathways into daughter nuclei during mitosis in early *Drosophila* embryos. J Cell Sci 82:155–172

Gazaryan KG, Nabirochkin SD, Tatosyan AG, Shakhbazyan AK, Shibanova EN (1984) Genetic effects of injection of Rous sarcoma virus DNA into polar plasm of early *Drosophila melanogaster* embryos. Nature 311:392–394

Germeraad S (1976) Genetic transformation in *Drosophila* by microinjection of DNA. Nature 262:229–231

Goldstein LSB, Fyrberg EA (1994) Methods in cell biology. Academic Press, New York
Gonzalez C, Glover D (1993) Techniques for studying mitosis in *Drosophila*. In: Fantes, Brooks (eds) The cell cycle. A practical approach. Oxford University Press, Oxford
Hiraoka Y, Minden JS, Swedlow JR, Sedat JW, Agard DA (1989) Focal points for chromosome condensation and decondensation revealed by three-dimensional in vivo time-lapse microscopy. Nature 16:293–296
Hoy MA (1994) Insect molecular biology. An introduction to principles and applications Academic Press, New York
Illmensee K (1972) Developmental potencies of nuclei from cleavage, preblastoderm, and syncytial blastoderm transplanted into unfertlized eggs of *Drosophila melanogaster*. Wilhelm Roux' Archiv Dev Biol 170:267–298
Kamdar P, Allem G, Finnerty V (1992) Transient expression of DNA in *Drosophila* via electroporation. Nucleic Acids Res 20:3526
Kellogg DR, Mitchison TJ, Alberts BM (1988) Behaviour of microtubules and actin filaments in living *Drosophila* embryos. Development 103:675–686
Kellogg DR, Field CM, Alberts BM (1989) Identification of microtubule-associated proteins in the centrosome, spindle, and kinetochore of the early *Drosophila* embryo. J Cell Biol 109:2977–2991
LaBonne SG, Sunitha I, Mahowald AP (1989) Molecular genetics of pecanex, a maternal-effect neurogenic locus of *Drosophila melanogaster* that potentially encodes a large transmembrane protein. Dev Biol 136:1–16
Minden JS, Agard DA, Sedat JW, Alberts BM (1989) Direct cell lineage analysis in *Drosophila melanogaster* by time-lapse, three-dimensional optical microscopy of living embryos. J Cell Biol 109:505–516
Oegema K, Withfield WGF, Alberts B (1995) The cell cycle-dependent localization of the CP190 centrosomal protein is determined by the coordinate action of two separable domains. J Cell Biol 131:1261–1273
Pawley JB, Erlandsen SL (1989) The case for low voltage high resolution scanning microscopy of biological samples. Scanning Microsc Suppl 3:163–178
Planques V, Warn A, Warn RM (1991) The effects of microinjection of rhodamine-phalloidin on mitosis and cytokinesis in early stage *Drosophila* embryos. Exp Cell Res 192:557–566
Preiss A, Rosenberg UB, Klienlin A, Seifert E, Jackle H (1985) Molecular genetics of Kruppel, a gene required for segmentation of the *Drosophila* embryo. Nature 313:27–32
Raff JW, Glover DM (1988) Nuclear and cytoplasmic mitotic cycles continue in *Drosophila* embryos in which DNA synthesis is inhibited by aphidicolin. J Cell Biol 107:2009–2019
Rubin GM, Spradling AC (1982) Genetic transformation of *Drosophila* with transposable element vectors. Science 218:348–353
Santamaria P (1986) Injecting eggs. In: Roberts DB (ed) *Drosophila:* a practical approach. IRL Press, pp 159–173
Scavarda NJ, Hartl DL (1984). Interspecific DNA transformation in *Drosophila*. PNAS 81:7515–7519
Shotton D, White N (1989) Confocal scanning microscopy:three-dimensional biological imaging. Trends Biochem Sci 14:435–439
Silhavy TJ, Beckwith JR (1985) Uses of Lac fusions for the study of biological problems. Microbiol Rev 49:398–418

Simpson P (1993) Flipping fruit flies:a powerful new technique for generating *Drosophila* mosaics. Trends Genet 9:227–228
Smith CL, DeLotto R (1994) Ventralizing signal determined by protease activation in *Drosophila* embryogenesis. Nature 368:548–551
Steller H, Pirrotta V (1985) Fate of DNA injected into early *Drosophila* embryos. Dev Biol 109:54–62
Struhl G, Basler K (1993) Organizing activity of wingless protein in *Drosophila*. Cell 72:527–540
Sullivan W, Minden JS, Alberts BM (1990) daughterless-abo-like, a *Drosophila* maternal-effect mutation that exhibits abnormal centrosome separation during the late blastoderm divisions. Development 110:311–323
Walker VK (1989) Gene transfer in insects. Adv Cell Culture 7:87–124
Wang S, Hazelrigg T (1994) Implications for bcd mRNA localization from spatial distribution of exu protein in *Drosophila* oogenesis. Nature 369:400–403
Warn RM, Flegg L, Warn A (1987) An investigation of microtubule organization and functions in living *Drosophila* embryos by injection of a fluorescently labelled antibody against tyrosinated alpha-tubulin. J Cell Biol 105:1721–1730
Whitfield WG, Millar SE, Saumweber H, Frasch M, Glover DM (1988) Cloning of a gene encoding an antigen associated with the centrosome in *Drosophila*. J Cell Science 89:467–480
Zuker CS, Mismer D, Hardy R, Rubin GM (1988) Ectopic expression of a minor *Drosophila* opsin in the major photoreceptor cell class: distinguishing the role of primary receptor and cellular context. Cell 53:475–482

Subject Index

Note: Page numbers with an F or a T indicate references to Figures and Tables, respectively.